# Advances in Geographical and Environmental Sciences

Advances in Geographical and Environmental Sciences synthesizes series diagnostigation and prognostication of earth environment, incorporating challenging interactive areas within ecological envelope of geosphere, biosphere, hydrosphere, atmosphere and cryosphere. It deals with land use land cover change (LUCC), urbanization, energy flux, land-ocean fluxes, climate, food security, ecohydrology, biodiversity, natural hazards and disasters, human health and their mutual interaction and feedback mechanism in order to contribute towards sustainable future. The geosciences methods range from traditional field techniques and conventional data collection, use of remote sensing and geographical information system, computer aided technique to advance geostatistical and dynamic modeling.

The series integrate past, present and future of geospheric attributes incorporating biophysical and human dimensions in spatio-temporal perspectives. The geosciences, encompassing land-ocean-atmosphere interaction is considered as a vital component in the context of environmental issues, especially in observation and prediction of air and water pollution, global warming and urban heat islands. It is important to communicate the advances in geosciences to increase resilience of society through capacity building for mitigating the impact of natural hazards and disasters. Sustainability of human society depends strongly on the earth environment, and thus the development of geosciences is critical for a better understanding of our living environment, and its sustainable development.

Geoscience also has the responsibility to not confine itself to addressing current problems but it is also developing a framework to address future issues. In order to build a 'Future Earth Model' for understanding and predicting the functioning of the whole climatic system, collaboration of experts in the traditional earth disciplines as well as in ecology, information technology, instrumentation and complex system is essential, through initiatives from human geoscientists. Thus human geoscience is emerging as key policy science for contributing towards sustainability/survivality science together with future earth initiative.

Advances in Geographical and Environmental Sciences series publishes books that contain novel approaches in tackling issues of human geoscience in its broadest sense—books in the series should focus on true progress in a particular area or region. The series includes monographs and edited volumes without any limitations in the page numbers.

Subhash Anand · Virendra Nagarale ·
Rajesh Kumar Abhay
Editors

# Emerging Geosustainability Transformations in India

Springer

*Editors*
Subhash Anand
Department of Geography
Delhi School of Economics
University of Delhi
New Delhi, Delhi, India

Virendra Nagarale
Department of Geography
S.N.D.T. Women's University
Pune, Maharashtra, India

Rajesh Kumar Abhay
Department of Geography
Dyal Singh College
University of Delhi
New Delhi, Delhi, India

ISSN 2198-3542 ISSN 2198-3550 (electronic)
Advances in Geographical and Environmental Sciences
ISBN 978-981-97-7992-5 ISBN 978-981-97-7993-2 (eBook)
https://doi.org/10.1007/978-981-97-7993-2



This Springer imprint is published by the registered company Springer Nature Singapore Pte Ltd.
The registered company address is: 152 Beach Road, #21-01/04 Gateway East, Singapore 189721, Singapore

# Preface

In today's world, where we face unprecedented environmental challenges, social inequalities, and economic uncertainties, adopting sustainable development practices is essential for ensuring a prosperous and equitable future for all. Sustainability is indeed the need of the hour. Sustainability incorporates the integration of economic growth, social equity, and environmental sustainability, and hence can lead to meeting the needs of the present without compromising the ability of future generations to meet their own needs. It recognizes the interconnectedness of human well-being and planetary health, emphasizing the importance of responsible resource management, social inclusion, and ecological resilience.

At its core, sustainability tries to address global issues such as climate change, biodiversity loss, poverty, hunger, health, and social injustice. By promoting sustainable practices in sectors such as energy, transportation, environment, urban planning, and education, the adverse impacts of human activities on the environment can be mitigated. Moreover, geosustainability is crucial for fostering inclusive and equitable societies where all individuals have access to education, healthcare, clean water, and economic opportunities. By prioritizing social equity and human rights, we can reduce inequalities, empower marginalized communities, and build sustainable communities capable of adapting to environmental and economic shocks.

In addition to its environmental and social benefits, sustainable development also offers economic advantages. Investing in renewable energy, green infrastructure, and sustainable agriculture can create jobs, spur innovation, and stimulate economic growth while reducing dependence on fossil fuels and promoting energy security. However, achieving sustainable development requires concerted efforts from governments, businesses, civil society, and individuals alike. It calls for bold leadership, innovative solutions, and collaborative partnerships to overcome barriers and drive systemic change. Governments must enact supportive policies, regulations, and incentives to promote sustainable practices and hold businesses accountable for their social and environmental impact. Businesses, in turn, must integrate sustainability into their operations, supply chains, and corporate strategies, recognizing that long-term success depends on responsible stewardship of natural and human resources.

At the same time, individuals can contribute to sustainable development through conscious consumption choices, lifestyle changes, and community engagement. By reducing waste, conserving energy, supporting sustainable products and services, and advocating for change, each of us can play a role in advancing the transition to a more sustainable and resilient world. In conclusion, sustainability is not just an option but a necessity in today's interconnected and interdependent world. It offers a pathway to addressing pressing global challenges, promoting social equity, and securing a prosperous future for generations to come. By embracing sustainable development principles and working together towards common goals, we can create a world where people and the planet thrive in harmony.

The book provides a series of chapters linked with each other and is related to the theme of the book. The chapters provide detailed insights into the emerging geosustainability transformation in India through various review papers, case study-based, and methodological papers. The contributors of the chapter include researchers, scholars, independent researchers, academicians, and well-established geographers. The book addresses national, regional, and local level issues related to sustainability of South Asia in general and India in particular. The volume is a reference book and worthwhile for disciplines like geography, planning, environment, rural development, public health, and policy. The book has been prepared under the Advances in Geographical and Environmental Sciences series of Springer as chapter contributors.

New Delhi, Delhi, India Subhash Anand
Pune, Maharashtra, India Virendra Nagarale
New Delhi, Delhi, India Rajesh Kumar Abhay

# Praise for *Emerging Geosustainability Transformations in India*

"A remarkable collection of brilliant papers".

—Prof. Pooran Chand Joshi, *Former Vice—Chancellor, University of Delhi, Delhi, India*

"This edited book on Geosustainability is a good collection of papers on various aspects of sustainable development".

—Prof. H. S. Sharma, *Formerly Professor and Head, Department of Geography, University of Rajasthan, Jaipur, Rajasthan, India*

"The geographical nature of sustainability is viewed by eminent scholars and adoptability of a number of concepts and methodologies are aptly explained".

—Prof. K. Kumaraswamy, *Emeritus Professor, Department of Geography, Bharathidasan University, Tiruchirapalli, Tamil Nadu, India*

"Realisation of economic vision, protection of fragile ecosystem and attainment of social equity as have been penned down with great endeavour and sincerity in the present work".

—Prof. Smita Bhutani, *Centre of Advanced Study in Geography, Panjab University, Chandigarh*

"From Gandhian philosophy on sustainability to UN sustainable development goals, the book diagnoses varied aspects of natural and human environments with a plea for Geosustainability. A must read for planners, policy makers and those who wish to save the Earth and save its natural and human resources".

—Prof. Sudesh Nangia, *Formerly Professor, CSRD, Jawaharlal Nehru University, New Delhi*

# Contents

# Editors and Contributors

## About the Editors

**Prof. Subhash Anand** is currently a professor at the Department of Geography, Delhi School of Economics, University of Delhi, India. He has 25 years of experience in teaching and research and has published eight books and more than 70 research papers, articles, and chapters in journals and books. He specializes in urban environmental issues, sustainability, solid waste management, and planning and geotourism. Dr. Anand is the vice president of the Indian Geographical Society (IGS) and the vice chair of the Commission on Geoheritage, International Geographical Union (IGU). He is editor of the Springer series on Advances in Geographical and Environmental Sciences and an associate editor-in-chief of the *International Journal of Geoheritage and Parks* as well as an editorial board member of many journals. He is Chairperson of governing body of Kalindi college, Treasurer in Dyal Singh college and member in various colleges of the University of Delhi and has supervised 14 Ph.D. and M.Phil. theses. Dr. Anand led the Indian delegation on the ICSSR-JSPS joint research program to Hiroshima. In addition, he has held an advanced training workshop, "Urban Spatial Planning in Response to Climate Change in Asia", in Taipei; and a Commonwealth Geographical Bureau (CGB) workshop, "Human Consequences of Climate Change", at the University of Colombo. He has traveled widely to participate in international conferences, seminars, and workshops and has delivered lectures in several countries.

**Prof. Virendra Nagarale** is a Senior Professor in the Department of Geography at S.N.D.T. Women's University Mumbai, Pune Campus (India). He has guided 14 M.Phil. and 12 Ph.D. students successfully and is currently guiding two Ph.D. students. He has published more than 35 research papers in reputed journals and presented more than 45 research papers in International, National, and other conferences. Professor Nagarale presented his research papers in countries like USA, France, Austria, China, Poland, Japan, Portugal, Russia, Norway, Ireland, etc. He has completed three research projects and presently working on one Major Research Project funded by ICSSR New Delhi. He is the recipient of "Prof. B. Hema Malini Eminent Geographer Award 2019" by National Association of Geographers, India (NAGI) in 2019, Prof. Pravin Saptrshi Best Teacher Award 2022–2023 by Maharashtra Bhugolshastra Parishad (MBP) and also recipient of Best Teacher Award honored by Pune Municipal Corporation Pune.

**Dr. Rajesh Kumar Abhay** is currently an Assistant Professor of Geography at Dyal Singh College, University of Delhi, Delhi. Dr. Abhay completed all his academic qualification (B.A.–Ph.D.) in Geography from the University of Delhi. His area of specialization includes Natural Resource Management, Urban Sustainability and Remote Sensing and GIS. He is the recipient of Prof. N. P. Aiyyar Young Geographer Award of National Association of Geographers, India (NAGI) held at Jaipur in 2014. He has also received an UGC Inter-University Centre Associate-ship for research at Indian Institute of Advanced Study (IIAS), Shimla, recently in September 2022. He has published five books, a Dictionary of Geography, 21 research papers, and 19 book chapters at national and international level. He is actively engaged in presenting research papers in national and international conferences.

## Contributors

**Rajesh Kumar Abhay** Department of Geography, Dyal Singh College, University of Delhi, New Delhi, India

**Sarfaraz Alam** Department of Geography, Institute of Science, Banaras Hindu University, Varanasi, Uttar Pradesh, India

**Subhash Anand** Department of Geography, Delhi School of Economics, University of Delhi, New Delhi, India

**Md Anisujjaman** Department of Geography, Sidho Kanho Birsha University, Purulia, West Bengal, India

**Ranjita Banik** Individual Researcher, Howrah, India

**Shabnoor Bano** Department of Geography, Delhi School of Economics, University of Delhi, Delhi, India

**Rosy Bansal** GSSDGS Khalsa College, Patiala, Punjab, India

**Santosh Bhailume** Department of Geography, KVN Naik Shikshan Prasarak Sanstha's Arts, Commerce, and Science College, Nashik, Maharashtra, India

**Purushottam V. Bhandare** Peace Economy Project, Chicago, USA

**Aritra Chakraborty** Department of Geography, Presidency University, Kolkata, India

**K. V. Chamar** Department of Geography, M.D. University, Rohtak, Haryana, India

**Dharmender Singh Chauhan** Department of Geography, University of Rajasthan, Jaipur, Rajasthan, India

**Shivam Chaurasiya** Department of Geography, Delhi School of Economics, University of Delhi, New Delhi, India

**Shanta Dutta** Department of Civil and Environmental Engineering, The Hong Kong Polytechnic University, Kowloon, Hong Kong, China

**Kapil Kumar Gavsker** Department of Geography, Institute of Science, Banaras Hindu University, Varanasi, India

**Shweta Ghimireh** Department of Geography, SNDT Women's University, Pune, Maharashtra, India

**Sourav Halder** P. G. Department of Geography, Prabhu Jagatbandhu College, Howrah, West Bengal, India;
Department of Geography, The University of Burdwan, Purba Bardhaman, West Bengal, India

**Monika Hans** Government PG College for Women Gandhi Nagar, Jammu, Jammu and Kashmir, India

**Md Sariful Islam** Media Lab, Massachusetts Institute Technology, Cambridge, MA, USA

**N. C. Jana** Department of Geography, The University of Burdwan, Purba Bardhaman, West Bengal, India

**Kunal Jha** Department of Geography, Institute of Science, Banaras Hindu University, Varanasi, India

**Yuvraj U. Kadam** Shivaji University, Kolhapur, Maharashtra, India

**Asmita Kanav** Department of Geography, Central University of Haryana, Mahendergarh, Haryana, India

**Snehal N. Kasar** K.V.N. Naik Shikshan Prasarak Sanstha's Arts, Commerce and Science College Nashik, Canada Corner, Nashik, Maharashtra, India

**Ashwini Kumar** Department of Geography, C.H.L. Government College, Chhara, Jhajjar, Haryana, India

**Harish Kumar** Department of Geography, Delhi School of Economics, University of Delhi, Delhi, India

**Jitendra Kumar** Department of Geography, Central University of Haryana, Mahendergarh, Haryana, India

**Mukesh Kumar** Department of Physics, Swami Shraddhanand College, University of Delhi, Alipur, Delhi, India

**Naresh Kumar** Jawaharlal Nehru University, New Delhi, India

**Indraj Kumar Kumawat** Department of Chemistry, Dyal Singh College, University of Delhi, New Delhi, India

**Mahima** Department of Geography, Delhi School of Economics, University of Delhi, Delhi, India

**Suvra Maji** Department of Geography, Pakuahat Degree College, Malda, West Bengal, India

**Sagar Mali** Vidya Prabodhini College of Commerce, Education, Computer and Management, Goa, India

**Vinod Kumar Mayala** Department of Economics, Hansraj College, University of Delhi, Delhi, India

**Asmita Mukherjee** Bijoy Krishna Girls College, Howrah, West Bengal, India

**Virendra Nagarale** Department of Geography, SNDT Women's University, Pune, Maharashtra, India

**Veena M. Nare** Department of English, Mahilaratna Pushapatai Hiray Arts, Science and Commerce Mahila Mahavidyalaya, Nashik, India

**Subhash N. Nikam** Arts, Science and Commerce College, Manmad, Nashik, India

**Deep Narayan Pandey** Special Centre for Disaster Research, Jawaharlal Nehru University, New Delhi, India

**Hemlata P. Patel** Department of Geography, Savitribai Phule Pune University Pune, Pune, India

**Amruta Pattnaik** EE, NIST University, Berhampur, Odisha, India

**Jyoti Paul** Department of Commerce, Dyal Singh College, University of Delhi, New Delhi, India

**Gandharva H. Pednekar** Give Grants, Bengaluru, India

**Venkatesh PrabhuGoankar** Parvatibai Chowgule College of Arts and Science, Margao, Goa, India

**Tejbir Singh Rana** Department of Geography, Shivaji College, University of Delhi, New Delhi, India

**Ashu Rani** Department of Geography, Government College Baund Kalan, District Charkhi Dadri, Haryana, India

**Neeraj Rani** Department of Geography, M.D. University, Rohtak, Haryana, India

**Shweta Rani** Department of Geography, Dyal Singh College, University of Delhi, New Delhi, India

**Usha Rani** Department of Geography, Shivaji College, University of Delhi, Delhi, India

**Shimul Roy** Department of Environmental Science and Resource Management, Mawlana Bhashani Science and Technology University, Tangail, Bangladesh

**Swati Sachdev** Department of Geography, Budge Budge College, affiliated to University of Calcutta, Kolkata, India

**Barnika Saha** Ufaber, Mumbai, India

**Nandkumar Sawant** Parvatibai Chowgule College of Arts and Science, Margao, Goa, India

**Arti Sharma** Department of Geography, Delhi School of Economics, University of Delhi, New Delhi, India

**Rashmee Sharma** Give Grants, Bengaluru, India

**Anamika Singh** St. Xavier's College, Jaipur, Rajasthan, India

**Rashmi Singh** Department of Geography, Miranda House, University of Delhi, Delhi, India

**Uma Shanker Singh** Department of History, Dyal Singh College, University of Delhi, New Delhi, India

**Ganesh Sonawane** Department of Geography, K. V. N. Naik Shikshan Prasarak Sanstha's Arts, Commerce, and Science College, Nashik, Maharashtra, India

**Gorakh Singh Teja** GSSDGS Khalsa College, Patiala, Punjab, India

**Piyush Telang** Department of Geography, SNDT Women's University, Pune, Maharashtra, India

**Namdev V. Telore** Department of Geography, Raja Shripatrao Bhagawantrao Mahavidyalaya, Satara, Maharashtra, India

**Deepak N. Thakare** Department of Geography, Mahilaratna Pushapatai Hiray Arts, Science and Commerce Mahila Mahavidyalaya, Nashik, India

**Swati Thakur** Department of Geography, M.M.H. College, Ghaziabad, India

**Tanmoy Roy Tusher** Department of Environmental Science and Resource Management, Mawlana Bhashani Science and Technology University, Tangail, Bangladesh

**Anil Yedage** Parvatibai Chowgule College of Arts and Science, Margao, Goa, India

**Rajendra B. Zolekar** K.V.N. Naik Shikshan Prasarak Sanstha's Arts, Commerce and Science College Nashik, Canada Corner, Nashik, Maharashtra, India

# Abbreviations

| | |
|---|---|
| ADDA | Asansol Durgapur Development Authority |
| AET | Actual Evapotranspiration |
| AHP | Analytic Hierarchy Process |
| AI | Aggregation Index |
| AMRUT | Atal Mission for Rejuvenation and Urban Transformation |
| ANN | Artificial Neural Networks |
| AWC | Annual Water Holding Capacity |
| BIS | Bureau of Indian Standards |
| BPL | Below Poverty Level |
| CDC | Centres for Disease Control and Prevention |
| CDP | Community Development Programme |
| CGWB | Central Groundwater Board |
| CRZ | Coastal Regulation Zone |
| DAHD | Department of Animal Husbandry and Dairying |
| DCHB | District Census Handbook |
| DJB | Delhi Jal Board |
| DPHE | Department of Public Health and Engineering |
| ED | Edge Density |
| EEZs | Exclusive Economic Zones |
| GCZMA | Goa Coastal Zone Management Authority |
| GDP | Gross Domestic Products |
| GIS | Geographical Information System |
| GSMC | Global System for Mobile Communication |
| GUTP | Green Urban Transport Policies |
| HEIs | Higher Education Institutions |
| HRIDAY | National Heritage City Development and Augmentation Yojana |
| HTL | High Tide Line |
| ICMR | Indian Council of Medical Research |
| IDPH | Illinois Department of Public Health |
| IFAD | International Fund for Agricultural Development |
| IJI | Interspersion and Juxtaposition Index |

| | |
|---|---|
| IPDS | Integrated Power Development Scheme |
| ISRO | Indian Space Research Organization |
| IUWM | Integrated Urban Water Management |
| JNNURM | Jawaharlal Nehru National Urban Renewal Mission |
| LGED | Local Government and Engineering Department |
| LSA | Land Suitability Analysis |
| LSI | Landscape Shape Index |
| LST | Land Surface Temperature |
| LULC | Land Use Land Cover |
| MADP | Mewat Area Development Project |
| MCDM | Multi-Criteria Decision Making |
| MDGs | Millennium Development Goals |
| MDS | Mewat Development Society |
| MGNREGA | Mahatma Gandhi National Rural Employment Guarantee Act |
| MLC | Maximum Likelihood Classifier |
| MPS | Mean Patch Size |
| MYRADA | Mysore Resettlement and Development Agency |
| NAAC | National Assessment and Accreditation Council |
| NDVI | Normalized Difference Vegetation Index |
| NLSI | Normalized Landscape Shape Index |
| NPI | Nemerow's Pollution Index |
| NRSC | National Remote Sensing Centre |
| NULM | National Urban Livelihood Mission |
| OA | Overall Accuracy |
| PA | Producer's Accuracy |
| PCA | Principal Component Analysis |
| PCM | Pairwise Comparison Matrix |
| PD | Patch Density |
| PET | Potential Evapotranspiration |
| RS | Remote Sensing |
| SCM | Smart Cities Mission |
| SDGs | Sustainable Development Goals |
| SEZs | Special Economic Zones |
| SHGs | Self-Help Groups |
| SHS | Solar Home Systems |
| SLM | Sustainable Land Management |
| SME | Small and Medium-Sized Enterprises |
| SNM | Strategic Niche Management |
| SPV | Special Purpose Vehicle |
| STS | Studies of Technology and Science |
| SVM | Support Vector System |
| TALC | Tourism Area Life Cycle |
| UA | User's Accuracy |
| ULBs | Urban Local Bodies |

| | |
|---|---|
| WASA | Water Supply and Sewerage Authority |
| WHO | World Health Organization |
| WLC | Weighted Linear Combination |
| WQI | Water Quality Index |

# Part I
# Introduction

# Chapter 1
# Geosustainability Transformations in India

## Introduction and Overview

**Subhash Anand, Rajesh Kumar Abhay, and Virendra Nagarale**

**Abstract** The volume is published under the Advances in Geographical and Environmental Sciences series of Springer as chapter contributors. The book tries to understand the transformation from the geographical as well as sustainability perspective. This volume consists of studies from various scholars across India. Geosustainability transformation represents a shift in societal values, behavior, system toward a more resilient, equitable, and sustainable future. At its centre, it involves transitioning from current unsustainable practices, which degrade the environment, exacerbate social inequalities, and compromise future generations' well-being, to more sustainable pathways that prioritize environmental stewardship, social justice, and economic prosperity. This transformation is imperative in light of pressing global challenges such as climate change, biodiversity loss, resource depletion, pseudo-urbanization, gender disparity, and social injustice, which threaten the sustainability and prosperity of human societies. All such issues are imperative to be addressed to achieve sustainability in the world in general and India in particular. The volume ends with several emerging research frontiers out of the issues discussed in the book.

**Keywords** Sustainable development · Education · Social · Resources · Health · Tourism

S. Anand (✉)
Department of Geography, Delhi School of Economics, University of Delhi, Delhi, India
e-mail: sanandpvs@gmail.com

R. K. Abhay
Department of Geography, Dyal Singh College, University of Delhi, New Delhi, India
e-mail: rkabhay@dsc.du.ac.in

V. Nagarale
Department of Geography, SNDT Women's University, Pune Campus, Pune, Maharashtra, India
e-mail: drnagarale@gmail.com

 
S. Anand et al. (eds.), *Emerging Geosustainability Transformations in India*, Advances in Geographical and Environmental Sciences,
https://doi.org/10.1007/978-981-97-7993-2_1

Sustainability transformation refers to the fundamental changes needed across various sectors of society to achieve sustainability goals. It involves transitioning toward more sustainable patterns of production, consumption, and governance to ensure the long-term well-being of both people and the planet (Patterson et al. 2017; Söderholm 2020). Key to sustainability transformation is the recognition of interconnectedness–the understanding that environmental, social, and economic systems are deeply intertwined and must be addressed holistically (Voulvoulis et al. 2022). It requires a departure from the prevailing linear, extractive economic model toward circular and regenerative approaches that mimic nature's cycles and minimize waste. Embracing principles of ecological integrity, such as biodiversity conservation, ecosystem restoration, and sustainable resource management, is essential for ensuring the long-term health and resilience of natural systems upon which all life depends (Niesenbaum 2019). Moreover, sustainability transformation necessitates a re-evaluation of societal values and priorities, moving away from a relentless pursuit of economic growth at the expense of social and environmental well-being toward more inclusive, equitable, and just societies (Hosseini 2023). This entails fostering cultures of care, cooperation, and solidarity that prioritize human flourishing, social equity, and community resilience over material consumption and individualistic pursuits. Empowering marginalized communities, promoting gender equality, and upholding indigenous rights are central to building more inclusive and participatory decision-making processes that ensure diverse voices are heard and respected.

At the heart of sustainability transformation lies innovation and technology, which play a pivotal role in driving systemic change and unlocking new possibilities for sustainable development (Martínez-Peláez et al. 2023). From renewable energy and clean technologies to sustainable agriculture and green infrastructure, technological innovations offer promising solutions for decoupling economic growth from environmental degradation and advancing the transition to a low-carbon, resource-efficient economy. However, technology alone is not sufficient; it must be coupled with social innovation, policy reform, and grassroots mobilization to create enabling conditions for transformative change (GoI 2020). Crucially, sustainability transformation requires bold and visionary leadership at all levels–from local communities and businesses to national governments and international institutions. Political leaders, policymakers, and business executives must demonstrate courage and foresight in setting ambitious sustainability goals, enacting supportive policies, and mobilizing resources for implementation. Collaboration and partnership are essential for harnessing collective action and leveraging diverse expertise, resources, and perspectives toward shared sustainability objectives.

Education and public awareness are also critical drivers of sustainability transformation, as they shape values, attitudes, and behaviors that influence individual and collective actions (Filho et al. 2018). Formal and informal education systems should integrate sustainability principles across curricula and promote critical thinking, systems literacy, and ethical decision-making skills. Public awareness campaigns, media outreach, and digital platforms can amplify messages of sustainability and inspire citizens to adopt more environmentally responsible lifestyles and consumption patterns (Borawska 2017). Ultimately, sustainability needs to be achieved

**Table 1.1** Content organization

| | |
|---|---|
| Part-I | Introduction (2 papers) |
| Part-II | Sustainable Urbanization and Policies (6) |
| Part-III | Geo-Environmental Sustainability and Tourism (7) |
| Part-IV | Rural Sustainability and Livelihood Security (5) |
| Part-V | Sustainability Through Education and Health Policy (7) |

through the approach of sustainable development (Washington 2015; Abhay and Rani 2023). It demands a fundamental shift in mindset from short-term thinking to long-term stewardship, from competition to collaboration, and from exploitation to regeneration. While the challenges ahead are daunting, the opportunities for positive change are immense. By embracing sustainability as a guiding principle and collective imperative, we can pave the way for a more just, resilient, and flourishing world for present and future generations.

The book is organized into 5 parts and 27 chapters. Various parts of the book are arranged, sequentially and theme-wise, so that a continuous understanding of transformations of various concepts can be understood. Among the five parts, the first is the introduction which includes two chapters. The second part explains the transformation through the urbanization process and evaluates their sustainability with policy implications. The third part deals with geo-environmental sustainability along with tourism a tourism-related case study. Both involve major sectors of the economy. The fourth part of the book elaborates on social and rural sustainability transformation and how their livelihood is secured through various means. The last section discusses various geo-sustainability issues related to education and health (Table 1.1). A detailed description of these sections is discussed below:

## 1.1 Part I Introduction

Part one provides introductory glimpses of the book. It conceptualizes the book formation and explains all sections of the book. The first chapter by Anand, Abhay, and Nagarale explains the meaning of geosustainability transformations and discusses how it requires the attention of scholars to address these challenges to achieve sustainability in India. The second chapter by Singh evaluates the role of Gandhian philosophy in achieving sustainability. The chapter also addresses the intellectual influence of Gandhi, his critique of industrialization and modernization along with his perspective on sustainable development. The author of the chapter concludes that the Gandhian philosophy of trusteeship has huge potential to act as the base of sustainability and sustainable development.

## 1.2 Part II: Sustainable Urbanization and Policies

Part II of the book addresses the geosustainability transformation in the field of sustainable urbanization through six chapters. The paper by Bhailume and Sonawane analyzes the urban growth using geospatial techniques. The study employs the Landsat satellite series data to present an in-depth geospatial investigation of urbanization patterns within three municipal corporations in Maharashtra, namely the Pune Municipal Corporation, Pimpri-Chinchwad Municipal Corporation and Nashik Municipal Corporation over the span of 2001–2021. The study investigates land use/land cover dynamics and urban growth and presents landscape metrics and develops several indexes. The authors address these urban transitions across the environmental entities (land use/cover) and argue for a growing conflict between environmental response and human progress. The next chapter by Pednekar, Bhandare, and Sharma evaluates the shifting perspectives in land acquisition policies and their impact on the urban development of Navi Mumbai region in the Maharashtra state. Authors explain that in a post-independence scenario, the urban centre in India, especially Mumbai, attracted migrants from rural areas and this led to the development of the urban fringe of Mumbai known as Navi Mumbai. The paper analyzes the developmental planning of the Navi Mumbai region, changes in land acquisition and rehabilitation and resettlement policies over the years and the role of stakeholders in development while asserting the gaps in implementation.

Another aspect of urbanization taken up by Kanav and Kumar examines green space planning which has a huge role in enhancing the quality of urban life, restoring ecological balance, and promoting the well-being of people. The chapter delves into the fundamental green space planning principles by highlighting the pressing need for sustainable urban development in Indian cities. The chapter critically examines six key green space planning principles, established through previous research that serve as a comprehensive framework for guiding sustainable urban development strategies; namely: connectivity, multi-functionality, multi-scale, integration, participation, and continuity. The author of the chapter concludes that green spaces can direct urban development by offering an opportunity for economic growth and environmental conservation if strategically planned, designed, and executed. On the other hand, the chapter by Halder and Jana analyzes the impact of migration and urbanization on sustainable development in coastal blocks of the Sundarbans region of West Bengal. The author tries to understand the relationships between population dynamics, migration, urbanization, vegetation index, and urban heat islands. The study observed the migration of many youthful populations due to adverse situations that indicate human brain drain from the study area. The authors also suggested relevant strategies to restore sustainability within the process of migration and urbanization in the study area. The next chapter by Maji and Alam also explores sustainable development opportunities in Asansol city located again in West Bengal state using the Eco2 city approach. The author argues that the city has undergone economic functional changes—mainly from a mining and transport town to an economically complex city. Due to the expansion of the industrial footprint, the problems of air and water pollution have become a major concern in the city. Against this backdrop, the chapter advocates for adopting the "Eco2 City" approach by integrating the economic and ecological concerns in the planning process and sustainable transformation of Asansol City. The last chapter of the section takes the section toward urban sustainability. Gavsker and Jha examine the performance, initiatives, and spatial contrasts of smart cities in the states of Punjab and Uttar Pradesh.

The authors suggest that to make urban places liveable, resilient and sustainable, smart solutions, technology-based applications, and advanced effective digital spaces need to be emphasized in urban planning. Under the Smart Cities Mission, three cities in Punjab and ten cities in Uttar Pradesh states of India are studied in this chapter. The present study analyzed urbanization and urban growth patterns and elucidated components, elements, and strategies of smart city planning; examined performance and unconventional initiatives introduced in making them smart urban spaces; and explored the significance of smart cities in urban sustainability and inherent challenges.

## 1.3 Part III: Geo-environmental Sustainability and Tourism

The third part of the book includes seven chapters and highlights geo-environmental sustainability and tourism. The first chapter of the section by Roy, Islam, Tusher, and Dutta summarizes the urbanization trend in Dhaka city, the status of water supply and sanitation services, existing infrastructural facilities, and significant challenges and prospects toward ensuring sustainable water supply and sanitation in light of the United Nation's sustainable development goal (SDG). The study shows that the rapid urbanization process in Dhaka city creates tremendous pressure on the city's infrastructures and essential services, including water supply and sanitation. The next study by *Rani, Mahima and Bano carried out and h*ydrological analysis and climatic variability of water balance in the Delhi Metropolitan Region. The study intends to estimate and calculate water balance based on the Thornthwaite-Mather water balance model for Delhi and to propose innovative measures to manage water during water deficit and water surplus months to ensure durability, self-reliance, and sustainability of water in the long run. The findings and recommendations of the chapter ensure the hydrological sustainability of the area. The next chapter is contributed by Singh, Kumar, Rani, and Kumar on groundwater quality assessment in the Jaipur city of Rajasthan state. Through the physicochemical, biological, and microbiological parameters-based analysis, this study evaluates the suitability of groundwater for drinking purposes. The findings underscore the need for effective water management strategies, including improved monitoring and treatment measures, to safeguard public health and ensure the sustainable utilization of groundwater resources in urban areas. Further, the study by Telore and Kadam applies morphometric analysis for groundwater prospects in the Koyna watershed region in the Maharashtra state.

The morphometric parameters like stream order, stream length, bifurcation ratio, drainage density, stream frequency, relief ratio, elongation ratio, circularity ratio, and compactness constant are used to study groundwater prospects. The study highlights the favorable groundwater prospects within the region, indicating a well-distributed network of streams contributing to groundwater availability.

The study by Zolekar, Patel and Kasar explores the methodological framework, and applications of land suitability analysis in agriculture, demonstrating its potential to analyse geosustainability transformations. Land suitability analysis has emerged as a hope in addressing this complex land use challenge. It serves as a valuable method for uncovering the inherent capabilities and potential of land for various purposes, providing valuable insights for sustainable land management.

The section also addresses geosustainability of scenic resources. Sawant, Goankar, Mali, and Yedage have studied coastal tourism and sustainability in Calangute and Anjuna Beach Destinations of Goa state. The study attempts to envisage the development of tourism and the dynamics of decadal changes and transformation of land use and land cover features over nearly three decades after globalization. The study concludes that the studied destinations have reached the stagnation stage of the Tourism Area Life Cycle and areas close to the beach have been converted into tourist services and commercial utilities. Another study by Nagarale, Telang, and Ghimireh evaluates the Impact of the Covid-19 Pandemic on tourism sustainability in Darjeeling District located in the eastern Indian state. The study explores Covid-19 implications for the environment, local communities, and the long-term viability of the tourism sector. The research finds that the decline in tourist inflow during the pandemic has led to a reduced carbon footprint, as transportation-related emissions, including those from tourist vehicles, decreased due to fewer visitors.

## 1.4 Part IV: Rural Sustainability and Livelihood Security

The fourth sections include five chapters and highlight geosustainability transformations through social and rural livelihood perspectives. The first chapter by Sharma and Rana establishes the link between technological advancement in cooking and its impact on women's quality of life for their sustainability. The study found that there is an alteration in the thinking of family members with each new addition of kitchen appliances in a household. With the advent of each new kitchen appliance, other family members also developed an interest in cooking food. But it also observed that the addition of new cooking appliances is given the least priority in the budget of middle- and lower-income households.

Rural transformation and sustainability are intertwined concepts crucial for the development and well-being of rural areas, which often play a significant role in national economies. The study by Sachdev, Chakraborty, Pandey, Anisujjaman, Thakur and Saha paper examines the nature of rural livelihood and the multi-faceted aspects of sustainability in general and especially in the Bankura district of West Bengal. The study argues that sustainable livelihood entails providing rural areas with physical, and economic knowledge, societal, and electronic connectivity in the contemporary era. The paper also focuses on one arena of livelihood management to promote sustainable rural development. In the next chapter, Mukherjee and Banik evaluate the role of shuttlecock manufacturing for a local sustainable economy in the Uluberia sub-division of the Howrah district in West Bengal. The paper focuses on the present situation of the industry and the socioeconomic condition of the people associated with the shuttlecock business.

Agricultural land and its sustainability have been analyzed by Kumar and Pattnaik. The author provides smart energy solutions based on solar irrigation systems for sustainable agriculture in India. The authors argue that the solution to growing energy demands in India lies in solar panels that are both efficient and affordable and Indian farmers need intelligent irrigation systems fuelled by solar energy. It has been suggested that the answer to address the energy issue for Indian farmers can be solved by integrated solar-powered water pumps.

The last chapter of the section, contributed by Kumar attempts to understand the concept of participation and the role of the state in promoting sustainable development in the Mewat Area Development Project in the Mewat region of Haryana state. The study aims to problematize the concepts of "participation" and "community" in the development process. The study concludes that the role of the state in community development is to create an enabling environment where communities can thrive, ensuring that all members have access to opportunities, resources, and services necessary for their well-being and prosperity. Collaboration between government, communities, and other stakeholders is essential for sustainable and inclusive development.

## 1.5 Part V: Sustainability Through Education and Health Policy

Sustainability through education and public health policy is important in fostering the well-being of individuals and communities while safeguarding the planet for future generations. Taking both into consideration, under this section, seven chapters have been listed. The first chapter contributed by Paul, Abhay and Kumawat on Higher Education Institutions (HEIs) and Sustainable Practices. The study tries to explore the areas in which HEIs work and what sustainable practices these HEIs are adopting related to social, economic, environmental, and governance dimensions. The results of the study are beneficial for the policymakers, university authorities, and others concerned to make sustainable plans for the higher education institutions in India. The next chapter by Rani and Chamar evaluates the gender disparities in the level of literacy in the socially deprived caste in the Jhajjar district of Haryana state. The study identified gender disparity in literacy rate and level of educational attainment up to the levels of post-graduation among the Balmiki caste. The study concludes, based on a sample survey, that the Jhajjar district has a higher literacy rate among the

Balmiki caste. The study concludes that there has been a growing awareness about the importance of education in promoting sustainable development and reducing poverty, among marginalized communities particularly the Balmiki caste in rural areas of Haryana.

Similarly, Chauhan and Singh analyze the relationship between literacy rate and sex ratio. It attempts to bring forth the correlation between the literacy rate and sex ratio of Ajmer district for sustainable education. The author argued that a sustainable approach to education will ensure quality in education, and gender equality and reduce inequalities through a change in the mindset of the societies. Educational attainment among females in rural India has been a subject of significant interest and concern due to its implications for gender equality, economic development, and public health. The study by Rani, Chamar, and Mayala examines the levels of educational attainment among females in rural India. The study brings out regional disparities in educational attainment among females across the country.

Public health is also important that has experienced geosustainability transformations in India. The study by Bansal, Teja, and Hans evaluates advances in public health policy for sustainable development in India. The author argues that the goal of public health is to prevent and control disease at the population level by coordinated efforts and well-informed decisions made by people, groups, the public and private sectors, and society. However, to achieve health fairness the government's intervention is essential. Nare, Nikam, and Thakare contribute another paper on public health policy. The paper examines case studies and examples of successful integration of public health policy and sustainable practices in various settings, including urban planning, transportation, food systems, and healthcare. It is concluded that there is a huge potential for a more holistic and integrated approach to public health policy that takes into account the long-term sustainability of our communities and planet.

The last chapter of the section and also of the book by Abhay, Anand, Nagarale, and Chaurasiya brings forth emerging research frontiers out of the issues discussed in the book. The paper highlights the emerging field of sustainable transitions, particularly through geographical lenses. It tries to address the gaps in existing literature by offering novel insights into the spatial dimensions of sustainability. Thus, this volume provides national-level information. The editors expect that this volume will serve as reference material for academicians, professionals, and students. The book is a combination of theoretical, case studies, reviews, and location-specific problems that will be helpful to the researchers in comprehending the geosustainability transformations in the country. Many chapters have raised new research questions and developed ideas that will generate interest among, academicians, researchers, and students.

## References

Abhay RK, Rani S (2023) Sustainable development: concept, components and history. In: Abhay RK, Pandey DN, Rani S (eds) Constraints in achieving sustainability of India. TERI, New Delhi, pp 15–28

Borawska A (2017) The role of public awareness campaigns in sustainable development. Econ Environ Stud 17(4):865–877. https://doi.org/10.25167/ees.2017.44.14

Filho WL, Raath S, Lazzarini B et al (2018) The role of transformation in learning and education for sustainability. J Clean Prod 199:286–295. https://doi.org/10.1016/j.jclepro.2018.07.017

GoI [Government of India] (2020) Science, technology, and innovation policy. Ministry of Science & Technology Department of Science & Technology, Draft STIP Doc 1.4 Dec 2020. https://dst.gov.in/sites/default/files/STIP_Doc_1.4_Dec2020.pdf. Accessed 10 Apr 2024

Hosseini SAH (2023) The well-living paradigm: reimagining quality of life in our turbulent world. Discov Glob Soc 1:19. https://doi.org/10.1007/s44282-023-00022-8

Martínez-Peláez R, Ochoa-Brust A, Rivera S et al (2023) Role of digital transformation for achieving sustainability: mediated role of stakeholders, key capabilities, and technology. Sustainability 15:11221. https://doi.org/10.3390/su151411221

Niesenbaum RA (2019) The integration of conservation, biodiversity, and sustainability. Sustainability 11(17):4676. https://doi.org/10.3390/su11174676

Patterson J, Schulz K, Vervoort J (2017) Exploring the governance and politics of transformations towards sustainability. Environ Innov Soc Trans 24:1–16. https://doi.org/10.1016/j.eist.2016.09.001

Söderholm P (2020) The green economy transition: the challenges of technological change for sustainability. Sustain Earth 3:6. https://doi.org/10.1186/s42055-020-00029-y

Voulvoulis N, Giakoumis T, Hunt C et al (2022) Systems thinking as a paradigm shift for sustainability transformation. Glob Environ Chang 75:102544. https://doi.org/10.1016/j.gloenvcha.2022.102544

Washington H (2015) Is 'sustainability' the same as sustainable development? In: Kopnina H, Shoreman-Ouimet E (eds) Sustainability: key issues. Routledge, London and New York, pp 359–376

## References

# Chapter 2
# Gandhian Philosophy and Sustainability of India: Relevance in the Globalized Era

**Uma Shanker Singh**

**Abstract** Post Industrial Revolution, the symbiotic association between man and the environment has become increasingly complex. With the expansion of industrialization, extraction of resources has been phenomenal thus affecting the ecosystem in unprecedented ways. Today, we require sophisticated systems to manage and sustain the complex process of industrialization. Society is at risk of using and adapting to technology-based products and processes that are highly sophisticated. Presently in the globalized world, environmental-related issues have become very important. Gandhian philosophy of development provides solutions to this and is similar to the concept of sustainability. Mahatma Gandhi's Hind Swaraj is seen as a critic of industrialization, urbanization, and modernization. Since the 1960s onwards the concept of development has been thoroughly debated. In this background, the notion of sustainability has now become a matter of concern within the international community. In this context, the Gandhian philosophy of trusteeship has been examined in this chapter which provides a historical basis for its understanding in the context of India.

**Keywords** Mahatma Gandhi · Trusteeship · Development · Industrialization · Hind Swaraj

## 2.1 Introduction

In the contemporary world, the human body, mind and heart are under great stress, disarray, and disharmony; primarily because of the insatiable material desire of humans. Such a relentless pursuit of worldly gains has corroded all values and the meaning of human life. In this process, human being have been deprived of the protection, love, and warmth of community life. But there is a far more serious dimension of human predicament. There is a state of total confrontation between man and nature which is threatening the survival of mankind. Rapid exhaustion of

U. S. Singh (✉)
Department of History, Dyal Singh College, University of Delhi, New Delhi, India
e-mail: umashanker.historian@gmail.com



S. Anand et al. (eds.), *Emerging Geosustainability Transformations in India*, Advances in Geographical and Environmental Sciences,
https://doi.org/10.1007/978-981-97-7993-2_2

non-renewable energy sources, climate change, environmental degradation, global warming, ozone layer depletion, and the impending threat of the melting of the ice in the Polar Regions are some of the major dimensions of the same crisis. What is more, all the ideological movements, like liberalism and Marxism in their variant forms, have failed to take man to the promised land of peace, prosperity, and plenty (Pradhan 2014).

In the above context, intellectually sensitive minds all over the world are seriously engaged in finding a sustainable model of development based on a simple, need-based (not want-based) way of living. This search has ended virtually at the doorstep of Mahatma Gandhi. There is a reason for it. Unlike many other thinkers, Mahatma Gandhi not only offered a critique of modern industrial civilization but suggested a viable and sustainable model of development, which was based on a new, simple, and environmentally friendly lifestyle. Some of his ideas sound so relevant not only for peace and harmony between humans and humans but also between humans and nature.

## 2.2 Mahatma Gandhi: Life Sketch

One of the greatest men of the nineteenth and twentieth centuries Mohan Das Karamchand Gandhi was born in a small town at Porbandar, on the western coast of India, presently a part of Gujarat state in 1969. On the advice of his family members (father Karamchand Gandhi and mother Putlibai), he sailed for England on 04 September 1888 to continue his legal education in England. After arriving in India in July 1891 he came to know that his mother had already passed away. As Rajkot was a small place, he moved to Bombay to start his legal practice.

In April 1893 he sailed for Durban (South Africa) after getting an offer for a legal engagement from a Muslim firm, which had a business interest in South Africa. South Africa became the social lab for Gandhian experiments. This was the place where Gandhi developed a strong mindset against Colonialism and Imperialism, and his strengths reached their potential. Even after many decades, scholars have shown great interest in Gandhi's life in South Africa starting from Nagindas Sanghavi's (Sanghavi 2006) to Ashwin Desai and Goolam Vahed's "*The South African Gandhi: Stretcher-Bearer of Empire*" (Kolge 2016). The Gandhian tools satyagraha, non-violence, fast, etc., were developed in South Africa especially when Mahatma Gandhi resided in Phoenix and the Tolstoy farm. On the advice of his political guru—Gopal Krishna Gokhale he sailed for India in December 1914. He was going back to his motherland for good after 20 years. He reached Bombay in January 1915 and was given a hero's welcome. In the words of Bhikhu Parekh: "In South Africa Gandhi acquired political skills and learned lessons, some of which served him well and others ill on his return to India" (Parekh 2010).

Later, he travelled to different parts of the country to acclimatize and understand the Indian situation without making any declaration and commitment. After completing his India tour, he decided to setup Kochrab Ashram on the bank of river

Sabarmati, near Ahmedabad on 25 May 1915, which was to be called *Satyagraha Ashram*. Soon he was invited by Pt. Madan Mohan Malviya to address the inaugural function of Banaras Hindu University to be held on 4 February 1916. Gandhi's speech created a stir when he bitterly criticized the British elite, including the Viceroy and the bejeweled princess of the Indian states, for their overindulgence in their showmanship and comforts. Meanwhile, he attended the Lucknow Congress (1916), where Tilak and Jinnah were the star performers resulting in the signing of the Lucknow pact (Pradhan 2014). After the death of Tilak on 01 August 1920, Gandhi started the first mass moment in India in the form of the Non-Cooperation Movement, and two other mass moments such as the Civil Disobedience moment in the 1930s and Quit India Moment in 1942 made Mahatma Gandhi a whole sole leader of Indian national movement. He was seen by the Indian masses as a charismatic leader who was the main force behind the Indian independence in 1947.

## 2.3 Intellectual Influence on Gandhi

In the opinion of most of the Indians, Gandhi is seen as merely a politician of national movement. But, he was more than that. Mahatma Gandhi's ideas and thoughts on different aspects of human civilization have been the subject of strong scholarly debates and continuous re-evaluation. Gandhian books and essays received a significant and sustained interest of scholars particularly in his experiences and different thoughts regarding nature and human civilization. Gandhian philosophy is multidimensional. His political ethics and supporting notions about man and the state are central to his teachings and practices. His philosophical views are influenced by India's civilizational resources. Gandhi's philosophy was influenced by the views of many worldly scholars. His readings included religious and moral literature including those of Plato's *Apology* and William Salter's *Ethical Religion (1889)*. Also, Gandhi revised these into his native language Gujarati. He also regards Socrates and tries to emulate his teachings.

In South Africa, three books influenced Gandhi deeply like Henry Thoreau's *On the Duty of Civil Disobedience (1847)*, a "masterly treatise", Tolstoy's *The Kingdom of God is Within You (1893)*. These books excited him and in these he claimed to have first discovered the principles of non-violence and love, and John Ruskin's *Unto This Last (1862)*, whose "magical influence" was a "turning point" in his life. This book was given to Gandhi by a British friend H.S.L. Polak which he read during his journey from Johannesburg to Durban in 1904. Ruskin's ideas on political economy influenced the principles of Gandhi's ashram organization. Being inspired by Ruskin, Gandhi began to live life in austerity and stayed at Phoenix Farm in Natal in 1904 and then at Tolstoy Farm outside Johannesburg in 1910 (Parekh 1997). Later in India, he established the Sabarmati Ashram in Gujarat and Sevagram Ashram in Maharashtra.

In England, he read about British and European law and politics, and also read the religious scriptures including the Old Testament. He also read about the religious traditions enshrined in the Hinduism and Buddhist scriptures. The works of Ruskin,

Thoreau, and Tolstoy had a significant intellectual influence on Gandhi. Gandhi spread his views through his writings especially in the *weekly Indian opinion.*[1] while he was in South Africa. When he returned to India, he edited newspapers/tabloid such *Navajivan, Harijan.* Gandhi was opposed to Plato's ideas of social organization as mentioned in *The Republic.* He instead preferred a minimalist role for the state. Gandhi believed in the inherent goodness of man and was a strong proponent of the art of persuasion. Hence the Gandhian movement remained entirely open. In modern India, the path of discourse was initiated by Raja Rammohan Roy and elaborated by Keshab Chandra Sen, Dayanand Saraswati, Swami Vivekananda, Rabindranath Tagore, and Mahatma Gandhi.

## 2.4 Gandhi's Critique of Industrialization and Modernization

Although Mahatma Gandhi was influenced by the West, he rejected Western civilization both as a model and as an inspiration. His seminal work, *Hind Swaraj* is the primer of his political philosophy (Gandhi 1910). His *Hind Swaraj* depicts his understanding of the chaos the modern civilization would usher in having witnessed the human devastation that industrialization had caused in England, he warned us of the impending dangers of an urban industrial society. He was baffled at the thought of India being heavily industrialized and its culture eroded through dehumanizing. He wrote in Young India (Gandhi 1928) that India should not follow the path of industrialization as the West. Gandhi was undoubtedly a visionary who could foresee the ills of industrialization and modernization. He was an early critic of the dehumanizing character of modern industrial civilization. It is in this context of new value orientation and the quest for human survival threatened by the environmental and ecological crisis that the re-discovering of Gandhi's warning of "industrialize and perish" has to be seen (Singh 1999). Gandhi's ideas can be compared with that of Rousseau's *Discourses on Inequality (1755).* Gandhi considered industrialization as a key cause of human misery.

This perspective solidifies as South Africa undergoes its period of industrialization. He applauds the significant boost in efficiency and the resulting improvement in living standards in the West, yet he disagrees with the notion that industrialization, as it is presently practiced, has truly contributed to the progress and well-being of human society. He disapproves of Western civilization. Similar to Rousseau, he objects to modern technology and industrialization, citing them as the cause of suffering and inequality. His attention is drawn to the noticeable decline in societal progress evident in Italy during the struggle for independence. In discussing this, he highlights the plight of the working class and the general populace, whose dreams were overlooked by the ruling elite, with Italy still under bondage. For Gandhi, true liberation is synonymous with the freedom of the working class and those who are impoverished. The progress seen in the West was out of step with Indian traditions and fell short of addressing India's actual needs, thereby obstructing any genuine progress or development at the individual level.

The genuine state would comprise of self-sufficient towns and communities based on truth and non-violence. Gandhi stands up to the foundational standards of the cutting edge world and, in their put, offers an idealized conception of conventional life in provincial India, which he sees as giving a participation to the complexity, realism, and destitution he identifies in present-day society (Tercheck 1998). Gandhi needed a free India that would not take after the Western way. This implies giving up apparatus, present day strategies of transportation, present day solutions and machine-spun cloth. In spite of the fact that he alters a few of his thoughts in this way like tolerating small-scale businesses and those businesses where work isn't valuable or alluring, he follows the in general push of his introductory arguments as enunciated within the Rear Swaraj.

The rise and modernization of a little Asian nation like Japan at the starting of the twentieth century blended passionate assumptions in India. Gandhi educated his supporters not to take after the Japanese demonstrate. Agreeing to him the solution to the Indian issue has got to come from inside India instead of bringing in outside thoughts and teaching. Gandhi's free India is one where the economy and nation would be distinctive from that of other advanced industrialized countries. This is since nations like India with huge work and large-scale unemployment and under-employment ought to limit the utilization of apparatus. He wants industrialization that would fulfill needs like nourishment, shield, wellbeing care, and fundamental instruction. In spite of the fact that cognizant of the indifferent and repetitive life that industrialization involves, he is prepared to acknowledge it in the event that it helps to fulfill the essential human necessities.

Gandhi's demeanor to the West varies significantly from other non-Western progressives. Mao Zedong, Ho Chi Minh, Che Guevara, and Frantz Fanon not as it were "rejected the western, political and financial mastery but too their traditional ways of life of their individuals and particularly the devout components that given the establishment for the ancient societies of Asia and Africa, supplanting them by western political shapes and by western technologies" (Woodcock 1971). Fanon's (1963) discernment appears a few curiously inconsistencies within the problem of third-world progressives. Fanon compellingly argues for decolonization within The Wretched of the Soil (1963) but is utterly scornful of the settler. He upbraids the Europeans virtually as war offenders. He respects European extravagance as shameful "for it has been established on slavery… which it comes straightforwardly from the soil and the sub-soil of the immature world". Unexpectedly in showing disdain toward this genuine prosecution, Fanon conceives of alter as it were with the crucial offer assistance of the Europeans and by dismissing innate ways. Gandhi, on the other hand, needs to keep the windows open whereas his feet are immovably settled in in his claim culture. His whole accentuation revolves around the recommendation that Indian issues ought to be unraveled by indigenous strategies and not by Western thoughts as these are inconsistent. It is this manufactured viewpoint, holding the most excellent of Western conventions and coordination it with the indigenous roots of the Indian conventions that produces Gandhi one of a kind. There's nothing only

Indian around his thoughts. He castigated Indians for not changing outdated social practices and traditions and focused on the significance of social equity and equity. A free India must be free from the present day political and financial institutions.

Gandhi's conception of Swaraj is distinctive from the one that's verbalized by the Western Marxist, Communist, and Liberals ideas. Just like the Marxists and the Communists Gandhi wants an libertarian society but contradicts their deterministic view of history and human nature. He wants, like Burke, to hold India's ancient legacy and modernizing anything is worth rescuing and valuable. He is a revolutionary, and a preservationist and however none of these for, he never misplaced his significantly progressive character (Bondurant 1967). Gandhi is basically concerned with modern issues and tries to discover arrangements that are both alluring and attainable. Gandhi was not in support of bourgeois patriotism classification in India. He emphasized the awareness quality of the masses so that masses may not be misused by the settler control or by any prevailing control. He did not need India to advantage at the cost of other countries. He considered humankind as one family which is why Gandhi rose above bourgeois patriotism.

Gandhi's thoughts speak to the fullest expression of the inborn roots of advanced Indian political theory. From Raja Smash Mohan Roy to Vivekananda, the journey has been for digestion of the Western thoughts and culture with that of the East. Within the twentieth century, Tagore and M.N Roy along with Gandhi depicted an independent evolution of political discourse. All three, established within the Western encounter, attempted to rise above it by starting a talk that's closer to the Indian reality. Gandhi gives a comprehensive evaluation of Western innovation and modernization by challenging the presumption and conclusion of a self-confident Orientalism, especially it sees almost the mediocrity of the East and the prevalence of the innovation of the West. Present day people were not tossed back this world upon themselves. One of the foremost tireless patterns in cutting edge philosophy... has been a comprehensive concern with the self, as recognized from the soul or individual or man in general... world alienation has been the trademark of modern age (Arendt 1958). Gandhi is one of the foremost pundits of modernity within the twentieth century. He sees innovation as unchecked, clearing absent everything that stands in its way because it tries to require "charge of the world" (Connolly 1988).

## 2.5 Sustainability and Sustainable Development: The Gandhian Approach

Nature gives human civilization with a complex life bolster framework, discuss, water, nourishment, and a reasonable climate for survival. It too gives the physical assets that are essential for the food of the economy. Nature has upheld and kept up life on Soil since time immemorial and ought to proceed to do so within the future. Sustainability encompasses three pillars—economic, environmental, and social, also represented as profits, planet, and people. However, humans have been interfering with the sustainability of the natural system through their activities, which is endangering various life forms.

Sustainability and sustainable development looks similar as both argues for consumption of natural resources faster than their replacement. However, there are different substantially. Sustainability is comparatively larger concept than the sustainable development. Sustainability includes social, economics, environment, and governance-related aspects. Sustainability is process-oriented that involves long-term improvement in the quality of life by taking into inter-generational and intra-generational equity.

The natural environment is very vital for sustainability and sustainable development. In the age of globalization, environmental issues have become a matter of concern. The impacts of global warming, depleting ozone layer, and biodiversity loss have made people aware about the increasing environmental challenges. Global warming and consequent climate change is a phenomenon in reality and is here to stay for a long. Though a non-traditional threat to human security but is more ferocious than traditional threats that could be contained given political will at the end of global leadership. Addressing issues of global warming is complex and difficult on most occasions beyond human imagination as it requires the role of many agencies: international legal regimes: individual governmental policy interventions; scientific innovations; international and national non-governmental organizations; and multilateral organizations are to name a few (Babu 2024). Therefore, environmental studies have become very important since the 1960s.

In France, the tradition related to environmental history has started since the arrival of the Annals ideology. In America, the history of the environment is accompanied by biologist Rachel Carson's(1962) work in the form of "The Silent Spring". It explains the negative impacts on environment due to overuse of pesticides. After the publication of this book, the campaign against pollution started worldwide. The systematic beginning of environmental history writing in India that also set the tone for future writings is invariably associated with Gadgil and Guha (1992). The author found out that during pre-colonial India, resource use was judicious and nature-based, and also among the society it was very cordial. According to them Verrier Elvin, IrawatiKarve, D.D Kosambi, and Radha Kamal Mukharjee were pioneers of human ecological studies in India (Gadgil and Guha 1992).

Guha (1999) argued that the main aim of the forest department which was set up in 1864 under a German expert, Dietrich Brandis was aimed at exploiting woods for the British Empire, and for this reason, they had established their right over forests by depriving the farmers of forest use. Due to this, there were protests at different places in the nineteenth century, which continued in the Gandhi era also. Grove et al. (2000) have challenged this argument. Guha (1999) has tried to fill the gap between pre-British and British times in his works. It is argued that the development of the tribal polity has a great effect on the environment. Sundar (1997) in her work has shown the tribal resistance in the Bastar region of present-day Chhattisgarh state against the exploitation of natural resources (Singh 2023).

Since the 1960s onwards the concept of development has been thoroughly debated. In this background, the notion of sustainability and sustainable development has become a matter of concern among the world community. In this background, the United Nations established a World Commission on Environment and Development (WCED 1987) headed by the Norwegian Prime Minister Gro Harlem Brundtland. The report of the commission also known as Brundtland Report, was entitled as "Our Common Future". It defines sustainable development as a "development that meets the needs of the present without compromising the ability of future generations to meet their own needs". Since then up to the meeting of COP-28, which took place on December 1, 2023 in Dubai, the capital city of UAE world is trying to address the issue of sustainability and sustainable development in the background of climate change and global warming.

In the above background, Mahatma Gandhi had developed the concept of trusteeship. It can be taken as an alternative to the capitalist mode of production in which industrialization was playing a vital role thereby destroying sustainability and sustainable development. After the fall of the USSR in 1991 communism as an experiment got a setback. Since 1990s a feeling among the world community has developed belief that capitalism is the only saviour for humanity and there is no alternative to it (Iyengar 2019). In this context understanding Gandhi's idea of trusteeship is important because it seems to be a ray of hope for the sustainability of human civilization. According to Gandhi ji both capitalism and communism were founded on violence. In capitalism, wealth creation produces savagery. Communism that talks almost value moreover suggests savagery. As against both schools of thought, the trusteeship rule stands on non-violence. The endeavors that society must make to attain this kind of value are not on savagery. In this background, for a feasible society, Gandhi contended that trusteeship stands better. Beneath trusteeship, it is accepted that the wealthy man will leave the ownership of riches which he will utilize when he fundamentally requires and sensibly fulfills his individual needs and after that act as a trustee for the leftover portion to be utilized for the society and by the society. "Absolute trusteeship is an abstraction like Euclid's definition of a point and is equally unattainable. But if we strive for it, we shall be able to go further in realizing a state of equality on earth, then by any method" (CWMG 1934).

The basic assumption behind the concept of trusteeship is the integrity and honesty of the trustee.

Many scholars have commented on the above model. For example, Prof. Dantwala believes that it is important to differentiate between ethical principles from the economic and business management model that could be deduced from it. Many scholars criticized the concept of trusteeship being not operational. Actually, the trusteeship is very much embedded in Indian traditional system. It is the value of the *aparigraha* non-acquisitive nature of the human being that has to be developed. This *aparigraha* is the foundational basis in trusteeship. *Aparigraha* is essential to human behavior and is also a part of behavior of an economic man.

Basically, trusteeship is based on the thought of ahimsa. The characteristic culmination of ahimsa is Satyagraha, that's, on the off chance that the affluent and the capitalist don't portion with their riches intentionally, at that point the weapon of satyagraha is to be utilized. When Gandhi ji was asked about the process of bringing about trusteeship then his answer was—persuasion and non-violence. When he was asked, if the trustee fails to behave as s trustee, would the state be justified in dispossessing them? His answer was yes. Creation and possession of wealth are justified in the scheme of trusteeship. During the textile mill strike in Ahmedabad Gandhi said, what I expect of you is that you should hold your riches as a trust to be used solely in the interest of those who sweat for you whose industry and labor you owe all your possession and property. "Since property becomes the cause of quarrels and tempts us to do wrong things, we should give it up, and so long as we are not ready to renounce it completely, we should, as its trustee, use it for the good of others and make the minimum use of it for our enjoyment," (CWMG 1923).

The other input of production is nature. Exhaustion and degradation of characteristic assets is comparatively a later phenomenon. Normal assets and inputs are utilized within the generation handle. As a corporate sector, consideration should be to play down extraction taken a toll or transportation taken a toll or on the off chance that there's a few outright shortage taken a toll of the characteristic asset. Gandhi was against the over-the-top and silly utilization of innovations. His concern was the sensible utilization of innovation. In modern speech, it can be treated as eco-friendly innovation that can be translated as trusteeship.

## 2.6 Conclusion

In the contemporary world, the humans are under great stress, disarray, and disharmony; primarily because of the insatiable material desire of humans. Such a relentless pursuit of worldly gains has corroded all values and the meaning of human life. In this process, humans have been deprived of the protection, love, and warmth of community life. But there is a far more serious dimension of human predicament. There is a state of total confrontation between man and nature which is threatening the very survival of humankind. Rapid exhaustion of non-renewable energy sources, environmental degradation, global warming, climate change, ozone layer depletion, and the impending threat of the melting of the ice in the Polar Regions are some of the major dimensions of the same crisis. After the industrial revolution (1750) different phases of industrialization lead to the spread of urbanization and modernization throughout the globe. Presently in the globalized world, environmental-related issues have become very important. In this background, Mahatma Gandhi's Hind Swaraj is seen as a critic of industrialization, urbanization and modernization. Since the 1960s onwards the concept of development has been thoroughly debated. In this background, the notion of sustainability and sustainable development has become a matter of concern among the world community. In this context, the Gandhian philosophy of trusteeship can act as the base of sustainability and sustainable development.

## Notes

1. Indian opinion was launched to develop better collaboration between the Europeans and Indians in South Africa. The paper was published in 4 languages and later in 2 to be discussed into public events and political, economic, ethical and religious ideas.

## References

Arendt H (1958) The human condition. University of Chicago Press, Chicago
Babu GK (2024) Editorial. World Focus 45:3
Bondurant JV (1967) Conquest of violence: the Gandhian philosophy of conquest. Princeton University Press, Princeton, New Jersey
Carson R (1962) The silent spring. Mariner Books, Boston
Connolly W (1988) Political theory and modernity. Basil Blackwell, Oxford
CWMG (1923) Collected works of Mahatma Gandhi. https://www.gandhiashramsevagram.org/gandhi-literature/mahatma-gandhi-collected-works-volume-26.pdf. Accessed 15 Feb 2024
CWMG (1934) Collected works of Mahatma Gandhi. https://www.gandhiserve.net/about-mahatma-gandhi/collected-works-of-mahatma-gandhi/059-19340916-19341215/. Accessed 15 Feb 2024
Fanon F (1963) The wretched of the Earth. Grove Press, New York
Gadgil M, Guha R (1992) The fissured land: an ecological history of India. Oxford University Press, New Delhi
Gandhi MK (1910) Indian home rule. The International Printing Press, Phoenix, Natal
Gandhi MK (1928) Young India 1927–28. S Ganeshan, Madras
Grove RH, Damodaran V, Sangwan S (2000) Nature and the orient: the environmental history of South and Southeast Asia. Cambridge University Press, New Delhi
Guha S (1999) Environment and ethnicity in India, 1200–1991. Cambridge University Press, Cambridge
Iyengar S (2019) A possibility in the eco-system of Swadeshi and Swaraj. Yojana 63:1–7
Kolge N (2016) Was Gandhi a racist? His writings in South Africa. Econ Pol Wkly LI(5):88–93
Parekh B (1997) Gandhi. Oxford University Press, Oxford
Parekh B (2010) Gandhi. Sterling Pub Co Inc., New York
Pradhan RC (2014). Integrating body, mind and heart: the Gandhian way. Trinity, New Delhi
Sanghavi N (2006) The agony of arrival, Gandhi South Africa years. Rupa, New Delhi
Singh S (1999) Global concern with environmental crisis and Gandhi's vision. APH Publishing Corporation, New Delhi
Singh US (2023) Environment, culture and sustainable development: a historical perspective. In: Abhay, RK, Pandey DN, Rani S (eds) Constraints in achieving sustainability of India. TERI Publication, New Delhi
Sundar N (1997) Subalterns and sovereigns: an anthropological history of Bastar, 1854–2006. Oxford University Press, New Delhi
Tercheck RJ (1998) Gandhi: struggling for autonomy. Vistaar Publications, New Delhi
WCED [World Commission on Environment and Development] (1987) Our common future: report of the world commission on environment and development. Oxford University Press, Oxford
Woodcock G (1971) Mohandas Gandhi. Viking Press, New York

# Part II
# Sustainable Urbanization and Policies

# Chapter 3
# Geospatial Analysis of Urban Growth

## A Comparative Study of Pune, Pimpri-Chinchwad and Nashik Municipal Corporations

**Santosh Bhailume and Ganesh Sonawane**

**Abstract** The burgeoning intricacy of urban landscapes has accentuated the necessity of comprehensive study. These are vital for illuminating sustainable growth paths, grasping socio-economic interplays, and permitting educated urban policy decisions. Guided by this imperative, the current research employs the Landsat satellite series data to present an in-depth geospatial investigation of urbanization patterns observed within three cardinal municipal corporations in Maharashtra, India, namely the Pune Municipal Corporation (PMC), Pimpri-Chinchwad Municipal Corporation (PCMC), and Nashik Municipal Corporation (NMC) over the span of 2001–2021. The findings unambiguously identify a significant augmentation in urban built-up regions, frequently at the expense of other land use classifications. Of the three, NMC experienced the most pronounced spatial increase, accounting for an additional 78.87 $km^2$, while PMC and PCMC witnessed additions of 64.14 $km^2$. and 40.17 $km^2$, respectively. In absolute spatial terms, PMC manifests as the central urban expanse with a sprawl of 189.34 $km^2$, succeeded by NMC at 137.51 $km^2$. and PCMC at 105.71 $km^2$. Furthermore, a period of increasing urban sprawl was evident across the municipalities during the 2006–2011 intervals. Spatial landscape metrics point to the type of urban landscape fragmentation, the compactness of cities, and the degree of sprawl characteristics. The results show that the built-up in the urban landscapes shows landscape fragmentation into small patches that happened on a large scale, indicating sprawl during the study period.

**Keywords** Urban growth · Land use/cover change · Landscape metrics · Geospatial analysis

S. Bhailume (✉) · G. Sonawane
Department of Geography, K. V. N. Naik Shikshan Prasarak Sanstha's Arts, Commerce, and Science College, Nashik, Maharashtra, India
e-mail: bhailumesantosh@gmail.com



S. Anand et al. (eds.), *Emerging Geosustainability Transformations in India*, Advances in Geographical and Environmental Sciences,
https://doi.org/10.1007/978-981-97-7993-2_3

## 3.1 Introduction

Urbanization is a pivotal element in human societal progression and represents a significant way humans influence the Earth's natural ecosystems (Tian et al. 2022; Liu et al. 2023). The mechanism of urbanization, characterized by the evolution and proliferation of urban locales, presents a compelling narrative in the annals of contemporary geographical and socio-economic research (Gavsker 2023). As conventionally delineated, urbanization refers to increasing population concentration within towns and cities and a marked shift from agrarian-based activities to more industrial and service-oriented vocations (Varquez et al. 2023; Alwedyan 2023; Aslam et al. 2023; Wiatkowska and Słodczyk 2021). Globally, the trajectory of urban growth delineates a dramatic shift in human settlement patterns. As affirmed by the United Nations' World Urbanization Prospects 2018, an estimated 55% of the world's inhabitants are currently constrained within the urban zones—a proportion slated to escalate to 68% by the year 2050 (UN 2018). This anticipated rise in urban dwellers underscores an impending alteration in the global demographic landscape. Most intriguingly, the pace of this urban surge is postulated to manifest with greater intensity in specific nations. In the contemporary epoch, the urbanization trajectories of Asian and African nations, particularly India, China, and Nigeria, have garnered significant scholarly attention. Collectively, these countries are anticipated to be responsible for approximately 35% of the global urban population augmentation from 2018 to 2050 (Manna et al. 2023; Latue and Rakuasa 2023; Dhanaraj and Anghadi 2022, Hailu and Mammo 2020).

Geospatial technology, anchored in Geographic Information Systems (GIS), remote sensing, and global positioning systems, has undergone a transformative journey (Mandal and Ghosh 2019; Baqa et al. 2021; Akubia and Bruns 2019; Dutta and Das 2019). Its advanced tools and techniques, such as high-resolution satellite imagery, real-time data collection, and sophisticated spatial analysis algorithms, offer unparalleled insights into urban morphology and dynamics. Notably, with its consistent, multi-temporal, and multispectral capabilities, the Landsat series is a sentinel in monitoring Land Use and Land Cover (LULC) alterations, making it a linchpin in urban-centric studies (Trinder and Liu 2020; Hamad 2020; Wahla et al. 2023).

In the domain of remote sensing and satellite imagery, Longley's (2005) incisive exploration of spectral imaging intricacies, Goodchild's (2011) resounding advocacy for the indispensability of Geographic Information Systems (GIS), and Jensen et al. (2009) methodological innovations for decoding remote sensing data have collectively revolutionized comprehension of urban landscapes' multi-layered intricacies. Transitioning to geospatial analytics and predictive modeling, Zeng and Guo (2023) pioneered the amalgamation of machine learning prowess with geospatial analytics. Magidi and Ahmed (2019) provided insights into the potential of predictive models for sustainable urban evolution, and Hamnett and Yuen (2019) highlighted innovative integration of hybrid predictive frameworks has collectively ushered in an era of redefining urban growth pattern analyses with heightened accuracy and foresight. Concurrently, in the arena of geo-visualization and 3D Urban Modeling, the contributions of Bodum and Kjems (2005) in demystifying the complexities of 3D urban cadastral registration, Elwood and Leszczynski (2018) pivotal discourse on digital geographies, and Batty (2013) seminal insights into the new science of cities, have collectively, transcended conventional paradigms. Also empowered urban planners and stakeholders with cutting-edge visual representation tools for informed decision-making (Dinda et al. 2021; Ray et al. 2023).

With the advancement of RS and GIS, geospatial technology provides greater flexibility to study various aspects of urbanization from local to worldwide (Ray et al.2023; Alam et al. 2020). The wide use of urban mapping and LULC analysis over different periods (Mishra and Rai 2020; Paul et al. 2021) provides excellent opportunities on a global scale. The introduction of new approaches and models for urban studies gained much attention from scholars and researchers; popular among them is the Cellular Automata (CA) modeling algorithm (Latueand Rakuasa 2023; Akubia and Bruns 2019; Aburas et al. 2016; Tong and Feng 2020; Moghadam and Helbich 2013). Further, there is Shannon's Entropy (Patra et al. 2022; Mohabey and Nongkynrih 2023; Bhailume 2021), SLEUTH model (Das and Jain2022; Liu et al. 2019), Landscape Metrics (Jaafari et al. 2020; Bosch et al. 2020; Flowers et al. 2020; Getu and Bhat 2021; Akin and Erdogan 2020) and so on. In recent years, landscape metrics have emerged as a new approach to studying the dynamics of urban growth. The present investigation employs land use/cover analysis and Landscape metrics for the study.

In the context of India, Maharashtra, with its burgeoning cities like Pune, Pimpri-Chinchwad, and Nashik, offers a microcosm reflective of these global trends. Each city's unique developmental trajectory accentuates the significance of a comprehensive geospatial analysis, especially in an era of rapid urban transformations. Against this backdrop, this study endeavors to harness the potency of the Landsat series data to fathom the urban expansion patterns of these pivotal municipalities over two decades. In doing so, it seeks to contribute to the rich tapestry of urban research, emphasizing the indispensability of geospatial analyses for sustainable urban futures and informed policy articulations.

## 3.2 Material and Methods

### *3.2.1 Study Area*

Encompassing a triad of distinct urban landscapes within Maharashtra, India, the study area encompasses the cities of Pune, Pimpri-Chinchwad, and Nashik. The geographical expanse of Pune spans from latitudes 18°24′16″ N to 18°37′18″ N and longitudes 73°44′58″ E to 74°0′12″ E, the territory of which covers 331.23 $km^2$. Adjacent to Pune, the Pimpri-Chinchwad region spans from latitudes 18°33′46″ N to 18°43′15″ N and longitudes 73°55′59″ E to 74°42′40″ covering 181.75 $km^2$ area and is characterized by dynamic administrative divisions and strategic positioning along the Mumbai-Pune National Highway. Further northwest lies Nashik, which extends from latitudes 19°54′40″ N to 20°05′08″ N and longitudes 73°41′08″ E to 73°54′22″ E, covering an extensive area of 259.10 km$^2$ (Fig. 3.1).

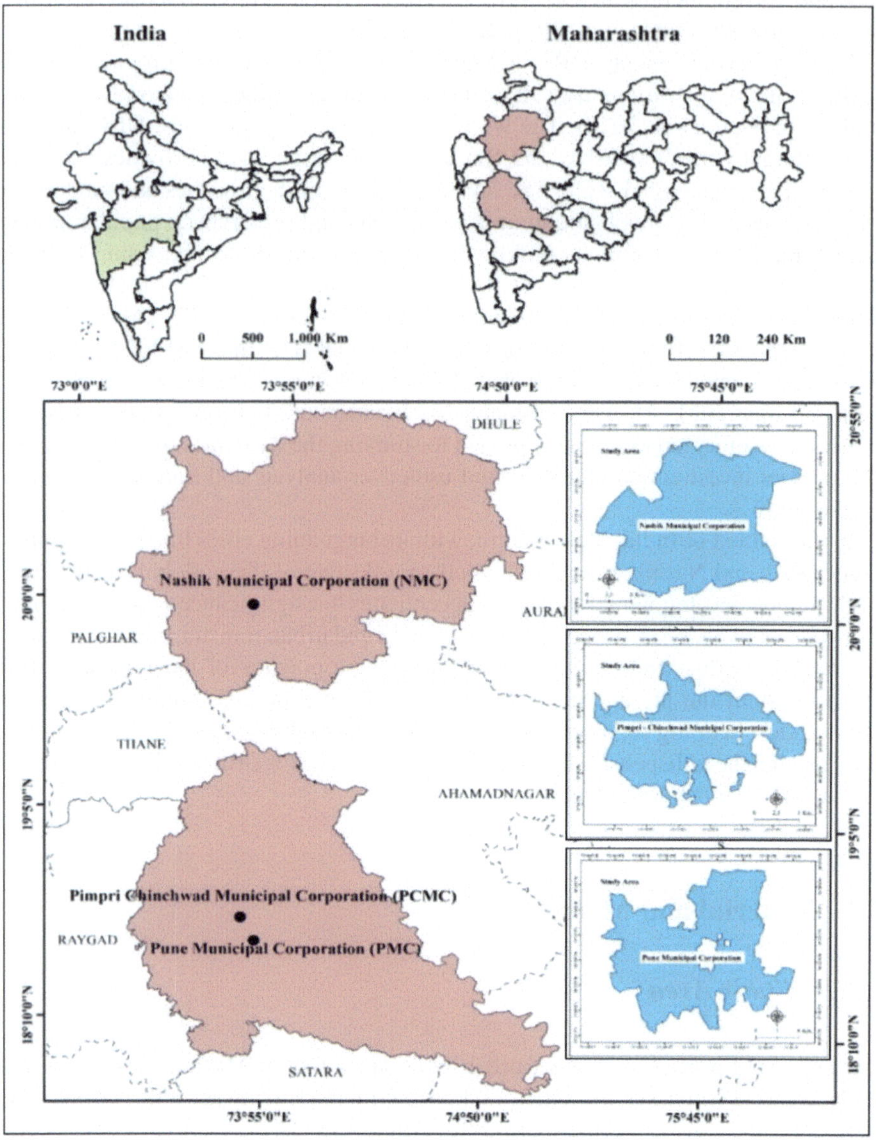


**Fig. 3.1** Location of study area

Nashik, Pune, and Pimpri-Chinchwad emerge as critical urban nuclei within Maharashtra, each offering distinctive contributions to the state's economic, educational, and cultural fabric. Nashik's historical legacy converges with its roles as an agricultural and industrial epicentre, rendering it both economically and spiritually significant. Pune's stature as an educational powerhouse magnetizes students nationally, bolstered by its diverse industrial and cultural panorama. Pimpri-Chinchwad's strategic positioning fuels its industrial advancement, rendering it an appealing locale for residential and commercial ventures. This triumvirate of cities, marked by their unique attributes, collectively furnishes a compelling canvas for geospatial analyses unraveling the intricate dynamics of urban expansion.

### *3.2.2 Data Acquisition and Processing*

The data employed for the present study was sourced from satellite imagery provided by the United States Geological Survey (https://earthexplorer.usgs.gov). The imagery was obtained through the utilization of the LANDSAT-7 Enhanced Thematic Mapper Plus (ETM+) and LANDSAT-8 Operational Land Imager (OLI) with Thermal Infrared Sensor (TIRS) sensors (Table 3.1). The dataset encompasses diverse acquisition dates across various paths. Specifically, the LANDSAT-7 ETM+ sensor contributed data from 12/03/2001, 18/02/2001, 22/03/2006, 07/01/2006, 15/03/2011, and 19/02/2011, characterized by a spatial resolution of 30 m. Furthermore, the LANDSAT-8 OLI-TIRS sensor offered imagery from 10/03/2016, 22/02/2016, 12/02/2021, and 28/01/2021. These meticulously curated datasets constitute the fundamental cornerstone for the rigorous geospatial analysis underpinning the study's investigation into urban expansion dynamics within the designated study area.

**Table 3.1** Source and characteristics of satellite imageries

| Satellite | Sensor | Path | Resolution (m) | Date | Cloud cover (%) |
|---|---|---|---|---|---|
| LANDSAT-7 | ETM+ | 147/46 | 30 | 12/03/2001 | <5 |
| | ETM+ | 147/47 | 30 | 18/02/2001 | <5 |
| | ETM+ | 147/46 | 30 | 22/03/2006 | <5 |
| | ETM+ | 147/47 | 30 | 07/01/2006 | <5 |
| | ETM+ | 147/46 | 30 | 15/03/2011 | <5 |
| | ETM+ | 147/47 | 30 | 19/02/2011 | <5 |
| LANDSAT-8 | OLI-TIRS | 147/46 | 30 | 10/03/2016 | <5 |
| | OLI-TIRS | 147/47 | 30 | 22/02/2016 | <5 |
| | OLI-TIRS | 147/46 | 30 | 12/02/2021 | <5 |
| | OLI-TIRS | 147/47 | 30 | 28/01/2021 | <5 |

The methodology employed in this study encompasses a series of rigorous image preprocessing and analytical steps. Initial image preprocessing involved Scan Line Error Removal to rectify any anomalies in the imagery. Subsequently, Atmospheric and Geometric Correction techniques were applied to ensure spatial accuracy and eliminate atmospheric interferences. Image Enhancement procedures were then executed to optimize visual clarity and information extraction (Batty 2013; Zeng and Guo 2023).

The next phase entailed Layer Stacking, wherein the corrected and enhanced images were combined to generate composite datasets. Spectral Signature Training was performed to derive spectral profiles of distinct land cover classes, enabling subsequent classification. For land cover classification, a two-tiered approach was undertaken. Firstly, the Maximum Likelihood Algorithm, renowned for its statistical robustness, was employed to classify the images into preliminary land cover categories. Subsequently, the Hybrid Classification Algorithm, amalgamating multiple classification methods, was applied to refine and enhance accuracy (Rajendran et al. 2020).

### *3.2.3 Land Use and Land Cover Classification*

In the context of our recent research, we have expressly adopted the LULC classification scheme outlined by the National Remote Sensing Centre (NRSC). The NRSC's systematic and hierarchical approach provides a comprehensive framework to interpret and categorize land and its myriad uses. By leveraging this well-established scheme, our study ensures consistency and comparability with other investigations, enriching the credibility of our findings and allowing for a more nuanced understanding of land patterns and transformations. Considering the characteristics of the study area, six distinct LULC classes were designated to evaluate urban growth (Table 3.2).

### *3.2.4 Landscape Metrics*

Based on a categorical, patch-based landscape description, landscape metrics were introduced in the late 1980s and incorporate measures from information theory and fractal geometry (Cardille and Turner 2017). Patches are described as homogeneous areas with a typical landscape feature of importance, such as "industrial land," "park," or "high-density residential area. Landscape metrics are used to measure the geographic variability of individual patches in situations where all patches belong to the same class and the landscape is a collection of patches. The metrics may be used to evaluate and characterize changes in the degree of spatial heterogeneity when applied to multi-scale or multi-temporal datasets. The value of these measures is calculated using the FRAGSTATS program. FRAGSTATS is a public domain spatial

**Table 3.2** Land use/cover classes for the present study

| Sr. no. | LULC class | Description |
|---|---|---|
| 1 | Agriculture Land | Encompasses all cultivable terrains, including both seasonal and temporal croplands |
| 2 | Built-up | Constitutes residential, industrial, commercial zones, and all other impervious terrains |
| 3 | Fallow land | Represents both temporary and permanent agricultural fallow lands |
| 4 | Vegetation | Comprises all forms of natural and urban vegetation |
| 5 | Wasteland | Areas that include open spaces, parks, playgrounds, scrublands, grasslands, pasturelands, etc. |
| 6 | Waterbody | Encompasses all water-related features |

measurement software that was created in the mid-1990s and has been constantly updated since then. FRAGSTATS supports many metrics at the class, patch, and landscape.

In the present research, parameters like Number of Patches (NP), Patch Density (PD), Edge Density (ED), Mean Patch Size (MPS), Landscape Shape Index (LSI), Normalized Landscape Shape Index (NLSI), Clumpiness Index (Clumpy), Aggregation Index (AI), percentage of Like Adjacencies (PLADJ), Interspersion and Juxtaposition Index (IJI), CONTAG were used. For analysis purposes, the geographical space of these three municipal corporations is divided into eight sectors based on the geographical quadrant.

### 3.2.5 *Accuracy Assessment*

A critical component of the methodology involved Accuracy Assessment, whereby the accuracy of the classification outcomes was rigorously evaluated through comparison with ground truth data. This assessment served to validate the classification results and gauge the reliability of the analysis. The accuracy assessment of the classified images pertaining to LULC classification is paramount. For the years 2001, 2006, 2011, 2016, and 2021, 207 Ground Control Points (GCPs) were meticulously gathered and executed via a systematic randomized sampling method (Fig. 3.2). To facilitate this, GCPs encapsulating all six distinguished LULC classifications were ascertained utilizing spectral signatures derived from supervised classification methodologies, archival visual data from Google Earth, in addition, ground truth points were used for the year 2021.

With the established GCPs serving as a foundational benchmark, the classified LULC was juxtaposed against the reference or ground truth imagery. Such a comparison was instrumental in determining the extent to which the computational classification was congruent with the actual terrestrial conditions. The fidelity of the

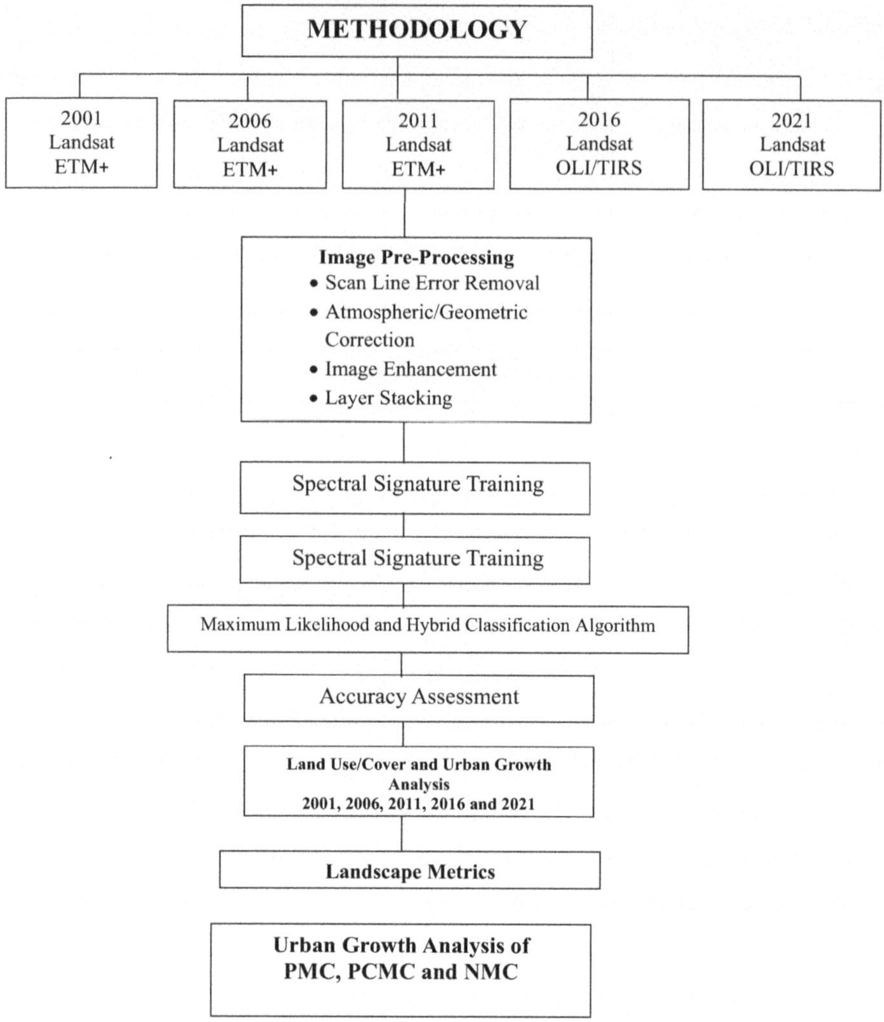


**Fig. 3.2** Methodological framework of the study

classification was appraised using metrics including the Producer's Accuracy (PA), the User's Accuracy (UA), and the Overall Accuracy (OA). Subsequently, a confusion matrix was meticulously constructed to delineate the accuracy associated with each LULC category. Conclusively, the exactitude of the LULC was articulated, employing a methodology resonant with that delineated in the source.

## 3.3 Results and Discussion

### *3.3.1 Land Use and Land Cover Change*

The LULC data, spanning two decades from 2001 to 2021, for the Pune Municipal Corporation (PMC), the Pimpri-Chinchwad Municipal Corporation (PCMC), and Nashik Municipal Corporation (NMC), presents an evolving paradigm of urban expansion and ecological transmutation. As illustrated in Tables 3.3, 3.4, and 3.5, the LULC metrics conspicuously reveal a burgeoning emergence of built-up areas across all three municipal domains—PMC (37.80–57.16%), PCMC (36.06–58.16%), and NMC (22.63–53.07%) (Fig. 3.3). This substantial increase points to population expansion, housing demand, and the corresponding need for expanded infrastructure and public services. Additionally, the rapid pace of urbanization brings crucial problems surrounding environmental effects, including increasing resource use and waste creation.

Contrasting this surge is a marked decline in agricultural land and vegetation, which reflects potentially worrisome environmental trajectories. PMC experienced a reduction in agriculture from 12.42 to 8.62% and vegetative land from 7.66% to 5.30%. Similarly, PCMC suffered declines from 12.55 to 9.10% for agriculture and from 10.90 to 7.69% for vegetation. In NMC, agriculture slightly reduced from 26.59 to 23.20%, while vegetation fell drastically from 4.28 to 2.65%. These losses confront problems about the loss of biodiversity, local food security, and broader ecological resilience.

The marginalization of fallow and wasteland categories further amplifies the complexity of land resource allocation. Their decline intimates a conversion—ostensibly into built-up areas—that may be economically advantageous in the immediate term yet harbors latent ecological detriment. In PMC, fallow land declined from 8.66% in 2001 to 4.23% in 2021, while wasteland plummeted from 29.65 to 21.21%. In PCMC, fallow land reduced from 7.03 to 4.88%, while wasteland decreased from

**Table 3.3** LULC classification of PMC

| LULC class | 2001 | | 2006 | | 2011 | | 2016 | | 2021 | |
|---|---|---|---|---|---|---|---|---|---|---|
| | Area ($Km^2$) | % | Area ($Km^2$) | % | Area ($Km^2$) | % | Area ($Km^2$) | % | Area ($Km^2$) | % |
| Built-up | 125.2 | 37.80 | 137.98 | 41.66 | 159.32 | 48.10 | 175.26 | 52.91 | 189.34 | 57.16 |
| Agriculture | 41.14 | 12.42 | 38.61 | 11.66 | 32.42 | 9.79 | 31.04 | 9.37 | 28.54 | 8.62 |
| Vegetation | 25.36 | 7.66 | 23.62 | 7.13 | 22.01 | 6.64 | 21.28 | 6.42 | 17.56 | 5.30 |
| Fallow land | 28.69 | 8.66 | 25.52 | 7.70 | 17.66 | 5.33 | 15.21 | 4.59 | 14.01 | 4.23 |
| Wasteland | 98.21 | 29.65 | 94.25 | 28.45 | 86.81 | 26.21 | 76.34 | 23.05 | 70.26 | 21.21 |
| Waterbody | 12.63 | 3.81 | 11.25 | 3.40 | 13.01 | 3.93 | 12.1 | 3.65 | 11.52 | 3.48 |
| Total | **331.23** | **100** | **331.23** | **100** | **331.23** | **100** | **331.23** | **100** | **331.23** | **100** |

*Source* Calculated by Authors

**Table 3.4** LULC classification of PCMC

| LULC class | 2001 | | 2006 | | 2011 | | 2016 | | 2021 | |
|---|---|---|---|---|---|---|---|---|---|---|
| | Area ($Km^2$) | % | Area ($Km^2$) | % | Area ($Km^2$) | % | Area ($Km^2$) | % | Area ($Km^2$) | % |
| Built-up | 65.54 | 36.06 | 75.86 | 41.74 | 88.32 | 48.59 | 94.49 | 51.99 | 105.71 | 58.16 |
| Agriculture | 22.81 | 12.55 | 21.88 | 12.04 | 17.47 | 9.61 | 17.41 | 9.58 | 16.54 | 9.10 |
| Vegetation | 19.81 | 10.90 | 16.69 | 9.18 | 15.91 | 8.75 | 14.04 | 7.72 | 13.97 | 7.69 |
| Follow land | 12.77 | 7.03 | 11.95 | 6.57 | 10.82 | 5.95 | 9.74 | 5.36 | 8.87 | 4.88 |
| Wasteland | 56.96 | 31.34 | 51.72 | 28.46 | 45.79 | 25.19 | 42.68 | 23.48 | 33.39 | 18.37 |
| Waterbody | 3.86 | 2.12 | 3.65 | 2.01 | 3.44 | 1.89 | 3.39 | 1.87 | 3.27 | 1.80 |
| Total | **181.75** | **100.00** | **181.75** | **100.00** | **181.75** | **100.00** | **181.75** | **100.00** | **181.75** | **100.00** |

*Source* Calculated by Authors

**Table 3.5** LULC classification of NMC

| LULC class | 2001 | | 2006 | | 2011 | | 2016 | | 2021 | |
|---|---|---|---|---|---|---|---|---|---|---|
| | Area ($Km^2$) | % | Area ($Km^2$) | % | Area ($Km^2$) | % | Area ($Km^2$) | % | Area ($Km^2$) | % |
| Built-up | 58.64 | 22.63 | 75.76 | 29.24 | 103.57 | 39.97 | 118.6 | 45.77 | 137.51 | 53.07 |
| Agriculture | 68.9 | 26.59 | 62.39 | 24.08 | 55.95 | 21.59 | 59.56 | 22.99 | 60.1 | 23.20 |
| Vegetation | 11.08 | 4.28 | 11.01 | 4.25 | 9.55 | 3.69 | 8.28 | 3.20 | 6.87 | 2.65 |
| Fallow land | 19.66 | 7.59 | 16.95 | 6.54 | 11.22 | 4.33 | 7.39 | 2.85 | 5.11 | 1.97 |
| Wasteland | 96.55 | 37.26 | 88.98 | 34.34 | 75.43 | 29.11 | 61.29 | 23.65 | 45.64 | 17.61 |
| Waterbody | 4.27 | 1.65 | 4.01 | 1.55 | 3.38 | 1.30 | 3.98 | 1.54 | 3.87 | 1.49 |
| Total | **259.1** | **100.00** | **259.1** | **100.00** | **259.1** | **100.00** | **259.1** | **100.00** | **259.1** | **100.00** |

*Source* Calculated by Authors

31.34 to 18.37%. In NMC, fallow land decreased from 7.59 to 1.97%, and wasteland decreased from 37.26 to 17.61%.

Interestingly, despite the enormous changes in land use, the proportion of aquatic bodies has mainly stayed consistent in each jurisdiction. This might either imply suitable water resource management methods or potentially a tenuous stability in environmental balance that could be prone to future change.

### *3.3.2 Land Use and Land Cover Change Dynamics*

In analyzing landuse transformations in the PMC, PCMC, and NMC regions over the two-decade span from 2001 to 2021, several salient trends and variations emerge that merit scientific scrutiny. All three regions manifest an unequivocal trend toward

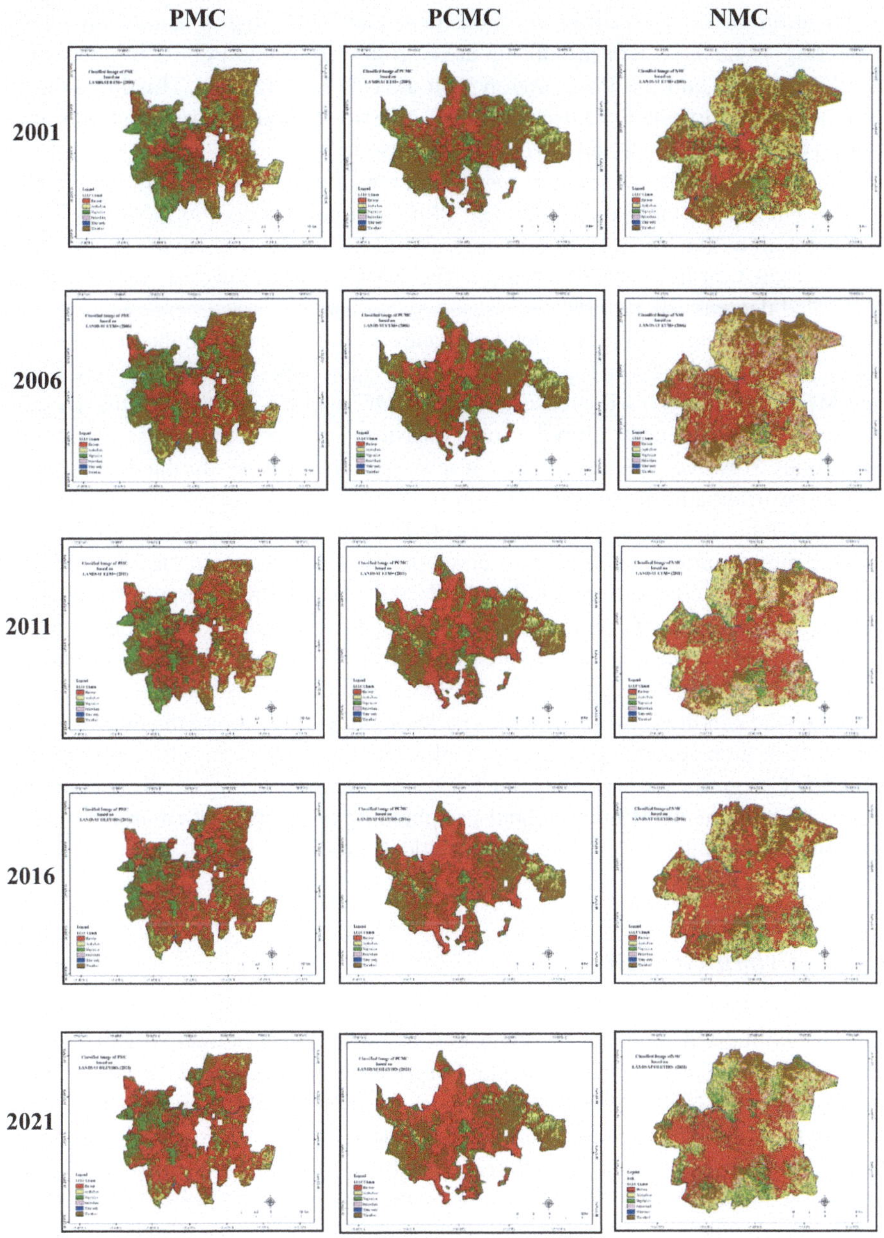


**Fig. 3.3** LULC analysis of PMC, PCMC, and NMC (2001–2021)

urbanization, albeit with differing intensities. The NMC region stands out with a staggering 134.5% growth in built-up areas, followed by the PCMC region with a 61.29% increase and the PMC region with a 51.23% increment. This hierarchical pattern in urban expansion raises pertinent questions about the regional pressures exerted on infrastructural and environmental resources.

Moreover, agricultural zones across all regions are on a downtrend, but the rate of contraction is notably disparate. The PMC region registers the most significant decline, with a 30.63% reduction in agricultural land, which is closely trailed by a 27.49% reduction in the PCMC region. The NMC region, however, shows a more moderated decrease of 12.77%.

Further, the degradation in vegetation zones is another area of concern but varies less significantly across the regions. The NMC region experienced a 38% reduction, the PMC region a 30.76% decrease, and the PCMC region a 29.48% decline. Despite the numerical differences, all point to a consistent environmental stressor affecting these zones. However, the underlying causative factors for these similar declines may vary and warrant additional investigation.

Follow land and wastelands have endured alarming contractions in all regions, with the most pronounced decline observed in the NMC region. In the NMC region, there is a 74.01% contraction in fallow lands and a 52.73% reduction in wastelands. In comparison, the PMC region reveals a 51.17% reduction in fallow lands and a 28.46% decline in wastelands. The PCMC region, on the other hand, experiences a 30.54% decline in fallow lands and a 41.38% reduction in wastelands.

Lastly, the condition of water bodies varies significantly among the three regions. The NMC region records a relatively modest decline of 9.37%, whereas the PMC and PCMC regions show declines of 8.79% and 15.28%, respectively. Such variability could indicate differential land-use planning or an array of environmental or anthropogenic influences affecting hydrological stability.

### 3.3.3 Urban Growth

The spatiotemporal analysis of urban expansion in Pune Municipal Corporation (PMC), Pimpri Chinchwad Municipal Corporation (PCMC), and Nashik Municipal Corporation (NMC) over the two decades from 2001 to 2021 presents intriguing insights into the dynamics of urbanization in these regions (Figs. 3.3, 3.4, and 3.5, Table 3.6). Both the absolute and relative growth metrics underscore a heterogeneous development pattern reflective of differing socio-economic and policy landscapes.

Nashik Municipal Corporation (NMC) consistently manifests a higher tempo of urbanization, culminating in an astonishing 134.5% growth rate over the 20 years. This is further corroborated by an average annual growth rate of 6.72%, which markedly outstrips its counterparts. In absolute terms, NMC's urban expanse increased by 78.87 km$^2$, indicative of significant land-use conversion, likely driven by a confluence of economic, demographic, and infrastructural factors.

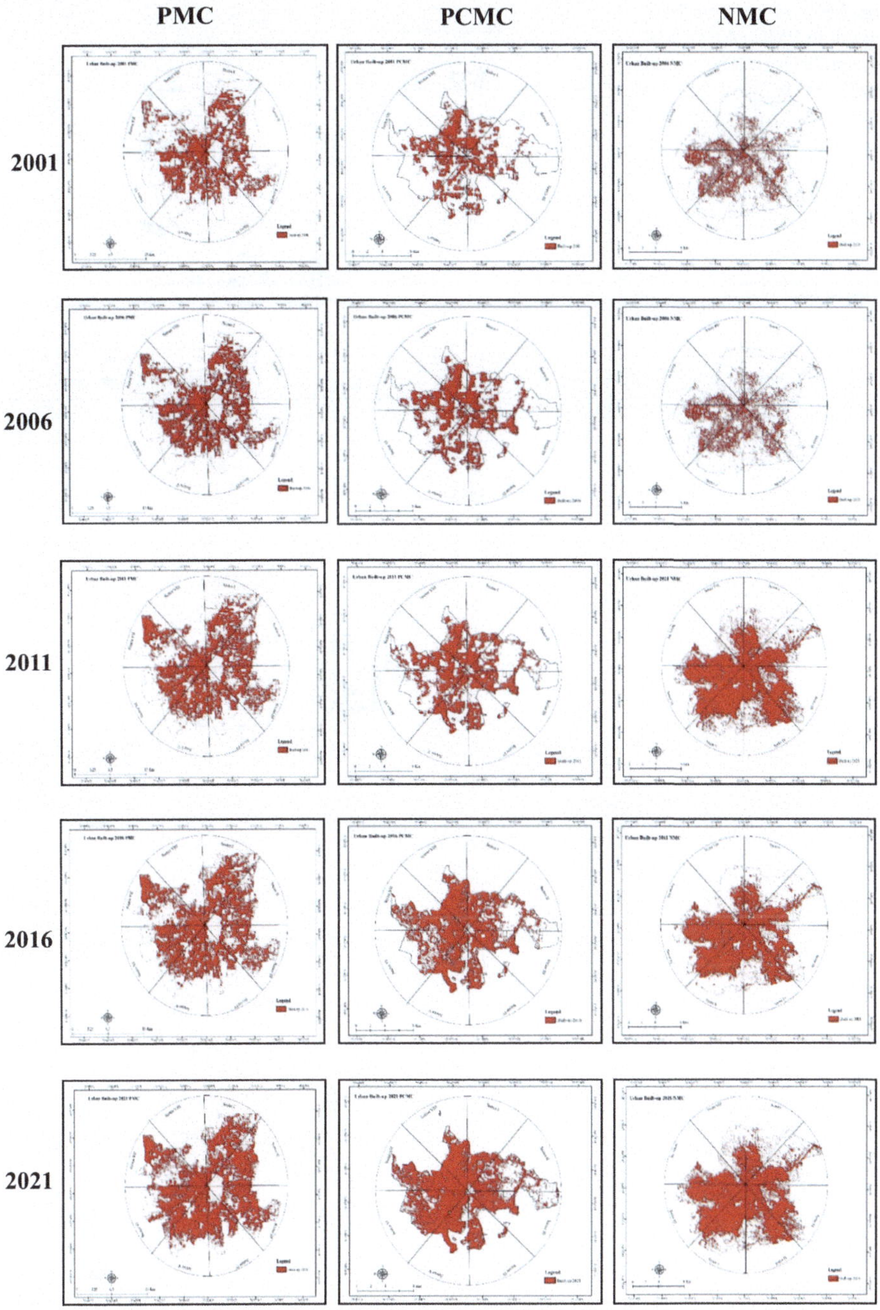


**Fig. 3.4** Urban growth in PMC, PCMC and NMC (2001–2021)

**Fig. 3.5** Composite urban growth in PMC (A), PCMC (B) and NMC (C) (2001–2021)

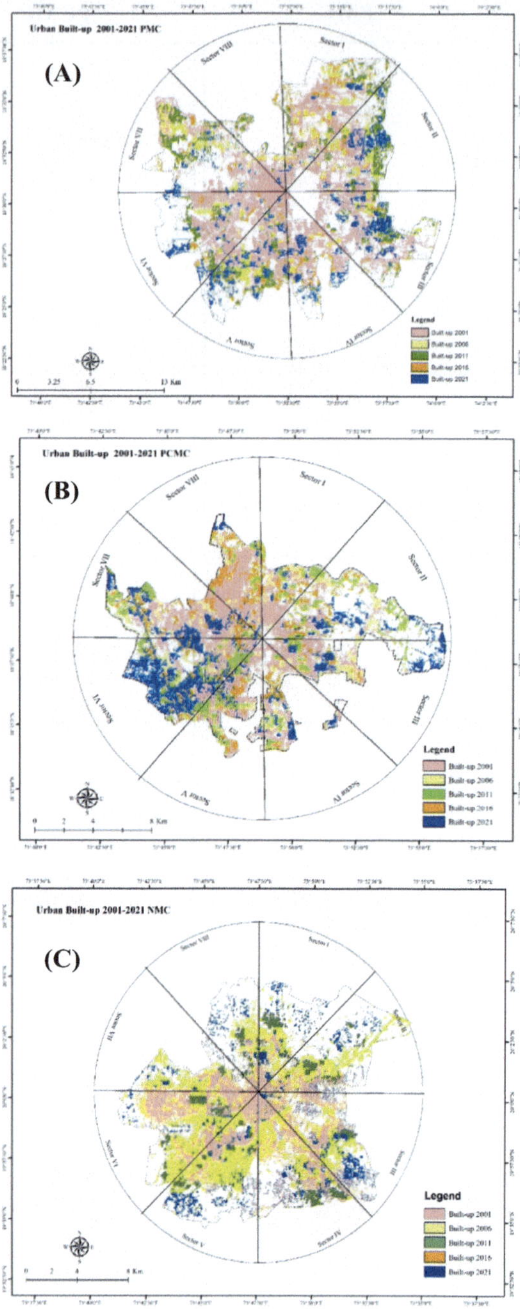

**Table 3.6** Urban growth in PMC, PCMC, and NMC

| Period | Parameter | PMC | PCMC | NMC |
|---|---|---|---|---|
| 2001–2006 | $Km^2$ | 12.78 | 10.32 | 17.12 |
| | Growth rate | 10.21 | 15.75 | 29.2 |
| | Annual growth rate | 2.042 | 3.15 | 5.84 |
| 2006–2011 | $Km^2$ | 21.34 | 12.46 | 27.81 |
| | Growth rate | 15.47 | 16.42 | 36.71 |
| | Annual growth rate | 3.094 | 3.284 | 7.342 |
| 2011–2016 | $Km^2$ | 15.94 | 6.17 | 15.03 |
| | Growth rate | 10.01 | 6.99 | 14.51 |
| | Annual growth rate | 2.02 | 1.38 | 2.92 |
| 2016–2021 | $Km^2$ | 14.08 | 11.22 | 18.91 |
| | Growth rate | 8.03 | 11.87 | 15.94 |
| | Annual growth rate | 1.606 | 2.374 | 3.188 |
| 2001–2021 | $Km^2$ | 64.14 | 40.17 | 78.87 |
| | Growth rate | 51.23 | 61.29 | 134.5 |
| | Annual growth rate | 2.561 | 3.064 | 6.72 |

In contrast, Pimpri Chinchwad Municipal Corporation (PCMC) follows a moderate yet sustained trajectory of urban growth, characterized by a 61.29% overall expansion and an average annual growth rate of 3.064%. The expansion in urban land, as denoted by an increase of 40.17 $km^2$, suggests a more measured pace of development, possibly influenced by planned urban policies and constrained by resource limitations. While not lagging far behind, Pune Municipal Corporation (PMC) registers a cumulative growth rate of 51.23% and an average annual growth rate of 2.561%, resulting in an overall expansion of 64.14 $km^2$. The relatively lower rates may reflect a mature urban setting with limited scope for rampant expansion, necessitating a focus on vertical development and infrastructural optimization.

It's important to highlight the temporal variability in growth patterns. The peak urban growth for NMC occurred during 2006–2011, with an annual growth rate of 7.342%. Conversely, PMC and PCMC showed peak growth during the same period with annual rates of 3.094% and 3.284%, respectively. These temporal fluctuations could be symptomatic of varying regional policy interventions, economic incentives, or demographic shifts and warrant further investigation.

### 3.3.4 Landscape Metrics

The present research employs a comprehensive array of landscape metrics to scrutinize the dynamics of urban growth under investigation. These metrics, outlined in Table 3.7, encompass significant parameters, including but not limited to Number of Patches (NP), Patch Density (PD), Edge Density (ED), Mean Patch Size (MPS), Landscape Shape Index (LSI), Normalized Landscape Shape Index (NLSI), Clumpiness Index (Clumpy), Aggregation Index (AI), Percentage of Like Adjacencies (PLADJ), Interspersion and Juxtaposition Index (IJI), and CONTAG. These metrics collectively provide a robust framework for quantitatively assessing various dimensions of landscape composition and configuration. For analytical rigour, the geographical expanse of the three municipal corporations has been discretely demarcated to facilitate an organized investigation into urbanization patterns. This meticulous spatial delineation enhances the capacity to discern evolving urban landscapes within specific regions, yielding deeper insights into the intricacies of urban growth dynamics.

#### 3.3.4.1 Number of Patches (NP)

The Number of Patches (NP), represents a total number of build-up patches in a given landscape. The analysis of the "Number of Patches" (NP) data for the landscape matrices of PMC, PCMC, and NMC reveals a consistent trend of landscape fragmentation in these urban areas from 2001 to 2021, all three municipal corporations—Pune, Pimpri-Chinchwad, and Nashik—exhibited a substantial increase in the number of patches. In PMC, the number of patches rose from 61,906 to 121,340, while PCMC saw a rise from 25,364 to 51,760 patches. Similarly, NMC witnessed an increase from 43,109 to 125,366 patches. This progression of elevated patch counts signifies a notable fragmentation of the urban landscapes. The transformation of larger land areas into smaller, distinct patches indicates ongoing urban expansion, changes in land use patterns, and increased development activities. As a result, these urban areas have become more heterogeneous in terms of land cover distribution.

#### 3.3.4.2 Patch Density (PD)

Patch Density (PD) serves as an indicator of landscape fragmentation, where it rises in correspondence with an increasing count of patches. The analysis of "Patch Density" (PD) data from 2001 to 2021 across PMC, PCMC, and NMC highlights a consistent pattern of increasing landscape fragmentation within these urban areas. In the context of PMC, the PD metric rises from 27.68 patches per km$^2$ in 2001 to 37.22 patches per km$^2$ in 2021. Notably, PCMC exhibits a comparable trajectory, with patch density increasing from 25.23 patches per km$^2$ in 2001 to 42.38 patches per km$^2$ in 2021. Similarly, NMC displays a parallel progression, with patch density increasing from 17.13 patches per km$^2$ in 2001 to 25.40 patches per km$^2$ in 2021. The overarching trend of augmented patch density values across these municipalities underscores a shared narrative of heightened landscape fragmentation. As patch density increases, the landscape becomes characterized by a larger number of smaller and more dispersed patches.

**Table 3.7** Landscape metrics for build-up in PMC, PCMC, and NMC

| *Number of patches (NP)* | | | | | | *Clumpiness Index (Clumpy)* | | | | | |
|---|---|---|---|---|---|---|---|---|---|---|---|
| | *2001* | *2006* | *2011* | *2016* | *2021* | | *2001* | *2006* | *2011* | *2016* | *2021* |
| PMC | 61,906 | 77,221 | 94,883 | 107,363 | 121,340 | PMC | 0.15 | 0.20 | 0.23 | 0.28 | 0.34 |
| PCMC | 25,364 | 31,592 | 40,197 | 46,415 | 51,760 | PCMC | 0.53 | 0.40 | 0.32 | 0.24 | 0.41 |
| NMC | 43,109 | 66,227 | 86,918 | 98,395 | 125,366 | NMC | 0.52 | 0.48 | 0.43 | 0.35 | 0.36 |
| *Patch density (PD)* | | | | | | *Aggregation Index (AI)* | | | | | |
| | *2001* | *2006* | *2011* | *2016* | *2021* | | *2001* | *2006* | *2011* | *2016* | *2021* |
| PMC | 27.68 | 29.84 | 32.92 | 35.07 | 37.22 | PMC | 46.25 | 52.25 | 57.75 | 62.5 | 66.62 |
| PCMC | 25.23 | 29.13 | 33.16 | 37.23 | 42.38 | PCMC | 45.08 | 49.62 | 50.56 | 54.5 | 58.25 |
| NMC | 17.13 | 18.02 | 21.35 | 23.82 | 25.40 | NMC | 66.88 | 63.38 | 60.38 | 57.17 | 36.55 |
| *Mean patch size (MPS)* | | | | | | *Percentage of like adjacencies (PLADJ)* | | | | | |
| | *2001* | *2006* | *2011* | *2016* | *2021* | | *2001* | *2006* | *2011* | *2016* | *2021* |
| PMC | 4.43 | 6.02 | 8.68 | 11.17 | 14.06 | PMC | 47.88 | 51.38 | 55.50 | 59.88 | 67.13 |
| PCMC | 12.26 | 15.89 | 19.57 | 21.77 | 28.52 | PCMC | 48.25 | 62.25 | 67.25 | 75.38 | 81.50 |
| NMC | 7.23 | 8.35 | 8.21 | 8.06 | 10.37 | NMC | 43.63 | 50.75 | 59.25 | 62.00 | 66.75 |
| *Edge density (ED)* | | | | | | *Interspersion and juxtaposition Index (IJI)* | | | | | |
| | *2001* | *2006* | *2011* | *2016* | *2021* | | *2001* | *2006* | *2011* | *2016* | *2021* |
| PMC | 72.50 | 92.32 | 107.19 | 135.56 | 172.39 | PMC | 65.45 | 70.64 | 74.84 | 79.12 | 82.85 |
| PCMC | 43.75 | 56.07 | 70.16 | 85.95 | 97.78 | PCMC | 46.61 | 55.25 | 70.08 | 73.94 | 78.74 |
| NMC | 99.99 | 192.11 | 165.54 | 187.05 | 211.35 | NMC | 29.05 | 35.95 | 39.41 | 28.19 | 24.64 |
| *Landscape shape Index (LSI)* | | | | | | *CONTAG* | | | | | |
| | *2001* | *2006* | *2011* | *2016* | *2021* | | *2001* | *2006* | *2011* | *2016* | *2021* |
| PMC | 34.47 | 44.63 | 56.30 | 67.84 | 76.12 | PMC | 70.95 | 73.01 | 75.49 | 79.40 | 70.95 |
| PCMC | 30.61 | 32.79 | 41.20 | 42.15 | 39.22 | PCMC | 55.39 | 58.95 | 63.95 | 67.20 | 70.55 |
| NMC | 41.14 | 61.00 | 128.73 | 75.50 | 83.39 | NMC | 61.83 | 57.23 | 68.48 | 71.52 | 66.94 |
| *Normalized landscape shape Index (NLSI)* | | | | | | | | | | | |
| | *2001* | *2006* | *2011* | *2016* | *2021* | | | | | | |
| PMC | 0.36 | 0.41 | 0.44 | 0.48 | 0.55 | | | | | | |
| PCMC | 0.60 | 0.56 | 0.52 | 0.50 | 0.47 | | | | | | |
| NMC | 0.25 | 0.43 | 0.46 | 0.43 | 0.45 | | | | | | |

### 3.3.4.3 Mean Patch Size (MPS)

"Mean Patch Size" (MPS) is a metric used in class or landscape analysis. It calculates the average area of patches by dividing the total patch area by the number of patches. Applying MPS to data from 2001 to 2021 in PMC, PCMC, and NMC urban areas reveals insights into the average size of built-up patches in these regions. In PMC, a consistent uptrend in MPS is observed, increasing from 4.43 in 2001 to 14.06 in 2021. Similarly, PCMC exhibits a parallel pattern, with MPS ascending from 12.26 in 2001 to 28.52 in 2021. This signifies a deliberate transition toward larger and potentially more nucleus patches. In contrast, NMC maintains relatively stable MPS values, ranging from 7.23 in 2001 to 10.37 in 2021, reflecting a consistent patch size distribution. Collectively, the trends in MPS provide insights into urban development approaches, illustrating the balance sought between urban expansion, land use efficiency, and ecological considerations in these municipalities.

#### 3.3.4.4 Edge Density (ED)

Edge density serves as an indicator for assessing landscape fragmentation and spatial diversity. Analyzing the data on edge density between 2001 and 2021 across PMC, PCMC, and NMC urban areas provides insights into the patterns of built-up edge concentration within these urban landscapes. Notably, for PMC, there is a consistent and substantial increase in edge density from 72.50 in 2001 to 172.39 in 2021. Similarly, PCMC displays a parallel trajectory, rising from 43.75 in 2001 to 97.78 in 2021, indicating increased habitat edges due to urban expansion. In contrast, the NMC edge density trajectory starts with a significant value of 99.99 in 2001, peaks at 192.11 in 2006, and gradually reduces to 211.35 in 2021. These trends collectively highlight the evolving urban landscapes, with elevated edge densities suggesting augmented habitat fragmentation and interface. The findings emphasize the significance of strategic urban planning to mitigate fragmentation's ecological effects, foster landscape connectivity, and ensure sustainable urban growth.

#### 3.3.4.5 Landscape Shape Index (LSI)

For PMC, there is a consistent upward trajectory in LSI values, increasing from 34.47 in 2001 to 76.12 in 2021. This indicates a continuous enhancement in landscape complexity and irregularity, possibly due to urban expansion and developmental activities that reshape the natural environment. Conversely, PCMC follows a fluctuating pattern with LSI values. Although there is a slight increase from 30.61 in 2001 to 42.15 in 2016, the value drops to 39.22 in 2021. This suggests a dynamic interplay between land development and preservation efforts within the urban fabric. NMC LSI values show a remarkable spike from 41.14 in 2001 to 128.73 in 2011, followed by a decline to 83.39 in 2021. This sharp rise and subsequent fall may reflect rapid urbanization followed by strategic planning initiatives to address landscape fragmentation. Overall, these trends underscore the intricate relationship between urbanization, development, and landscape configuration, urging the need for comprehensive urban planning strategies that balance growth with ecological sustainability.

#### 3.3.4.6 Normalized Landscape Shape Index (NLSI)

For PMC, the NLSI values consistently increase from 0.36 in 2001 to 0.55 in 2021. This signifies a continuous enhancement in landscape complexity and diversity, suggesting that while patches are growing larger, their shapes are becoming more varied and irregular. PCMC, on the other hand, presents a fluctuating pattern, with NLSI values starting at 0.60 in 2001, declining to 0.50 in 2016, and then further to 0.47 in 2021. This indicates shifts in the balance between patch size and shape complexity. NMC NLSI values show a rise from 0.25 in 2001 to 0.46 in 2011, followed by relative stability around 0.43 in subsequent years. This suggests an initial effort to enhance landscape complexity, followed by a more consistent configuration. Overall, these trends underscore the intricate interplay between urbanization, development, and landscape structure, emphasizing the importance of holistic urban planning strategies that consider both patch size and shape intricacy in achieving sustainable and ecologically balanced urban landscapes.

#### 3.3.4.7 Clumpiness Index (Clumpy)

The Clumpiness Index analysis reveals distinct trends in the patch distribution patterns of Pune, Pimpri-Chinchwad, and Nashik Municipal Corporations. For PMC, there is a consistent and steady increase in Clumpiness Index values from 0.15 in 2001 to 0.34 in 2021. This signifies a notable shift toward a more clustered arrangement of habitat patches over time, indicating potential efforts to consolidate patches into more cohesive groupings. PCMC, in contrast, displays a fluctuating trend, with the Clumpiness Index values initially decreasing from 0.53 in 2001 to 0.24 in 2016 before rebounding to 0.41 in 2021. This suggests varying degrees of dispersion and clustering within the patch configuration. Similarly, NMC Clumpiness Index values show a decrease from 0.52 in 2001 to 0.36 in 2021, indicating a trend toward more dispersed patches.

#### 3.3.4.8 Aggregation Index (AI)

The Aggregation Index analysis unveils distinct patterns in the patch distribution of Pune, Pimpri-Chinchwad, and Nashik Municipal Corporations. For Pune, there is a consistent and gradual increase in Aggregation Index values, rising from 46.25 in 2001 to 66.62 in 2021. This signifies a notable shift toward more clustered patch arrangements over time, potentially indicating initiatives to concentrate habitat patches in closer proximity. Pimpri-Chinchwad showcases a steady upward trajectory in Aggregation Index values, moving from 45.08 in 2001 to 58.25 in 2021. This suggests an ongoing trend toward increased patch clustering, aligning with strategies to foster spatial cohesion. In contrast, Nashik's Aggregation Index values display an initial high of 66.88 in 2001, followed by a gradual decline to 36.55 in 2021. This could signify a shift in urban planning priorities toward a more dispersed arrangement of habitat patches.

#### 3.3.4.9 Percentage of like Adjacencies (PLADJ)

The PLADJ analysis unveils discernible patterns in the arrangement of similar land cover types in Pune, Pimpri-Chinchwad, and Nashik Municipal Corporations. For Pune, there is a consistent upward trend in PLADJ values, increasing from 47.88 in 2001 to 67.13 in 2021. This signifies a continuous augmentation in the prevalence of like land cover types being adjacent, potentially indicating urban planning efforts to cluster similar land uses together. Pimpri-Chinchwad demonstrates a similar pattern, with PLADJ values rising from 48.25 in 2001 to 81.50 in 2021. This highlights a pronounced increase in the adjacency of like land cover types, reflecting strategies that promote spatial cohesiveness of similar land uses. Nashik, meanwhile, follows suit with a rise in PLADJ values from 43.63 in 2001 to 66.75 in 2021, indicating a consistent trend toward clustering similar land cover types.

#### 3.3.4.10 Interspersion and Juxtaposition Index (IJI)

The IJI analysis elucidates discernible trends in the juxtaposition and arrangement of distinct land cover types within Pune, Pimpri-Chinchwad, and Nashik Municipal Corporations. For Pune, there is a steady increase in IJI values, rising from 65.45 in 2001 to 82.85 in 2021. This signifies an ongoing effort to enhance the mingling of different land cover types, potentially indicative of urban planning strategies that promote functional interactions among diverse land uses. Pimpri-Chinchwad also displays a noteworthy ascent, with IJI values progressing from 46.61 in 2001 to 78.74 in 2021. This highlights a pronounced increase in the intertwining of various land cover types, suggesting strategies to foster spatial diversity and interconnectivity. Nashik, however, showcases a varying trend with IJI values initially increasing from 29.05 in 2001 to 39.41 in 2011, followed by a decline to 24.64 in 2021. This may indicate shifts in urban development dynamics and land use practices that have influenced the arrangement of different land cover categories.

#### 3.3.4.11 CONTAG

The CONTAG analysis unveils distinctive patterns in the spatial clustering and connectivity of diverse land cover types in Pune, Pimpri-Chinchwad, and Nashik Municipal Corporations. For Pune, there is a relatively consistent trajectory in CONTAG values, starting at 70.95 in 2001, peaking at 79.40 in 2016, and returning to 70.95 in 2021. This suggests a stable tendency of certain land cover types to cluster together in contiguous patches. Pimpri-Chinchwad follows a similar pattern, with CONTAG values increasing from 55.39 in 2001 to 70.55 in 2021. This signifies an incremental enhancement in the cohesion and contiguity of different land cover categories. Nashik, meanwhile, displays fluctuations, with CONTAG values starting at 61.83 in 2001, decreasing to 57.23 in 2006, rising to 71.52 in 2016, and settling at 66.94 in 2021. These variations indicate shifts in the spatial arrangement and connectivity of land cover types over time, likely influenced by urban development dynamics.

## 3.4 Conclusions

The comprehensive urban growth analysis for Pune Municipal Corporation (PMC), Pimpri-Chinchwad Municipal Corporation (PCMC), and Nashik Municipal Corporation (NMC) from 2001 to 2021 reveals a compelling narrative: the transformative force of urban expansion has irreversibly altered the landscape across these municipalities. While these urban transitions have varied across the environmental entities (land use/cover), they all point to a growing conflict between environmental response and human progress. In terms of rapid urban expansion, NMC, which has seen a 134.5% surge in built-up areas, the pace of change is indicative of both robust economic activities and a rapidly growing population. On the other hand, PCMC's more moderate, yet steady, 61.29% urban growth rate. PMC, with its slower but substantial expansion, represents the transition between older and emerging urban areas, suggesting that its future growth may focus more on vertical development and infrastructural optimization rather than sprawling outward. The landscape metrics suggest rapid evolution of the landscape where older urban landscapes suggest maturity, and the new emerging patches suggest peripheral urban sprawl.

**Acknowledgements** The author extends heartfelt gratitude to the Indian Council of Social Science Research, New Delhi, for the generous funding provided through the IMPRESS research grant scheme (Grant No. IMPRESS/P2832/2018-19/ICSSR) that supported the substantial research project. Additionally, sincere appreciation is conveyed to K.V.N. Naik Shikshan Prasarak Sanstha's Arts, Commerce, and Science College Nashik for their unwavering encouragement and support throughout this research endeavor.

## References

Aburas MM, Ho YM, Ramli MF et al (2016) The simulation and prediction of spatio-temporal urban growth trends using cellular automata models: A review. Int J Appl Earth Obs Geoinf 52:380–389. https://doi.org/10.1016/j.jag.2016.07.007

Akın A, Erdoğan MA (2020) Analyzing temporal and spatial urban sprawl change of Bursa city using landscape metrics and remote sensing. Model Earth Syst Environ 6:1331–1343. https://doi.org/10.1007/s40808-020-00766-1

Akubia JE, Bruns A (2019) Unravelling the frontiers of urban growth: spatio-temporal dynamics of land-use change and urban expansion in greater Accra metropolitan area, Ghana. Land 8(9):131. https://doi.org/10.3390/land8090131

Alam A, Bhat MS, Maheen M (2020) Using Landsat satellite data for assessing the land use and land cover change in Kashmir valley. GeoJournal 85:v1529-1543. https://doi.org/10.1007/s10708-019-10037-x

Alwedyan S (2023) Monitoring urban growth and land use change detection with GIS techniques in Irbid City, Jordan. Int Rev Spatial Plann Sustain Dev 11(1): 253–275. https://doi.org/10.14246/irspsd.11.1_253

Aslam RW, Shu H, Yaseen A (2023) Monitoring the population change and urban growth of four major Pakistan cities through spatial analysis of open source data. Ann GIS 29(3):355–367. https://doi.org/10.1080/19475683.2023.2166989

Baqa MF, Chen F, Lu L et al (2021) Monitoring and modeling the patterns and trends of urban growth using urban sprawl matrix and CA-Markov model: a case study of Karachi. Pakistan. Land 10(7):700. https://doi.org/10.3390/land10070700

Batty M (2013) The new science of cities. MIT Press, Cambridge, MA

Bhailume SA (2021) An assessment of land use land cover change and urban growth of Nashik city using geospatial techniques. Appl Ecol Environ Sci 9(2): 225–235. https://doi.org/10.12691/aees-9-2-14

Bodum L, Kjems E (2005) Focus for 3D city models should be on interoperability: not verisimilarity! In: Proceedings of UDMS'06: 25th urban data management symposium. Urban Data Management Society pp 9–1

Bosch M, Jaligot R et al (2020) Spatio-temporal patterns of urbanization in three Swiss urban agglomerations: insights from landscape metrics, growth modes and fractal analysis. Landscape Ecol 35:879–891. https://doi.org/10.1007/s10980-020-00985-y

Cardille JA, Turner MG (2017) Understanding landscape metrics. In: Gergel S, Turner M (eds) Learning landscape ecology. Springer, New York, pp p45-63

Das S, Jain GV (2022) Assessment and prediction of urban expansion using CA-based SLEUTH urban growth model: A case study of Kolkata Metropolitan area (KMA), West Bengal. India. J Indian Soc Remote Sens. 50(12):2277–2302. https://doi.org/10.1007/s12524-022-01602-y

Dhanaraj K, Angadi DP (2022) Land use land cover mapping and monitoring urban growth using remote sensing and GIS techniques in Mangaluru, India. GeoJournal 87(2):1133–1159. https://doi.org/10.1007/s10708-020-10302-4

Dinda S, Chatterjee ND, Ghosh S (2021) An integrated simulation approach to the assessment of urban growth pattern and loss in urban green space in Kolkata, India: a GIS-based analysis. Ecol Indic 121:107178. https://doi.org/10.1016/j.ejrs.2021.02.001

Dutta I, Das A (2019) Application of geospatial indices for detection of growth dynamics and forms of expansion in English Bazar urban agglomeration, West Bengal. J Urban Manag 8(2):288–302. https://doi.org/10.1016/j.jum.2019.03.007

Elwood S, Leszczynski A (2018) Feminist digital geographies. Gend Place Cult 25(5):629–644. https://doi.org/10.1080/0966369X.2018.1465396

Flowers B, Huang KT, Aldana GO (2020) Analysis of the habitat fragmentation of ecosystems in Belize using landscape metrics. Sustainability 12(7):3024. https://doi.org/10.3390/su12073024

Gavsker KK (2023) Urban growth, changing relationship between biophysical factors and surface thermal characteristics: a geospatial analysis of Agra city, India. Sustain Cities Soc 94:104542. https://doi.org/10.1016/j.scs.2023.104542

Getu K, Bhat HG (2021) Analysis of spatio-temporal dynamics of urban sprawl and growth pattern using geospatial technologies and landscape metrics in Bahir Dar, Northwest Ethiopia. Land Use Policy 109:105676. https://doi.org/10.1016/j.landusepol.2021.105676

Goodchild MF (2011) Scale in GIS: an overview. Geomorphology 130(1–2):5–9. https://doi.org/10.1016/j.geomorph.2010.10.004

Hailu A, Mammo S (2020) Dynamics of land use, land cover change trend and its drivers in Jimma Geneti District, Western Ethiopia. Land Use Policy 99:105011. https://doi.org/10.1016/j.landusepol.2020.105011

Hamad R (2020) A remote sensing and GIS-based analysis of urban sprawl in Soran District, Iraqi Kurdistan. SN Appl Sci 2(1):1–9. https://doi.org/10.1007/s42452-019-1806-4

Hamnett S, Yuen B (eds) (2019) Planning Singapore: the experimental city. Routledge, London

Jaafari S, Shabani AA, Moeinaddini M et al (2020) Applying landscape metrics and structural equation modeling to predict the effect of urban green space on air pollution and respiratory

mortality in Tehran. Environ Monit Assess 192:1–15. https://doi.org/10.1007/s10661-020-08377-0
Jensen JR, Hodgson ME, Garcia-Quijano M et al (2009) A remote sensing and GIS-assisted spatial decision support system for hazardous waste site monitoring. Photogramm Eng Remote Sens 75(2):169–177. https://doi.org/10.14358/PERS.75.2.169
Latue PC, Rakuasa H (2023) Analysis of land cover change due to urban growth in central Ternate district, Ternate city using Cellular Automata-Markov Chain. J Appl Geospat Inf 7(1): 722–728. https://doi.org/10.30871/jagi.v7i1.4653
Liu R, Dong X, Wang X et al (2023) Relationship and driving factors between urbanization and natural ecosystem health in China. Ecol Indic 147:109972. https://doi.org/10.1016/j.ecolind.2023.109972
Liu Y, Li L, Chen L et al (2019) Urban growth simulation in different scenarios using the SLEUTH model: a case study of Hefei, East China. PLoS ONE 14(11):e0224998. https://doi.org/10.1371/journal.pone.0224998
Longley P, Goodchild MF, Maguire DJ el al (2005) Geographic information systems and science. 2nd ed. Wiley, West Sussex, London
Magidi J, Ahmed F (2019) Assessing urban sprawl using remote sensing and landscape metrics: a case study of city of Tshwane, South Africa (1984–2015). Egypt J Remote Sens Space Sci 22(3):335–346. https://doi.org/10.1016/j.ejrs.2018.07.003
Mandal J, Ghosh N (2019) Urban growth dynamics and changing land-use land-cover of megacity Kolkata and its environs. J Indian Soc Remote Sens 47:1707–1725. https://doi.org/10.1007/s12524-019-01020-7
Manna H, Sarkar S Hossain M et al (2023) Modeling and predicting spatio-temporal land use land cover changes and urban sprawling in Kalaburagi City Corporation, Karnataka, India: a geospatial analysis. Model Earth Syst Environ, 1–24. https://doi.org/10.1007/s40808-023-01814-2
Mishra PK, Rai A (2020) Land use and land cover change detection using geospatial techniques in the Sikkim Himalaya, India. Egypt J Remote Sens Space Sci 23(2):133–143. https://doi.org/10.1016/j.ejrs.2019.02.001
Moghadam HS, Helbich M (2013) Spatiotemporal urbanization processes in the megacity of Mumbai, India: a Markov chains-cellular automata urban growth model. Appl Geogr 40:140–149. https://doi.org/10.1016/j.apgeog.2013.01.009
Mohabey DP, Nongkynrih JM (2023) Urban growth trend analysis of proposed Greater Silchar City, India, using landscape metrics and Shannon entropy model. Environ Dev Sustain, 1–32. https://doi.org/10.1007/s10668-023-03681-4
Patra PK, Behera D, Goswami S (2022) Relative shannon's entropy approach for quantifying urban growth using remote sensing and GIS: a case study of Cuttack city, Odisha, India. J Indian Soc Remote Sens 50(4):747–762. https://doi.org/10.1007/s12524-022-01493-z
Paul S, Saxena KG, Nagendra H et al (2021) Tracing land use and land cover change in peri-urban Delhi, India, over 1973–2017 period. Environ Monit Assess 193:1–12. https://doi.org/10.1007/s10661-020-08841-x
Rajendran GB, Kumarasamy UM, Zarro S et al (2020) Land-use and land-cover classification using a human group-based particle swarm optimization algorithm with an LSTM classifier on hybrid pre-processing remote-sensing images. Remote Sens 12(24):4135. https://doi.org/10.3390/rs12244135
Ray R, Das A, Hasan MSU et al (2023) Quantitative analysis of land use and land cover dynamics using geoinformatics techniques: a case study on Kolkata Metropolitan Development Authority (KMDA) in West Bengal, India. Remote Sens 15(4):959. https://doi.org/10.3390/rs15040959
Tian Y, Tsendbazar NE, von Leeuwen E et al (2022) A global analysis of multifaceted urbanization patterns using Earth Observation data from 1975 to 2015. Landsc Urban Plan 219:104316. https://doi.org/10.1016/j.landurbplan.2021.104316

Tong X, Feng Y (2020) A review of assessment methods for cellular automata models of land-use change and urban growth. Int J Geogr Inf Sci 34(5):866–898. https://doi.org/10.1080/13658816.2019.1684499

Trinder J, Liu Q (2020) Assessing environmental impacts of urban growth using remote sensing. Geospatial Inf Sci 23(1):20–39. https://doi.org/10.1080/10095020.2019.1710438

United Nations, Department of Economic and Social Affairs (2018) Revision of World Urbanization Prospects. https://www.un.org/en/desa/2018-revision-world-urbanization-prospects

Varquez ACG, Dong S, Hanaoka S et al (2023) Evaluating future railway-induced urban growth of twelve cities using multiple SLEUTH models with open-source geospatial inputs. Sustain Cities Soc 91:104442. https://doi.org/10.1016/j.scs.2023.104442

Wahla SS, Kazmi JH, Tariq A (2023) Mapping and monitoring of spatio-temporal land use and land cover changes and relationship with normalized satellite indices and driving factors. Geol Ecol Landsc, 1–17. https://doi.org/10.1080/24749508.2023.2187567

Wiatkowska B, Słodczyk J (2021) Spatial-temporal land use and land cover changes in urban areas using remote sensing images and GIS analysis: the case study of Opole, Poland. Geosciences 11(8):312. https://doi.org/10.3390/geosciences11080312

Zeng T, Guo Z (2023) Tempo-spatial landslide susceptibility assessment from the perspective of human engineering activity. Remote Sens 15(16):4111. https://doi.org/10.3390/rs15164111

# Chapter 4
# Perspective Shift in Land Acquisition Policies and Its Impact on the Urban Development of Navi Mumbai Region, India

**Gandharva H. Pednekar, Purushottam V. Bhandare, and Rashmee Sharma**

**Abstract** The interactions between development projects and the communities around them are complex and dynamic. The communities can be affected by the projects as well as influence them in various ways. Urban areas often face the challenge of accommodating the growing population and meeting their needs without compromising the infrastructure and the quality of life. This requires careful planning and regulation of urban structures. New urban areas can emerge as a result of economic and social factors. One such example is Navi Mumbai, the satellite city of Mumbai located near the west coast of India, which was developed after independence due to the influx of migrants from rural areas to Mumbai. The urbanization process in India is different from that of the developed countries. The current plan of developing the rural areas around the upcoming Navi Mumbai International Airport through an aerotropolise model, known as Navi Mumbai Airport Influence Notified Area (NAINA), is a major urban development project undertaken by CIDCO. This paper examines the planning and development of the Navi Mumbai region, the evolution of land acquisition and rehabilitation and resettlement policies over time and the role of stakeholders in development while identifying the gaps in implementation. The paper also proposes effective measures for the future development of urban centres in India.

**Keywords** Urbanization · Sustainability · Policy · Inclusive development · Urban planning

G. H. Pednekar (✉) · R. Sharma
Give Grants, Bengaluru, India
e-mail: me.gandharva@gmail.com

P. V. Bhandare
Peace Economy Project, Chicago, USA

 
S. Anand et al. (eds.), *Emerging Geosustainability Transformations in India*, Advances in Geographical and Environmental Sciences,
https://doi.org/10.1007/978-981-97-7993-2_4

## 4.1 Introduction

The first wave of globalization resulted in the formation of large, dominant cities such as New Delhi, Kolkata, Mumbai, Rio de Janeiro, Jakarta, and Nairobi. These cities, which led the urban hierarchies in their respective countries, served as administrative centers and channels for the export of raw materials from the developing world. Concurrently, this period saw the emergence of dualistic societies, characterized by a vast rural population living in poverty (mirrored in the cities' informal sectors), under the control of a small ruling class comprised of colonialists, foreign businesspeople, and local oligarchs (McGee 1997). Differences in the timing and processes of decolonization, which occurred earlier in Latin America, have resulted in varied urbanization patterns. Latin America has achieved urbanization levels comparable to those of developed countries. In contrast, Asia and Africa are just beginning the 'accelerated' phase of urban transition from rural to urban populations. In each of these continents, migration and endogenous growth in large primate cities and their surrounding industrial locations are expanding urban areas, leading to the formation of what are now referred to as extended metropolitan regions.

Most people in the world, especially in Asia, Africa and Latin America, still live in places where networks are mainly local and personal; these networks may be vital for their survival (McGee 1997). McGee also argues that globalization is accompanied by an increase in urbanization and the formation of global and sub-global systems of highly connected cities. If the current trends of urban concentration continue, the developing world will witness the rise of mega-urban regions as major components of their urban systems. Although some argue that this is the second phase of globalization in developing countries, after the first one marked by their incorporation into the colonial system, the urban consequences are equally significant. Hudalah et al. (2007) describe the specificity of peri-urbanization in East-Asian countries in terms of its dependence on the metropolitan centres, capital accumulation and dynamic coexistence of urban and rural livelihoods. Robertson and Pinstrup-Andersen (2010) reported that land rich developing nations are sanctioning the transfer of land rights toward urbanization. This involves land acquisition for the industrial setups by foreign investments. In India, rapid urbanization has led to the growth of peri-urban areas and thus the conversion of rural land into urban settings. Land acquisition and urbanization have become one of the most contentious policy issues in relation to spatial development planning (Panda 2010).

In India, urbanization has been increasing since the beginning of the twentieth century. The census data on India's rural–urban composition indicate a consistent increase in urbanization, particularly in the latter half of the twentieth century. The urban population percentage, which was merely 11% in 1911, rose slightly to 11.3% in 1921 and gradually to 14% by 1941. In 1951, a broader definition of urban areas led to a significant rise in the urban population percentage to 17.6%. However, with a more stringent definition in 1961, the percentage saw a modest increase to 18.3%. A new urban unit definition was adopted in the 1971 census and retained in the 1981 census.

The proportion of India's urban population has grown from 25.8% in 1991 to 27.8% in 2001. By 2011, the urban population reached 377 million, accounting for 31.16% of the total population. Additionally, the number of towns in India has seen a steady increase, rising from 1,627 in 1951 to 3,060 in 1971, 3,126 in 1971, 4,029 in 1981, 5,166 in 2001, and 7,935 in 2011 (Table 4.1).

In the late nineteenth century, several cities emerged as transportation hubs due to the railways, hill stations, and cantonments for the ruling elite, as well as market towns for exporting agricultural products during the first wave of globalization (Heitzman 2008). Additionally, small industrial clusters for specialized products developed, particularly in Tamil Nadu, within the irregular quadrilateral formed by Coimbatore, Salem, Tiruchirapalli, and Madurai (Ghurye 1974). Following Independence in 1947, the Indian government established several new cities for state-level political administration, such as Chandigarh, Bhubaneswar, and Gandhinagar, as well as industrial townships. However, planning policies at that time exhibited an anti-urban bias, with a disproportionate emphasis and funding on rural development projects (Ramachandran 1989). The liberalization of the economy in 1991 marked another significant milestone, as India opened up to the global economy, attracting foreign investment directed toward specific sectors and cities (Shaw 2012). With the introduction of neoliberal economic policies and reductions in public expenditure on rural development, cities began to be seen as "engines of growth" capable of attracting national and global business and investment, thereby contributing to the broader economic growth of the country (Mitra and Mehta 2011; Siddiqui 2014). Achieving this vision necessitated improvements to existing infrastructure and the development of new infrastructure. It was acknowledged that the public sector lacked sufficient funds, which opened the door for the private sector to become more significantly involved in urban development.

In 2005, the most substantial urban development effort since its independence was embarked upon by the Indian government, known as the Jawaharlal Nehru National Urban Renewal Mission (JNNURM) (Ministry of Housing and Urban Affairs 2024). This initiative integrated all prior funding from the central government and urban development programs, including the National Slum Development Programme (NSDP), Swarna Jayanti Sahakari Rozgar Yojana, Valmiki-Ambedkar Awas Yojana (aimed at providing housing for the socially disadvantaged urban population), and the National Transport Policy. The JNNURM encompassed elements such as the privatization and commercialization of essential services like water provision, sanitation, and waste management through collaborations between public and private entities, which also introduced user charges. It involved the liberalization of the land and property market by revoking the Urban Land (Ceiling and Regulation) Act (Ministry of Housing and Urban Affairs 1976) and modifying the Rent Control Acts. It also included the establishment of a robust mortgage market in conjunction with 100% foreign direct investment in housing and real estate, simplification of land-use conversion rules, property tax reforms and reduction in stamp duties, financial and administrative overhaul of municipalities, rationalization and outsourcing of services offered by the Urban Local Body (ULB), implementation of e-governance,

**Table 4.1** Urbanization and urban growth in India, 1871–2011

| Year | Number of towns and cities | Number of cities/UAS with million + population | Urban population (million) | Total population (million) | Urbanization % | Annual inter-censal urban population growth rate % | Annual inter-censal total population growth rate % | Share of major 10 cities as % of urban population | Share of major 10 cities as % of total population | Urban sex ratio | Total sex ratio |
|---|---|---|---|---|---|---|---|---|---|---|---|
| 1871 | | 1 | 18 | 212 | 8.7 | | | | | | 944 |
| 1881 | | 1 | 20 | 214 | 9.3 | 0.75 | 0.08 | | | | 962 |
| 1891 | | 1 | 22 | 234 | 9.4 | 1.03 | 0.92 | | | | 962 |
| 1901 | 1,916 | 1 | 26 | 238 | 10.8 | 1.63 | 0.19 | 17.3 | 1.9 | 910 | 972 |
| 1911 | 1,908 | 2 | 26 | 252 | 10.3 | 0.04 | 0.56 | 19.4 | 2 | 872 | 964 |
| 1921 | 2,048 | 2 | 28 | 251 | 11.2 | 0.8 | −0.03 | 19.8 | 2.2 | 846 | 955 |
| 1931 | 2,220 | 2 | 33 | 279 | 12 | 1.77 | 1.05 | 19 | 2.3 | 838 | 950 |
| 1941 | 2,427 | 2 | 44 | 319 | 13.9 | 2.81 | 1.34 | 22 | 3 | 831 | 945 |
| 1951 | 3,060 | 5 | 62 | 361 | 17.3 | 3.52 | 1.26 | 24.4 | 4.2 | 860 | 946 |
| 1961 | 2,700 | 7 | 79 | 439 | 18 | 2.37 | 1.98 | 26 | 4.7 | 845 | 941 |
| 1971 | 3,126 | 9 | 109 | 548 | 19.9 | 3.29 | 2.24 | 26.5 | 5.3 | 858 | 930 |
| 1981 | 4,029 | 12 | 159 | 683 | 23.3 | 3.87 | 2.23 | 25.7 | 6 | 880 | 934 |
| 1991 | 4,689 | 23 | 218 | 846 | 25.7 | 3.16 | 2.16 | 25.4 | 6.5 | 894 | 926 |
| 2001 | 5,161 | 35 | 286 | 1,029 | 27.8 | 2.78 | 1.97 | 26 | 7.2 | 900 | 933 |
| 2011 | 7,935 | 53 | 377 | 1,211 | 31.2 | 2.8 | 1.64 | 24.6 | 7.7 | 929 | 943 |

*Source* Census of India, 1951–2011

and the promotion of the private sector and private credit-rating agencies over elected civic bodies.

## 4.2 Development of New Settlements

Urban expansion strategies have been implemented across Asia during the 1950s and 1960s in response to the growing urban population. One common method of urban expansion, which also facilitated the controlled growth of large metropolitan cities, was the establishment of satellite towns. Asian countries such as Japan, Hong Kong, Singapore, South Korea, and China have implemented strategies to control urban growth and decentralize populations. Japan passed the Capital Region Development Law in 1956 (Nakamura and White 1988) to create a greenbelt around Tokyo, but demands from residents led to suburban development. Satellite towns were established to absorb people from inner Tokyo, though some struggled to attract residents. Income disparities persisted, leading to the development of 'technopolises'.

Hong Kong and Singapore successfully decentralized populations through the construction of new residential towns and public housing, reducing core area populations. South Korea's attempt with the satellite town of Seongnam faced challenges due to insufficient job opportunities (Wang and Tan 1981; Yeung 1988). China focused on building satellite towns around Shanghai, with later towns constructed farther out. These self-contained communities reduced commuting distances (Fung and Freeberne 1981). Overall, the goal across these countries was to alleviate overcrowding in core areas, promote regional development, and address income disparities. While success varied, the strategies aimed to create sustainable and balanced urban growth.

In India, efforts have been made to manage urban growth by constructing satellite towns around the major metropolitan areas. This includes six ring towns surrounding Delhi, four encircling Hyderabad, three new towns near Chennai, two around Kolkata, and one outside of Mumbai.

## 4.3 Development of Urban Fringe

From 1991 to 2011, India's urban population expanded from 217 million, accounting for 27.5% of the total population, to 377 million, representing 31.16% of the total population. In urban areas, 43% of the population resides in cities or urban agglomerations (UAs) with a population of over 1 million. The number of such UAs has grown from 35 in 2001 to 53 in 2011. The three largest UAs, also known as megacities with populations exceeding 10 million, are Greater Mumbai UA with 18.4 million residents, Delhi UA with 16.3 million, and Kolkata UA with 14.1 million. Chennai UA and Bangalore UA follow these with populations of 8.7 million and 8.5 million respectively. During the first decade of economic reforms, there was a significant increase in the population of these megacities, but this growth has considerably decelerated in the period from 2001 to 2011 (Census of India 2011).

The transformation of erstwhile villages into urban areas and the redefinition of rural regions as urban, due to the expansion of city limits, have been key factors in urban development from 2001 to 2011 (Bhagat 2012). Rural areas are transitioning into what's known as urban fringe areas, where urban and rural activities and institutions exist side by side. These areas are characterized by the dynamic exchange of commodities, capital, natural resources, people, and pollution, and a variety of processes that amplify the connections between urban and rural areas (Marshall and Randhawa 2017). A significant portion of the urban poor have been either displaced or relocated to the peripheries of the city (Bhan 2014). The majority of the population in these urban fringe areas, which includes residents of slums, unauthorized colonies, and villages, lack basic services such as water, sanitation, and health, leading to serious public health problems.

Urbanization increasingly influences the distribution of the population toward higher income categories. As a result, urban regions have a larger proportion of people in lower middle-income, middle-income, upper-middle-income, and higher-income groups compared to rural areas. In this context, the urban fringe is viewed as a transitional area between the city and the countryside. It's not a distinct region, but rather a broad territory characterized by a mix of features and phenomena, largely driven by activities within the urban zone (Nottingham and Liverpool Universities 1998). Defining these areas is challenging due to the inherent difficulties in conceptualizing both rural and urban worlds.

The concept of "urban fringe" was initially coined by T.L. Smith in 1937 to refer to the developed region immediately beyond a city's official boundaries. Pastalan (1967) offered a functional interpretation of the fringe as a transitional space where rural land use gradually transitions to urban land use. Oosthwizen (1969) conducted research on the rural–urban fringe and proposed alternative terms for the suburban area, including "rural suburbs" and "quasi-urban" areas or townships.

A study conducted by Rao Prakasa and Tweari (1979) on the Bangalore city region underscored the significance of the Distance-Decay effect in shaping the zonal structure of the city region based on distance. The growth of urban areas also influences the occupational structure of the people in the surrounding regions. In this context, Seetharaman (1984) conducted a noteworthy study, suggesting that small towns serve as crucial links between the scattered rural population and medium to large urban centers by providing essential goods and services. These small towns are a significant part of the National Settlement System. Furthermore, Phadke and Sita (1981) attempted to examine the spatial pattern of urban influence in Bombay, concluding that accessibility, rather than geographical distance, is a key factor in the dissemination of urban characteristics.

## 4.4 History of Law of Land Acquisition in India

The initial legislation regarding property acquisition in India was introduced through the Bengal Regulation I of 1824. This law, which was enforced across all provinces under the presidency of Fort William, established guidelines for government officials to acquire land or other immovable assets at a fair price for public purposes such as roads and canals.

In 1850, during the development of the Railways, these were classified as public works and the aforementioned regulation was extended to include them. Concurrently, in the Madras presidency, Act XX of 1852 was enacted to ease the process of land acquisition for public use. The concept of 'public purpose' was expanded in 1850 to encompass Railways. It's noteworthy that until 1857, there was no uniform law governing land acquisition across all of British India.

### *4.4.1 Land Acquisition Act, 1894*

The Land Acquisition Act of 1894 is the principal law in India that governs the process by which the Government can procure land for a public purpose (LAA 1894). This Act outlines five key steps for the mandatory acquisition of privately owned land for a public purpose: (1) a preliminary notification of acquisition must be issued under Sect. 4.4; (2) the public has a thirty-day window following this notification to file objections; (3) the government's intent to acquire the land must be announced under Sect. 4.6; (4) affected parties must be invited to submit compensation claims, and the compensation amount must be determined; and finally, (5) once the compensation is paid, the government can take possession of the land (Acharya 1988).

Prior to India's independence, the issue of land acquisition was at the heart of numerous disputes and policy stalemates. Since its inception, the Land Acquisition Act of 1894 has been the subject of contentious debates and controversies. Despite several rounds of amendments aimed at addressing its shortcomings, the Act has not been successful in resolving key issues related to land acquisition. These include forced acquisitions, the vague definition of "public purpose", rampant misuse of the "urgency clause", compensation matters, lack of transparency in the acquisition process, and the involvement of communities whose land is being acquired. Moreover, the Act does not include provisions for rehabilitation and resettlement packages for farmers. Furthermore, poor implementation and ineffective administration at the grassroots level have exacerbated the distress and suffering of the people.

Land acquisition has evolved into a lucrative opportunity for numerous intermediaries. This is due to the stringent regulations and intricacies associated with land acquisition in the present day, which offer substantial arbitrage opportunities for these individuals. There have been instances in India where poor management of land acquisition issues has resulted in cost overruns and even project abandonment. Conversely, there are also examples where effective management of the issue has been beneficial for both the industry and those displaced by the acquisition. For instance, when the Punjab government sought to acquire 860 acres of fertile land in Mansa District, Punjab, for the development of a Thermal Power Project, it led to local unrest. The owners of approximately 160 acres of land have declined the offered compensation of Rs. 23 lakh per acre, insisting on continuing their farming activities and asserting that the land rate is about Rs. 50–60 lakh per acre.

Dhingra (2022) undertook a detailed historical analysis of the laws, ordinances, and legal cases related to land acquisition to study the political economy of land acquisition in India. She posited that while the emphasis of India's land acquisition laws is on the logic of law-making, they also present a conflict regarding fair compensation for dispossession due to infrastructure development in the country. She pointed out a deficiency in the modeling of state power within the political economic conditions. The Land Acquisition Act in India is a key land reform measure that supports infrastructure development. Besley and Burgess (2000) contended that land reforms are primarily linked to poverty reduction. However, despite infrastructural development, the issue of poverty continues to persist, largely due to delays in disbursing compensation. Recent land acquisition has been a consequence of the government's growth-focused initiatives, and a similar model has been in place since the economic reforms of 1991 (Mukerji 2017). Mathur's examination of displacement caused by development is relevant in the context of rapid urbanization, and he asserts that such displacement primarily stems from land acquisition programs in India, further exacerbated by unsuccessful resettlement programs (Mathur 2013).

### *4.4.2 Maharashtra State Land Acquisition Act*

The Land Acquisition Act in India, also known as the Right to Fair Compensation and Transparency in Land Acquisition, Rehabilitation, and Resettlement (R&R) Act, 2013. Key provisions of the Land Acquisition Act, 2013 include Social Impact Assessment (SIA), Consent of affected families, and finally compensation and rehabilitation of the impacted individual/community (R&R 2013).

The state of Maharashtra has adhered to the LA Act of 1894 and subsequently the RFC & TLARR, 2013 for the purpose of land acquisition. It has implemented numerous amendments to the Centre's RFC & TLARR, 2013 to expedite development projects within the State. In terms of land valuation, it references the ready reckoner value established under the Maharashtra Stamp Act (59 of 1958) and the Maharashtra Stamp (Determination of True Market Value of Property) Rules, 1995 as the basis for determining the market value of land. If the land is procured for urbanization, 20% of the developed land will be set aside and offered to the landowner at a price equivalent to the acquisition and development costs.

### 4.4.3 Maharashtra Regional Town Planning Act, 1966

The Maharashtra Regional Town Planning Act of 1966 is a significant piece of legislation that guides and controls development in urban regions as well as areas with urbanization potential. This Act serves as a primary model for town and country planning in India and has been subsequently adopted by other states such as Gujarat, Himachal Pradesh, Tamil Nadu, and Madhya Pradesh. Key features of this Act encompass provisions for regional planning, development plans, town planning schemes, and implementation strategies like land acquisition, transfer of development rights, and plot reconstitution methods. The Act also covers financial aspects (Finance account and Audit) and governance for plan enforcement (Planning authority). However, the Act has been criticized for lacking a specific time frame for the development plan's preparation by the planning authority and its approval by the State Government, reservations on private lands often exist only on paper, and development plans are often idealistic without considering the planning authority's capacity to implement them.

### 4.4.4 Right to Fair Compensation and Transparency in Land Acquisition and Resettlement and Rehabilitation Act, 2013 (RFC & TLARR ACT, 2013)

The RFC & TLARR Act 2013 is applicable to a broad spectrum of initiatives including Special Economic Zones (SEZs), Exclusive Economic Zones (EEZs), Intensive Agriculture Development Projects (IADPs), defense initiatives, Public–Private Partnerships, Government Projects, and sectors such as irrigation, power, infrastructure projects, and government educational institutions. The Act stipulates that there is no minimum land requirement for the implementation of rehabilitation and resettlement provisions, and it excludes an emergency clause, except in cases related to national security or natural disasters. This Act introduces two significant changes to the traditional Doctrine of Land Acquisition: (1) it imposes a duty to resettle and rehabilitate those affected by land acquisition; and (2) it mandates the involvement of the affected individuals through transparent processes that include public disclosure, consultation, and consent. These new obligations are in addition to the three established principles of serving a public purpose, ensuring fair compensation, and providing a mechanism for judicial resolution of disputes.

This act has led to the establishment of a new administrative structure. This includes the National Monitoring Committee, State Monitoring Committees, a Commissioner for Resettlement and Rehabilitation (R&R) in each state, R&R Committees at the Project Level, an Administrator for each project, and an Expert Group for evaluating the Social Impact Assessment (SIA) and Social Impact Management Plan (SIMP). Additionally, a Collector and Land Acquisition & R&R Authority have been instituted for dispute resolution, replacing the traditional role of Civil Courts.

In the process of modifying the Preliminary Notification (PN), Final Declaration (FD), and Award Enquiry, the New Act maintains procedures that are largely similar to those in the Land Acquisition (LA) Act. The time frame between PN and FD remains unchanged from the LA Act, at one year. However, the duration between FD and Award has been shortened from two years in the LA Act to one year. Despite this reduction, the overall time from the start to the end of the land acquisition process has extended from three years to five years. This includes six months for Social Impact Assessment (SIA), up to one year from the date of Environmental Clearance (EC) for issuing PN, an additional year after Displacement Notice (DN), and a further eighteen months after the Award is issued to finalize Resettlement Infrastructure before the land can be taken possession of.

The New Act (RFC & TLARR ACT, 2013) has outlined certain actions as non-cognizable offenses and has established corresponding penalties. It also restricts the direct acquisition of land beyond certain thresholds for any reason without the prior consent of the Collector or government. The Preliminary Notification (PN) is required to disclose details such as the public benefit of the proposed project, the necessity for the displacement or impact on individuals, a condensed report of the Social Impact Assessment's (SIA) key findings, and the contact information for the designated Project Administrator for additional inquiries, including information about the land intended for acquisition.

Contrary to the LA Act, the New Act introduces certain preparatory measures at the Preliminary Notification (PN) stage itself to address future challenges and legal disputes. These measures include mandatory updating of land records (including survey records) within a two-month period following the PN, resolution of forest land title issues concerning Scheduled Tribes and Other Forest Dwellers, cessation of all transactions, alienation, developments, and divisions of lands specified in the PN. It also involves the collection of data on tenants, mortgages, and rentals for the three years prior to the PN to safeguard tenants and easements, etc., and revising values under the Stamp Act for compensation determination. However, the two-month timeframe set for the completion of these tasks appears to be insufficient.

The similarities between the LA Act and the RFC & TLARR Act conclude with the notification for consultation and social impact assessment. Additionally, the RFC & TLARR Act incorporates local self-governance and the Gram Sabha of the impacted region. A Social Impact Assessment (SIA) is mandated to examine six key aspects: (i) Minimal Land Requirement; (ii) Definition of Public Interest; (iii) Absence of Alternative Land; (iv) Social Impact and Associated Costs; (v) Extent of Land Acquisition; (vi) Scale of Displacement including homes, communities, and significant roads. The agency responsible for SIA is tasked with creating a Social Impact Management Plan (SIMP) to mitigate adverse social effects. This assessment must be finalized within a six-month period following its announcement. Details regarding both SIA & SIMP should be disseminated publicly, followed by an organized public discussion on these documents as well as on the specifics of the proposed area. An expert panel is to be formed to evaluate the reports on SIA and SIMP, although their recommendations are not obligatory for governmental decision-making. Exemptions from conducting an SIA are granted for irrigation projects and urgent land acquisitions provided an Environmental Impact Assessment (EIA) has been completed.

## 4.5 Recent Land Acquisition in Maharashtra

In the last two decades, a significant portion of land that has been acquired is for agricultural use, predominantly designated for irrigation initiatives. These projects have utilized approximately 76% of the total land acquired in this timeframe, in contrast to the mere 14% used by industrial sectors. Land allocation for road infrastructure comprised about 4.4%, while railway expansion required only 1.74% as shown in Fig. 4.1. A small fraction, amounting to 0.67%, was set aside for hospital construction and another 3.3% fell into an assorted category labeled "others." Out of the total land acquisition of nearly 192,000 ha throughout the state, a substantial portion—nearly 37%—originated from Vidarbha (Kakodkar 2015). Following closely was the drought-prone Marathwada area contributing to 28%. The Konkan region saw minimal land acquisition activity with just an 8.6% share over these two decades. This data underscores the magnitude of farmland acquisitions across various state regions. Specifically, in Vidarbha's Amravati division, all acquired land spanning over an area of approximately 25,999 ha was used for agriculture purposes. In North Maharashtra as well, a high percentage (96.5%) represented agricultural acquisitions; similarly high percentages were noted in Konkan (90%) and Marathwada (94%). However, details regarding agricultural land acquisitions within Nagpur's sixth revenue division were not disclosed to state authorities (Kakodkar 2015).

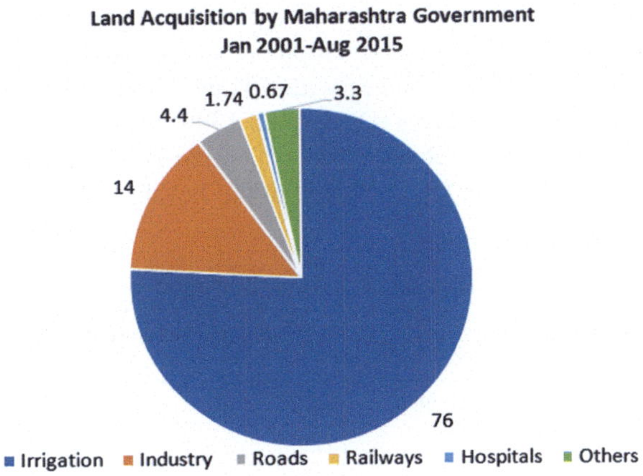


**Fig. 4.1** Land acquisition by Maharashtra Government for various purposes. *Source* Kakodkar (2015)

### 4.5.1 Land Acquisition in Navi Mumbai

The development of a city on the scale of Navi Mumbai required a significant amount of land acquisition. This process began well before the Navi Mumbai project was initiated. In the 1960s, the Maharashtra Industrial Development Cooperation (MIDC) established two industrial estates in the Trans-Thane Creek (TTC) and Taloja, both of which are located within the boundaries of Navi Mumbai. By 1988, MIDC had acquired 2388 ha in TTC and 893 ha in Taloja, affecting a total of 3632 families (MIDC 1987–88). In the case of Navi Mumbai, on February 5, 1970, the state government, under the LA Act of 1894, issued a notification for the acquisition of privately owned land spanning 86 villages and measuring 15,954 ha. An additional 2800 ha of land from 9 more villages were notified and added in August 1973. These villages were situated in the Thane and Raigarh districts, and the notification impacted 19,400 families or approximately 117,000 individuals (CIDCO 1973). At the taluka level, 29 villages in Thane taluka, 38 in Panvel taluka, and 28 in Uran taluka were affected by the notification. Of the total project area of 34,400 ha, private land constituted 16,567 ha or 56.3%, government land accounted for 10,137 ha, and saltpan land made up 2720 ha or 9.2%. The remaining area was comprised of MIDC land, municipal areas, and the gaothans (villages). The government's acquisition notification affected only the private land and the saltpan land. Since the built-up areas of the villages or the gaothans were not included in the acquisition, the villagers retained their homes but lost their lands, which were their primary source of livelihood.

At the time of the project's notification, the land was registered under the names of 24,878 landowners (Deshmukh1999). By the fiscal year 1998–99, property procurement had been completed from 19,875 owners/households, while notification had been issued for the properties of the remaining 5,003 households. As indicated in Table 4.2, by the fiscal year 1998–99, three-quarters of private property and over half of state-owned land had been appropriated for Navi Mumbai's construction. CIDCO (City and Industrial Development Corporation of Maharashtra) identified that out of a total of 95 villages within Navi Mumbai's proposed limits, land acquisition impacted 52 villages (Deshmukh 1999) (Fig. 4.2). The villages not included in this count are those abandoned (14), situated within MIDC regions (8), designated as green zones (16), and located in the Jawaharlal Nehru Port Terminal area (5).

**Table 4.2** Proposed and actual land acquisition for Navi Mumbai

| Type of land | Proposed land acquisition in 1970 | Actual land acquisition till 1998–99 |
|---|---|---|
| Private agricultural land | 16,677 | 11,960 |
| Saltpan land | 2720 | 2686 |
| Total private land | 19,397 (100%) | 14,646 (75.5%) |
| Government land | 10,137 (100%) | 5289 (52.2%) |

*Source* Deshmukh (1999)

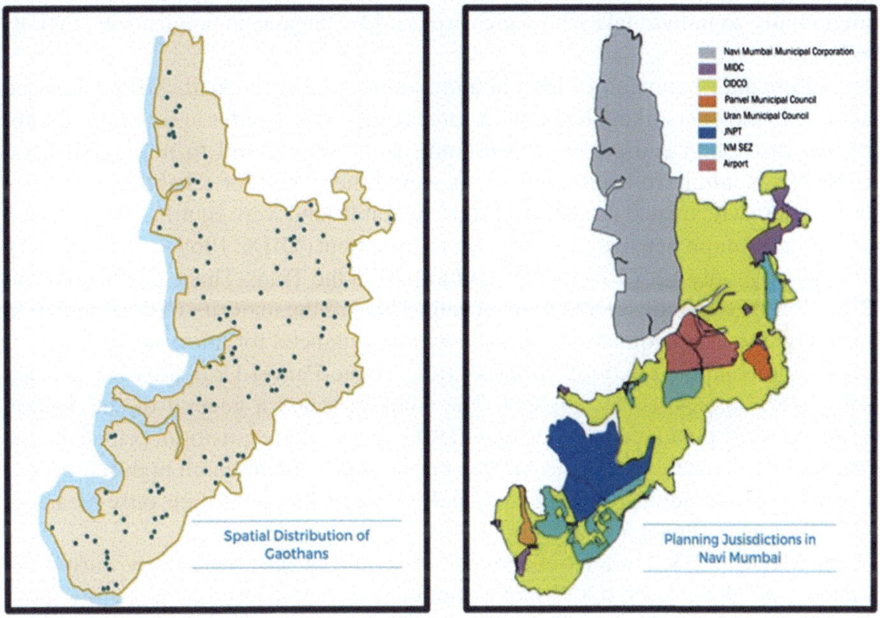


**Fig. 4.2** Distribution of gaothans and planning agencies jurisdiction in Navi Mumbai. *Source* http://cidco-smartcity.niua.org

In addition to the families directly impacted by the government's land acquisition, there were several households that didn't own any land but relied on it for their livelihood. This includes individuals like saltpan workers and agricultural laborers. Village artisans, who provided farming tools and other services to farmers, were also affected. Some individuals depended on fishing in Thane Creek for their livelihood. The building of the new city, which involved the acquisition of agricultural land and the reclamation of marshes along the creek, also had an impact on these people's incomes. However, since they made up only a small fraction (5 to 10%) of the affected families, they did not receive separate cash compensation (Deshmukh 1999).

## 4.6 MIDC and CIDCO Variation in Policies

Between 1964 and 1988, the Maharashtra Industrial Development Corporation (MIDC) saw an eightfold increase in its land holdings, from 2,406 to 22,263 ha. Reflects a concerted effort to reduce the adverse effects of this land acquisition. Strategies included prioritizing the use of government-owned non-arable land and steering clear of fertile farmland to limit the displacement of farmers (MIDC 1965–66). During the mid-60 s, MIDC lease agreements incorporated a specific provision that mandated industrialists leasing MIDC land to offer employment opportunities preferentially to individuals who were displaced by these land acquisitions (MIDC 1965–66).

A significant proportion of land holdings were relatively small. MIDC deduced that these landowners likely had other sources of income besides agriculture. Therefore, the disruption caused by transitioning from agricultural to non-agricultural professions would have been minimal, as would their need for rehabilitation. Only those farmers who owned more than 5 acres of arable land were significantly affected and required comprehensive assistance for resettlement (MIDC 1969–70) (Table 4.3). Consequently, only 12.5% of the agriculturalists in the Trans-Thane Creek area and 6.8% in Taloja were considered to need help. This left the majority of the population affected by the land acquisition without any pressing need for intervention.

A socio-economic study of villages in the Thane-Panvel-Uran area was carried out by TISS, sponsored by CIDCO. The study revealed a general dissatisfaction among the local population regarding CIDCO's property acquisition, primarily due to negative past interactions with MIDC. Furthermore, the research highlighted that for numerous individuals, the targeted land for acquisition was their only means of financial support (Punekar and Desai 1970).

Like MIDC, CIDCO faced challenges in acquiring land, especially during the initial stages. However, CIDCO had a comprehensive strategy for relocating those displaced and improving the villages impacted. Even with these measures for relocation, acquiring land was a struggle and by 31 August 1980, a decade following the initial announcement, merely 14.45% of the privately notified land had been secured. Meanwhile, close to 40% of the land notified by the government had been obtained (Table 4.4) (Verma 1985).

The data further indicates that the majority of the land notified for acquisition was private agricultural land, accounting for 56.3%, with over half of this land (58.34%) situated in the Panvel tahsil. Of the total private agricultural land acquired, which amounted to 2397.03 ha, 62% was obtained through award declarations, 8.9% via the urgency clause, and a mere 23.13% through voluntary surrender, predominantly from Panvel tehsil (90%) (Refer to Table 4.4). The process of land acquisition for the construction of Navi Mumbai became more streamlined post-1980, and by 1989, a significant 13,352 ha, or 80.59% of the private land, had been secured (CIDCO 1989).

**Table 4.3** Agriculturalists affected by MIDC land acquisition by size of landholdings, 1969–70

| MIDC area | Total acreage | Number of agriculturalists by holding size | | | | Total agriculturalists |
|---|---|---|---|---|---|---|
| | | <1 acre | 1–5 acres | 5–10 acres | >10 acres | |
| Trans Thane Creek | 4936 | 790 (40.45) | 917 (46.95) | 168 (8.60) | 78 (3.99) | 1953 (100) |
| Taloja | 451 | 151 (44.54) | 165 (48.67) | 14 (4.13) | 9 (2.65) | 339 (100) |

*Source* MIDC (1969–70)

**Table 4.4** Notified lands and their acquisition in Navi Mumbai

| Particulars | Area in ha | | | |
|---|---|---|---|---|
| | Thane | Panvel | Uran | Total |
| *Land notified* | | | | |
| Government land | 4599 | 3853 | 1685 | 10,137 |
| Private agricultural land | 2214 | 9666 | 4688 | 16,567 |
| Saltpan land | 878 | 14 | 1828 | 2720 |
| Total | 7691 | 13,532 | 8201 | 29,424 |
| Land acquired | | | | |
| Government land | 3057.68 | 857.97 | 85.39 | 4001.04 |
| *Private agricultural land* | | | | |
| By declaration of awards | 658.33 | 850.50 | - | 1508.82 |
| By urgency clause | 0.18 | 30.85 | 183.04 | 214.07 |
| By voluntary surrender | 55.60 | 607.45 | 11.08 | 674.14 |
| Total private agricultural land | 714.10 | 1488.81 | 194.12 | 2397.03 |
| Total | 3771.78 | 2346.78 | 279.51 | 6398.07 |

*Source* Verma (1985)

As per Verma (1985), until 31 August 1980, the cost for non-farming land ranged between Rs. 1 and Rs. 12 for each $m^2$ in the districts of Thane and Panvel. For farming land, the price spanned from Rs. 1000 to Rs. 7500 per acre in areas like Panvel and Uran. Due to local protests and negotiations, there was a gradual increase in these compensation rates. In a 1979 agreement, it was decided that the least amount for an acre would be set at Rs. 15,000, ensuring that with additional ex-gratia payments, the total compensation would equal Rs. 15000 for every acre. Deshmukh (1999) states that during the '70 s, CIDCO adjusted its compensation rates for land owned by locals from Rs. 2 per $m^2$ up to Rs. 12 per $m^2$—equivalent to an increase from Rs. 8000 to Rs. 48,000 per acre.

### *4.6.1 Rehabilitation Measures for PAPs of Navi Mumbai Project*

In order to prepare local individuals for job opportunities, intensive training programs were established in specially created technical institutes for a variety of technical professions and jobs related to agriculture. A number of individuals were trained to become bus drivers. CIDCO's employment service maintained a strong connection with industries, businesses, and government offices, and was successful in securing jobs for many local individuals (CIDCO 1995).

The gaothans were improved by providing essential services such as piped water, access roads, drainage systems, septic tanks, enhanced educational facilities, and healthcare services. CIDCO implemented a variety of programs, such as providing grants for the construction of school rooms, community centres, and sanitation facilities. This allowed the villagers to have access to urban-standard services within the sectors of Navi Mumbai.

The Project Affected Persons (PAPs) were offered a range of benefits including alternative job opportunities, contractual employment, allocation of shops and stalls, business loans, permissions for quarry operations, complimentary vocational education, and scholarships for educational and training programs. CIDCO initiated a comprehensive program to resettle those affected by the project in Navi Mumbai. Approximately 19,400 households, representing around 117,000 individuals, were already residing in 95 gaothans (villages). These gaothans, although located within the project's boundaries, were not subject to land acquisition. Instead, they were enhanced by providing essential services such as piped water, access roads, adequate drainage systems, septic tanks, improved educational facilities, and healthcare services. The rehabilitation program is designed to ensure that at least one individual from each household has the chance to earn an income as profitable as what their land could yield. It also aims to provide a stable income from investments for those unable to work, in addition to offering them loans for small businesses. To prepare the local populace for job opportunities, intensive training programs have been launched in newly established technical institutes for a variety of technical skills and jobs related to farming. Numerous individuals have also received training to become bus and scooter operators. CIDCO's placement service maintains an active connection with industries, commercial entities, and government agencies, achieving notable success in securing employment for many locals. There are also initiatives that offer scholarships, financial aid, and subsidies for educational support of local students and enhancements of village amenities (CIDCO 1973). The main goal of this program was to provide the displaced communities with means to earn non-farm incomes. A secondary aim was "to acquaint people with the forecasted environment and possibilities within a new urban context" (Deshmukh 1999).

In response to public demand, CIDCO had to formulate strategies for returning a portion of the developed land, in addition to establishing new physical and social infrastructure in the villages. The two schemes that were launched for this purpose are the Gaothan Expansion Scheme (GES) and the Twelve and Half Percent Scheme.

### *4.6.2 Gaothan Expansion Scheme (GES)*

Initiated in 1986, the Gaothan Expansion Scheme (GES) mandates that a village must set aside 10% of its expropriated land for developmental purposes, which is then redistributed to the villagers. Within this allocated 10%, half is designated for the villagers' use while the remaining half facilitates the construction of infrastructure such as roads and social amenities, as well as providing open areas. The size of plots provided to Project Affected Persons (PAPs) varies between 100 sq. m. and 500 sq. m. The Gaothan Expansion Scheme permitted various activities such as residences, retail shops, schools, community centres, religious places, clinics, essential public services, and more. It also allowed for stables, traditional household industries, parks, fishing-related activities, public conveniences, service industries, and limited warehousing. However, the scheme had limited beneficiaries, and within four years, only 27 ha of land was allotted across seven villages. Plot development was also slow under the scheme. The GES was closed in 1990 (CIDCO 2015).

### *4.6.3 Twelve and a Half Percent Scheme (12.5% Scheme)*

The 12.5% scheme, initially introduced in India, aimed to compensate landowners whose land had been acquired but who had not accepted compensation by a certain date. It later extended to include those who had accepted inadequate compensation. Under the scheme, landowners were given back 12.5% of the acquired land, with 30% reserved for social facilities and public utilities. The net allotment was 8.75% of the original land, allowing for development with 1.5 FSI and 15% commercial use. The scheme became fully functional in 1994, with CIDCO disbursing over 750 ha of land (CIDCO 2015). It had no plot size limit and provided additional areas if needed. The schemes offered numerous benefits. Property owners were involved as stakeholders in the urban development, receiving improved compensation and resettlement options. This approach boosted the availability of homes in Navi Mumbai via independent business initiatives, particularly benefiting the middle and lower-income brackets. It also motivated the Project-Affected Persons (PAPs) to take on developer roles, which diminished resistance to land acquisition. However, some drawbacks were noted, such as the emergence of an alternative housing market to CIDCO's offerings that was more affordable but of inferior quality and higher density. Consequently, this led to two separate models of urban development within Navi Mumbai, where differing standards resulted in significant disparities in living conditions for its inhabitants.

### *4.6.4 Issues Faced by CIDCO During Land Acquisition*

Despite amendments and improvements in the Land Acquisition Act, certain aspects are critically important. There are certain provisions of the act have been diluted or bypassed through subsequent amendments, reducing the safeguards for affected communities, and compromising their rights.

In the first ten years of Navi Mumbai's development, there was strong opposition from the local population, leading to acrimonious dealings with the government. In 1970–72, the state police force was required to keep the situation under control due to various forms of protest and even physical assaults on CIDCO officials. This opposition caused delays in the land acquisition process. The agitations gained regional and national support as in the 1980s, CIDCO began reselling developed plots at higher prices, causing the local population to feel cheated. To address the ongoing agitations, the 12.5% scheme was initiated in 1986, providing some relief to the Project-Affected Persons (PAPs) (CIDCO 1992).

CIDCO encountered significant and sustained resistance from Project Affected Persons (PAPs), primarily due to the method of land acquisition in Navi Mumbai. This process excluded the Gaothan (village) area, which is in the middle of a developing urban landscape. Initially, the PAPs either did not gain any immediate advantages or were reluctant to accept any, fearing it might imply agreement. Furthermore, previous experiences with land seizures by different entities had already caused them difficulties, leading to a more distrustful attitude toward CIDCO.

## 4.7 The Way Forward

The importance of communication in the execution of land acquisition is emphasized due to the involvement of various government departments, agencies, consultants, and experts at various stages of LA&RR. The acquiring authority is required to establish transparent communication with other stakeholders promptly to prevent overlap, disputes, or delays. The process also includes direct engagement with the impacted individuals and their grassroots governance structures. The RFC & TLARR Act mandates several Gram Sabhas to ensure efficient and prompt management of communication. The Act also requires that affected individuals give their knowledgeable agreement for all PPP and PP projects it implies that they must be fully informed about the project's specifics, benefits and drawbacks, compensation offered to them, and relevant contact authorities.

A key aspect of communication involves informing stakeholders about their rights and the processes to access them. This approach also addresses numerous public complaints. If such issues are not addressed, they can escalate into significant reasons for industrial action and demonstrations. Both Public and Private Sector enterprises often include Corporate Social Responsibility (CSR) initiatives at the inception and throughout the execution of new projects. These initiatives, along with R&R compensation plans, can be executed through robust communication plans to facilitate seamless project execution.

Compulsory land acquisition for public purposes leads to conflicts. Adequate preparation and effective management are essential for successful project completion. Early resolution is crucial to prevent escalating conflicts, but if not possible, conflicts can be managed through education, communication, and keeping policymakers informed.

Navi Mumbai was initially conceptualized as an independent township and not a satellite of Mumbai, hence was not designed with strong transport links in mind. The early stages of development by the CIDCO were marked by a lack of innovative land procurement strategies and poor rapport with local communities, resulting in delays and increased costs. Despite a sluggish start during its first twenty years, the subsequent decades saw Navi Mumbai flourish in terms of economic development and population growth. CIDCO encountered challenges in maintaining accurate land acquisition records, while PAPs struggled with legal inheritance issues. Both, local residents and developers sought to manipulate CIDCO for greater financial gain, leading to unauthorized occupation of land near traditional villages and designated areas, causing disorganized expansion in parts of Navi Mumbai. Drawing on past lessons, CIDCO has introduced initiatives like the Navi Mumbai International Airport (NMIA) and the Navi Mumbai Airport Influence Notified Areas (NAINA), which build upon previous schemes but with explicit regulations established from the outset. To improve future operations, there is a call for more transparency in determining eligibility and allocation procedures. Rapid implementation of schemes will ensure enhanced communication and conflict management among CIDCO and future PAPs.

## References


Acharya BP (1988) The instruments of urban land policy in India: opportunities for future direction. Environ Plan A 20(11):1427–1447. https://doi.org/10.1068/a201427

Besley T, Burgess R (2000) Land reform, poverty reduction, and growth: evidence from India. Q J Econ 119(2):389–430

Bhagat RB (2012) A turnaround in India's urbanization. Asia-Pac Popul J 27(2):23–39

Bhan G (2014) The impoverishment of poverty: reflections on urban citizenship and inequality in contemporary Delhi. Environ Urban 26(2):547–560

Census of India (1951–2011) Final population totals. Office of the Registrar General and Census Commissioner, India, Ministry of Home Affairs, New Delhi

CIDCO [City and Industrial Development Corporation of Maharashtra] (1992) Twenty years of development. Belapur

CIDCO [City and Industrial Development Corporation of Maharashtra] (1973) The draft development plan. Government Press, Bombay

CIDCO [City and Industrial Development Corporation of Maharashtra] (1989) New Bombay: an outline of progress. Mumbai

CIDCO [City and Industrial Development Corporation of Maharashtra] (1995) Sustainable urban development: case study of New Bombay. Belapur

CIDCO [City and Industrial Development Corporation of Maharashtra] (2015) Highlights of socio-economic survey from 1995 to 2010, CIDCO. Smart Newsletter, Jan–Mar 1(1)

Deshmukh SB (1999) The affected villagers and their rehabilitation. Seminar on Navi Mumbai. Tata Institute of Social Sciences, 16–17 February, Mumbai, p 42

Dhingra N (2022) Political economy of law, efficiency and adverse 'inclusion': rethinking land acquisition in India. Rev Evol Polit Econ 3:379–403. https://doi.org/10.1007/s43253-022-00083-6

Fung KI, Freeberne A (1981) Shanghai. In: Pacione M (ed) Problems and planning in third world cities. Routledge, London, pp p250-287

Ghurye GS (1974) Cities of India: their growth and location. In: Rao MSA (ed) Urban sociology in India. Orient Longman, Hyderabad, pp 178–216

Heitzman J (2008) The city in South Asia. Routledge, London and New York

Hudalah D, Winarso H, Woltjer J (2007) Peri-urbanisation in East Asia: a new challenge for planning? Int Dev Plan Rev 29(4):315–332. https://doi.org/10.3828/idpr.29.4.4

Kakodkar P (26 Oct 2015) In 15 years, Maharashtra government acquired land 4.4 times size of Greater Mumbai. The Times of India. https://timesofindia.indiatimes.com/india/in-15-years-maharashtra-government-acquired-land-4-4-times-size-of-greater-mumbai/articleshow/49532213.cms

LAA (1894) Land Acquisition Act of 1894, Government of India. http://dolr.nic.in/dolr/LandAcquisitionAct1894.asp. Accessed 15 Mar 2022
Marshall F, Randhawa P (2017) India's peri-urban frontier: rural-urban transformations and food security. International Institute of Environment and Development (IIED). https://www.iied.org/10794iied. Accessed 15 Mar 2022
Mathur HM (2013) Displacement and resettlement in India: the human cost of development. Routledge, London
McGee T (1997) Globalisation, urbanization and the emergence of sub-global regions. In: Watters RF, McGee T (eds) Asia Pacific: new geographies of the Pacific Rim. Hurst and Company, London
MIDC [Maharashtra Industrial Development Cooperation] (1965–66) Annual reports
MIDC [Maharashtra Industrial Development Cooperation] (1969–70) Annual report
MIDC [Maharashtra Industrial Development Cooperation] (1987–88) Annual reports
Ministry of Housing and Urban Affairs (1976) The urban land (Ceiling and Regulation) Act, 1976 No. 33 OF 1976. https://mohua.gov.in/upload/uploadfiles/files/ulcra_1976.pdf
Ministry of Housing and Urban Affairs (2024) Jawaharlal Nehru national urban renewal mission. https://mohua.gov.in/cms/jawaharlal-nehru-national-urban-renewal-mission.php
Mitra A, Mehta B (2011) Cities as the engine of growth: evidence from India. J Urban Plan Dev 137(2):171–183
Mukerji S (2017) Land acquisition in contemporary India: the growth agenda, legislation and resistance. Indian J Public Adm 63(1):85–103. https://doi.org/10.1177/0019556117689851
Nakamura H, White JW (1988) Tokyo. In: Dogan M, Kasarda JD (eds) Mega cities: the metropolitan era, vol II. Sage Publication, Beverly Hills, CA, pp 123–156
Oosthwizen AJG (1969) Delimitation of urban area. J Geogr, 437–450
Panda PK (2010) Land conversion and urbanization in India: snippets from literature. Jharkhand J Dev Manag Stud 8(3):4057–4066
Pastalan LA (1967) Towards an operational definition of the rural urban fringe. Indian Social Bulln 15:223
Phadke VS, Sita K (1981) Spatial patterns of urban impact in a metropolitan region—a case study of Bombay. Ann Natl Assoc Geogr India 1(1):18–28
Punekar SP, Desai KG (1970) Metro centre villages: a survey. Tata Institute of Social Sciences, Mumbai
R&R [Rehabilitation and Resettlement Act] (2013) Right to fair compensation and transparency in land acquisition, Rehabilitation and Resettlement Act, 2013. https://www.indiacode.nic.in/bitstream/123456789/2121/1/A2013-30.pdf. Accessed 15 Mar 2022
Ramachandran R (1989) Urbanization and urban systems in India. Oxford University Press, New Delhi
Rao Prakasa VLS, Tweari VK (1979) The structure of an Indian metropolis: a study of Bangalore. Allied Publishers, New Delhi
Robertson B, Pinstrup-Andersen P (2010) Global land acquisition: neo-colonialism or development opportunity? Food Sec 2:271–283. https://doi.org/10.1007/s12571-010-0068-1
Seetharaman SP (1984) An approach to the understanding of the occupational structure of small towns in Tanjore District in Tamil Nadu. Deccan Geogr 23(1):31–38
Shaw A (2012) Metropolitan city growth and management in post-liberalized India. Eurasian Geogr Econ 53(1):44–62
Siddiqui K (2014) Higher education in the era of globalization. Int J Humanit Soc Sci 3(2):9–32
Smith TL (1937) The population of Louisiana: its composition and changes. Louisiana Bulletin No. 293, Louisiana State University and Agricultural and Mechanical College, Agricultural Experiment
Verma HS (1985) Bombay: New Bombay and metropolitan region: growth processes and planning lessons. Concept Publishing House, Delhi

Wang LH, Tan T (1981) Singapore. In: Pacione M (ed) Problems and planning in third world cities. Croom Helm, London, pp p218-249

Yeung Y (1988) Great cities in Asia. In: Dogan M, Kasarda JD (eds) A world of giant cities: the metropolitan era, vol 1. Sage Publication, Beverly Hills, CA, pp 123–156

# Chapter 5
# Demystifying Green Space Planning Principles for Sustainable Urban Development in Indian Cities

**Asmita Kanav and Jitendra Kumar**

**Abstract** Urbanization and rapid population growth in Indian cities have brought about significant challenges in maintaining a balance between economic development and environmental sustainability. Green spaces are known to play a vital role in enhancing the quality of urban life, restoring ecological balance, and promoting the well-being of people. Although the benefits of urban green space are well established, researchers and policymakers are still debating their implementation strategies. Research has developed green infrastructure planning principles to assist professionals with planning techniques to aid in these conversations. Thus, this chapter aims to delve into the fundamental green space planning principles that are essential for achieving sustainable urban development. We begin by highlighting the pressing need for sustainable urban development in Indian cities, where issues such as air and water pollution, congestion, and inadequate lifestyles have become pervasive. In response to these challenges, we critically examine six key green space planning principles, established through previous research that serve as a comprehensive framework for guiding sustainable urban development strategies; namely: connectivity, multi-functionality, multi-scale, integration, participation, and continuity. These principles enable green space planning to interweave into a city's master plan. Through a comprehensive review of the literature and case studies, this research provides a simple understanding of these green space planning principles and their specific relevance to sustainable development. By demystifying these principles, the chapter equips urban planners, policymakers, and stakeholders with a valuable toolkit to navigate the intricate challenges of urbanization and promote sustainable living environments that prioritize human well-being as well as ecological balance.

**Keywords** Green space · Urban planning · Sustainability · Indian cities

A. Kanav · J. Kumar (✉)
Department of Geography, Central University of Haryana, Mahendergarh, Haryana, India
e-mail: jitendra@cuh.ac.in

A. Kanav
e-mail: reachtoasmita@gmail.com

 
S. Anand et al. (eds.), *Emerging Geosustainability Transformations in India*, Advances in Geographical and Environmental Sciences,
https://doi.org/10.1007/978-981-97-7993-2_5

## 5.1 Introduction

India's urbanization is accelerating as the country shifts from a predominantly rural to an urban society. The 31.8% increase in urban population from 2001 to 2011 contrasts sharply with the 17.6% increase in India's total population (Census of India 2011). It is estimated that by 2050, 53% of India's population will reside in urban areas, adding a whopping 416 million people to Indian cities (UNDESA 2018).In 2011, the median age of the country was just 24.9, much younger than China's 36.8 years. Although projections report that the average age of the country is on the rise, India population remains younger than most European, American and even other higher-income developing countries. This young workforce allows for the opportunity for a surge in investments given a large consumer market. People gravitate toward urban areas in search of employment, education, improved infrastructure, and a variety of services. Thus, such investments are typically generated by the country's urban areas, resulting in the densification of urban centers (Kumar 2015).

According to the United Nations, India hosts 17.4% of the world's urban population, with five megacities housing more than ten million people (UNDESA 2018). By 2030, two more cities, Ahmedabad and Hyderabad are expected to cross the 10 million mark. Delhi, with its 22.7 million inhabitants, is now the second largest urban agglomeration in the world. Mumbai and Kolkata, ranked 7th and 10th, respectively, house 19.7 million and 14.4 million people. Urbanization and rampant urban expansion have diminished urban green spaces and made cities more susceptible to the dangers of climate change. The alteration in the patterns of the southwest monsoon, the severity of seasonal variations, and the increased incidence of temperature anomalies persist as key concerns. Changes in the land cover patterns of major urban centers have led to a shift in the local climatic conditions (Yang et al. 2012). Moreover, poor air quality, urban heat island effect, and acute water scarcity also deteriorate the quality of city life. As research predicts longer, more frequent and more severe heat waves, India shall have to encounter critical challenges in the face of climate change and upholding sustainable development (Singh et al. 2021a). While India grapples with rising carbon emissions and climatic abnormalities, a lack of budget allotted to mitigation limits the potential for environmental improvement. Concurrently, rapid population expansion and a lack of public involvement limit the efficacy of governmental actions.

According to the World Health Organization, diseases tied to air pollution claim the lives of about seven million people worldwide annually (WHO 2015). As a developing country with a growing population, India suffers from severe air pollution and contains nine of the world's ten most polluted cities (WHO 2016). The rising levels of air pollution in Indian cities over the last several decades, as well as the resultant human health consequences, have received considerable scholarly attention in recent years.

It is well established that climate change affects the planet's temperature, rainfall patterns, and amount of insolation. Furthermore, the increased frequency and severity of floods, storms, torrential rains, droughts, and other severe weather phenomena brought on by climate change impact millions of people and claim thousands of lives. Ali et al. (2014) studied the impact of climate change in 57 Indian cities and found that 33 of them are expected to see an increase in intense rainfall events and subsequent flooding from 2036 to 2060. In the remaining 24 cities, a decline in precipitation is expected, indicating a greater likelihood of drought. A rise in the frequency of intense rainfall events was found in Mumbai, which causes flooding in the city. Mumbai is said to have the largest population at risk from coastal flooding, making it the world's most vulnerable city to climate change (Murray et al. 2012). According to Ranger et al. (2011), Mumbai's risk of flooding has been heightened by anthropogenic changes to its geography, particularly due to increasing impervious surfaces and a shortage of drainage networks that discharge excess water into the sea.

Given the health problems it poses to city dwellers, air pollution is a major problem in Indian megacities, where pollutant levels are frequently above permitted thresholds. Air pollution exposes India's urban centers to unsafe and hazardous circumstances (Dutta et al. 2021). Over the past few decades, urban air pollution has been a continuously troubling environmental challenge in India's megacities (Faheem et al. 2021). Singh et al. (2021b) studied air quality in the five Indian megacities during a period of six years, from 2014 to 2019. Although the monthly mean PM2.5 concentration has declined for all cities over the years, it was determined that Delhi was the most polluted city in the world, followed by Kolkata, Mumbai, Hyderabad, and Chennai. As a developing country, India faces increased air pollution on account of a lack of proper infrastructure, such as adequate roads, efficient public transport and planned industrial land allocation (Rumana et al. 2014). The average vehicular speed is declining on crowded city roads, which raises automobile emissions and air pollution levels. Dutta et al. (2021) find that uncontrolled population increases and industries are responsible for air pollution and resultant health problems in Indian cities. Furthermore, intensive human activities in these urban regions are exacerbating the issue.

Heat-related illnesses in cities, such as skin cancer, heat stroke, heat exhaustion, diarrhoea, and heat-induced fatal heart attacks, are exacerbated by climate change (Bai et al. 2014; Orimoloye et al. 2019). Dehydration, heat cramps, heat exhaustion, heat stroke, eye and skin conditions are other heat-related human health consequences (Dutta and Chorsiya 2013). The urban heat island effect is a significant result of climate change. Mumbai, one of India's megacities, witnessed a 2.4 °C rise in average temperatures between 1881 and 2015 (NASA 2015). The UHI impact has been observed to cause extreme heat-related discomfort in numerous Indian cities (Kikon et al. 2016; Kaur and Pandey 2020; Shukla and Jain 2021). Ramachandra and Kumar (2010) recorded a 2°–2.5 °C rise in Bangalore's local temperature in the previous decade. Veena et al. (2020) find that all metropolitan cities in India display the urban heat island effect, which can be fatal for people with serious respiratory conditions. According to statistical data published in research by Dutta and Chorsiya (2013), more than 600 people lost their lives to heat waves in 2013 in India. In cities, children's susceptibility to respiratory infections and slum dwellers' sensitivity to seasonal severity have skyrocketed.

Urban lifestyles, which are heavily reliant upon automobile usage, tend to make the residents even more vulnerable to threats. Lifestyle diseases like diabetes, hypertension, obesity, certain types of cancer, infertility, and cardiovascular diseases are caused by a lack of physical exercise and unhealthy dietary habits coupled with exposure to pollutants. Mathur and Mascarenhas (2019) found that 65% of urban population is physically inactive, with women tending to be more lethargic than men. Rural men report about five times more physical activity than urban men, whereas rural women describe being as many as seven times more physically active than urban women (Yadav and Krishnan 2008).

One strategy for building climate-resilient cities is to integrate climate adaptation into urban development planning (Sharma and Tomar 2010). Improving urban green areas or urban forests is one method that can alleviate the negative consequences of urbanization in a sustainable fashion, making cities more appealing and pleasant to live in. India's central and state governments are recognizing the value of urban green areas in cities.

## 5.2 Urban Green Spaces: Bridging the Gap Between Economic Development and Environment Conservation

Urban green spaces are an essential component of sustainable urban planning. Sustainability is ubiquitously understood as an amalgamation of social, environmental, and economic development (Purvis et al. 2019) (Fig. 5.1). Green spaces can yield tangible economic benefits by increasing property values, boosting tourism, and creating jobs. Simultaneously, they contribute significantly to environmental conservation by protecting biodiversity, enhancing air and water quality, and aiding in climate change mitigation. Social benefits such as improved physical health, community cohesion, and stress reduction can also be derived from green spaces (Woolley et al. 2004). Moreover, green spaces enrich the urban landscape aesthetically and give the residents a sense of place identity.

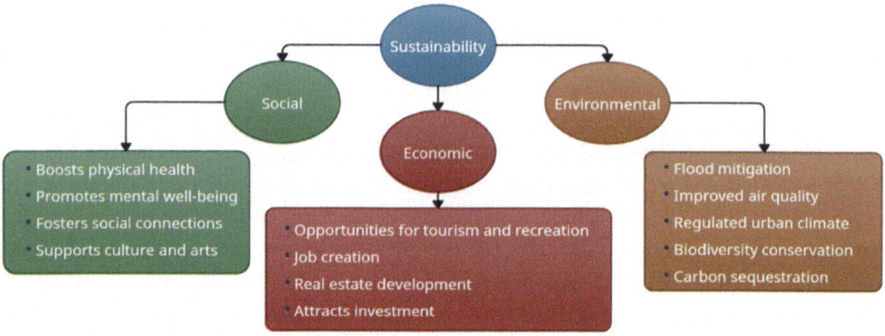


**Fig. 5.1** How urban green spaces serve the components of sustainable development

The presence of urban green spaces is critical, especially in a developing country like India with high population density and extreme levels of air pollution. Green spaces are a vital source for managing the urban microclimate (Gunawardena et al. 2017), reducing flood risks (Bai et al. 2018), and facilitating active lifestyles (Richardson et al. 2013). They also play a significant role in air purification, which enhances the quality of the air (Diener and Mudu 2021). Furthermore, by controlling the surface temperatures of urban areas, UGSs aid in lessening soil erosion, noise pollution, and energy consumption.

Urban India's green cover has suffered due to population growth; the UN-recommended threshold of 9 $m^2$ of green space per capita is far beyond Chennai's meagre 0.81 $m^2$ (NIUA 2022) and Mumbai's 1.24 $m^2$ (Udas-Mankikar 2020). Ramachandra et al. (2016) modeled the urban land cover trajectory of four Indian cities: Ahmedabad, Bhopal, Hyderabad, and Kolkata, finding that built infrastructure is expected to expand toward buffer regions of cities. Vegetation cover in Ahmedabad is expected to drop by approximately 50% in the decade of 2020. During the same period, Kolkata's vegetation cover may drop from 8.31 to 3.37%. Similarly, the cities of Bhopal and Hyderabad will follow a similar trajectory of increasing built-up areas and diminishing green spaces. As housing quarters become more vertical, the residential garden space is diminishing, exacerbating environmental degradation. In Kozhikode, there has been a noticeable decline in the area of residential gardens between 2000 and 2010 (Gangopadhyay and Balooni 2012). The loss in the garden area was found to be dependent upon household income, increasing from 2.43 $m^2$ for low-income households to 47.35 $m^2$ for high-income ones. The evident shift in preferences toward garden space suggests that due to changing lifestyles, urban green cover is perceived as less valuable. Ramachandra and Kumar (2010) reported that in Bangalore, there had been a 76% decline in vegetation cover over the period of 37 years. The city has lost such a great deal of urban open spaces to unchecked development that permanent changes have taken place in its local hydrology and drainage systems (Sudhira et al. 2004).

The challenge of applying green space principles to urban planning is not merely related to the complexity of integrating ecological concepts into urban planning. As stated by McPhearson et al. (2014), it is more concerned with the difficulties in understanding and planning for resilient and sustainable urban systems. Cities are places that need to have adaptable strategies to decrease risks and take advantage of useful tools to lessen the susceptibility of urban systems to climate change.

Efficient UGS planning involves undertaking innovative approaches and various interventions at different levels of the planning process. To fulfill this objective, several authors have tried to identify different planning principles for green spaces to help practitioners during planning procedures, including implementation and management. These principles, which are predominantly based on geography and landscape ecology, try to incorporate ecological, social, and economic concerns into the decision making and implementation processes in urban areas. However, due to the ambiguity of green space planning principles, practitioners, and decision-makers struggle to understand their true benefits and the best practices for its implementation and management at local level. Monteiro et al. (2020) recognized the following eight principles of green infrastructure planning through a meta-analysis of the literature: connectivity, multi-functionality, applicability, integration, diversity, multi-scale, governance, and continuity. These planning principles were further corroborated by Monteiro et al. (2022) while assessing the priority principles of urban planners in the Lisbon metropolitan area and in 2023 by Mosissa et al. while replicating the study in Addis Ababa. Some of these principles have been repeatedly mentioned throughout the literature, whereas others are relatively newer ideas developed after analyzing experiences of cities and a deep analysis of ecosystem services. As there is no set of pre-established planning principles (Hansen and Pauleit 2014), we recognize the following six principles as most relevant and frequently mentioned: connectivity, multi-functionality, multi-scale, integration, participation, and continuity.

## 5.3 Connectivity

Connectivity is a fundamental principle in green space planning which involves strategically designing urban landscapes to establish seamless links between various green spaces. This concept emphasizes the development of interconnected routes, trails, and corridors (Ahern 2013) that allow the inhabitants to move easily and access green spaces from all parts of the city. Furthermore, this idea encompasses functional integration, biodiversity protection, social engagement, environmental resilience, and overall well-being in addition to physical access. Connectivity between natural landscapes is also important in addition to network links between urban parks and green belts, so they can be incorporated into the urban environment. The rationale behind the adoption of this principle is that the stability of individual green areas and the urban landscape as a whole is significantly higher when it is directly tied to the natural environment native to that area (Liu et al. 2022). As a result, urban built infrastructure should be intermittent and discontinuous, with a continuous system of green areas. The city will then be able to serve as a system of recreational services for the populace. The connectedness of green spaces is crucial to the preservation of vital ecological processes and services as well as the health of people and wildlife (Benedict and McMahon 2002). Ensuring connectivity ascertains that the benefits of green spaces are maximized. The connectivity is boosted between: people and the built environment; people and a contiguous part of nature; various flora and fauna species to cross over and form an interconnected web; and between urban green spaces and natural landscape.

Connectivity promotes active transportation modes such as walking and cycling by connecting parks, gardens, and recreational places, promoting healthier lives (Tzoulas et al. 2007) and decreasing urban congestion, adding to a more sustainable, lively, and organized cityscape (De Ridder et al. 2004). Social cohesion is developed as an interconnected network of UGSs that provides public spaces for individuals to visit and engage. Appropriate expenditure on a city's green spaces is vital for attracting private investment and businesses to create jobs. With a connected system, a person can embark on a journey with only natural landscapes surrounding them throughout the whole venture, which are all interconnected by pedestrian networks.

To create functionally connected green spaces, planning, and design professionals may use the fundamental concepts of landscape ecology: patch, corridor, and matrix (Rouse and Bunster-Ossa 2013). In an urban matrix, green corridors should connect patches of green space to preserve and reduce fragmentation of original vegetation cover. Investigating these physical and functional linkages between various green spaces can also improve ecological connectedness by restoring and conserving urban ecology by controlling water flow or climatic functions. Connectivity can also be achieved by planting along gas and electricity easements, key roadways, and rail corridors.

## 5.4 Multi-functionality

The capacity of green infrastructure to provide various services at the same time is referred to as multi-functionality. Green spaces ought to be multi-functional to maximize communal, environmental, and financial benefits (Hansen and Pauliet 2014). If effectively designed, green spaces can accomplish several goals. Firstly, they can influence and optimize the growth of urban areas in a sustainable manner and shape their identity. By managing water runoff and rising temperatures, green spaces can assist in mitigating the effects of climate change. Furthermore, they also serve as indispensable places for fostering a connection to nature, as people can access open skies, engage in sports, improving the visitor experience and quality of life for all. However, multi-functionality must be balanced alongside the need to conserve and strengthen the local environment (Von Haaren and Reich 2006), which may include certain areas set aside for exclusive use by indigenous wildlife or to safeguard vulnerable plant species.

As a green space planning requisite, multi-functionality stresses the design and usage of green spaces in such a way that they serve a variety of functions, broaden their array of benefits, and boost their efficiency within the urban setting (Ioja et al. 2014). This idea acknowledges that green spaces may go beyond their usual functions as mere areas of rest and rejuvenation and are deliberately planned to serve multiple functions, thus serving as a solution for different urban challenges while enhancing the urban quality of life. It is based on the variety of benefits offered by green spaces, whether environmental, economic, or social. These benefits result from the diverse and overlapping actions offered by green systems. The multi-functionality concept recognizes that green spaces can play a dynamic role in urban life, addressing numerous problems at the same time and improving the overall quality of the urban environment. The concept encourages urban planners and green space designers to achieve the greatest value for the people they are responsible for by integrating seemingly different aims such as flood management, energy efficiency, and enhanced public health outcomes. Effectively incorporating the multi-functionality concept includes not only improving the synergies between serving multiple functions but also avoiding any conflict over land use (Demuzere et al. 2014).

It also optimizes land use by accommodating diverse activities within a single area. For example, a park might feature spaces for recreation, cultural events, stormwater management, and habitat preservation. By serving multiple purposes, green spaces can contribute to economic growth through tourism, events, and recreational activities. This can lead to increased local spending and job creation in related industries. Integrating ecological functions within green spaces, such as rain gardens, urban forests, and wildlife habitats, enhances urban biodiversity, improves air and water quality, and supports overall ecological health. Multifunctional green spaces can mitigate the impacts of climate change by providing shading, reducing heat island effects, and absorbing rainwater, thus minimizing flooding risks and enhancing urban resilience.

One example of multifunctional green spaces is urban farms, which not only contribute to local food production but also provide opportunities for learning for visitors. Similarly, rooftop gardens in densely populated urban areas aid in alleviating heat waves while serving as spots for recreation. Community gardens also foster relationships among people while supporting indigenous plant species. Green spaces may greatly contribute to the creation of sustainable, resilient, and prosperous cities by including multiple functions in their design and maintenance. Planners must establish a thorough grasp of user wants and preferences in order to comprehend the multi-functionality requirements. The quality of space being multi-functional can cause conflicts if public opinion is not gathered (Ioja 2014).

## 5.5 Multi-scale

Green spaces are necessary at various scales or sizes (van Herzele and Wiedemann 2003). The multi-scale principle in green space planning emphasizes the creation of green spaces of different sizes that cater to different functions, ranging from small neighborhood parks to larger urban parks, green corridors, and forests. This principle recognizes that a variety of green spaces of different scales contribute to a more balanced, accessible, and inclusive urban environment, meeting the diverse needs of residents and fostering a sense of connection to nature.

Sub-city parks are not a replacement for the countryside or neighborhood play areas. Providing small-scale green spaces within neighborhoods ensures that residents have easy access to nature for relaxation, recreation, and social interaction without the need to travel long distances. The concept of multi-scale is integrally tied with multi-functionality (Demuzere et al. 2014) as different scales of green spaces can offer varying functions, such as tot lots for children's play, community gardens for local food production, and larger urban parks for organized sports and cultural events. Multi-scale planning also allows for the creation of connected networks of green spaces that facilitate pedestrian and cycling movement throughout the city, contributing to better urban mobility and accessibility. It can help in supporting ecosystems and biodiversity, which may not be achievable with a single large-scale green area.

Examples of the multi-scale principle include a mix of community gardens, playgrounds, and pocket parks within neighborhoods, mid-sized parks that cater to local communities and larger urban parks that serve entire city regions. Moreover, linear parks and greenways, that can traverse various scales, connecting neighborhoods with larger green spaces, contribute to a holistic urban green network. Although research has not delved into interrelationships that may be present among green spaces at different spatial scales, the majority of the authors assume that green spaces of different sizes play distinct roles (Demuzere et al. 2014; Gupta et al. 2016). Large expanses of woodland at the city perimeter may have value to the urban population for weekend excursions, and smaller parks in the inner city may be purposeful for everyday recreation. At different spatial scales, green spaces can be included in the following ways:

- At the sub-city scale, green spaces may help promote development and achieve objectives of urban renewal, rural land preservation, and enhanced possibilities for people walking, biking, and using public transit. Examples of UGSs can be zoos, stadiums, or urban forests.
- At the district scale, large parks, botanical gardens, and greenways into urban forests and greenery along major streets can be used in planning policies and capital investments.
- At the community scale, gathering places like community centers, and bus stops can be made greener to promote loitering in public spaces, thereby increasing social interactions.
- At the neighborhood scale, small and frequent parks with play facilities can provide everyday recreational opportunities for the residents within walking distance.
- At the housing scale, green walls covered with vines, household gardens, and green rooftops can help alleviate the negative impacts of high-density housing.

To ensure that private spaces are adequately greened, city codes and regulations can be developed accordingly (Rouse and Bunster-Ossa 2013). They can include incentives for green approaches, requirements for water-runoff management and rainwater harvesting, or setting aside protected spaces for environmental conservation.

## 5.6 Integration

Integration, as a green space planning principle, emphasizes the seamless blending of green spaces into the broader urban setting, ensuring that they are not isolated elements but rather integral components that enhance the overall functionality of the city. The principle of integration suggests combining green space with urban growth and grey infrastructure. Incorporation of green spaces in major infrastructure projects can serve as a catalyst for sustainable development (Tzoulas et al. 2007). Globally, there is a shift toward more multi-functional grey infrastructure that resembles nature, offers essential ecosystem services, and encourages healthy and active living. Integration ensures that the planning of green spaces is linked with other urban infrastructure design plans, such as residential and commercial complexes, transportation, or water management systems in order to create an optimum urban master plan.

Integration considers green spaces as another integral part of urban infrastructure, recognizing their potential to complement grey infrastructure or even replace it (Hansen et al. 2017). For example, a city with a cultural and historic background can transform its image by integrating green spaces into its artistic elements. Green spaces surrounding buildings of architectural or historical value, sculptures, murals, museums, or performance arenas enhance the appeal and cultural relevance of these areas and draw in more tourists. In turn, green space in itself can act as a tourist attraction and boost real estate and investment.

It is important to create a common goal by acknowledging the structural and operational synergies between urban green space and other infrastructure like housing, water management, transportation networks, and waste disposal (Kim and Tran 2018). Through collaboration between all administrative departments and applying expert advice from several disciplines and sectors, an integrated methodology can be devised. The city should allow for the use of publicly held assets like land near abandoned railway lines, water bodies, drainage, and pipeline networks for plantation and recreational use. To maximize the possibility of building green infrastructure along transportation corridors, coordination between city planners and road and rail authorities should be optimized. Integration at an earlier stage also ensures that the development process is unobstructed and urban renewal objectives are achieved. Ensuring a consistent source of funding is essential for the maintenance, upkeep, and improvement of green spaces. Thus, integrating their planning with long-term urban financial goals guarantees that green spaces remain vibrant and well-maintained.

## 5.7 Participation

Participation means involving stakeholders in the planning process for green spaces. It recognizes that sustainable and successful green spaces meet the needs, preferences, and aspirations of people residing in that area (Hansen et al. 2016). It is considered that the facilities and restrictions of a location, as well as the activities that may be done there, are most important in people's perception of green spaces. In contrast to most professional planners, the public defines not just parks and gardens as green spaces, but riversides, shrubs, farms, woods, golf courses, and cemeteries are all included in the term green space. Furthermore, locations that are not legally managed for "recreation", such as green fingers produced by river valleys and abandoned land are frequently more highly regarded by recreational users (Nicol and Blake 2000).

The principle of participation calls for an inclusive planning process, opens to participation by diverse population groups, and caters to their needs and preferences. It involves stakeholders directly or indirectly in the creation and management of neighborhood, local, district, and regional green space plans and actions.

When a broad group of people participate in the planning process, better ideas generally emerge. Embracing diversity broadens the understanding of user's needs and subsequent knowledge, thoughts, and perspectives, resulting in equitable and inclusive strategies. Community engagement as well as participation across government agencies, particularly those at the state and municipal levels is advised. Developing a network of green infrastructure necessitates the involvement of several agencies and user groups. Long-term planning of green networks throughout urban areas will benefit from shared knowledge and resources. A city should involve the stakeholders from the infancy stage of green projects and keep them engaged throughout the project's lifespan. Plans for green spaces should identify and balance the interests of several stakeholders in order to maximize their utilization. It should be ensured that equitable access to green spaces is taken into account; thus, the values, requirements, motivations behind visits, uses, and barriers to entry of various demographic groups should be studied (Korpilo et al. 2022). To encourage the usage of underutilized green spaces among local communities, designers should create inclusive play areas that are accessible to all members of the community, regardless of ability, age, gender, or culture.

## 5.8 Continuity

A significant setback in green space projects has been the lack of post-implementation monitoring or practical measures of the ecosystem services and functions they promise to deliver (Monteiro et al. 2020). Continuity underscores the importance of creating a cohesive monitoring framework of green spaces so that they can be designed, managed, and maintained over the long term. This principle recognizes that the success and effectiveness of green spaces rely on their sustained presence and functionality within the urban landscape, ensuring that they remain valuable assets for generations to come. It involves developing comprehensive and adaptable long-term plans that consider the evolving needs of the city and its residents. These plans help guide decisions regarding the establishment, expansion, and enhancement of green spaces. Engaging local communities in the planning and management of green spaces fosters a sense of ownership and responsibility, contributing to the continual use and maintenance of these spaces. Examples of continuity in green space planning include the establishment of dedicated funds for ongoing maintenance and improvements, the integration of green space plans into broader city master plans, and the implementation of community stewardship programs that encourage local residents to take an active role in the long-term care of their green spaces.

For green space projects to thrive, they require constant investment, maintenance, and updates (Kim and Tran 2018). If green space provisions are provided in a city's master plan, governments must be able to share new information about the projects, their aims, achievements, and future prospects on a regular basis. Thus, to ascertain their success, green space plans must include a well-defined assessment procedure and periodic updates on the progress of the intended green initiatives. Regular updating of these plans is also important as green spaces should evolve to accommodate changing urban dynamics, demographics, and environmental factors. Monitoring and evaluation of the performance and usage of green spaces, allowing for adjustments and improvements based on observed trends and feedback.

The frequent monitoring and reporting of strategic plans are made possible by integrating the supply of green spaces into the city's strategic planning framework. Standard review procedures for urban plans, for instance, might include a "green space annual monitoring report" to track progress on objectives and evaluate activities undertaken to achieve them. If every urban plan includes an associated execution plan that specifies priority and schedule for activities in the short, medium, and long term, management of green spaces can be done regularly and effectively.

## 5.9 Conclusion

India has to combine its current environmental efforts with green space initiatives to develop a robust economic strategy. The idea of embracing climate change adaptation and mitigation measures to produce "improved human wellbeing and social equity while significantly reducing environmental risks and ecological scarcities" is where the notion of a green economy emerged (Udas-Mankikar and Driver 2021). Urban green spaces are a way to achieve urban resilience and sustainability when integrated with urban planning. Despite the growth in literature on sustainability in recent decades, there is a lack of integration of the concept with urban development. We believe that by controlling urban transformational processes and planning diligent actions to enhance urban recovery capability, sustainability might be more readily attained. As a result, key components of the framework include process change, receiving feedback from evaluations, and preparing active and opportunistic solutions.

Unfortunately, urban plans are not easy to modify since creating them requires plenty of time and their crafting is based on a lot of data. Cities that are transitioning to these changes face a major threat from this slower pace, which coexists with fast changes in urban growth, technological innovation, and possible damages from climate change. The conversion of these ambitious objectives into local goals and actionable plans is the challenging part. Many cities have difficulties with implementation, such as a lack of financing, difficulty obtaining land, low levels of public involvement, and administrative fragmentation.

India's urban planning laws are superannuated and employ strict strategies through land use regulations and building codes. Indian cities must adopt dynamic urban planning that takes environmental changes into account if they are to adapt to an evolving component like green spaces. The public sector must provide more comprehensive and transparent digital interfaces, such as GIS mapping, to facilitate real-time monitoring and assessment of green spaces. Management of green space necessitates striking the correct balance between environmental and economic sustainability, as well as ensuring the safety of the people. Accordingly, successful green space projects in developed countries have incorporated public opinion into their planning processes (Korpilo et al. 2022). People will better understand green initiatives and be encouraged to take a leadership role in their planning, formulation, implementation, and monitoring through active participation and engagement with the government, planners, and policymakers. Green spaces have the capacity to direct urban development by offering an opportunity for economic growth and environmental conservation if they are strategically designed, created, and maintained. Such a strategy would provide multiple opportunities for blending together urban growth, health goals, and environmental preservation.

## References

Ahern J (2013) Urban landscape sustainability and resilience: the promise and challenges of integrating ecology with urban planning and design. Landscape Ecol 28:1203–1212. https://doi.org/10.1007/s10980-012-9799-z

Ali H, Mishra V, Pai DS (2014) Observed and projected urban extreme rainfall events in India. J Geophys Res Atmos 119:12621–12641. https://doi.org/10.1002/2014JD022264

Bai L, Ding G, Gu S et al (2014) The effects of summer temperature and heat waves on heat-related illness in a coastal city of China, 2011–2013. Environ Res 132:212–219. https://doi.org/10.1016/j.envres.2014.04.002

Bai T, Mayer AL, Shuster WD et al (2018) The hydrologic role of urban green space in mitigating flooding (Luohe, China). Sustainability 10(10):3584. https://doi.org/10.3390/su10103584

Benedict MA, McMahon ET (2002) Green infrastructure: smart conservation for the 21st century. Renew Resour J 20(3):12–17

Census of India (2011) Registrar general and census commissioner, ministry of home affairs, Government of India

Demuzere M, Orru K, Heidrich O et al (2014) Mitigating and adapting to climate change: multi-functional and multi-scale assessment of green urban infrastructure. J Environ Manag 146:107–115. https://doi.org/10.1016/j.jenvman.2014.07.025

De Ridder K, Adamec V, Bañuelos A et al (2004) An integrated methodology to assess the benefits of urban green space. Sci Total Environ 334:489–497. https://doi.org/10.1016/j.scitotenv.2004.04.054

Diener A, Mudu P (2021) How can vegetation protect us from air pollution? A critical review on green spaces' mitigation abilities for air-borne particles from a public health perspective-with implications for urban planning. Sci Total Environ 796:148605. https://doi.org/10.1016/j.scitotenv.2021.148605

Dutta P, Chorsiya V (2013) Scenario of climate change and human health in India. Int J Innov Res Dev 2(8):157–160

Dutta S, Ghosh S, Dinda S (2021) Urban air-quality assessment and inferring the association between different factors: a comparative study among Delhi, Kolkata and Chennai Megacity of India. Aerosol Sci Eng 5:93–111. https://doi.org/10.1007/s41810-020-00087-x

Faheem M, Danish M, Ansari N (2021) Impact of air pollution on human health in Agra district. Int Res J Educ Technol 1(6):5–18

Gangopadhyay K, Balooni K (2012) Technological infusion and the change in private, urban green spaces. Urban for Urban Green 11(2):205–210. https://doi.org/10.1016/j.ufug.2011.12.003
Gupta K, Roy A, Luthra K, Maithani S (2016) GIS-based analysis for assessing the accessibility at hierarchical levels of urban green spaces. Urban for Urban Green 18:198–211. https://doi.org/10.1016/j.ufug.2016.06.005
Gunawardena KR, Wells MJ, Kershaw T (2017) Utilising green and blue space to mitigate urban heat island intensity. Sci Total Environ 584:1040–1055. https://doi.org/10.1016/j.scitotenv.2017.01.158
Hansen R, Pauleit S (2014) From multifunctionality to multiple ecosystem services?: a conceptual framework for multifunctionality in green infrastructure planning for urban areas. Ambio 43:516–529. https://doi.org/10.1007/s13280-014-0510-2
Hansen R, Rall E, Chapman E et al (2017) Urban green infrastructure planning: a guide for practitioners. GREENSURGE. https://ign.ku.dk/english/green-surge/rapporter/D5_3_Urban_GIP_-_A_guide_for_practitioners.pdf. Accessed 15 Dec 2023
Hansen R, Rolf W, Santos A et al (2016) Advanced urban green infrastructure planning and implementation—innovative approaches and strategies from European cities. GREEN SURGE Report D5.2. https://doi.org/10.13140/RG.2.1.4813.2243. Accessed 15 Dec 2023
Ioja CI, Gradinaru SR, Onose DA et al (2014) The potential of school green areas to improve urban green connectivity and multifunctionality. Urban for Urban Green 13(4):704–713. https://doi.org/10.1016/j.ufug.2014.07.002
Kaur R, Pandey P (2020) Monitoring and spatio-temporal analysis of UHI effect for Mansa district of Punjab, India. Adv Environ Res 9:19–39. https://doi.org/10.12989/aer.2020.9.1.019
Kikon N, Singh P, Singh SK et al (2016) Assessment of urban heat islands (UHI) of Noida city, India using multi-temporal satellite data. Sustain Cities Soc 22:19–28. https://doi.org/10.1016/j.scs.2016.01.005
Kim HW, Tran T (2018) An evaluation of local comprehensive plans toward sustainable green infrastructure in the US. Sustainability 10(11):4143. https://doi.org/10.3390/su10114143
Korpilo S, Kaaronen RO, Olafsson AS et al (2022) Public participation GIS can help assess multiple dimensions of environmental justice in urban green and blue space planning. Appl Geogr 148:102794. https://doi.org/10.1016/j.apgeog.2022.102794
Kumar J (2015) Metropolises in Indian urban system: 1901–2011. Eur J Geogr 6(3):41–51
Liu Y, Huang TT, Zheng X (2022) A method of linking functional and structural connectivity analysis in urban green infrastructure network construction. Urban Ecosyst 25(3):909–925. https://doi.org/10.1007/s11252-022-01201-2
Mathur P, Mascarenhas L (2019) Lifestyle diseases: keeping fit for a better tomorrow. Indian J Med Res 149(Suppl-1):S129. https://doi.org/10.4103/0971-5916.251669
McPhearson T, Hamstead ZA, Kremer P (2014) Urban ecosystem services for resilience planning and management in New York city. Ambio 43:502–515. https://doi.org/10.1007/s13280-014-0509-8
Monteiro R, Ferreira JC, Antunes P (2020) Green infrastructure planning principles: an integrated literature review. Land 9(12):525. https://doi.org/10.3390/land9120525
Monteiro R, Ferreira JC, Antunes P (2022) Green infrastructure planning principles: identification of priorities using analytic hierarchy process. Sustainability 14(9):5170. https://doi.org/10.3390/su14095170
Mosissa ST, Zhongwei S, Tsegaye WH et al (2023) Prioritization of green infrastructure planning principles using analytic hierarchy process: the case of Addis Ababa. Urban for Urban Green 85:127965. https://doi.org/10.1016/j.ufug.2023.127965
Murray V, McBean G, Bhatt M et al (2012) Case studies: managing the Risks of Extreme events and disasters to advance climate change adaptation. A Special Report of Working Groups I and II of the Intergovernmental Panel on Climate Change (IPCC). Cambridge University Press, Cambridge, UK, and New York, USA, pp 487–542

NASA [National Aeronautics and Space Administration] (2015) GISS surface temperature analysis. Goddard Institute for Space Studies. https://data.giss.nasa.gov/gistemp/station_data_v4_globe/. Accessed 15 Dec 2023

NIUA [National Institute of Urban Affairs] (2022) CSCAF in monitoring urban green spaces. https://niua.in/c-cube/blog/content/cscaf-monitoring-urban-green-spaces. Accessed 15 Dec 2023

Nicol C, Blake R (2000) Classification and use of open space in the context of increasing urban capacity. Plan Pract Res 15(3):193–210. https://doi.org/10.1080/713691902

Orimoloye IR, Mazinyo SP, Kalumba AM et al (2019) Implications of climate variability and change on urban and human health: a review. Cities 91:213–223. https://doi.org/10.1016/j.cities.2019.01.009

Purvis B, Mao Y, Robinson D (2019) Three pillars of sustainability: in search of conceptual origins. Sustain Sci 14:681–695. https://doi.org/10.1007/s11625-018-0627-5

Ramachandra TV, Aithal BH, Shivamurthy V et al (2016) Modelling and visualization of urban trajectory in 4 cities of India. In: 32nd annual in-house symposium on space science and technology, ISRO-IISC space technology cell. Indian Institute of Science, Bangalore

Ramachandra TV, Kumar U (2010) Greater Bangalore: emerging urban heat island. Geospatial World 10:89–95

Ranger N, Hallegatte S, Bhattacharya S et al (2011) An assessment of the potential impact of climate change on flood risk in Mumbai. Clim Change 104:139–167. https://doi.org/10.1007/s10584-010-9979-2

Richardson EA, Pearce J, Mitchell R et al (2013) Role of physical activity in the relationship between urban green space and health. Public Health 127(4):318–324. https://doi.org/10.1016/j.puhe.2013.01.004

Rouse DC, Bunster-Ossa IF (2013) Green infrastructure: a landscape approach. American Planning Association Planning Advisory Service Report Number 571. https://caeau.com.ar/wp-content/uploads/2018/11/46.GREEN-INFRAESTRUCTURE.pdf. Accessed 10 Nov 2023

Rumana HS, Sharma RC, Beniwal V, Sharma AK (2014) A retrospective approach to assess human health risks associated with growing air pollution in urbanized area of Thar Desert, western Rajasthan, India. J Environ Health Sci Eng 12:1–9. https://doi.org/10.1186/2052-336X-12-23

Sharma D, Tomar S (2010) Mainstreaming climate change adaptation in Indian cities. Environ Urban 22(2):451–465. https://doi.org/10.1177/0956247810377390

Shukla A, Jain K (2021) Analyzing the impact of changing landscape pattern and dynamics on land surface temperature in Lucknow city, India. Urban for Urban Green 58:126877. https://doi.org/10.1016/j.ufug.2020.126877

Singh C, Madhavan M, Arvind J et al (2021a) Climate change adaptation in Indian cities: a review of existing actions and spaces for triple wins. Urban Climate 36:100783. https://doi.org/10.1016/j.uclim.2021.100783

Singh V, Singh S, Biswal A (2021b) Exceedances and trends of particulate matter (PM2. 5) in five Indian megacities. Sci Total Environ 750:141461. https://doi.org/10.1016/j.scitotenv.2020.141461

Sudhira HS, Ramachandra TV, Jagadish KS (2004) Urban sprawl: metrics, dynamics and modelling using GIS. Int J Appl Earth Obs Geoinf 5(1):29–39. https://doi.org/10.1016/j.jag.2003.08.002

Tzoulas K, Korpela K, Venn S et al (2007) Promoting ecosystem and human health in urban areas using Green Infrastructure: a literature review. Landsc Urban Plan 81(3):167–178. https://doi.org/10.1016/j.landurbplan.2007.02.001

Udas-Mankikar S (2020) Formulating open-space policies for India's cities: the case of Mumbai, ORF Occasional Paper, 241

Udas-Mankikar S, Driver B (2021) Blue-green infrastructure: an opportunity for Indian cities. ORF Occasional Paper, 317

UNDESA [United Nations Department of Economic and Social Affairs] (2018) World Urbanization Prospects: The 2018 Revision. https://esa.un.org/unpd/wup/. Accessed 10 Nov 2023

Van Herzele A, Wiedemann T (2003) A monitoring tool for the provision of accessible and attractive urban green spaces. Landsc Urban Plan 63(2):109–126. https://doi.org/10.1016/S0169-2046(02)00192-5

Veena K, Parammasivam KM, Venkatesh TN (2020) Urban heat island studies: current status in India and a comparison with the international studies. J Earth Syst Sci 129:1–15. https://doi.org/10.1007/s12040-020-1351-y

Von Haaren C, Reich M (2006) The German way to greenways and habitat networks. Landsc Urban Plan 76:7–22. https://doi.org/10.1016/j.landurbplan.2004.09.041

Woolley H, Rose S, Carmona M et al (2004) The value of public space: how high quality parks and public spaces create economic, social and environmental value. Commission for Architecture and the Built Environment. https://www.designcouncil.org.uk/fileadmin/uploads/dc/Documents/the-value-of-public-space1.pdf. Accessed 10 Nov 2023

WHO [World Health Organization] (2015) China: country health profile. http://www.who.int/countries/chn/en/. Accessed 10 Nov 2023

WHO [World Health Organization] (2016) Global urban ambient air pollution database. https://www.who.int/phe/health_topics/outdoorair/databases/cities/en/. Accessed 12 Nov 2023

Yadav K, Krishnan A (2008) Changing patterns of diet, physical activity and obesity among urban, rural and slum populations in north India. Obes Rev 9(5):400–408. https://doi.org/10.1111/j.1467-789X.2008.00505.x

Yang B, Zhang Y, Qian Y (2012) Simulation of urban climate with high resolution WRF model: a case study in Nanjing, China. Asia-Pac J Atmos Sci 48:227–241. https://doi.org/10.1007/s13143-012-0023-5

# Chapter 6
# Impact of Migration and Urbanization on Sustainable Development in Selected Coastal Blocks of Sundarbans, West Bengal, India

**Sourav Halder and N. C. Jana**

**Abstract** Ecological diversity, inter-regional, and intra-regional migration, and growth of urbanization with population dynamics highlight significant footprints in the Sundarbans region of South 24 Paraganas District, West Bengal, India. According to Census reports of India, 310 million people were the total number of migrants and 98 million people migrated through different push and pull factors into a new residence in the decade 1991–2001. In 2020–21, 12.2% of people migrated due to loss of job/closure of limit/ lack of employment opportunities and 15.7% due to health-related reasons, as per the migration reports of India. Sundarbans region going through climo-morphological hazards, which directly impact human living and livelihood and trigger to initiate the process of migration. The urban growth rate increased from 13.31% in 1991 to 25.610% in 2011 and observed rapid urbanization in several blocks of the Sundarbans region. In this chapter, an attempt has been made to delineate and analyze the diversification of ecological factors, the process of migration, and the growth of urbanization to reach and ensure SDG1, SDG8, and SDG9–goals.

**Keywords** Ecological diversity · Climo-morphological hazards · Migration · Urbanization · Sustainable development

S. Halder
P. G. Department of Geography, Prabhu Jagatbandhu College, Howrah, West Bengal, India

S. Halder · N. C. Jana (✉)
Department of Geography, The University of Burdwan, Purba Bardhaman, West Bengal, India
e-mail: jana.narayan@gmail.com

 
S. Anand et al. (eds.), *Emerging Geosustainability Transformations in India*, Advances in Geographical and Environmental Sciences,
https://doi.org/10.1007/978-981-97-7993-2_6

## 6.1 Introduction

Sundarbans is characterized by dynamic climate and hydro-geomorphological phenomena with high ecological diversity and human-nature conflict. This region is enriched by geographical, anthropological and heritage resources. Fertility and mortality, migration is the essential indicator of population change (Banerjee and Das 2021). To understand the "law of migration" (1885) of Ernst Ravenstain (1885), neoclassical economic theory has been the dominant approach that involves improving wages and economic stability to make the perception for most of the twentieth century. The inadequacy of job opportunities and low wages is the important factors of migration in Sundarbans (Mistri and Das 2020). The reason behind the migration for male migrants was dominated by employment reasons in rural and urban areas. In India, 29 and 56% of rural and urban male migrants migrated due to employment (NSSO 2007–2008). From 1904 to 2016, Sundarbans lost 450 sq km (Dasgupta et al. 2020). Lohachora, Bedford, and Suparibhanga islands were consumed by the River Hugli (Lower course of River Ganges) in the recent past and Ghoramara island is slowly eroding significantly in the last five decades, 50% of it has already submerged due to enhanced land erosion and sea level rise. Jammu and Mousumi's islands shrunk by 75% and 20% respectively in their area (Garg et al. 2021). Approximately 4.5 million people live in the Bengal Delta, in 54 islands out of 102 Islands. The Deltaic region was populated in 273–232 BC which is found from historical findings. According to Indian Meteorological Department, the frequency of severe cyclones and high-intensity cyclones increased by 26% in the Indian Sundarbans during the last century and Post monsoon tropical disturbances in the Bay of Bengal will increase by 50% by the year 2050 and will contribute to form high-intensity cyclonic phenomenon due to climate change (Garg et al. 2021). Different climatic hazards especially cyclonic storms cause storm surges and the consequence of heavy floods and a larger impact on sustainability on the land and increasing level of salinity in fresh water and land in Sundarbans (Halder and Jana 2022). From 1920 to 2021 increasing frequency of climatic hazards like tropical cyclones, and storm surges, exerted a wider impact on upper land and underwater natural habitats in case of irrespective of ecological succession. The growth of the urban population in the Sundarbans region is rapid. As per the census report of 2001 and 2011, the growth of the urban population is 18.38%, and high growth in Canning-I, Canning-II, Mathurapur-I, Jaynagar-I, Jaynagar-II and Basanti blocks of Sundarbans, South 24 Paraganas, West Bengal. The increasing pressure on the urban sector is concentrated within the urban setup, but the immense pressure drags people to other surrounding regions. On the other hand, human activities in the urban sector cause environmental depletion through overutilization of resources that accelerate air pollution, overconsumption of water resources, and problems like urban floods. The complex river network of Hooghly, Muriganga, Battala, Hatania-Doania, Saptamukhi, Chunkuri, Sundarika-Doania, and Chunnai, rivers dominated the delta plain of tropical climate along with the high proportion of fine-grained sediment comprises organic-rich mud with plant debris largely derived from mangrove vegetation (Paul et al. 2021). The present study has tried to appraise migration and rapid growth of urbanization and their impact on climate change with the inherent relationship of sustainable growth to ensure SDG1, SDG8, and SDG9 goals in Canning-I and Canning-II. Blocks Sundarban, South 24 Paraganas, West Bengal in India.

## 6.2 Historical Background of the Study Area

From a historical-structural point of view, migration is related to the structural change in the world scenario, especially the function of globalization, transforming economies, and new forms of production (Massey et al. 1993, Saha et al. 2020). Sundarbans lies between India and Bangladesh with only 19% of India's share. The Sundarbans is enriched by the diversification of a large number of flora and fauna (Halder and Jana 2022). Migration is an important livelihood strategy, and at least one member of three-fourths of the households migrates for job opportunities (Mistri and Das 2020). Sundarbans stretched between 21°27′ to 22°30′N and 89°03′ to 90°00′ E. The present study is concerned with Canning-I and Canning-II C.D. blocks (22°31′ N and 88°66′ E) (Fig. 6.1). Due to the different climatic hazards induced by forced migration, an element of external impact dominates, including threats to life and livelihood arising from natural factors (Samling et al. 2015). Environmental factors play an important role in shaping rural–urban migration. Immense pressure on the urban setup is likely distributed by increasing population rate, changing land use patterns, and newly formed services. The growth of urban population increases rapidly in Canning-I and II blocks through the comparative study between different Census reports (1991–2011). It is observed through the 2001 Census that Canning-I and II blocks had no urban population but in the present-day context, it is observed that Canning-I and II have eight and one census towns, respectively, which indicates the rapid growth of urban population in this geographical region. The total population of Canning-I was 304,724 and the decadal population growth is 24.57%. The significance is negative growth of the rural population at −25.80% and there is no urban population share in 2001 as such but in 2011 the urban population share was 40.44%. In Canning-II C.D block, the total population was 252,253 and rural and urban population growth was 23.15 and 4.43% with the comparison between the 2001 and 2011 census reports. The study tries to explore migration, urbanization, and their impact on sustainability toward reaching sustainable development goals.

## 6.3 Objectives of the Study

This chapter tries to analyze the migration, urbanization, and sustainable development in Sundarbans that is very significant in terms of the high growth of urbanization. The main objectives are to explore the reasons for migration and their impact and to make a comprehensive assessment of urbanization and its impact on sustainability.

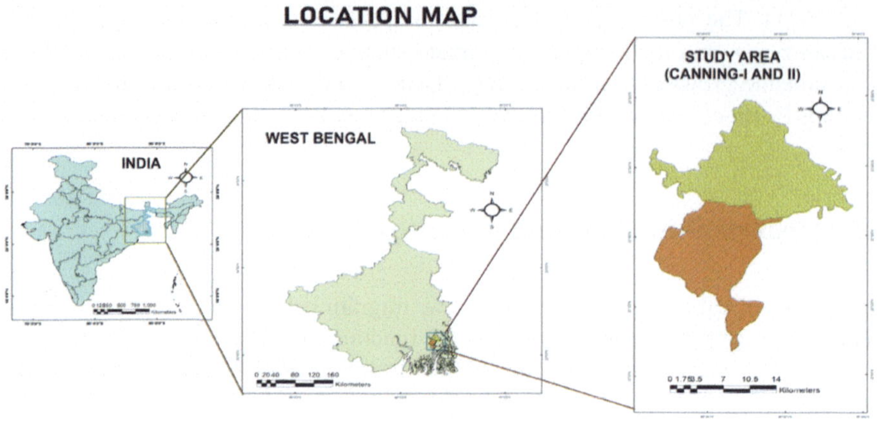


**Fig. 6.1** The study area. *Source* Authors

## 6.4 Database

The study is designed with an empirical approach and the work goes through by using both primary and secondary data. Both Primary and secondary data analyses are done by quantitative and qualitative methods. Primary data were collected through field surveys with the help of probability and non-probability sampling techniques to study the nature of migration and urbanization. The sample sites were selected from the urban areas eight census towns of Canning-I and one Census town in Canning-II block. 100 respondents from each block were taken through the stratified and snowball sampling methods. The database of the study has been collected from different census years from 1951 to 2011 for urban population growth. The figures have been prepared using Multi-year Landsat 4–5 and Landsat 8 DEM for the detection of Normalized Difference Vegetation Index (NDVI) change from 1992 to 2023 in the study area. Landsat 8 DEM was used for the observation of Land Surface Temperature (LST) in the study area (Table 6.1). For the detection of cyclonic phenomena, which augments the possibility of migration from the geographic region, we tried to analyze 52 years (1970–2022) of the formation of depression, cyclonic storms, and severe cyclonic storms in the Bay of Bengal and landfall on the South 24 Paraganas Coast, West Bengal, India. Indian Meteorological Department (IMD), Kolkata, and weather online portal and NOAA Climatic and IMD cyclone E-Atlas data). Other secondary data have been collected from the Handbook, reviewing several research articles, and government reports.

Arc GIS 10.8.2 software was used for the preparation of the NDVI and LST maps. For the preparation of different thematic maps, MS Excel and MS Word were used.

**Table 6.1** Specification of LANDSAT images (1992 and 2023) used in the present study

| Satellites | Raw | Path | Date of acquisition | Scene centre time | Bands | Sensors | Resolutions |
|---|---|---|---|---|---|---|---|
| Landsat 5 | 045 | 138 | 08/03/1992 | 03:55:21 | 5 | TM | 30 m |
| Landsat 8 | 045 | 138 | 07/04/2023 | 04:31:04 | 9 | OLI | 30 m |

*Source* USGS Earth Explorer

## 6.5 Methodology

The study goes through a pragmatic research design with different sequential and interlinked steps through the use of primary and secondary databases.

### *6.5.1 Preparing a Primary Data Table for Delineating the Factors of Migration of Canning-I and Canning-II Blocks*

The data table is prepared using primary surveyed data. Analyzed the determining factors of migration and distribution of migrants in inter-districts, intra-districts, and intra-states and thereafter analyzed the reasons for migration. Explaining analytically with the help of a graphical representation of urbanization in India (1961–2011), the annual exponential growth rate of India (1971–2011), the growth of urban population in India (1901–2011), decadal variation of the urban population in South 24 Paraganas (1901–2011), level of urbanization in South 24 Paraganas (1951–2011), decadal growth of urban population in South 24 Paraganas (1911–21 to 2001–2011), Trends of urbanization in South 24 Paraganas (1961 to 2011), the urban population of Canning-I block (2011), number of census towns in south 24 Paraganas (2001 and 2011), number of census towns in Canning-I and Canning-II blocks.

### *6.5.2 Computation of NDVI of the Study Area (1992 and 2011)*

NDVI of Canning-I and Canning-II is prepared by the use of 1992 and 2011 USGS DEM Data.

For 1992, Band 3 and Band 4 from Landsat 4–5 TM C2 L1 were used and for 2011, Band 4 and Band 5 from Landsat 8–9 OLI/TIRS C2 L1 were used.

**Formula for Landsat 4–5 TM C2 L1**

$$NDVI = \left(Band4 - Band3 \middle/ Band4 + Band3\right)$$

**Formula of Landsat 8–9 OLI/TIRS C2 L1**

$$NDVI = \left(Band5 - Band4 \middle/ Band5 + Band4\right)$$

### 6.5.3 *Estimation of LST*

LST is estimated to explain temperature from the thermal band and its preparation for measuring the impact of the built-up area.

**Formula**

$$L(\lambda) = (LMAX \; - LMIN) \big/ 255 \times DN + LMIN$$

*Where,*

L(λ) = Spectral radiance; LMIN = Spectral radiance of DN value; LMAX = Spectral radiance of DN value; DN = Digital Number.

Spectral radiance is converted into the top of atmosphere brightness temperature in Kelvin.

**Formula**

$$T = K2 \big/ In(K1/R) + 1$$

*Where*

K1 = Calibration Constant 1; K2 = Calibration Constant 2; R = Radiance values W/m2 SR μm; T = Surface Temperature.

Land Surface Emissivity (LSE) is estimated by the use of NDVI value.

Kelvin Temperature is converted into degrees Celsius.

TB = T–273.

### 6.5.4 *Computation of Urban Heat Island (UHI)*

UHI Index has been prepared with the help of LST in both Canning-I and Canning-II blocks in 2023 and both maps were analyzed after reflecting the value from low to high.

**Formula**

$$UHI = (LST - LSTm)/SD$$

*Where*

UHI = Urban Heat Islands; LST = Land Surface Temperature; LSTm = The mean temperature of the LST in the study area; SD = Standard Deviation of Temperature.

## 6.6 Results and Discussion

Sundarbans deltas are coastal complexes characterized by the interplay between Rivers, lands, and oceans and highly influenced by the combination of rivers. The study observed migration of many youthful populations due to the adverse situations which indicates human brain drain from this region.

### *6.6.1 Migration*

The migration of population refers to the transfer of population from one place to another temporarily or permanently and it means the mobility of people from one place to another in case of short-term and long-term migration. An important factor initiates significant change in population growth, size, and structure of population from the perspective of geographic region. Intra-state and inter-state migration is an ancient and significant phenomenon in the case of West Bengal (Banu 2016). The population density of West Bengal increased by 13.84% in between 2001 and 2011 (Census of India 2001 and 2011). Significantly, a high number of the population has migrated from the urban areas of West Bengal to the urban areas of other states, while in the case of in-migration to West Bengal from other states, most of the rural inhabitants of other states migrated to the urban areas of West Bengal (Sarkar 2019). The net migration decreased in West Bengal from 1.86 to 1.44 in 2001 and 1.02in 2011 (Banerjee and Das 2021). Migration in Sundarbans occurs due to various reasons. Sundarbans faced inter-state migration (49%), inter-district migration (48%), and 4% of migrants preferred intra-district migration. It is very relevant that after Kolkata (32.0%), the mobility of migration occurs in Tamilnadu (13.0%), Karnataka (7.0%), Kerala (6.5%), Gujrat (6.0%), other States (15.5%), and Bardhaman (5%), other districts (11.0%) and south 24 Paraganas(4.0%) respectively (Mistri and Das 2020).

#### 6.6.1.1 Territorial Factors and Impact of Migration on Population Change

The incidences of migration phenomenon constitute due to different circumstances, it can be classified into natural and anthropogenic factors. Petersen (1958) classified based on two main criteria those are conservation and innovation and based on the classification of the driving force of migration, into nature (ecological push), state (migratory policy), norms (aspiration), and others (social movement). A few determinants, which accelerate the process of migration can be classified into economic, demographic, environmental, social, political, and geographical. In 2009, just after the occurrence of the severe cyclonic hazard Aila, inter-state migration (around 64%) over inter-district (32%) and intra-district (3.0%) migration constituted in Sundarbans (Mistri 2012) (Fig. 6.2). It is understood from the previous study and the result of the recent study that climatic hazards and the consequences of storm surges and saline water ingression speed up vulnerability, which induces the rate of migration.

The devastating nature of severe tropical cyclones on the West Bengal Coast or near the Bangladesh coast such as Sidr (2007), Aila (2009), Fani (2019), Bulbul (2019), Amphan (2020), Yash (2021) and Mocha (2023) brought significant changes of land use and land cover (Halder and Jana 2022), loss of human life, human property and influence on the ecological imbalance to a greater extent. The present study shows that from 1970 to 2022, the West Bengal coastal region was affected by more than 35 depressions and 14 Cyclonic storms and extremely severe cyclonic storms. Especially Canning-I and Canning-II blocks were highly affected by more than five Cyclonic storms and extremely severe cyclonic storms which initiated the human migration phenomenon from this region (Table 6.2) 0.31.05 and 50.32% BPL (Below Poverty Level) populations live in Canning-I and II blocks (Census of India 2011). It is observed that 80% of Canning-I and 90% of Canning-II blocks (DasGupta and Shaw 2014) have primitive residential infrastructure and this is an important reason that they are highly impacted by natural calamities. Canning-I block is situated in the transitional zone of Sundarbans mangroves and the southeast region of Matla River and this area is highly affected by tidal flooding due to excessive sedimentation and clogging of the riverbed. While the Canning-II block has relatively low coastal exposure.

#### 6.6.1.2 Inter and Intra-Regional Impact of Population Dynamics on Migration

Through the primary survey with the use of multiple sampling techniques, it is observed that climatic hazards, hydro-geomorphological hazards (especially in the Canning-I block), higher wage rates, job opportunities are the key factors of migration and better health facilities and other basic amenities, academic facilities, the higher scope of business, etc., are the secondary factors (Fig. 6.3). It is observed through the analysis of primary data and comparative study between finding factors, better job opportunities (32.4%), and higher wage rate (38.6%) are the most relevant factors, and climatic hazard (12.5%) and hydro-geomorphological hazard (8.2%) are also very significant for migration. Due to the different push and pull factors (Ravenstein 1885) inter-districts, intra-districts and intra-state migration occurred. Due to several natural and anthropogenic perspectives induced intra-districts (especially in Kolkata, Howrah, Bardhaman, Hoogli), inter-districts (Diamond Harbour and near Kolkata),

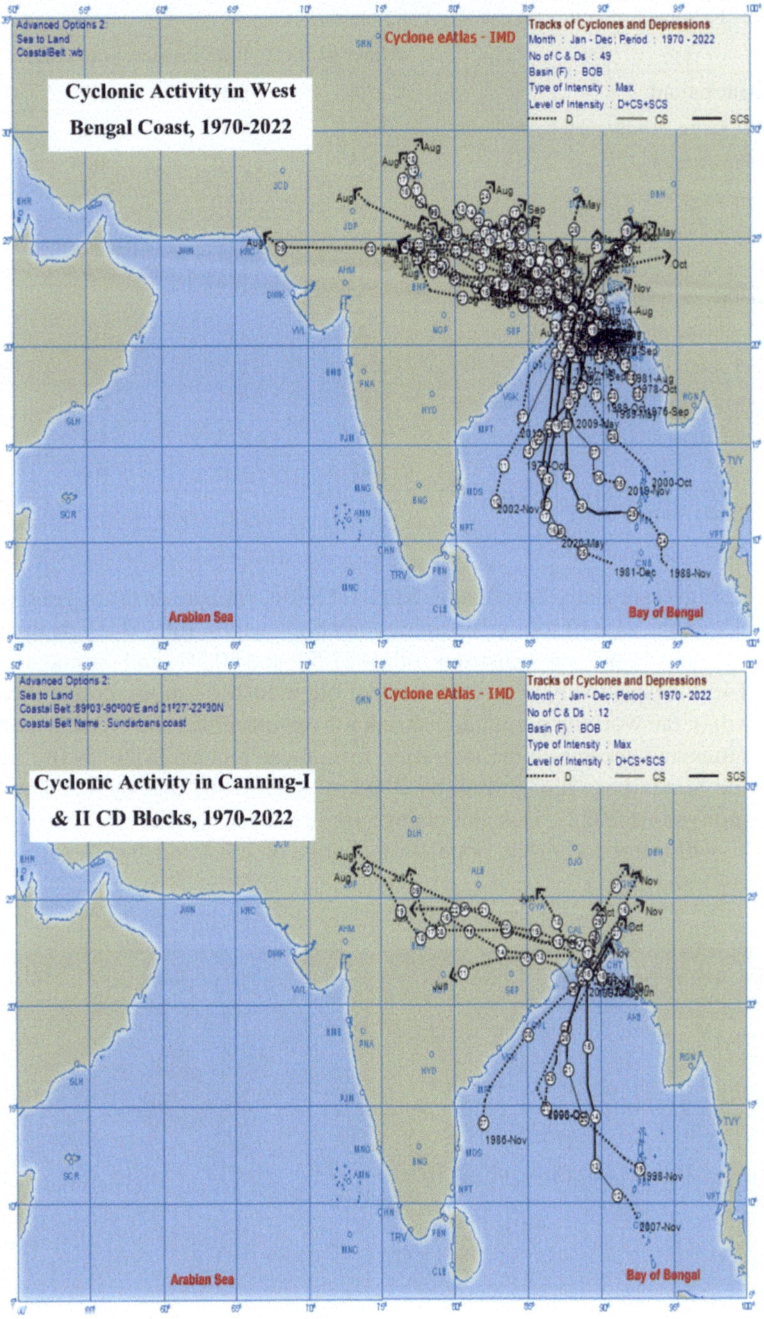


**Fig. 6.2** Cyclonic activity in West Bengal coast, 1970–2022. *Source* Cyclone eAtlas-IMD

**Table 6.2** Factors of migration in canning-I and canning-II blocks (%)

| Factors | Inter-districts | Intra-districts | Intra-states | Total |
|---|---|---|---|---|
| 1. Climatic hazard | 25.4 | 8.5 | 2.5 | 12.5 |
| 2. Hydro-geomorphological hazard | 5.2 | 4.5 | 1.4 | 8.2 |
| 3. Job opportunities | 27.5 | 34.4 | 38.5 | 32.4 |
| 4. Higher wage rate | 30.5 | 41.6 | 46.8 | 38.6 |
| 5. Family reason | 0.6 | 2.3 | 0.2 | 1.2 |
| 6. Better health facilities and other basic amenities | 0 | 1.2 | 0.4 | 0.8 |
| 7. Academic facilities | 0.2 | 2.5 | 1.5 | 1.5 |
| 8. Higher scope of business | 3.5 | 1 | 2.8 | 2.3 |
| 9. Others reason | 2.1 | 2 | 2.5 | 0.7 |
| 10. Do not know | 5 | 2 | 3.4 | 1.8 |
| Total | 100 | 100 | 100 | 100 |

*Source* Primary Survey 2023

and intra-state (especially Tamilnadu, Kerala, Gujrat, Maharashtra) migration from this blocks. In the case of job opportunities and high wages, Intra-districts migration share 34.4% and 41.6%, inter-districts follow 27.5% and 30.5%, and intra-states share 38.5 and 46.5, respectively (Fig. 6.4). Lack of big industries, mono-crop cultivation, and growth of the working population shrink job opportunities and create pseudo-job opportunities and accelerate the migration from these blocks. In 2015–16, the total area of the Sundarbans had shrunk by 210 km$^2$ since 1967 and by 451 km$^2$ since 1904 (Anonymous 2022). Decadal change by residence in West Bengal is a total of 13.8%, with rural and urban population change of 7.7% and 29.7% respectively (Census of India 2011).

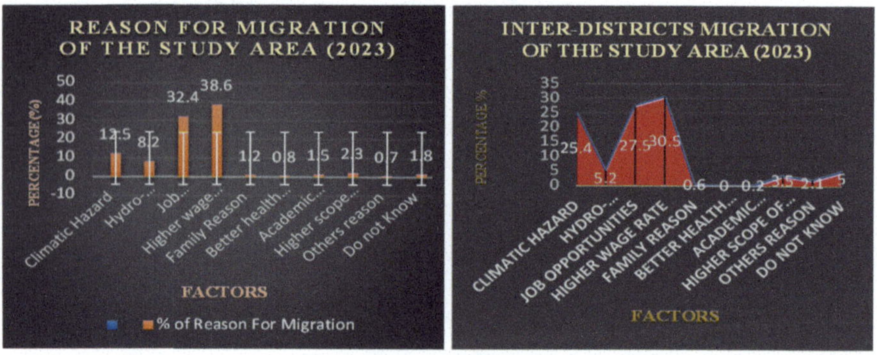


**Fig. 6.3** Migration scenario. *Source* Primary survey 2023

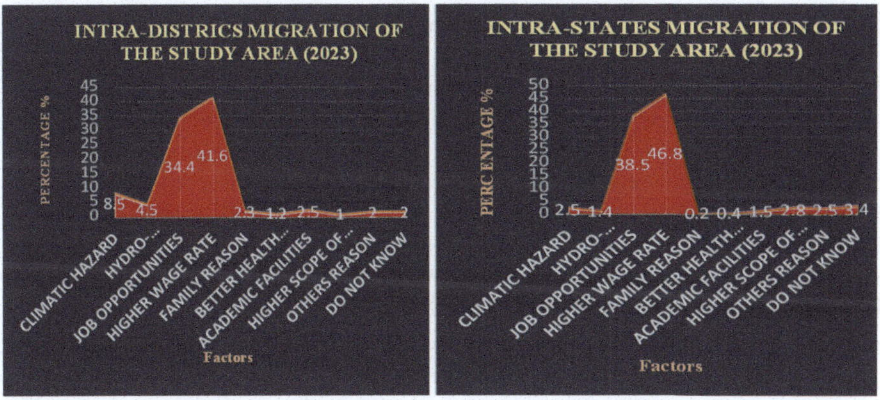


**Fig. 6.4** Intra-district and intra-state migration. *Source* Primary survey 2023

## 6.6.2 *Urbanization*

The process of urbanization is determined as an "engine of inclusive economic growth" (Katke 2016). The increasing ratio of the urban population over the period brings significant changes to the loss of natural components. Growth of urban expansion induced land use and land cover changes on the earth's surface which had a greater impact on forest cover, cultivated land, and water areas, contributing to an imbalanced ecosystem and massive impact by the process of urbanization (Das and Das 2019). The development of systematic environmental protection was necessary with continuous urban development. Economic progress is linked with the environment, so environmental restoration became essential (Gohar 2016).

### 6.6.2.1 Urban Population Dynamics and Trends of Urbanization

Global urbanization has been continuing in Asia, Africa, and Latin America. The first municipal corporation in India was set up at Madras in 1687–88, after that Mumbai and Calcutta municipal corporations were established in 1726. After the independence, especially after the 1960's the growth of urbanization in India increased rapidly.

The process of urbanization is the main driving force of higher productivity and higher economic growth. Mathur (2005) observed that urban growth was driven by substantial growth of urban population share and extended the share of employment manufacturing and service sector after post-liberalization. According to the comparative analysis of census data from 1951 to 2011, it is observed that the growth of urbanization increased rapidly. It shows that in 1961 India shared 17.97% urban population while in 2011 urban population increased to 31.16%. The study shows that in 1981, the annual exponential urban growth rate was high (3.79%) and comparatively low in 2001 (2.75%). Growth of the economy, educational development, infrastructural development, trade, industrialization, connectivity, road density, clustering of the service sector, and population growth are the key determinants of the growth of urbanization in India as well as in West Bengal (Fig. 6.5).The process of urbanization consists of social transformation from traditional rural socio-cultural societies to modern, industrial, and urban communities. It is observed that through the analytical evaluation from the 2001 to 2011 census year data, there is continuous growth of urbanization but the study shows that after 1931 urbanization growth was very progressive. According to the census, the share of the urban population in 1901 was only 12.8%, and after independence, with very significant growth of the urban population, in 2011 the urban population share increased to 31.89% in West Bengal. Urban population growth accelerates due to the different pull factors and urban agglomeration is a continuous process that initiates urban spreads constituting a town and its adjoining urban outgrowth (Datta 2006) (Fig. 6.6). It is observed that the decadal variation of the population in South 24 Paraganas is very significant, in 1921 the decadal variation was 6.54%, which is the lowest, and in 1971 the decadal variation in population was highest at 34.24%. The percentage urban population in South 24 Paraganas is increasing rapidly. The study states that the level of urbanization was only 4.6% in 1951 and 2011 it increased to 25.61% but the most relevant finding is that from 2001 to 2011 the level of urbanization increased by 9.91%. Decadal growth of the urban population is lowest in 1921–31 (2.4%) and the study observed significantly more than double the growth of the urban population in relation to 1941–51 42%) to 1951–61 (98%) and between 1991–2001 (42.9%) to 2001–2011 (92.2%) (Figs. 6.7 and 6.8).The spread of urbanization in South 24 Paraganas was negative in 1991 (−5.024%) and highest in 2011 (6.284). Thirteen blocks of South 24 Paraganasare included in the Sundarbans region namely Canning-I, Canning-II, Mathurapur-I, Mathurapur-II, Jaynagar-I, Jaynagar-II, Kultali, Basanti, Gosaba, Namkhana, Kakdwip, Patharpratima, Sagar C.D. Blocks.

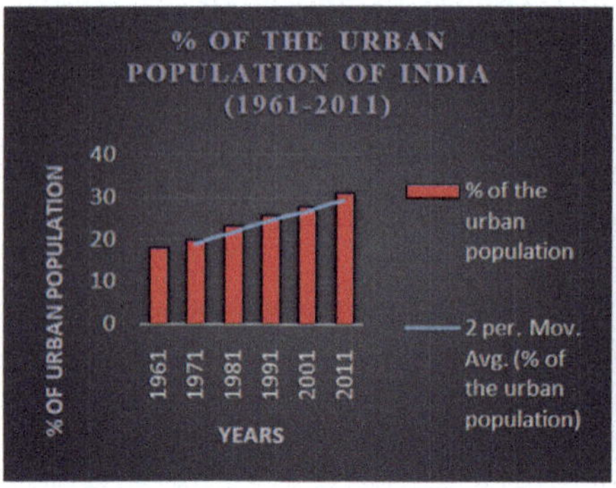


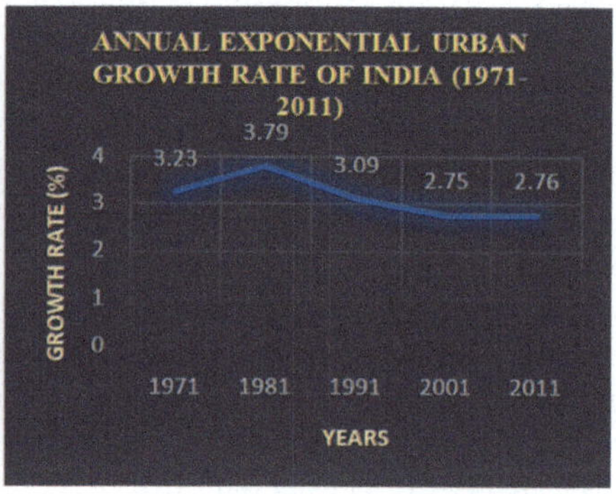


**Fig. 6.5** Urban population scenario of India. *Source* Census Report of India, 1961–2011

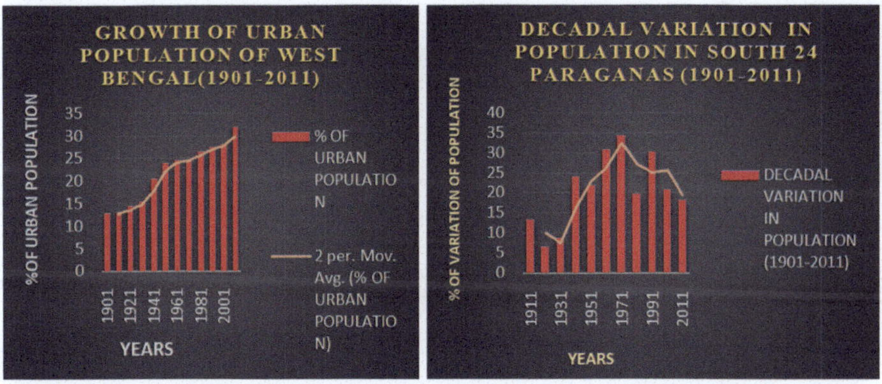


**Fig. 6.6** Urban population scenario in West Bengal. *Source* Census Report of India, 1961–2011

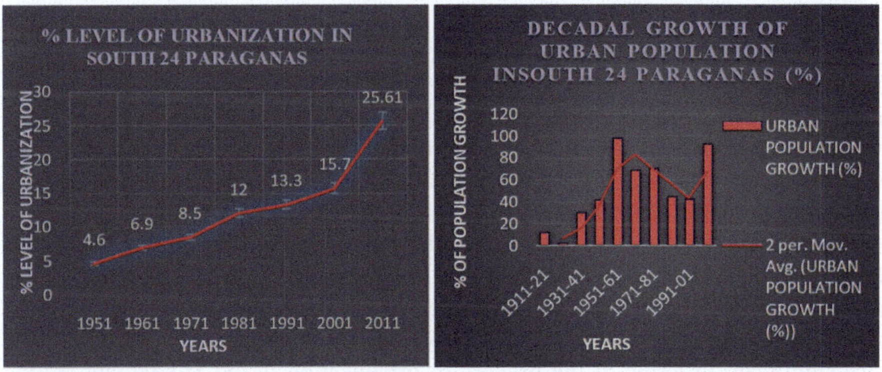


**Fig. 6.7** Urbanization and decadal growth in 24-Parganas(S) *Source* Census Report of India, 1911–2011

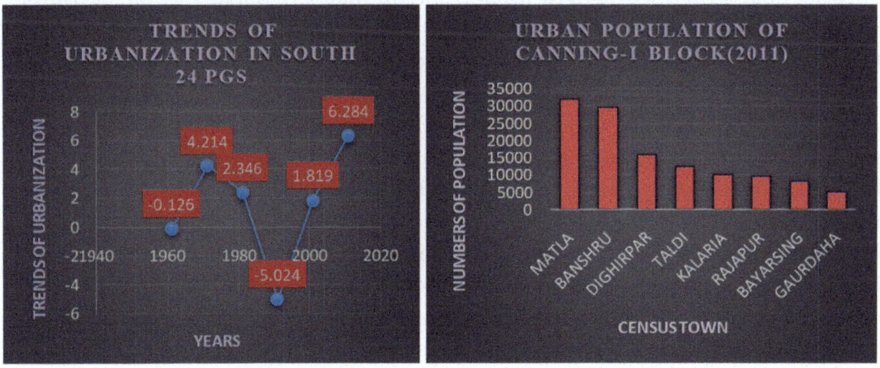


**Fig. 6.8** Urbanization scenario of 24-parganas (S) and canning-I. *Source* Census of India, 1951–2011

**Fig. 6.9** Census towns of 24-parganas (S) and canning-I and II. *Source* Census of India, 2001–2011

The study observed that no urbanization process started till 2001, but as per 2011 census, in six blocks of the Sundarbans region such as Canning-I (8), Canning-II (1), Mathurapur-I (4), Jaynagar-I (6), Jaynagar-II (2) and Basanti(1) C.D. Blocks under South 24 Paraganas, urbanization process started and twenty-two Census Town (CT) were established. Canning-I has the highest number of census towns among the total blocks of South 24 Paraganas' Sundarbans. Matla census town denoted the highest number of urban population and Gourdaha Census town denoted the lowest urban population and urban population share of 2011 was 40.44% whereas the urban population share was nil in 2001 in Canning-I C.D Block. In 2011, the urban population share in the Canning-II block was 4.43%, and only one census town Makhal Tala (11,192) whereas no urban population was recorded in 2001, which is very significant in terms of the process of urbanization in the Sundarbans region (Fig. 6.9).

NDVI and LST maps are important indicators of land use and land cover changes (Hussain et al. 2023). Growth of urbanization initiates frequently in the peri-urban areas which is a result of economic development. It is observed from the NDVI, 1992 that the built-up area of the Canning-I block was only 9.66% and sparse vegetation share was 44.17% but dense vegetation was very low only 6.36%. In Canning-II block built-up area was 14.85% and sparse vegetation was 25.92% and dense vegetation was 9.54%. But significantly in 2023 built-up area of Canning-I increased by more than 31% and Canning-II increased by 55.95% which clearly indicated the growth of urbanization in the studied blocks of Sundarbans (Figs. 6.10 and 6.11). LST is the surface radiation and longwave radiation energy balance which is the significant determinant of climate, environment, and biochemical models (Hussain et al. 2023). Heat island is processed from the arrangement of the urban environment and human-made warming. When the growth of urbanization replaces natural LULC the concentration of Built-up areas and surfaces absorb and retain heat (Fig. 6.12). The result of UHI shows that both the blocks by remote sensing and analysis are affected by NDVI. Urban heat island refers to those urban areas experiencing increased temperature due to human activities and the built environment reflecting heat from the built-up area,

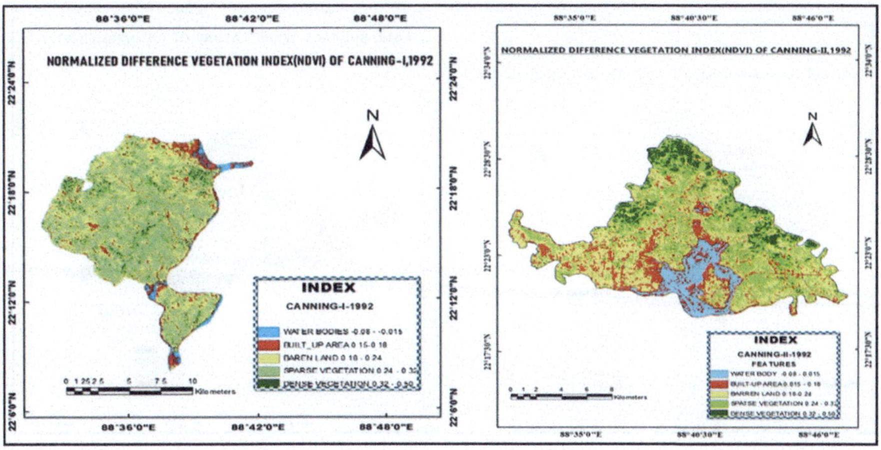


**Fig. 6.10** NDVI of canning-I and II, 1992. *Source* Authors using Landsat 4–5 TM C2 L1 data

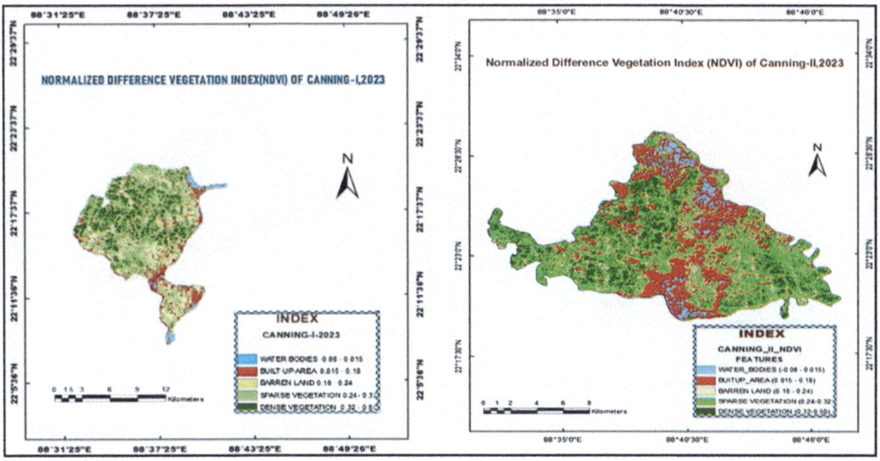


**Fig. 6.11** NDVI of canning-I and II, 2023. *Source* Authors using Landsat 8–9 OLI/TIRS C2 L1 data

and decreased vegetation covers of the study area (Figs. 6.13 and 6.14) environment. The study indicates that the reason is the modification of the land surface.

### 6.6.3 Sustainability

In the Asia–Pacific region, population growth and economic development induced an increase in environmental pollution. The Sundarbans is characterized by complex

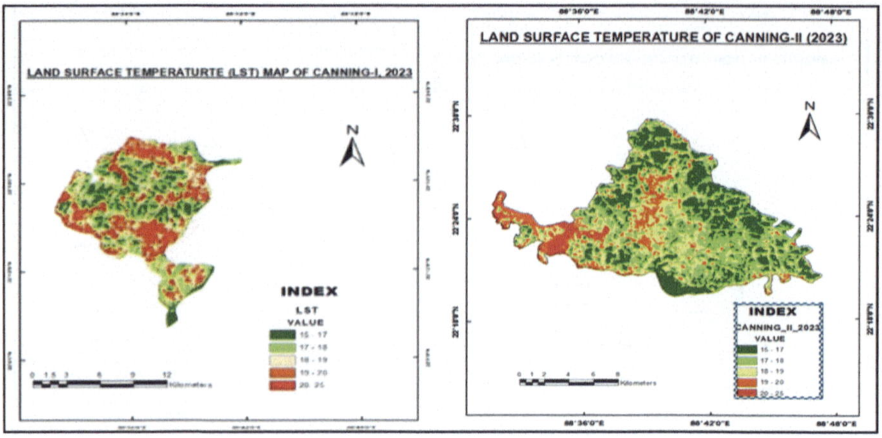


**Fig. 6.12** LST of canning-I and II, 2023. *Source* Authors using Landsat 8–9 OLI/TIRS C2 L1 data

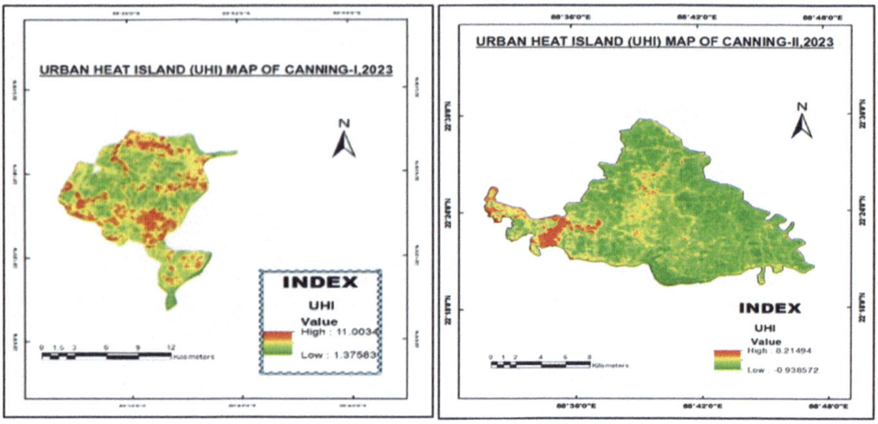


**Fig. 6.13** Urban heat Island of canning-I and II. *Source* Authors using Landsat 8–9 OLI/TIRS C2 L1 data

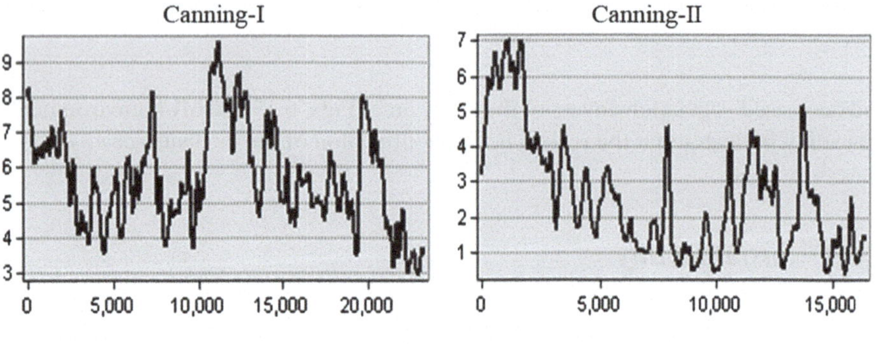


**Fig. 6.14** Urban heat Island graph of canning-I and II

biodiversity and is enriched by large numbers of flora and fauna. The transformation from rural to urban or the growth of urbanization induced a larger impact on land use and land cover change. Due to the growth of urbanization, Canning-I and Canning-II block faced four major challenges to restore sustainability, such as rapid growth of population, loss of agricultural productivity due to natural calamities, loss of natural resources, and the emergence of ecological imbalance. Research findings in the present context have shown that the changes in the growth of population in rural and urban areas, vegetation cover, LST, and growth of urbanization are closely related to environments and anthropogenic activities. The study has shown that the total population growth of South 24 Paraganas Sundarbans is very significant, with the comparison between the 2001 and 2011 census reports and that is 25.10%. The total population growth in Canning-I was 24.57% with the share of urban and rural population growth being 40.43% and −25.80%. In terms of Canning-II C.D. Block, the total population growth was 4.43% and rural population growth was 23.15%. As per the result of the NDVI value of 1992 and 2023, the study observed that the growth of the Built-Up area in Canning-I and Canning-II showed a significant decrease in natural vegetation cover and an increase in LST, indicating threats to Sustainability. UHI indicates the adverse effects on human health, energy consumption, ecological impact, as well as air quality.

## 6.7 Suggestions and Conclusion

Sustainability, migration, and urbanization are complex debatable topics from different aspects. The study observed the migration of many youthful populations due to the adverse situations that indicate human brain drain from this region. Canning-I and Canning-II are important C.D. Blocks of the Sundarbans region, where the process of urbanization started after 2001 but the rate of increase is very rapid. So, there is a need to formulate strategies and ground-level implementation of the same to restore sustainability and the ultimate development of these blocks. The ecosystem, natural habitat, and the co-relation between different ecosystems are highly impacted by the growth of population, the process of urbanization, structural built-up area, lifestyle modification, etc. Environmental degradation can be initiated by the occurrence of new economic opportunities and the new growth of urbanization (UN 2021). Relevant strategies can be adopted to restore sustainability within the process of migration and urbanization of the study area which are mentioned in Fig. 6.15. The Government should adopt necessary environment-friendly capacity-building techniques with the help of local inhabitants to restore and increase the natural vegetation for controlling the rapid urban growth and urbanization, toward achieving sustainable development goals.

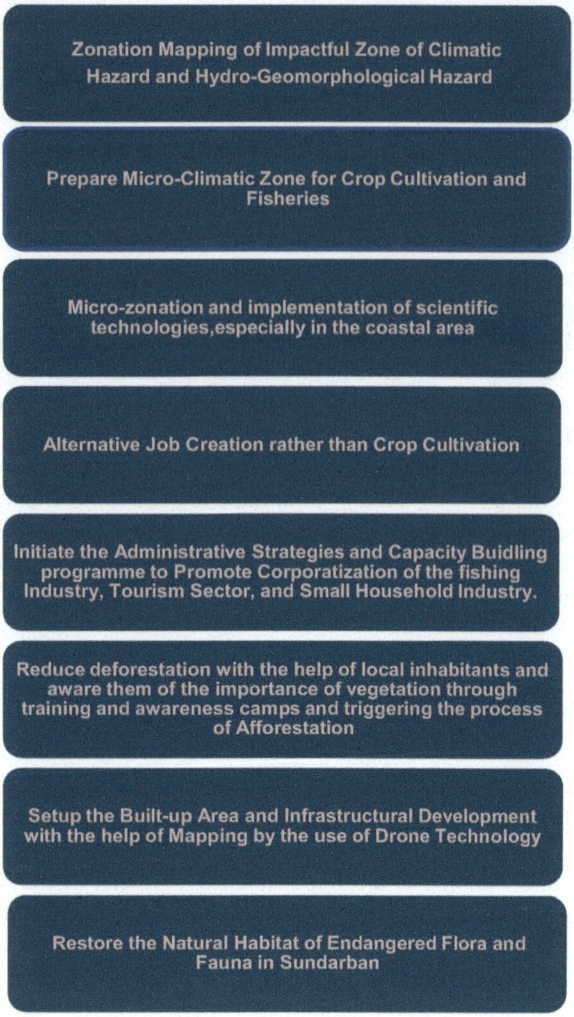


**Fig. 6.15** Suggested relevant strategies to restore sustainability within the process of migration and urbanization

## References

Anonymous (2022) Climate justice: loss and damage in the Sundarbans. Zero Carbon Analytics. https://zerocarbon-analytics.org/archives/justice/loss-and-damage-in-the-sundarbans. Accessed 15 Feb 2024

Banerjee P, Das B (2021) Regional disparity in patterns of out-migration from West Bengal: evidences from census data. Demography India 50(2):131–151

Banu S (2016) Trend and pattern of internal migration and socio-economic transformation in West Bengal. Dissertation. Aligarh Muslim University

Census of India (2011) Population Census 1901–2011. Office of the Registrar General and Census Commissioner, Delhi, India

Cyclone eAtlas-IMD (1970–2022) Tracks of cyclones and depressions over north Indian Ocean. http://14.139.191.203/AboutEAtlas.aspx

Das M, Das A (2019) Dynamics of urbanization and its impact on urban ecosystem services: a study of a medium-sized town of West Bengal, Eastern India. J Urban Manag 8:420–434

DasGupta R, Shaw R (2014) Participatory planning for enhancing community resilience in mangrove rich Indian Sundarbans. Kyoto University, Kyoto, Japan, p 42. https://rajibshaw.org/wpRS/wp-content/uploads/2018/09/Sundarban-report.pdf. Accessed 15 Feb 2024

Dasgupta S, Wheeler D, Sobhan MI et al (2020) Coping with climate change in the Sundarbans: lessons from multidisciplinary studies. Int Dev Focus. World Bank, Washington, DC. https://doi.org/10.1596/978-1-4648-1587-4

Datta P (2006) Urbanisation in India. In: Regional and sub-regional population dynamic population process in urban areas European population conference

Garg R, Zade D, Basu J et al (2021) Climate-induced displacement and migration in India: case studies from West Bengal, Maharashtra, Odisha, Uttarakhand and Bihar. https://cansouthasia.net/wpcontent/uploads/2021/02/Migration_India_20_02_2021.pdf Accessed 10 Sep 2023

Gohar A (2016) Urbanization and sustainable development: evolution and contemporary challenges. J Civ Eng Arch Res 3(12):1813–1825

Halder S, Jana NC (2022) Impact of climate change on sustainability in selected coastal blocks of Sundarban region, West Bengal. In: Mor S, Madan S, Das M (eds) Sustainability driven initiatives in Indian higher education institutions. Excel India Publishers, New Delhi

Husain MA, Kumar P, Gonencgil B (2023) Assessment of spatio-temporal land use/cover change and its effect on land surface temperature in Lahaul and Spiti, India. Land 12(7):1294. https://doi.org/10.3390/land12071294

Katke MS (2016) Urbanization in India: challenges and current issues. Int J Creat Res Thoughts 4(1):247–253

Massey DS, Arango J, Hugo G, Kouaouci A, Pellegrino A, Taylor JE (1993) Theories of international migration: a review and appraisal. Popul Dev Rev 19(3):431–466. https://doi.org/10.2307/2938462

Mathur OP (2005) Impact of globalization on cities and city-related policies in India. In: Richardson HW, Bae CHC (eds) Globalization and urban development. Advances in Spatial Science. Springer, Berlin, Heidelberg, pp 43–58. https://doi.org/10.1007/3-540-28351-X_4

Mistri A, Das B (2020) Environmental change, livelihood issues and migration: Sundarban biosphere reserve. Springer, Singapore, India

Mistri A (2012) Natural resource dependence and migration from Sundarban biosphere reserve: case study of a remote Island. Dissertation, Jawaharlal Nehru University

NSSO (National Sample Survey) (2007–08) Employment, unemployment and migration particulars India, 2007–2008. 64th round-Schedule 10.2. https://catalog.ihsn.org/index.php/catalog/1907. Accessed 15 Feb 2024

Paul A, Ghosh M, Sardar J et al (2021) Rapid land use and land cover changes in response to environmental variables: a study on Namkhana Abad of Indian Sundarban. Int J Adv Res Dev 6(2):8–18

Petersen W (1958) A general typology of migration. Am Sociol Rev 23(3):256–266. American Sociological Association, Stable. http://www.jstor.org/stable/2089239

Ravenstein EG (1985) The laws of migration. J Stat Soc Lond 48(2):167–235

Saha S, Goswami R (2020) Destinations of male outmigration and their drivers in Indian Sundarban. Space Cult, India 8(1):111–142. https://doi.org/10.20896/saci.v8i1.601

Samling CL, Das S, Hazra S (2015) Migration in the Indian Bengal delta and the Mahanadi delta: a review of the literature. DECCMA Working Paper, Deltas, Vulnerability and climate change: migration and adaptation, IDRC Project Number 107642

Sarkar R (2019) Recent changing patterns of migration and spatial patterns of urbanization in West Bengal: a demographic analysis. South-Asian J Multidiscip Stud 4(1):46–56

United Nations (2021) Department of economic and social affairs population division. New York

# Chapter 7
# Toward a Sustainable Development of Asansol City, West Bengal: An 'Eco2 City' Approach

**Suvra Maji and Sarfaraz Alam**

**Abstract** Cities in developing countries are becoming increasingly susceptible to environmental crises due to adverse impacts of activities related to urban growth and urbanization. The environmental problems caused by these activities render the city dwellers vulnerable and cities unsustainable. To tackle these problems, the World Bank has advocated the "Eco2 City" approach (i.e., developing simultaneously an economic and ecologically sustainable city) for urban planning, development, and management. It considers the ecological city as the economic city. The present paper focuses on the sustainable development of Asansol, the second-largest city in West Bengal. The total population of the city has grown nearly seven times in the last sixty years and the areal extent expanded over 20 times. The city has undergone economic functional changes–mainly from a mining and transport town to an economically complex city. In the process experienced profound implications in land use and environmental sustainability. The emerging land use pattern of the city is characterized by the expansion of built-up area and consequent shrinkage of the area under vegetation, agriculture, and water bodies. Further, due to the expansion of the industrial footprint, the problems of air and water pollution have become a major concern in the city. Due to the discharge of untreated effluents by industries directly or indirectly into the main rivers, water pollution has become a major concern. Further, in the absence of soak pits in the city, domestic sewage flows through open drains, resulting in unhygienic and unclean surroundings. Against this backdrop, the chapter advocates for adopting the "Eco2 City" approach by meaningfully integrating the major economic and ecological concerns in the planning, development, and management of Asansol City.

**Keywords** Urban · Sustainability · Ecology · Economy · Land use · Pollution

---

S. Maji
Department of Geography, Pakuahat Degree College, Malda, West Bengal 732138, India

S. Alam (✉)
Department of Geography, Institute of Science, Banaras Hindu University, Varanasi, Uttar Pradesh 221005, India
e-mail: sarfaraz05@bhu.ac.in

 
S. Anand et al. (eds.), *Emerging Geosustainability Transformations in India*, Advances in Geographical and Environmental Sciences,
https://doi.org/10.1007/978-981-97-7993-2_7

## 7.1 Introduction

A city is viewed as a well-defined territory where a large number of people from diverse socioeconomic backgrounds and geographical origins come to live and work for their growth and prosperity. It is also characterized by diversity in terms of socioeconomic and cultural activities. "The city is one of the highest pinnacles of human creation ... Through agglomeration, cities have the power to innovate, generate wealth, enhance the quality of life, and accommodate more people within a smaller footprint at lower per capita resource use and emissions than any other settlement pattern" (OECD 2006). A city acts as a nerve center of regional and national economies and functions as an engine of growth (Turok and Mcgranahan 2013). For these reasons, people from immediate and distant rural areas are attracted to these cities. Rural poor migrate to cities to earn money to meet their livelihood and escape the vicious cycle of poverty. On the other hand, wealthy people from villages are attracted to cities for their high quality of lifestyle and services (e.g., health and education).

The processes of movement of people from rural to urban areas contribute to urbanization, which is a major driver of a city's development. Urbanization is a multidimensional process. It includes the movement of people in the city, changes in social and economic structures, as well as the generation of agglomeration economies as a result of economic clustering. Urbanization promotes economic growth because it generates operational economies for businesses, improves the quality of life for citizens, and generates immense tax revenue for governments (Tripathi 2019). For these qualities, cities always remain dynamic and keep on growing and expanding with time.

The developed countries of the world are highly urbanized. In contemporary times, it is developing countries which are experiencing a high rate of urban growth. Cities in developing countries continue to be attractive places for people due to the growing gap in opportunities and services between urban and rural areas. They offer opportunities as well as quality of life. But most of the poor migrants end up living in ever-proliferating slums and squatters in cities of developing countries. Further, the growing consumerism in cities generates mountains of solid wastes and causes pollution of water bodies. The growing traffic in cities causes air pollution. Many cities in developing countries face rampant water crises for drinking and domestic use. Thus, cities in developing countries are becoming increasingly susceptible to environmental problems due to the adverse impact of activities related to urban growth and Urbanization, such as population growth, rising consumerism, industrialization, expansion of the built environment, and depletion and degradation of the environmental resource base. The environmental problems caused by these activities render the city dwellers vulnerable and cities unsustainable. Given these, cities of developing countries are facing severe sustainability challenges. Therefore, a question arises about their sustainable future.

The efforts to achieve urban sustainability give rise to a dilemma. Urbanization is essential for the economic prosperity of developing countries. According to OECD

(2006), 'No country has grown to middle income without industrializing and urbanizing. None has grown to high income without vibrant cities. The rush to cities in developing countries seems chaotic, but it is necessary. On the other hand, it is also true that the rapid Urbanization process in developing countries produces adverse effects on environmental resources, such as water, air, soil, land, flora, and fauna. It also challenges disposing of vast quantities of solid waste and treating polluted water. What is even more frightening is the increasing vulnerability of many urban centers to the consequences of climate change, including flooding, landslides, heat waves, and droughts (UNEP 2018). The UNEP (2018) further states, "Many cities are already struggling with environmental degradation, traffic congestion, inadequate urban infrastructure, and a lack of basic services, such as water supply, sanitation, and waste management. Environmental footprints of cities are quite alarming and can threaten the natural resources required to sustain the economic development and poverty alleviation rates." The persistence of these environmental challenges in urban habitats will negatively impact regional and national economies. Therefore, even purely from the interest of regional and national economy, the effective tackling of urban environmental challenges is essential. To tackle these problems, the World Bank (2011a) has advocated the "Eco2 City" approach (i.e., developing simultaneously a more economical and ecologically sustainable city) for urban planning, development, and management. It considers the ecological city as the economic city. The underlying idea is that the sustainable development of a city is achievable only when its major ecological and economic concerns are considered in an interdependent and integrated manner.

However, tackling urban environmental challenges requires heavy investment in environmentally clean technology and the employment of substantial human resources. In other words, a strong economy is needed to fight urban environmental challenges and achieve a sustainable city goal. Urbanization is one of the means by which economic growth can be achieved in developing countries. Thus, even though the chaotic Urbanization in developing countries causes environmental problems in cities, it is true that these ecological problems can be tackled by increasing the pace of urbanization. Experiences from urban centers of Europe, the USA, Australia, Japan, South Korea, and Singapore indicate that cities will become environmentally friendly in the long run. It is because urbanization results in economic growth which in sine qua non for investment in environmental-friendly technologies and funding of human resources.

Indian cities face similar problems which are experienced by the cities of other developing countries. According to the World Bank (2011b), 'Urbanization in India has expanded rapidly as increasing numbers of people migrate to towns and cities in search of economic opportunity. Slums now account for 1/4 of all urban housing.' In some Indian cities, more than half of their population lives in slums and squatters. Most cities have poor infrastructure and their environmental health (congestion, air and sound pollution, pollution and depletion of surface and groundwater, etc.) is deteriorating, resulting in diminishing health and productivity of urban dwellers and declining quality of life. It is in this context that the need for sustainable development in Indian cities arises.

This chapter focuses on the sustainable development of Asansol, the second largest city in West Bengal. Though Asansol initially emerged as a mining and transport town in the 1860s, it has witnessed a phenomenal rate of growth and expansion since the independence of the country. According to the Census of India (2011), the total population of the city has grown nearly seven times in the past 60 years (from 76,277 in 1951 to 56,3917 in 2011). During the same period, the areal extent of the city expanded over 20 times, from 6.43 sq. km. in 1951 to 127.874 sq. km. in 2011 (Census of India 2011). The city has undergone changes in its economic functional characteristics–from mainly a mining and transport town to an economically complex city–with profound implications on its land use and environmental sustainability. Its emerging land use pattern is characterized by the expansion of built-up area and consequent shrinkage of the area under vegetation, agriculture, and water bodies. Further, due to the development of the industrial footprint, the problems of air and water pollution have become major concerns in the city. The discharge of untreated effluents by industries directly or indirectly into the main rivers has caused water pollution. Further, in the absence of soak pits in the city, domestic sewage flows through open drains, resulting in unhygienic and unclean surroundings. Against this backdrop, the chapter advocates for adopting the "Eco2 City" approach by meaningfully integrating the major economic and ecological concerns in the planning, development, and management of Asansol City.

## 7.2 Concept of Eco2 City

The concept of sustainable cities needs to take into consideration environmental, economic, and social parameters. However, many scholars have approached the sustainable development of the city by considering either environmental, morphological, or social parameters in a non-integrated manner. For example, Kennedy et al. (2007) consider that sustainable cities are the only ones in which the flow of energy and material resources and the disposal of wastes do not exceed the environmental capacity. Jabareen (2006) has examined sustainable cities in the context of urban forms. He considered four types of sustainable urban forms: the neotraditional development, the urban containment, the compact city, and the eco-city. As mentioned above, investment in clean technologies and human resources can improve urban environmental quality. Therefore, economic prosperity is essential for adopting environmentally sustainable practices. In social sciences, sustainable cities are often described in terms of the goal of 'social sustainability' when social equity or justice is evident in a spatial setting (Ruzek 2015). Thus, a sustainable city may be defined as an economically prosperous city with a good quality environment, social equity, and justice.

The Eco2 city concept is the most recent approach to sustainable urban development. However, the concept of eco-cities is not new. The Industrial Revolution caused rapid urbanization in Europe. Since the Industrial Revolution was based on the logic of capitalism, i.e., profit motif, it had no regard for protecting environmental quality.

Urban settlements during the Industrial Revolution were highly polluted and poorly managed, leading to frequent outbreaks of diseases. Therefore, during this period, many people felt the need to live in harmony with nature. However, during the late 1980s, the discourse on sustainable development gave rise to the eco-city concept. The concept became a preferred response to sustainable development challenges. An architect, Richard Register, is widely credited for coining the term 'eco-city' in 1987, who defined it as 'an urban environmental system where inputs (of resources) and outputs (of waste) are minimized' (Register 2002; Bibri 2020).

Eco2 Cities is an approach to sustainable urban development. It aims at the development of cities in ecologically and economically sustainable and socially inclusive ways. According to Cities Alliance (2023), 'Eco2 Cities enhance their citizens' well-being through integrated urban planning and management that harnesses the benefits of ecological systems while protecting them for future generations. They also enhance their resource efficiency while reducing pollution and waste. This improves their economic competitiveness and resilience, strengthens their fiscal capacity, and provides significant benefits to the poor. 'Eco2 Cities are defined as cities that create economic opportunities for their citizens in an inclusive, sustainable, and resource-efficient way, while also protecting and nurturing the local ecology and global public goods, such as the environment, for future generations (World Bank 2011a). Thus, the three pillars of sustainable development of a city are the environmental quality of the city, economic prosperity of people, and social equity between people (Fig. 7.1). These elements can be thought of as an integrated whole. They reinforce each other in complex ways.

The concept of Eco2 Cities is based on certain principles. According to Hiroaki Suzuki et al. (2010), the Eco2 framework has been structured around the following four fundamental principles:

- **A city-based approach**. It enables local governments to lead a process that takes into account their specific circumstances;

**Fig. 7.1** Eco2 city approach to sustainable development

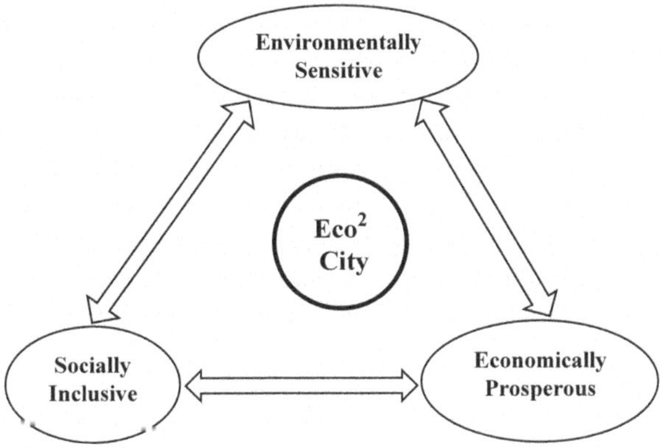

- **An expanded platform for collaborative design and decision-making**. It accomplishes sustained synergy by coordinating and aligning the actions of key stakeholders;
- **A one-system approach**. It enables cities to realize the benefits of integration by planning and managing the whole urban system; and
- **An investment framework that values sustainability and resiliency**. It incorporates lifecycle cost–benefit analysis, the value of all capital assets (manufactured, natural, human, and social), and a broader scope of risk assessments for resilience.

Thus, Eco2 City may be considered the most recent approach to address the current problems of the cities around the world.

## 7.3 Materials and Method

### *7.3.1 The Study Area*

The Asansol City ($86^0 53' 11''$ E to $87^0 2' 40''$ E and $23^0 45' 00''$ N to $23^0 37' 56''$ N) is located in the lateritic land of the northern (left) bank of river Damodar in the heart of the triangle formed by joining the rivers Damodar, Barakar and Ajoy in the West Burdwan district of West Bengal. The city is located on the lower Chotanagpur Plateau, which occupies most of Jharkhand. While the Dhanbad district in Jharkhand lies on the western side, the Durgapur subdivision lies on the eastern side of the city. To the south, the Purulia and Bankura districts share a boundary with the Damodar River. The Birbhum district lies to the north. Asansol is the second most populous city in West Bengal after the Kolkata Metropolitan Area. It is spread over an area of 127.874 $km^2$. It has a population of 5,64,491 persons and a population density of 4431 persons/$km^2$ (Census of India 2011).

Asansol is traditionally a coal mining and industrial city as well as one of the busiest commercial centers in India. Its economy was primarily dependent on its steel, coal, and railways. However, in recent times, trade and commerce have become very important. The city has a well-developed transport network connected by roads and railways from different parts of the country. It lies on the central urban axis of the major iron and steel industry, coal mines, and commercial hub on Howrah-Asansol-Mughalsarai (now Deen Dayal Upadhya Nagar)-Delhi and Asansol-Adra-Kharagpur-Bhubaneswar-Chennai South-Eastern Railways. It is also situated on the crossing point of NH-19 (Grand Trunk Road) and NH-14 (National Highway) roads. The NH 19 has been broadened as part of the Golden Quadrilateral project.

### *7.3.2 Objectives*

This chapter broadly focuses on the sustainable development of Asansol, the second-most populous city of West Bengal. The first part discusses the concept of a sustainable city with particular reference to Eco2 City. It then examines the economic characteristics of the city and its ecological and societal challenges. The chapter advocates for adopting the "Eco2 City" approach for sustainable development of the city. In this regard, it argues for the sustainable development of a city by integrating its ecological and societal concerns into the background of the salient features of its economic structure.

### *7.3.3 Data Sources and Methodology*

The chapter is mainly based on secondary sources of data and published literature. Data for this chapter has been collected from the Census of India and the Municipal Corporation of Asansol. In particular, data on the distribution of industries in the city has been taken from Chambers of Commerce & Industries, Asansol, Raniganj & Durgapur, NSSO (FOD), Government of India. Data from the *Disaster Management Plan of Paschim Bardhaman District* (West Bengal Disaster Management and Civil Defence Department 2019) has been used to show the distribution of chemical and industrial hubs and inflammable industries in the city. A map showing the distribution of industries (Fig. 7.2) has been prepared by Arc GIS 10.1 software based on Google Earth Image, 2019, and a personal survey. In addition, the chapter uses concepts and theories from various articles and research papers. It applies the idea of Eco2 City, developed by the World Bank, to identify the key challenges in making Asansol a sustainable city. It uses the Eco2 City approach for the sustainable development of the city.

## 7.4 Results and Discussion

### *7.4.1 Economic Base of Asansol City*

Asansol evolved as an urban center in the second half of the nineteenth century. Initially, it developed as a coal mine township. Subsequently, a complex network of railways was developed to transport coal from this belt. Soon, Asansol became a vital transport township in eastern India. In the next phase, the city witnessed the development of industries and commercial activities. Thus, the economic base of Asansol City stood on three pillars—coal mining, railways, and iron steel. The Eastern Coalfields Company is one of the largest employers in the region. Railways significantly contributed to Asansol's economy and are credited with developing the

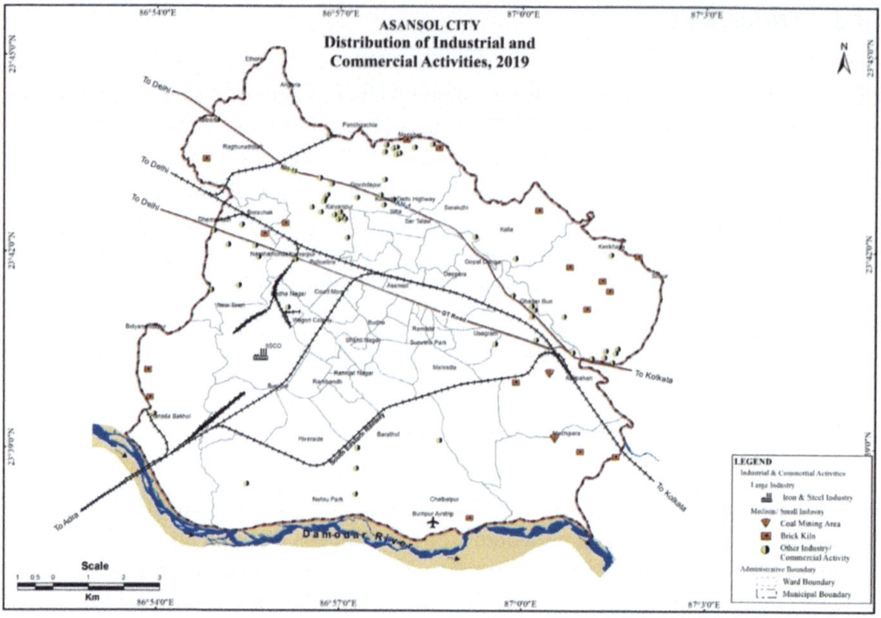


**Fig. 7.2** Distribution of industrial and commercial activities of Asansol city, 2019. *Source* Prepared by authors using arc GIS 10.1 software based on google earth image, 2019 and Personal Survey

city in the late nineteenth century (Chatterjee 2016). Since Asansol is located in the coal belt, many industries developed in the city and its surroundings and still play a major role in its economy. Due to large coal deposits in the city and the deposits of other minerals in the surrounding areas, and the development of a dense network of railways in the region and beyond, the iron and steel industry flourished in the past. However, at present, the industrial base of Asansol City is very much diversified. It has several large, medium, and small-scale industries.

#### 7.4.1.1 Large-Scale Industries

The leading industries in Asansol City are iron and steel manufacturing, railway wagon manufacturing, cement industry, and hardware machinery. The city and its surroundings have a large number of large industrial plants such as the IISCO Steel plant, Chittaranjan Locomotive Works, Hindustan Cables Limited, Disergarh Power Supply, DVC, Burn Standard, Reckitt and Coleman, some cement plants such as Burnpur Cement, Asansol Cement, Lafarge Cement India Pvt. Ltd., I-core Cement and refractories units such as Kabita Refractories (Table 7.1). Iron and steel industries are the most important heavy metallurgical industry in the city. Coal, iron ore, dolomite, and ferroalloys are among the raw materials available in this region. Iron ore and coal are the two most important bulky raw materials for the iron and steel industry. It is not necessary to bear the transport costs for at least one raw material in this industry, namely, coal. The adjacent state of Jharkhand supplies the remaining raw materials. Water for the Burnpur steel plant is drawn from the Damodar River.

**Asansol Steel Castings Pvt. Ltd.**, a limited liability company established in the year 1984, is engaged in manufacturing and supplying a wide range of industrial castings. The company manufactures diverse products such as MN-Steel Casting, Plain Carbon Steel Casting, High-Alloy Steel Casting, Special Steel Castings, and Stainless-Steel Casting. A production unit is also operated by a multinational company named Reckitt & Benckiser in the city. A foreign liquor company, the McDowell, has a production plant on the city's western side. Asansol City has evolved as a hub of chemical and inflammable industries (Table 7.2).

**Table 7.1** List of large-scale industries/public sector undertakings in Asansol and its surroundings

| Sl. No | Name of the unit | Products |
|---|---|---|
| 1 | IISCO steel plant | Pig iron, Z-type sheet piling, wire rod mill, bar mill, spring steel, low-alloy steel, etc |
| 2 | Burn standard (earlier Indian standard wagon) | Railway wagons: tanker, hopper, Casnub bogies, steel castings, wagon components, etc |
| 3 | Burnpur cement ltd | Three types of cement-OPC, PPC, and PSC |
| 4 | Legacy co-operative services Pvt. Ltd | Residential and commercial builders, interiors, and smart home construction |
| 8 | Kami machinery | Machinery manufacturing and hardware dealers |
| 9 | Vidhi international | Supplier of machinery-tiles machinery, grinding machinery, flour mill plant, bricks machinery, etc |
| 10 | Tungnath hydraulic corporation | Filters for air filters, hydraulic cylinders, loop filters, trade and supply of hydraulic |
| 11 | Desai trading co | Medical equipment wholesalers, pharmaceutical medicine wholesalers, and oxygen gas distributors |
| 12 | A to Z engineering works | Trader, manufacturer, and supplier of railway parts, oil-filled equipment, elevators, tie plates, screw spikes, etc |
| 13 | Radha SoamiJcb earthmovers spare parts | Manufacturer and supplier of JCB brakes, chains, gears, hooks, resistors, rings, etc |
| 14 | Royal refractories Pvt. Ltd | Manufacturer of fire bricks, grinding services, and motor clay |
| 15 | Technomech enterprise | LED light dealers, fan, switch, cable, geyser repair and dealer |
| 16 | Asansol steel casting Ltd | MN-steel casting, plain carbon steel casting, high alloy steel casting, etc |
| 17 | Sahara super cement | Cement |

*Source* MSME-Development Institute, (2018)

**Table 7.2** List of chemical and inflammable industries

| Sl. No | Name of the factory | Location |
|---|---|---|
| 1 | Ajudhia distillery (A unit of NIC Ltd.) | M-2, A.D.D.A. Industrial corporation |
| 2 | Anupam detergent & chemicals Pvt. Ltd. Industrial estate | B-30, ADDA industrial estate, Kanyapur |
| 3 | Asansol bottling & packaging Co. Pvt. Ltd | Barapukuria, Kalla |
| 4 | Asansol poly fabs Pvt. Ltd | G.T. Road, Gopalpur, Opposite-FCI Godown |
| 5 | Asansol pump station Indian oil corporation Ltd | P.O.-Kalla central hospital |
| 6 | Mcdowell & Co. Ltd. (Carew & Co. Ltd.) | G.T. Road, Narsamuda |
| 7 | Burnpur poly fabs Pvt. Ltd | Kamalpur, Mithani |

*Source* West Bengal Disaster Management and Civil Defence Department, (2019)

#### 7.4.1.2 Medium and Small-Scale Industries

There are many registered forest-based industrial units, including sawmills, furniture-making centers, and bidi-making centers (Table 7.3). Some of the sawmills are Laxmi Saw Mills and Timber Co. (Ward-26), Satyanaryan Saw Mill (Ward 6), Shree Ram Saw Mill (Upper Chelidanga, Ward-5), Jwalamukhi Saw Mill (Ward18), Jai HinglajSaw Mill (Ward-42), Jyoti Saw Mill in Ragania para (Ward 17), Jai Bhawani Saw Mill (Ward21). Kanyapur Industrial Hub, located in the northern part of Asansol, was developed for medium and small-scale industries. Additionally, the Asansol Durgapur Development Authority (ADDA) provides land for entrepreneurs to develop medium to small-scale industries. These industries are Flow Control Equipment Pvt. Ltd., Anupam Detergent and Chemicals Ltd., United Air Express, Renew Industries Pvt. Ltd., Medicott Industries, Deal Magsil, Lofty Paints and Chemical Works, Angel Bakery, Hazra Industries Interior Pvt. Ltd., Ashiana Cement, Paul Industries, Pharma Manufacturing, McNally Bharat Engineering Ltd., Megacity Iron and Steel Pvt. Ltd., Aspa Plast Tech Pvt. Ltd.

#### 7.4.1.3 Agro-Based Industries

Agro-based industries in Asansol City include the food processing industries, cold storage facilities, oil mills, and flour mills. There are two cold storages which mainly store potatoes, fruits and vegetables. These two cold storage are Durga Cold Storage in Benarsi Lal Kedia Road and Bedia Cold Storage in Domohani Railway Colony. Food processing in the city includes preparing bakery items, such as bread, biscuits, cakes, pastries, cookies, snacks, muffins, etc. Four food processing industries, namely, Angel Bakery, Arti Bakery, Morish Bakery, Payel Food Products Pvt. Ltd., and Shreeji Flour Mills Pvt. are located in Ward 31 at Kalyanpur. A flour mill, Ananda Flour Mill, is located at Gopalpur.

**Table 7.3** List of medium-scale enterprises

| Sl. No | Name of the unit | Products |
|---|---|---|
| 1 | Amba fly ash bricks | Fly ash bricks |
| 2 | Anish international | Cement, castable cement, grey cement, white cement |
| 3 | Sai brick field | Cement hollow bricks, fly ash bricks |
| 4 | Subha brick traders | Red bricks |
| 5 | AK ceramics | Tiles, marble, sanitaryware dealers |
| 6 | MPG alloy steel Pvt. Ltd | Manufacture of basic iron and Steel |
| 7 | Nanak refractories | Red bricks manufacture |
| 8 | Alfa wire product | Super enamelled copper wire |
| 9 | Royal refractories pvt | Ceramic and refractories |
| 10 | Arti bakery pvt. Ltd | Bakery products-cakes, pastries, cookies, snacks, muffins, biscuits, etc |
| 11 | Om roller flour mill | Manufacture of bajra atta, maida atta, and wheat atta |
| 12 | Ashiana cement Pvt. Ltd | Cement |
| 14 | ATC international Pvt. Ltd | Manufacture of cement asbestos products |
| 15 | Bhawani biscuit Co | Biscuits, cookies, health supplement biscuits |
| 16 | Binapani oil mill | Mustard oil |
| 17 | Hanuman cement Ltd | Manufacture, exporter, supplier and retailer of cement, quartz, potassium feldspar, magnesite, etc |
| 18 | Bedia cold storage Pvt. Ltd | Storage of foods at different temperatures |
| 19 | Ferro scrap nigam Ltd. Burnpur | Manufacturer, supplier of slag, metal slag, and raw metal ore |
| 20 | Kachhi patel saw mill | Ushagram, G.T. Road east |

*Source* MSME-Development Institute, (2018)

### *7.4.2 Commercial Activities*

The spatial pattern of Asansol City reveals that markets are generally intermixed with the residential areas. Commercial activities have recently developed rapidly along both sides of the GT Road. The city's core, the Central Business District (CBD), has a large number of retail businesses and commercial activities. The CBD has many outlets selling manufactured goods such as clothes, readymade garments, footwear, cosmetics, jewelry, mattress, hardware items, electrical and electronic items, decorative goods, plastic materials, and groceries. All three types of commercial spaces are observed in the city: wholesale markets, retail markets, and shopping complexes.

Wholesale and specialized commercial areas dealing with heavy and bulky articles are located in the central commercial zone, i.e. Munshi Bazar. Bulky and perishable goods markets, including vegetable markets, are located in Joynagar, Munshi Bazar. Fish markets are located at Munshi Bazar, Fish Patty, and Court More. Wholesale fruit markets are located in Ganja Gali, Tajpal Lane, and Burnpur Station Road. The electronics and automobile wholesale markets are located at ADDA market on Pathak Bari Road, Narayan Market near Raha Lane, Sabana Market at Ushagram and Pucca Bazar. Cloth wholesale markets are located at B.B. Road and Leighton Street while Grocery wholesale at Pathak Bari Road and Munshi Bazar. Medicine wholesale shops are located at Raha Lane. Most of these areas were developed in the 1960s; therefore, the buildings are old and dilapidated. Shops are arranged haphazardly along narrow and highly congested lanes.

The retail commercial area usually deals with local products and consumer goods from wholesale. With the population increase, commercial facilities have dispersed from a single core to multiple centers in the city. The retail markets are mainly concentrated in Durga Market at GrijaMore, Yadav Market at Pathak Bari, Kamal Market at Hutton Road, Gani Market at Leighton Street, Arghya Commercial Complex at Rabindra Nagar Road, and Anannya Shopping Complex at Sen Raleigh Road. Besides commercial areas, neighborhood shopping centers are also emerging in residential areas of the city. In general, these shops cater to the daily needs of the residents. Many of these shops are scattered throughout the city and constitute mixed-use land.

Shopping malls are the most recent form of doing business. A shopping mall is a large structure with various shops and an air-conditioned building with multiple floors (Arslan and Ergener 2023). It is regarded as a place for life and entertainment and is usually located in the city center. Nowadays, malls are not only places to shop, but to enjoy social, cultural, and entertainment activities as well (Nebati and Ekmekci 2019). There has been a profound change in the marketing system over the last two decades. In recent years, many supermarkets and hypermarkets have been established in the city (e.g., Galaxy Mall, Sentrum Mall, Spencer's Mall, Vishal Mega Mart, Reliance Jio, and V Mart). Adoption of Western shopping styles and behaviors is now quite common in Asansol City as well. *Galaxy Mall* was the first shopping mall that developed in Asansol City. It covers an area of 3 lakh square feet and is located on Burnpur Road, opposite Chitra Cinema Hall. It is one of the city's most visited shopping malls. Many major retailers, such as Big Bazar, Reebok, Turtle, Chique, Hoffman, John Players, Cantabil, Gini and Jony, Archies, Eylex, and Himalaya Opticals, are located in this mall. *Sentrum Mall* is located in Shristi Nagar, near Ramakrishna Mission. It is spread over 250,000 square feet of retail space. *Reliance Market* is a departmental store, located in the Kalyanpur Housing Complex. Other shopping malls are *Mohan Family Store*, *Bazar Kolkata*, and *Geeta Fashion*, which are located in the heart of the city, at Munshi Bazar Road.

### 7.4.3 Urbanization of Asansol City

Urbanization is a complex process and is influenced by many factors. Martin and Ottaviano's (1999) study shows that economic growth and geographic agglomeration as "mutually self-reinforcing processes". They suggest that spatial agglomeration of economic activities in a region drives its growth. As a sector expands, new companies tend to locate close to its origins. It is because of economic growth that a place attracts other economic activities there, which in turn, cause agglomeration. The development of industrial and commercial activities has led to economic development and spatial expansion of Asansol City. In 1951 Asansol was a mono-centric town then; the number of wards increased from 7 to 25 from 1951 to 1981. The economic changes also led to a rise in urban growth and expansion during the next decade. The area of the city increased. The number of wards also increased from 25 to 50 with the expansion of its area. Slowly, the city transformed from mono-centric to multicentric. Asansol City has dramatically experienced the expansion of built-up areas in the past 20 years due to economic growth and development. A large number of informal economic sectors have come up in different parts of the city. In particular, due to the rapid increase in population and expansion of urban areas, Asansol industrial belt has witnessed a rapid expansion of informal activities. The urbanization of Asansol has followed the classic urban development model of the Western countries, i.e., urbanization is proceeded by industrialization.

The urban growth in the Asansol and surrounding region is a combined impact of the emergence and growth of some new wards as well as the increase in the population of these towns. Auddya (2019) believes that the history of the Asansol region is a history of evolution from a forest-based subsidence economy to a mining and industrial economy, which in turn, transformed the socio-economic and cultural characteristics as well as the built environment of this region. Since the colonial period, a hierarchy of urban settlements has developed in the Asansol region based on coal, transport work, steel, and linkage industries.

A study of the urbanization trajectory of Asansol City reveals an interesting trend. The population of the city grew by 35.51% between 1891 and 1901 and 77.05% between 1901 and 1911 (Table 7.4). The phenomenal population growth during these decades was due to the development of the railway works at Asansol. The opening of the Bengal-Nagpur Railway and increasing activities in the coal and iron fields attracted large numbers of migrants to Asansol town from Manbhum, Bankura, and Bihar (Peterson 1910). Asansol has been continuously growing at a steady pace since then. Between 1891–1901, Asansol was a Class IV town. From 1901 to 1911, its population increased to 21,919 persons, resulting in a change in the status of Asansol from a Class IV to a Class III town, and this status continued for the next two decades. The 1921 and 1931 census figures show a decline in population growth rates (20.90 and 18.06%, respectively) due to the large-scale spread of the epidemic. Aquantumjump in Asansol City's population occurred in the 1930s. Between 1931 and 1941, the population increased by 78.34% in the inter-censual period. Its total population in 1941 was 55,797. As a result, the status of Asansol changed from a class III to a class II town in 1941 (Table 7.4).

Since 1941, the population has increased continuously and steadily until 1971 (Table 7.4). In 1961, its population crossed the one lakh mark. The city became a Class I town with a population of 103,405. However, its population growth rate declined to 17.57% in 1981. But again, the 2001 census indicates an abrupt jump with a percentage increase of 81.34%. The highest population growth was between 1991–2001. The city experienced a record population growth rate during this decade (81.34%) against 42.98% during the previous decade (1981–1991). As can be seen in Fig. 7.3, the city experienced a brisk pace of population growth (i.e., 47.64%) between 1991 and 2011. Thus, within 70 years (between 1951 and 1971), the population of Asansol grew sevenfold. In addition, to natural growth, migration from rural to urban areas and Urbanization were the main factors contributing to high growth. Asansol has experienced large-scale industrialization, which has resulted in large-scale migration of people from not only West Bengal but also from Uttar Pradesh and Bihar. The Indian Iron and Steel Company, Eastern Coal Field Ltd., the aluminium industry at J.K. Nagar, Alloy Steel Plant, Chaittaranjan Locomotive Works, and the copper industry at Hindusthan Cables attracted migrants to settle in the Kulti, Asansol, and Raniganj regions.

Due to rapid population growth (caused by the natural increase as well as in-migration) and the rapid development of economic activities in the city, the demand for built-up spaces for residences, shops, commercial activities, government offices, factories, hotels, restaurants, recreation centers, and playgrounds has increased with

**Table 7.4** Decadal growth rate of population and changes in the urban centre class of Asansol city, 1891–2011

| Year | Population | Decadal variation (Persons) | Decadal growth rate (%) | Class of urban centre of Asansol |
|---|---|---|---|---|
| 1891 | 11,000 | – | – | IV |
| 1901 | 14,906 | 3906 | 35.51 | IV |
| 1911 | 21,919 | 7013 | 47.05 | III |
| 1921 | 26,499 | 4580 | 20.90 | III |
| 1931 | 31,286 | 4787 | 18.06 | III |
| 1941 | 55,797 | 24,511 | 78.34 | II |
| 1951 | 76,277 | 20,480 | 36.70 | II |
| 1961 | 103,405 | 27,128 | 35.57 | I |
| 1971 | 155,968 | 52,563 | 50.83 | I |
| 1981 | 183,375 | 27,407 | 17.57 | I |
| 1991 | 262,188 | 78,813 | 42.98 | I |
| 2001 | 475,439 | 213,251 | 81.34 | I |
| 2011 | 563,917 | 88,478 | 18.61 | I |

*Source* Primary Census Abstract 1901–2011, Census of India (2011), New Delhi

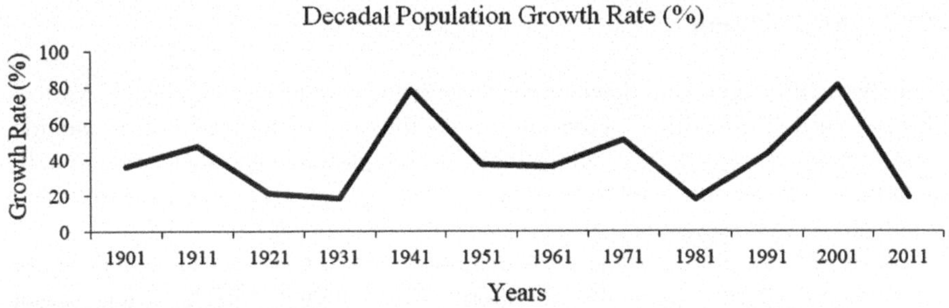


**Fig. 7.3** Decadal population growth rate of Asansol, 1901–2011. *Source* Primary Census Abstract 1901 to 2011 Burdwan district, West Bengal

time. This demand for the land is met by conversion from one form of land use to another. Accelerated economic growth for a region led to the urbanization of Asansol, which in turn led to a decline in the quality of the natural resource base of the city. Besides, the rapid growth of urban areas has adversely affected the city's landscape and its surroundings due to the conversion of forest and agricultural land into construction land and other urban uses.

## *7.4.4 Ecological Challenges of the City*

In developing countries, mining activities, industrialization, and urbanization are major threats to environmental challenges. Asansol owes its growth to mining, transportation, and industrial activities. Due to large-scale mining, the expansion of the transportation network, rapid industrialization, and urbanization, Asansol faces many environmental challenges. Some of these challenges are discussed below.

### 7.4.4.1 Water Pollution

Damodar is the major river in Asansol. Some other rivers are Nunia, Dihika, Damraand Gharui. Several *nallas* existed in the city and its surroundings besides these rivers. The primary sources of water pollution in Asansol City are industries and domestic. Industries discharge untreated effluents directly or indirectly into the main rivers. According to a study by the state government, the amount of water pollution from various processes of industries is about 40,000 KLD, and from municipal sewage arising out of the residential establishment is about 20,000 KLD (West Bengal Pollution Control Board 2015a). A large quantum of effluent generated in the industrial area at Kanyapur also reaches the river Damodar via the Damra Canal. Another study informs that industrial waste effluents are directly thrown into the Damodar River water without treatment (Singh et al. 2014).

#### 7.4.4.2 Air Pollution

Air quality in this region has deteriorated largely due to industrial expansion. Various industries in this industrial region emit air pollutants, such as carbon monoxide, sulfur dioxide, nitrogen oxide, chlorofluorocarbon, particulate matter, ammonia, fly ash particles, smoke, dust, and toxic materials causing serious problems surrounding the industrial environment.

According to the pollution analyzer consultant of WBPCB (West Bengal Pollution Control Board 2015b), air quality monitoring was conducted in the winter season covering the period of November 2010 to March 2015 at the Asansol-Durgapur industrial area, the concentration of Suspended Particulate Matter (SPM) in the area was 1127.69 $\mu g/m^3$. This value far exceeded the industrial and residential standard of SPM (360 $\mu g/m^3$) set by the National Ambient Air Quality Standard, 2009 (CPCB 2010). The locality surrounding the Indian Iron and Steel Company steel plant (i.e., Burnpur) is a critically polluted area of the city. Increasing vehicular traffic and vehicle numbers have exacerbated air and noise pollution. Another study states that, in an industrial city like this, heavy trucks contribute considerably to vehicular pollution. Several commercial and hiring vehicles may use contaminated fuel, which increases air pollution (Chatterjee 2017).

#### 7.4.4.3 Fly Ash

Coal-fired thermal power stations generate much fly ash as a result of their combustion. Fly ash is so fine and light that it remains airborne for a long time, contributing to air pollution. Steel plants and heavy metal industries handle large quantities of raw materials. A considerable amount of dust is generated when these materials are transported, loaded, reclaimed, sorted, and crushed. Furthermore, these industrial units emit sulfur, nitrogen, and carbon oxides, along with yellow–brown smoke and fumes. A dense, smoky environment prevails around the Asansol-Durgapur Industrial Complex (Chatterjee 2017).

#### 7.4.4.4 Shrinkage of Vegetation Cover

The conversion of forests for other purposes is a serious ecological issue in Asansol City. Deforestation in Asansol is strongly correlated with rapid industrialization and Urbanization. In the previous few decades, the area around Asansol has seen a rise in industrialization and Urbanization. Decadal satellite remote sensing data showed a three percent drop in vegetation-covered areas. In 2001, the forest cover area was 12% of the total area, but in 2019, the forest cover decreased to 9.4%.

On the other hand, built-up areas, open-cast coal mining, and industrial zones had fast growth. For example, in ward number 31, Shristi Nagar, an elite-class residential cum commercial area, was developed by clearing a 26-hectare vegetation-covered area. As a result of the rapid increase in built-up area and loss of green space, the surface temperature in Asansol City has risen.

#### 7.4.4.5 Shrinkage of Water Bodies

Water bodies and wetlands are a significant part of the urban ecosystem. Currently, cities face the challenge of rapid unplanned urbanization, which makes their role even more critical. There are many examples of shrinkage of water bodies. The continued disposal of waste materials from the IISCO plant in a nearby pond has converted the pond into a regular dumping site. In Ward-32, there were two small ponds in 2001, but now those ponds have been replaced by the Reliance Market Shopping Mall.

#### 7.4.4.6 Land Degradation

Rapid and unplanned expansion of cities has resulted in land degradation. The city's growth has led to a massive demand for construction materials, mainly bricks. The brick kilns have caused severe damage to agricultural land with their soil digging and quarrying activities in the city's peripheral area. The agricultural land has been converted into a wasteland or a ditch when the brick kilns are abandoned. These are the leading causes of land degradation in the city.

#### 7.4.4.7 Shrinkage of Agricultural Land

Asansol Sub-division has a relatively small percentage of its land used for farming (36%) purposes. The proportion of agricultural land is higher in periphery areas than in central regions of the city. In certain places, it isn't easy to grow crops in specific places since the coal seam is so near the surface. Asansol City and its surroundings have radically changed over the past few years due to the development of new industries, expansion of built-up areas, and shift in the occupational structure of people. The area under agriculture has declined from 61.73 sq. km (i.e., 48.27% of the total area) in 2001 to 37.88 sq km (i.e., 29.62% of the total area) in 2019.

#### 7.4.4.8 Increasing Slum Pockets

With industrialization, slums have mushroomed in Asansol City and the share of people living in its slums have increased over time. As per the 2011 Census, 31% of people live in slums in the statutory towns of West Bengal. According to the 2001 Census, 1,58,946 people lived in slums in Asansol City. It comes to around 33.43% of the city's overall population. In the Census 2011, the share of people living in slums to the total population of Asansol Municipal Corporation increased to 36.86% (ADDA 2023). In many of these areas, slums and informal settler communities are expanding at an alarming rate. There are 214 slums spread across 14 wards in the Asansol City area. In addition to their deteriorating environmental conditions, most of these slums are devoid of any public lavatories or recreational opportunities. Slums are pockets of agglomerations of urban poor and the outward symptoms of urban poverty.

#### 7.4.4.9 Solid Wastes

A suitable solid waste management system for the AMC region is the only way to reduce environmental damage. Proper site selection is required for the disposal of municipal solid waste, so as not to affect the environment in any way. Almost the entire city of Asansol lacks a proper system for handling garbage. In 14 wards, door-to-door garbage collection has started. The Asansol Municipal Corporation has planned to start an appropriate Solid Waste Management (SWM) System in this ward on an experiment basis, but the separation of solid wastes has not started yet. The people of the remaining wards throw garbage in bins or at open places according to their convenience. The municipal waste is finally disposed of at two dumping grounds, one beside NH2 and the other near IISCO. The environment is degrading under this arrangement because of land, water, and air pollution.

For the disposal of Municipal Solid Waste (MSW), a suitable area is needed. A site at Bansra, about 18 km from the Asansol City boundary, has been identified for this purpose. Presently, the Asansol Municipal Corporation is preparing to implement a scheme based on DPR (Detailed Project Report) and PPP (Public Private Partnership).

### *7.4.5 Eco2 City Approach to Sustainable Development*

It is apparent from the above discussion that the growth and expansion of Asansol City since its origin has been possible due to economic growth. In particular, the role of mining, transportation, and industries in the development of the city has been crucial. In recent years, the role of markets and commercial activities in the development of the city has become quite profound. It is also true that many of the ecological problems of the city are due to these economic activities. Moreover, the fruits of economic development have not reached equally to all communities of the city. But, it would not be prudent to take drastic measures for ecological preservation and socioeconomic equity of the city, which could hinder its economic growth. Instead, the city's economic growth should be considered an integral part of ecological and socioeconomic sustainability pursuits.

A significant part of the city population does not have access to adequate sanitation, so water-borne diseases are prevalent. Therefore, it becomes imperative to preserve its water bodies. To ensure a better quality of life, a holistic and integrated approach to managing and restoring these bodies is proposed by the Asansol City Development Plan (2006).

- Sewage should not be diverted to water bodies.
- Dragline machines can be used to desilt and deepen water bodies.
- It is not acceptable to dump garbage on water bodies.
- Aquatic plants and fish can be introduced to prevent mosquito breeding to improve water quality.
- Water harvesting involves collecting and storing rainwater on the surface or in sub-surface aquifers before it disappears as surface runoff.
- Discharge of effluents into rivers after treatment to the appropriate standards.

The proportion of people living in slums in the city is very high and is growing. People living in slums contribute significantly to the city's economy, but they face abject poverty and lack access to even the most essential amenities and resources. Dilapidated environmental conditions and the absence of leisure facilities characterize slums in Asansol. Asansol City Development Plan (2006) has proposed the following strategy for the slums of the city.

- Improvements in sanitation and living conditions in existing slums and squatter communities through the provision of low-cost dwellings and other physical infrastructure;
- Provision of health care services, such as those for families, children, and those in need of immunizations;
- Construction of new recreational and community halls;
- Provision of elementary and secondary schooling for children;
- Protecting and enhancing the existing housing supply in economically disadvantaged areas;

Transportation is critical to the sustainability of a city. Improvements to the city's current traffic and transportation infrastructure are needed to produce changes envisioned in Asansol's Vision 2025 (ADDA 2015). Asansol City Development Plan (2006) emphasizes the need to improve connectivity within the city, both east and west and north and south. The regional road network is undergoing expansion and improvement. A new bypass alignment has been proposed to ease the adverse outcomes of ribbon growth. Asansol City's future will depend on the progress of urban transportation infrastructure. Asansol City Development Plan (2006) suggests that road network development measures included an improved road surface quality, an expansion of chosen linkages, a geometrical enhancement of crossings, the addition of parking facilities, and the adoption of an efficient traffic circulation and management plan. The following measures have been proposed by the Asansol City Development Plan (2006) to improve city transportation problems.

- Develop bypass facilities;
- Develop regional bus terminals along with NH-2;
- Relocation of Regional bus terminal facilities from the GT Road to NH-2;
- Intra-city bus/minibus facilities.

The infrastructure development plan of Asansol City is also necessary for its sustainable economy and ecology. The Asansol City Development Plan (2006) has made the following suggestions for improving the general infrastructure of the city.

- Provision of effective drainage systems for all major commercial and residential areas in Asansol Municipal Corporation, as well as low-lying areas;
- Water connection at the household level in every house in Asansol City with an adequate water supply throughout the year;
- Maintaining an adequate sewerage system and sewage treatment plant to minimize the pollution of natural receiving water bodies;
- Fitting of streetlights along all major roads and public institutions to ensure safety and security;
- Extending of road network with roads of suitable surfaces and widths while maintaining the existing roads.

As part of the proposed Land Use and Development Control Plan, the Asansol City Development Plan (2006) has prepared a comprehensive and flexible framework for future land use.

The cultural and natural heritage add value to the richness of the city. Preserving the uniqueness of Asansol by developing its natural environment and cultural tradition would be crucial. In this regard, a vertical garden of 1250 square feet and a fountain were inaugurated in 2019. They have become a landmark of the city. The Durand Institute (now Vivekananda Institute, established in 1878), Eastern Railway Boy's High School, Durand Mixed Primary School, and Subhas Institute have been renovated.

In this regard, the Land Use and Development Control Plan-2025 for the Asansol Sub-division is an ambitious and forward-moving document (ADDA 2015). It has been rationally prepared to consider the city's strengths and weaknesses. The document outlines the key opportunities and threats in the city's development process. Some of the key features of this document are as follows.

- It plans to develop Asansol as a futuristic city.
- It addresses the environmental concerns of the city.
- A city is a part of the region. The document proposes urban–rural linkages and city-satellite town linkages in the planning.
- The document does not consider slums as a hindrance to the city's development. It calls for improvement in the quality of life of people in the slums.
- The document emphasizes intra-urban linkages and proposes north–south links.
- The document is mindful of the ribbon-like development of the city, which can cause problems in the future. Hence, it suggests the construction of link roads and the development of settlements along those roads.
- The document emphasizes issues of public spaces and proposes to create more public spaces in the city.

However, a critical reading of the Master Plan 2025 reveals many flaws from the point of view of the idea of the city and measures suggested to improve the city (ADDA 2015). Some of these are given below.

- The Master Plan over-emphasizes the physicality of the city and ignores its social and geographical aspects.
- The importance of digital technology has not been adequately emphasized.
- Many castes and communities, such as Scheduled Castes, Scheduled Tribes, and Minorities, inhabit the city. These groups are socially and spatially marginalized in the city. The report is silent on the issue of their socio-spatial marginalization.
- The Master Plan is silent on the issue of inter-ward disparities in the availability of facilities and amenities.
- Many ponds and water bodies in the city are vanishing. Interestingly, many new water bodies have been created due to soil digging for the brick kiln industry. How do they control brick-kiln industries or minimize their impact on local geomorphology and atmosphere? The Master Plan is largely silent on these issues.
- The booming brick kiln industries in the peripheral wards of the city are causing many problems related to land use and environmental degradation. Still, the Master Plan does not provide suggestions to improve this situation.
- Illegal coal mining is taking place in some of the peripheral wards of the city. The Master Plan does not spell out the strategies to tackle this problem.
- The Master Plan is silent on stormwater that causes floods in many low-lying city localities, such as Dildar Nagar, Lower Chelidanga, Gopalnagar, Mahua Dangal, Shanti Nagar, and West Apcar Garden.
- The city's green cover has declined over a period of time, about which the Master Plan does not provide a concrete plan.
- The Master Plan neglects the declining agricultural land of the city and the rejuvenation of urban agriculture.

It is important to note that implementing the Master Plan on the ground level is relatively poor. One of the reasons could be that while the plan is very lofty, the city administration does not have the resources to implement it. The Master Plan will be completed in 2025, but most of its recommendations are yet to be implemented.

## 7.5 The Way Forward

In the backdrop of the above discussion, the following suggestions are proposed for the development of Asansol City on a sustainable basis.

- Construction of a ring road through the open land in the southern part of the city;
- Construction of a wide road linking the northern and southern areas of the city with bridges over GT Road, AH-1, and the railway track;
- Construction of Rail Over Bridge in two places-at Manoj and Court Road–to reduce congestion and improve accessibility;
- Construction of rural link roads with GT Road and AH 1 to facilitate urban–rural linkages;
- Shifting of the bus depot to the northwest part of the city;

- Afforestation on barren land along the banks of the Damodar River and the railway track;
- Making landfilling mandatory for existing brick kiln areas;
- Shifting of industries from the core and congested areas of the city;
- Imposing restrictions on new starting brick kilns and coal mining within the Asansol City boundary;
- Expansion and development of the city without disturbing the existing stormwater drains, and streams;
- Renovation of existing ponds and water bodies;
- Shifting of dumping yards beyond the boundary of the Asansol Municipal Corporation;
- Development of another industrial hub in wards 31 and 50.

## 7.6 Conclusions

The present-day Asansol City is a result of different phases of its evolution ever since it emerged as a mining town in the late nineteenth century. The domination of specific types of economy characterizes each phase. In the present times, Asansol City has evolved into a complex and dynamic space under the impact of varied forces. The economic base of the city is mining, transportation, industries, and businesses. These economic activities have influenced the development of Asansol City in varying ways and degrees. At present, Asansol City is the second largest city in West Bengal. It has expanded considerably since independence. Its economy and society have diversified. Economic development has increased urbanization and resulted in urban sprawl. However, the city is facing many ecological and social challenges. Most of these challenges are a result of activities related to its growth and expansion (e.g., population growth, rising consumerism, industrialization, expansion of the built environment and depletion and degradation of the environmental resource, waste generation, pollution, inequality, etc.). To meet these challenges, there is a need for more investment in cleaner technologies so that economic activities become ecologically more sustainable. Investment is also needed in the field of waste management, slum upgradation, and provision of basic amenities and facilities in deprived areas. Undoubtedly Asansol City has evolved into an important economic centre in eastern India; however, its ecological and social challenges can be tackled through its further economic development. In other words, a vibrant economy is vital for tackling urban environmental challenges and achieving a sustainable city goal. As urban centers are engines of growth, further economic development of Asansol City will be crucial for achieving its sustainable development goals.

## References


ADDA [Asansol Durgapur Development Authority] (2006) City development plan: Asansol urban area: Asansol-Durgapur-Raniganj-Jamuria-Kulti, Asansol. https://dokumen.tips/download/link/cdp-asansol.html. Accessed 30 July 2023

ADDA [Asansol Durgapur Development Authority] (2015) Land use and development control plan-2025 for Asansol sub-division. http://addaonline.in/wp-content/uploads/2017/07/Report-LUDCP-for-Asansol-Sub-division1.pdf. Accessed 5 Jan 2023

Cities Alliance (2023) Eco2 cities: a model for sustainable urban development. https://www.citiesalliance.org/resources/publications/project-case-studies/eco2-cities-model-sustainable-urban-development Accessed 7 Jul 2023

Arslan HD, Ergener H (2023) Comparative analysis of shopping malls with different plans by using space syntax method. Ain Shams Eng J 14(9):1–10. https://doi.org/10.1016/j.asej.2022.102063

Asansol-Durgapur Development Authority (2023) Land use and development control plan for Durgapur (Andal, Pandabeswar, Faridpur-Durgapur and Kanksa blocks and Durgapur municipal corporation). https://www.wburbanservices.gov.in/upload_file/circular/Draft_LUDCP_ADDA_PC_170.pdf. Accessed 28 July 2023

Auddya R (2019) Urbanization in Asansol: an impact assessment. Int J Adv Sci Res Manag 4(5):223–227

Bibri SE (2020) The eco-city and its core environmental dimension of sustainability: green energy technologies and their integration with data-driven smart solutions. Energy Inform 3(4):1–26. https://doi.org/10.1186/s42162-020-00107-7

Census of India (2011) District census handbook: Barddhaman, village and town wise. Directorate of Census Operations, West Bengal, Kolkata

Chatterjee U (2017) Environmental pollution in Durgapur-Asansol industrial complex: issues and concerns. Int J Acad Res Dev 2(4):378–381

Chatterjee C (2016) Transition of informal sector: a case study of Durgapur-Asansol region, West Bengal. Unpublished Thesis, Department of Economics, The University of Burdwan, West Bengal

CPCB [Central Pollution Control Board] (2010). National ambient air quality standard (2009) Ministry of Environment and Forest, Government of India, New Delhi

Jabareen YR (2006) Sustainable urban forms: their typologies, models and concepts. J Plan Educ Res 26(1):38–52

Kennedy C, Cuddihy J, Engel-Yan J (2007) The changing metabolism of cities. J Ind Ecol 11(2):43–59. https://doi.org/10.1162/jie.2007.1107

Martin P, Ottaviano GIP (1999) Growing locations: industry location in a model of endogenous growth. Eur Econ Rev 43(2):281–302. https://doi.org/10.1016/S0014-2921(98)00031-2

MSME-Development Institute (2018) District industrial profile (2017–2018): Paschim Bardhaman. Ministry of Micro, Small and Medium Enterprises Government of India. www.msmedikolkata.gov.in/uploads/2021/03/districtprofiles/2017-18/PASCHIM%20BARDHAMAN. Accessed 26 Dec 2022

Nebati EE, Ekmekci I (2019) A proposal of novel performance criteria development for shopping malls. J Polytech 22(2):495–507. https://doi.org/10.2339/politeknik.470617

OECD (2006) Competitive cities in a global economy. OECD Territorial Reviews, Paris

Peterson JCK (1910) Bengal district gazetteers, Burdwan. The Bengal Secretariat Book Depot, Kolkata

Register R (2002) Eco-cities: rebuilding cities in balance with nature. Berkeley Hills Books, Berkeley

Ruzek W (2015) The informal economy as a catalyst for sustainability. Sustain 7(1):23–34. https://doi.org/10.3390/su7010023

Singh A, Deo B, Singh SP (2014) Risk analysis on the use of Damodar river water for drinking purposes. Int J Curr Eng Technol 4(1):405–410. https://inpressco.com/wp-content/uploads/2014/03/Paper72405-410.pdf

Suzuki H, Dastur A, Moffatt S, Yakubi N, Maruyama H (2010) Eco2 cities: ecological cities as economic cities. The World Bank, Washington, DC

Tripathi S (2019) Do economic reforms promote urbanization in India? Asia-Pac J Reg Sci 3(7):647–674. https://doi.org/10.1007/s41685-019-00117-8

Turok I, McGranahan G (2013) Urbanization and economic growth: the arguments and evidence for Africa and Asia. Int Inst Environ Dev (IIED 25(2):465–482. https://doi.org/10.1177/0956247813490908

UNEP (2018) Sustainable cities. https://www.unep.org/regions/asia-and-pacific/regional-initiatives/supporting-resource-efficiency/sustainable-cities. Accessed 7 Jul 2023

West Bengal Disaster Management and Civil Defence Department (2019) Disaster Management Plan of Paschim Barddhaman District, 2019–20. http://wbdmd.gov.in/writereaddata/uploaded/DP/DPPaschim%20Bardhaman20500. Accessed 28 July 2023

West Bengal Pollution Control Board (2015a) Environmental pollution abatement action plans for the industrial clusters in West Bengal: action plan for Asansol. https://cpcb.nic.in/displaypdf.php?id=Q0VQSS1Bc2Fuc29sX1JlcG9ydC5wZGY. Accessed 16 Feb 2023

West Bengal Pollution Control Board (2015b) Implementation of the action plan for critically polluted area, Asansol. https://cpcb.nic.in/displaypdf.php?id=QXNhbnNvbF9BdWdfMjAxNS5wZGY. Accessed 30 Jul 2023

World Bank (2011b) Indian Urban Challenges. https://www.worldbank.org/en/news/feature/2011/07/04/indias-urban-challenges. Accessed 7 Jul 2023

World Bank (2011a) Eco2 cities: a guide for developing ecologically sustainable and economically viable cities. https://www.worldbank.org/en/news/press-release/2011/11/08/eco2-cities-guide-for-developing-ecologically-sustainable-and-economically-viable-cities. Accessed 7 Jul 2023

# Chapter 8
# Towards Urban Sustainability: Assessing Performance, Initiatives and Spatial Contrasts of Smart Cities of Punjab and Uttar Pradesh

**Kapil Kumar Gavsker and Kunal Jha**

**Abstract** Urbanization has been shaping economies, standards of living and the environment in the global south. With a rapid pace of urbanization, there is increasing demand for better access to basic amenities, housing, quality and resilient infrastructure and energy. At the same time, the adverse effects on the environment, resources, and urban commons are also visible. Addressing such concerns requires measures of sustainable development at national, sub-national/regional scales particularly in urban planning and urban design. To make urban places liveable, resilient and sustainable, smart solutions, technology-based applications, and advanced effective digital spaces are being emphasized in urban planning. The Smart Cities Mission covering one hundred cities, launched in the country in the middle of the past decade offers immense promises and opportunities in making cities smart, integrated and sustainable. However, whether smart cities would lead to achieving the agenda of urban sustainability or smart cities remains a partially addressed query. Under the Mission, there are three cities in Punjab and ten cities in Uttar Pradesh states of India. Based on secondary data sources and relevant information, the present study analyses urbanization and urban growth patterns in the selected cities of Punjab and Uttar Pradesh; elucidates components, elements and strategies of smart cities planning; examines performance and unconventional initiatives introduced in making them as smart urban spaces; and study explores the significance of smart cities in urban sustainability and inherent challenges.

**Keywords** Liveable · Resilient · Smart cities · Urbanization · Urban sustainability

---

K. K. Gavsker (✉) · K. Jha
Department of Geography, Institute of Science, Banaras Hindu University, Varanasi, India
e-mail: kkgavsker@bhu.ac.in



S. Anand et al. (eds.), *Emerging Geosustainability Transformations in India*, Advances in Geographical and Environmental Sciences,
https://doi.org/10.1007/978-981-97-7993-2_8

## 8.1 Introduction

The contemporary urbanization process and the rising population concentration in the urban centres have posed unprecedented challenges-growth, development, basic amenities, infrastructure, planning and good governance towards the realization of sustainable urban habitat. The United Nations (2014) stated that "cities drive economic and social development as hubs of commerce, transportation, communication and government". The world is highly heterogeneous in that of the urbanization process and the growth of large cities and the quality of urban life. Different regions are urbanizing at different paces and rates in the world. Mathur et al. (2021) noted that "urbanization in India as in several other countries is a complex phenomenon; moreover, it is taking place under conditions of low per capita income and substantial governance and infrastructure and service deficits and is, therefore, far more complex".

The countries with higher levels of urbanization are home to a small proportion of the urban population to that of the world. In fact, a huge urban population is now living in the countries of the developing world and emerging economies. Xie et al. (2019) argued that "this rapid urbanization is being accompanied by the agglomeration of the urban population, the utilization of urban land, and severe industrial emissions leading to high ambient air pollution". In addressing these challenges and rising concerns, the idea of a smart city has become an approach in recent urban planning and a change in policy-making to make cities better than what they are- more sustainable and smart. In fact, "the concept of what constitutes a smart city is vague, and ranges from a narrow definition of technology to one that considers the quality of life more broadly across a range of areas, such as education, the environment, safety and governance" (The Economist Intelligence Unit 2016).

The idea of smart cities has come to light as a possible means of resolving multiple issues through the use of technology and data-driven strategies. The OECD (2020) elaborated that "the smart city" concept initially referred to initiatives that use digital and ICT-based innovation to improve the efficiency of urban services and generate new economic opportunities in cities". The concept of a smart city has evolved over time and crucial aspects have been identified and its scope widened. However, the literature on smart cities is growing and studies deal with multiple dimensions and elements as are highlighted in the definitions and research on such a city. Toli and Murtagh (2020) noted that "their diversity ranges from what elements a city needs to encompass to be deemed as smart, to what resources it needs to employ, what characteristics it needs to present and what are the smart city's goals, purpose and scope". Since its inception somewhere in the nineties, the concept of a smart city has been in practice in different ways, in different forms and in different regions and countries. So, there has been a transition in the way a *smart city* is imagined, designed and practiced in the world. The launch of the 11th UN Sustainable Development Goal (SDG) *making cities inclusive, safe, resilient and sustainable* has contributed towards an upsurge in its utility in urban planning and development.

In countries of Asia, "urbanization is fuelling economic growth, but the breakneck pace has left many cities struggling to provide adequate housing, infrastructure, and services to meet the needs of a surging population" (McKinsey Global Institute 2018). In South Asia, cities have been subject to economic, social and spatial transformation and change during different reigns and many of the contemporary cities have a common colonial past. Region's major cities-those with political, economic and cultural significancehave always played a great role in the development and progress of the region. Bhatikar and Sami (2021) unraveled concerns that "rapid urbanization over the next two or three decades will lead to an explosion in the demand for livelihoods, urban renewal, infrastructure and real estate development, food, water, energy, housing and digital technology". Cities of the region require addressing the increasing demand for urban services, and also vulnerable to a variety of socio-economic issues of inequality, poverty, health and social welfare and environmental factors such as environmental degradation. Gavsker (2020a) noted that "the growing cities of the third world experience an unimaginable challenge and crisis and create a distinctive form of geographies".

Thus, India is one of the important countries and emerging economies of South Asia. To respond, address and direct contemporary urban development and planning process, India has launched the Smart Cities Mission (SCM) to fulfill the needs of urban growth and overcome the challenges of major cities (SCM 2015). As initially understood, urban regions that use cutting-edge technology, data analytics, and digital infrastructure to improve the effectiveness, liveability, and sustainability of urban centres are conceptualized as smart cities. These cities seek to ponder upon a set of multi-necessity problems by integrating key areas, including energy, transportation, waste management, and governance. The Mission in India aims to improve resource management, improve quality of life, and advance environmental sustainability by integrating several sectors, including energy, transportation, waste management, and governance. Through the deployment of cutting-edge technology and data-driven decision-making, smart cities serve as crucial facilitators in accomplishing sustainability goals. Thus, the evaluation of ambitions, development projects, outcomes and geographical variations is timely and crucial. The present study aims to comprehend the efforts, performance and geographical factors of smart cities in Punjab and Uttar Pradesh. Both states have contributed to an agriculture-based economy and significant transitions thereafter in the sectoral share of the economy. However, the state of Uttar Pradesh falls in the lower socio-economic development and has lower level of urbanization in comparison to that of Punjab. As per the Census of India 2011, the level of urbanization was 37.5% and 22.3% in Punjab and Uttar Pradesh states respectively. The per capita income in 2012–13 was reported at Rs. 84,526 and Rs. 33,482 in Punjab and Uttar Pradesh respectively. This has increased to Rs. 1,51,491 and Rs. 65,431, respectively, in 2019–20. However, both states are significant as has been witnessed by their high concentration of population, agricultural productivity and growing manufacturing activities. In this wider context, whether smart cities would lead to realizing urban sustainability or smart sustainable cities remains a partially addressed query. Under the SCM three cities are selected in Punjab and ten cities in Uttar Pradesh (Fig. 8.1). The present study based on secondary sources of

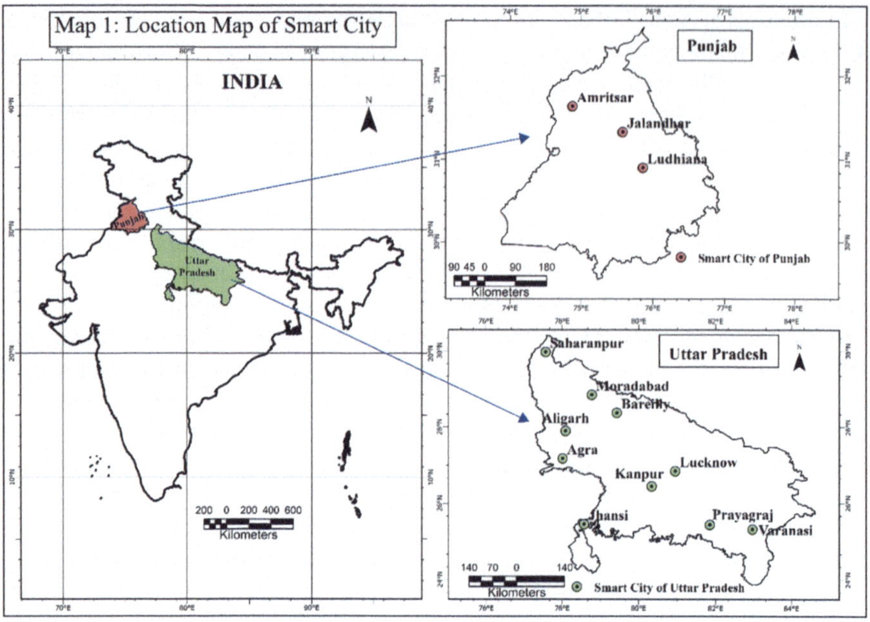


**Fig. 8.1** Location of smart cities of Punjab and Uttar Pradesh. *Source* Prepared on the basis of list of smart cities, smart city mission, Ministry of Urban Development, Government of India, New Delhi

data and relevant information, attempts to address the following key objectives: (i) to analyze urbanization and urban growth patterns of the select cities in Punjab and Uttar Pradesh; (ii) to elucidate components, elements, and strategy of *smart* (cities) planning of cities of these states, (iii) to examine the performance of these smart cities and the unconventional initiatives introduced in making them smart urban spaces, and (iv) to explore the significance of smart cities in tackling urban sustainability concerns and inherent challenges.

## 8.2 Data Source and Methodology

The methodology applied in the study is to evaluate and assess the performance of the cities of Punjab and Uttar Pradesh covered under the SCM. The study is based on secondary sources of data and information collected and retrieved from the official records and implementing agencies. Thus, it is a comparative study of two states-Punjab and Uttar Pradesh. The reason for selecting these states is that Uttar Pradesh is witnessing a continuation of state machinery under the same political organization-led government (the Bhartiya Janata Party), whereas, the state of Punjab has seen a drastic change in the political sphere over the years (from Congress to

Aam Aadmi Party) and, consequently, this way a change in the state machinery in the practice is inevitable. Whether there is any interesting reflection on the progress and nature of implementation of projects under SCM in the context of a *status quo* and *any change* in the political sphere, is explained in the study. The present analysis is framed in this wider circumstance of political aspirations and economic growth which is connected to raise urban habitat standards and overall quality of urban life. However, the deriving point of the present study is the level of urbanization and the number of projects undertaken in both states and how they have a kind of similarity and difference. Thus, the study critically assesses: (i) the total number of projects undertaken; (ii) the present status of a number of projects in implementation stage; (iii) reforms under the SCM; and iii) budget and finance improvements at local levels: urban local bodies (ULBs).

## 8.3 Conceptual Background

The world is now more urban than rural and major urban centres are growing at unprecedented rates and India is experiencing this change at a rapid scale. The United Nations (UN) noted that "Between 1950 and 2018, the urban population of the world grew more than four-fold, from an estimated 0.8 billion to an estimated 4.2 billion" (WUP 2019). The world's urban population is projected to increase from 4.2 billion in the year 2018 to 6.7 billion in the year 2050, and about 90% of this growth will take place in Africa and Asia (Mahendra et al. 2021). Cities have been centres of such growth processes accompanying economic development and technological progress. This is reflected in as "Cities create wealth, generate employment and drive human progress by harnessing the forces of agglomeration and industrialization" (World Cities Report 2016). According to the World Cities Report (2016), "Over the last two decades, cities have emerged as the world's economic platforms for production, innovation and trade. Urban areas offer significant opportunities for both formal and informal employment, generating a sizeable share of new private sector jobs". But, most of the global cities particularly the cities of the Global Urban South have been haphazardly growing and experiencing urban expansion. Urbanization here has been driven by an unsustainable development process that poses severe challenges to urban planning. From an environmental standpoint, the existing urbanization model promotes sprawling low-density suburbs, primarily influenced by private interests rather than public welfare. This model results in high energy consumption and significant contributions to climate change (Cohen 2013; World Cities Report 2016). Rapid urbanization and urban development have fuelled many challenges. According to Arku and Marais (2021), "megacities of the Global South have major socio-economic and environmental problems, such as in increase in slum-dwellers, air and water pollution, resource depletion, and rising urban poverty". They are facing multiple challenges and attempts are now being made to make cities more *sustainable, inclusive, safe and resilient*. The challenges can be overcome only by urban planning and focussing on the development of specific cities at the local and regional level. Arku and Marais (2021) further explain, "the local level is the level closest to people, their needs and their knowledge. It is the level of implementation of sustainable development policies in the form of urban infrastructure, basic services and land use and mobility planning". At present, the concept of *'smart cities'* has gained momentum in advanced countries already and it's in the initial phase in most of the emerging economies such as in India as it targets to address technology applications related weaknesses and infrastructure and service challenges of specific cities and provide Information and Communication Technology (ICT) enabled strategic planning and a new system in place.

## 8.4 Making Urbanization "Smart"

The Smart Cities Readiness Guide (2015) of the smart cities council noted that the smart city can be defined as "...city uses ICT to enhance its liveability, workability and sustainability. In simplest terms, there are three parts to that job: collecting, communicating and 'crunching". This indicates that smart cities are cities with robust networks of ICT integrated with the day-of-day life of city dwellers and promoting internet-based administration and governance. The origin of the concept of smart cities has several dimensions. The term smart city is believed to have appeared in the mid-1990s. Hollands (2008) believes that the term "smart cities" was used in a self-congratulatory manner, with cities declaring themselves as smart without substantial evidence or achievements to support it. Eremia et al. (2017) believed that the concept of the smart city is rooted in the emergence of the term urbanism in 1898 and argued that the term smart city gained momentum in 2009 in the discourse. They opined that initially, the term sustainable city was used but later with the foundation of European Digital Cities (1996–1999), the term digital cities gained momentum. The term *smartness* is seen as politically neutral compared to the term *sustainability*. Because of this, people in countries where the terms "sustainability" and "greenness" were linked to liberal or progressive policies, were more open to using different variations of the term 'smart' including terms like smart city, smart growth, smart development (see, Eremia et al. 2017; Moir et al. 2014). Thus, the idea of smart cities rapidly gained momentum in the urban discourse. Major cities around the globe tried to achieve *smartness,* though it was not properly defined. Technology emerged as a major foundation for the rise of smart cities. In this way, "Smart cities involve the creation of new relations between technology and society. According to this vision, urban infrastructures and everyday life in cities are optimized through technologies provided by IT companies (Söderström et al. 2014). Such cities can be understood as a way in which the life of people is enhanced in multiple ways by utilizing digital technologies and infrastructures. It acts as a bridge between technology and the people by connecting them and providing the required services and smooth delivery of basic amenities. There are several advantages associated with using technology in the smart cities. Using ICT in smart cities can have several advantages. It can help save resources like energy and water, decrease the amount of carbon dioxide released into the environment, and make better use of the current infrastructure (Harrison and Donnelly 2011). Smart cities are aimed at making cities inclusive, where the services by the government are dispensed to the citizens in better and more advanced ways. According to Veselitskaya et al. (2023), "The main factors which form the integrative structure of a smart city are: city management, technology, municipal government administration, political program, communications, economics, infrastructure and environment".

India's SCM drove its idea from the present process of making urbanization smart worldwide, the smart city plans are focused on strengthening technology-based city management and e-governance, promoting the economy and building a new digital image of the city concerned. The idea of a smart city initially emerged as the cities with technological advancement. The concept of smart cities has evolved over time through various factors, including advancements in technology, urbanization challenges, and the need for sustainable and efficient urban development. Advancements in technology, particularly in the field of ICT, have to play a significant role in the growth and development of smart cities. The increasing connectivity, availability of data, and the potential for automation and optimization offered by these technologies presented opportunities for transforming urban areas. The challenges posed by rapid urbanization, such as overcrowding, inadequate infrastructure, traffic congestion, environmental concerns, and inefficient public services, also prompted the need for new approaches such as in urban development and planning. Governments and policymakers started exploring innovative solutions to address these challenges effectively. Though the term Smart City was first used by IBM in 2008 when it used the 'smarter planet and smart city concept', there are some indications of using innovative ways for city development in India. The first plan which was at the centre of the development of the city was the Mega City Program of the Government of India (2012). Chakravorty (1996) stated that "It is an attempt to shore up infrastructure in five of the six largest metropolitan regions in India (Bombay, Calcutta, Madras, Hyderabad and Bangalore), using innovative financing mechanisms, and emphasizing cost recovery". In times to come, further, significant initiatives focusing on technology-based urban development evolved and one such models was the Bangalore Agenda Task Force (BATF) which played a significant role in initiating smart governance in the city way back in the 2000s. It was a public–private partnership initiative established in 1999 in Bengaluru City (then Bangalore), India. The BATF aimed to transform Bengaluru into a world-class city by addressing governance and infrastructure challenges. While the BATF's focus was not explicitly on making cities smart, its efforts in initiating smart governance and utilizing technology to enhance urban services were precursors to the broader concept of smart urban development and governance in India. It focused on initiatives such as e-governance, citizen-centric services, and public–private partnerships to improve governance and service delivery in Bengaluru. Such efforts and initiatives laid the foundation for later discussions on smart urban development and management, and the implementation process as it has found a place in SCM in the country. These and several other similar initiatives and largely aimed at the facelift of the megacities and cities of national and regional importance. However, all this was greatly influenced by the backdrop that "as a result of 1991s, opening of Indian economy and 1992s, 73rd and 74th Constitutional Amendments, suggesting decentralization of power between the central and local authorities, Indian economic growth got a new dimension, also encompassing urban development in the following years" (Rai 2012).

In the middle of the 2000s, the flagship program the Jawaharlal Nehru National Urban Renewal Mission (JNNURM) was launched for a period of seven years with an outlay of Rs.66.085 crores with the "aim is to encourage reforms and fast track planned development of identified cities. Focus is to be on efficiency in urban infrastructure and service delivery mechanisms, community participation, and accountability of ULBs/ Parastatal agencies towards citizens" (JNNURM 2005). It had two sub-missions: (a) urban infrastructure and governance administered by the Ministry of Urban Development; and (b) basic services for urban poor administered by the Ministry of Housing and Urban Poverty Alleviation of the Government of India. Thus, key sectors and projects included were: water supply and sanitation, solid waste management, urban transportation, development of slums, and projects to ensure basic services to the urban poor. The study carried out on the progress of JNNURM by the Public Policy Research Centre (2014) argued that "the implementation-architecture of JNNURM has several lacunae, few of which have been taken care of but some still continue. These include failure to mainstream urban planning, the top-down process of finance, bypassing state governments and land issues, etc. The quality of implementation of JNNURM could have been better had these issues been taken care of". The experiences gained and challenges faced from the past initiatives have influenced the approach to be taken in urban planning, particularly in the formulation of future-oriented SCM in India.

## 8.5 Urbanization: Spatial Patterns and the Urban Agglomerations

India is one of the rapidly developing countries as exhibited in India's Gross Domestic Product (GDP) growth rate in the last 10 years, which has been at an average of 6–7%. Indian cities contribute to about two-thirds of the economic output of the nation. The country has just crossed the rank of second most populous country and become the most populous one in the world. As per the Census of India (2011), there were 377.10 million urban dwellers living in 7933 cities, urban agglomerations, urban outgrowths and towns. This exhibits tremendous urban progress in the country during the post-independence period. It is estimated that by 2030, more than 400 million people will be living in such urban places in India.

Trends in urban population, number of towns and urban agglomerations in the country in relation to the states of Punjab and Uttar Pradesh have been represented in Table 8.1. The share of the urban population in the country was 19.91% in 1971, which increased to 23.34% in 1981. It further increased to 25.72% in 1991, 27.86% in 2001. Ultimately, it reached to 31.16% in 2011. Urbanization in Punjab has been showing higher levels as compared to the country and to that of the state of Uttar Pradesh. In Punjab, the percentage of the urban population was 23.76% in 1971, which was higher than that of Uttar Pradesh. This share grew to 27.7% in 1981, and a further 29.54% in 1991. By 2001, the level of urbanization in Punjab was 33.91% which grew to 37.29% in 2011. This trend has been above the national average throughout census years. In terms of the size of the urban population in Punjab, this increased from 3.22 million in 1971 to 4.65 million in 1981. This further increased to 5.99 million by 1991 and 8.26 million by 2001. The Census of India 2011 reported a 10.33 million urban population in the state of Punjab (Table 8.1). The number of towns and urban agglomerations in the state has increased from 108 in 1971 to 217 by 2011. The compound annual growth rate of the urban population is of relevance here, in 1971 it was 2.27% which jumped to 3.75% in 1981. It came down to 2.58% in 1991 and again a slight jump to 3.24% in 2001. It slumped to 2.35% by 2011 (Census of India 1971–2011).

Uttar Pradesh is larger in both size of urban population and area compared to Punjab. The state has a high number of urban dwellers but its percentage share remained below the national average all these years. In Uttar Pradesh, the share of the urban population was 14% in 1971, which increased to 18% in 1981. In 1991, it further increased to 19.8%. It was reported at 20.78% in 2001 and 22.27% by 2011 (Fig. 8.2). Simultaneously, the number of urban centres has increased from 256 in 1971 to 915 in 2011in Uttar Pradesh. The number of towns and urban agglomeration in Uttar Pradesh is higher as compared to that of Punjab. In 1971, out of a total of 2476 urban agglomerations in India, the share of Uttar Pradesh was 10.33% (i.e., 256) and Punjab had only 4.36% (i.e., 108). In 1981, there were 3301 urban agglomerations in India, the share of Uttar Pradesh was 18.11% (i.e., 598) and Punjab had 4.05 (i.e.134). During the next decade, the number of urban agglomerations increased to 3696 in India with a 17.07% share of Uttar Pradesh (i.e., 631) and Punjab contributed to 3.24% (i.e.120). In 2001, India had 5161 urban agglomerations with 670 in Uttar Pradesh and 157 in Punjab contributing 12.98% and 3.04%, respectively.

The number of cities and urban agglomerations has increased to 7933 in India with 915 urban agglomerations in Uttar Pradesh and 217 urban agglomerations in Punjab contributing 11.53% and 2.73%, respectively. The analysis reveals that the share of urban centres to the total urban centres of the country has been declining over the decades in both states. *The Town and Country Planning Department* (2021) of Uttar Pradesh noted that "although, Uttar Pradesh has the largest urban system in the country with 630 municipalities however, it ranks 23rd in the level of urbanization. There are marked regional imbalances in the level of urbanization in the state. As per the Census 2011, the Western Region with 32.45% urban population is the most urbanized, and Eastern Region with 13.40% urban population is the least urbanized. Urban population of Central and Bundelkhand Regions stands at 20.06% and 22.74%, respectively". The large cities hold a significant share of the urban population in both states of Punjab and Uttar Pradesh. However, the fact reveals a higher concentration of urban population, i.e. 58.05% in 2011 is found in Class I cities in Punjab, whereas in the case of Uttar Pradesh the share of the urban population is indeed found in large cities, i.e., 61.04 in 2011 and remaining in medium and small towns. The urbanization

**Table 8.1** Urbanization trends in India, Punjab and Uttar Pradesh (1971–2011)

| Year | India | | | Punjab | | | Uttar Pradesh | | |
|---|---|---|---|---|---|---|---|---|---|
| | Urban population (million) | Urban population (%) | Number of towns and urban agglomerations | Urban population (million) | Urban population (%) | Number of towns and urban agglomerations | Urban population (million) | Urban population (%) | Number of towns and urban agglomerations |
| 1971 | 109.11 | 19.91 | 2476 | 3.22 | 23.76 | 108 | 11.63 | 14.00 | 256 |
| 1981 | 159.46 | 23.34 | 3301 | 4.65 | 27.70 | 134 | 18.74 | 18.00 | 598 |
| 1991 | 217.18 | 25.72 | 3696 | 5.99 | 29.54 | 120 | 25.97 | 19.80 | 631 |
| 2001 | 286.12 | 27.86 | 5161 | 8.26 | 33.91 | 157 | 34.51 | 20.78 | 670 |
| 2011 | 377.10 | 31.16 | 7933 | 10.33 | 37.29 | 217 | 44.49 | 22.27 | 915 |

*Source* Census of India (1971–2011)

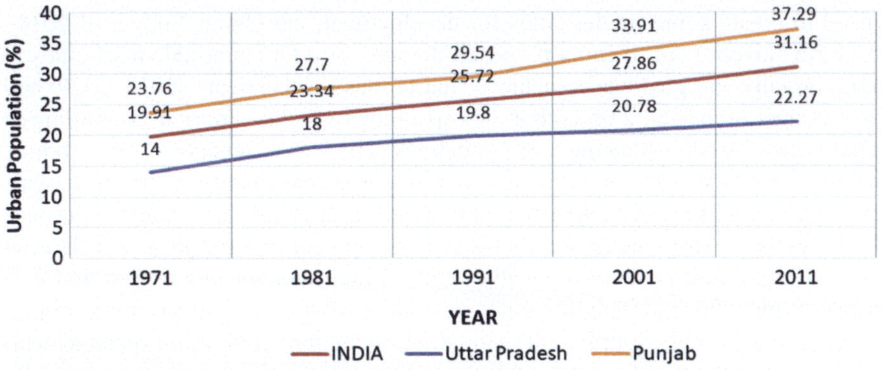


**Fig. 8.2** Comparative urbanization trends (1971–2011). *Source* Census of India (1971–2011)

process reciprocals the economic growth of cities and the region they are located in and thus, "cities play a significant role in social, economic, political and ecological systems on the earth surface" (Gavsker 2020b).

## 8.6 India's Smart Cities Mission

In rapidly urbanizing India, cities are being fuelled by economic and technological growth and expanding demographic size. The NITI Aayog (2021) noted that "India is the second largest urban system in the world with almost 11% of the total global urban population living in Indian cities. In absolute numbers, the urban population in India is more than highly urbanized countries/regions across the globe". However, substantial urbanization and urban growth remain unplanned, haphazard, lopsided, and ill-governed. Indian cities' cores are highly congested and prone to several pollution-led challenges. *Greenpeace's* study revealed that in Punjab $PM_{10}$concentrations in Amritsar, Jalandhar, Ludhiana, Mandi Gobindgarh, Khanna and Bhatinda were, respectively, 184, 151, 139, 130, 122 and 111 $\mu g/m^3$ for the year 2015. In Uttar Pradesh $PM_{10}$concentrations in Ghaziabad, Bareilly, Allahabad, Kanpur, Agra, Lucknow, Varanasi (average of $PM_{10}$levels from October 2015 to September 2016 was 228 $\mu g/m^3$ Varanasi) and Sonebhadra were, respectively, 258, 240, 250, 201, 186, 169, 145 and 132 $\mu g/m^3$ for year 2015 (Greenpeace 2017).

The schemes for urban development and rejuvenation have been undertaken in India, but these plants failed to fulfil the required development benchmarks. For example, the Mega City Program of the Government of India failed due to a shortage of funding throughout the ten-year period of the scheme (Chakravorty 1996, Hoelscher 2016). The JNNURM met the same fate of shortage of proper funding. It faced criticism due to delays in funding, incomplete transfer of authority, and inefficiencies in urban land acquisition procedures to carry out the development projects in the cities (GoI 2012, Weinstein et al. 2014; Hoelscher 2016). Realizing a need to create some model cities for development, the Union Budget of 2014–15 of the Government of India proposed the idea of a few hundred smart cities in India. Initially, the goal was to achieve smart cities for five years (2015–2020) and had a budget of around 2.05 lakh crores spanning over this duration. According to Smart Cities Mission-Statement & Guidelines (2015) "the objective is to promote cities that provide core infrastructure and give a decent quality of life to its citizens, a clean and sustainable environment and application of 'Smart' Solutions. The focus is on sustainable and inclusive development and the idea is to look at compact areas, and create a replicable model which will act like a light house to other aspiring cities". It includes major elements such as: (i) adequate water supply; (ii) assured electricity supply; (iii) sanitation, including solid waste management; (iv) efficient urban mobility and public transport; (v) affordable housing, especially for the poor; (vi) robust IT connectivity and digitalization; (vii) good governance, especially e-Governance and citizen participation; (viii) sustainable environment; (ix) safety and security of citizens, particularly women, children and the elderly; and (x) health and education. The smart city strategy has key components including: (i) pan-city initiatives in which at least one smart solution is applied city-wide; and (ii) development areas step-by-step: *retrofitting; redevelopment* and *greenfield development*. The promising cities were supported to propose their strategies based on these components. The guidelines laid down some comprehensive features of development. It included promoting mixed land use, expanding housing opportunities, creating walkable paths, promoting variable transport options, making citizen-friendly governance, giving identity to the city and applying smart solutions in the city development. With the selection of cities based on their proposal competition, the *Special Purpose Vehicle* (SPV) has been formed in such cities. The SPV is responsible for planning, appraising, approving, releasing of funds, operating, managing, monitoring and evaluating the smart cities. The SPV will be formed under the Companies Act, 2013 with 50% shareholding with the State/UT and fifty %of shareholding with concerned Urban Local Bodies (ULBs). The Ministry of Urban Development will use the funds provided only for the assigned mandate. Other projects will be carried out by the convergence of funding from various State and Union Government sponsored projects. The central level scheme converged for the funding of SCM are *Atal Mission for Rejuvenation and Urban Transformation* (AMRUT), *Integrated Power Development Scheme* (*IPDS*), National Heritage City Development and Augmentation Yojana (HRIDAY), National Urban Livelihood Mission (NULM) Swachh Bharat Mission, Digital India, Make in India, Pradhan Mantri Awas Yojana (*Shelter for All*), Solar City Mission, etc. In Punjab, convergence of projects has been of which were under the various offices and departments such as the Punjab Water Supply & Sewerage Board Project, Punjab Restructured Accelerated Power Development and Reform Program (R-APDRP), Punjab Nirman Scheme, Department of Sports, Department of Irrigation, etc. Whereas in Uttar Pradesh, convergence was done of schemes and departments such as Uttar Pradesh New and Renewable Energy Development Agency (*UPNEDA*), Nagriya Sadak Sudhar Yojana, Department of Tourism, District Urban Development Agency (DUDA), etc. There will be a national, a state and a city-level body providing guidance and directions to the plan.

In this wider context, the cities selected for the SCM of Punjab and Uttar Pradesh are under study here.

Though the primary focus of the SCM is people and the better well-being of people, technology and smart technical and digital solutions are the way forward. The major objective is to promote *cities* that can provide good infrastructure; a clean environment; the application of smart solutions; and a better quality of life for the citizens. The method of selection of cities is based on equal weightage it gives to the urban population in the State and the number of *statutory towns* in the State concerned. Conditions precedent included in Stage 1 of the City Challenge Competition: State's undertaking to make the city smart; inter-departmental task force consisting of parastatal bodies, urban local bodies, and organizations and urban development authorities; elected city council's resolution to make the city smart; and consultations held with residents on city development priorities. Scoring criteria consisted of: existing service levels (25 points); institutional systems/capacities (15 points); self-financing (30 points); and past track record and reforms (30 points). The selected under Stage 1, prepared Smart City Plans for Stage 2 of the competition for prioritizing cities for financing. Thus, the Stage 2 criteria for evaluation of Smart City Plans included: City Level Evaluation (30 points); and Proposal Level Evaluation (70 points). In different rounds of the selection process, important cities were selected under the SCM in both the states of Punjab and Uttar Pradesh. Overall in India, the total allocated investments for the SCM stood at Rs. ~205,018 crores (US$ 27.60 billion) as of March 2021. Of the total investments, 5,614 projects worth ~Rs. 173,018 crores (US$ 23.29 billion) have been tendered, work orders have been issued for 4,912 projects worth ~Rs. 139,851 crores (US$ 18.83 billion) and 2,420 projects worth ~Rs. 40,152 crores (US$ 5.40 billion) have been completed as of March 2021 (MoHUA, 2019).

Selected cities of Punjab are Amritsar, Jalandhar and Ludhiana. Among these cities, Amritsar is known for its religious, cultural and heritage significance. Ludhiana is known for the hosiery and machine tool industry and Jalandhar for small and medium manufacturing and commercial enterprises. The cities of Uttar Pradesh are Agra, Aligarh, Kanpur, Lucknow, Varanasi, Moradabad, Saharanpur, Prayagraj (Allahabad), Jhansi and Bareilly. Agra is well known for its historic value and the UNESCO Heritage-the monument Taj Mahal. Cities of Varanasi and Prayagraj are known for their cultural and religious significance, Jhansi for its historic and strategic values as well as, public sector undertakings and industrial and commercial activities. There are important industrial urban centres like Aligarh for lock industries, Ludhiana for the hosiery and machine tool industry, and Kanpur for the leather industry. All selected cities under SCM are the most important ones in several dimensions and emerging as urban hotspots in the contemporary process of urbanization as well in these states.

Table 8.2 presents aspects of cities selected under the SCM including urban population, share to state's urban population, urban households having access to electricity, urban households having access to tap water, urban households having access to sanitation, urban households having access to housing and percentage of slums in urban population. Punjab's Amritsar and Ludhiana contribute the highest share to the state's

**Table 8.2** Aspects of cities selected under SCM in Punjab and Uttar Pradesh, 2011

| Aspects | | Urban population | Share to state's urban population | Percent of households having electricity | Percent of households having tap water supply | Percent of households having sanitation facility | Percent of households having housing | Percent of slums population |
|---|---|---|---|---|---|---|---|---|
| Cities | | | | | | | | |
| Punjab | Amritsar | 1,159,227 | 11.15 | 93.6 | 68.8 | 59.8 | 95.3 | 26.88 |
| | Jalandhar | 862,886 | 8.30 | 99.3 | 69.5 | 98.0 | 96.8 | 19.14 |
| | Ludhiana | 1,618,879 | 15.57 | 98.9 | 92.0 | 97.7 | 94.5 | 15.91 |
| Uttar Pradesh | Agra | 1,585,704 | 3.56 | 95.8 | 62.9 | 88.3 | 95.8 | 30.52 |
| | Aligarh | 874,408 | 1.97 | 86.7 | 45.7 | 91.3 | 92.8 | 41.98 |
| | Bareilly | 904,797 | 2.03 | 89.3 | 61.0 | 95.9 | 95.8 | 17.02 |
| | Jhansi | 547,638 | 1.23 | 91.7 | 60.3 | 82.3 | 95.4 | 19.68 |
| | Kanpur | 2,768,057 | 6.22 | 90.0 | 45.3 | 86.1 | 94.1 | 19.71 |
| | Lucknow | 2,817,105 | 6.33 | 92.6 | 74.6 | 90.2 | 94.8 | 27.40 |
| | Moradabad | 887,871 | 2.00 | 86.8 | 55.8 | 95.4 | 92.8 | 10.00 |
| | Prayagraj | 1,168,385 | 2.63 | 95.0 | 92.5 | 93.0 | 94.1 | 11.68 |
| | Saharanpur | 705,478 | 1.59 | 93.6 | 61.2 | 93.3 | 95.4 | 33.06 |
| | Varanasi | 12,01,815 | 2.70 | 94.2 | 75.3 | 94.8 | 86.1 | 33.86 |

*Source* Based on District Census Handbooks of Punjab and Uttar Pradesh, 2011 urban population, i.e., 11.15% and 15.57%, respectively (Table 8.2). In Uttar Pradesh cities of Kanpur and Lucknow contribute 6.22% and 6.33% to the total urban population, respectively. Amritsar and Jalandhar of Punjab select cities still face challenges in providing safe and tap water supply to citizens. Except for Prayagraj, the rest of the cities face the worst challenges in Uttar Pradesh.

### *8.6.1 Status of Cities Under the SCM*

The various projects have been funded by the SCM and some with convergence of the existing plans (Table 8.3). With the SCM, cities of Punjab have progressed on the path of development. Several large projects of the smart city of Amritsar are still in the work order stage. Major projects in the city under this stage are the construction of a water treatment plant (1156 Crores), a gas pipeline in the city (450 Crores), the construction of smart roads (126 Crores), the development of green mobility (108.33 Crores), etc. Improvement of sanitation conditions, boosting local identity, providing access to open space, etc. are some of the projects completed under the SCM. In the city of Jalandhar, some projects have been finished under the SCM such as the installation of rooftop solar in schools (Rs. 17.83 Crores), junction improvements (21.68 Crores), development of smart classrooms in government schools (6.53 crores), etc. The major projects of the city of Jalandhar are in the work stage. This includes the construction of the Integrated Command and Control Centre (116.94 Crores), the construction of Burlton Park Sports Hub (250 Crores), the bulk water supply scheme (808 Crores), and several other projects. Major projects in the city of Ludhiana are in progress. These major projects consist of STP rehabilitation (650 Crores), rejuvenation and landscaping of Budha Nalla (550 Crores), the establishment of industrial manufacturing innovation, incubation and skill development centre (292.98 Crores), providing compactor hand carts for sweeping (60 Crores), etc. Others include Smart school gradation (6.28 Crores), retrofitting of the Sarabha Nagar market (15.88 Crores), and waterfront development (4.74 Crores).

Similar progress is being reported from the state of Uttar Pradesh. The most important work completed under the SCM in Agra is the construction of the Integrated Command and Control Centre (ICC) for traffic management. This project incurred a cost of Rs. 287.09 Crores. The other high-cost projects included the pilot 24 × 7 water supply (160.41 Crores), Water supply system (142.6 Crores), Beautification & Street scaping of Fatehabad Road (116.46 Crores), and rehabilitation of key roads. Other important projects included the development of heritage walks, the development of street vending zones, the upgradation of schools and municipal health centres, etc. Attention has been paid to utilising the funds from different projects also by the convergence method. The city's progress on smart cities shows its alignment with the proposed plan.

In the city of Kanpur, several key projects are launched. The city has developed a smart city operation centre with a cost of Rs. 389.87 Crores, which includes the Integrated Traffic Management System, smart parking solutions, city surveillance

**Table 8.3** Total spending in various projects in select smart cities

| Type of project | Total Spendings[a] in various projects (in crores) | | | | | | | | | | | |
|---|---|---|---|---|---|---|---|---|---|---|---|---|
| | Punjab | | | Uttar Pradesh | | | | | | | | |
| | Amritsar | Ludhiana | Jalandhar | Agra | Kanpur | Aligarh | Bareilly | Saharanpur | Lucknow | Jhansi | Varanasi | Moradabad |
| Integrated command and control center and master system integrator (MSI) | 109.68 | 352.52 | 116.94 | 422.26 | 289.87 | 54.17 | 44.52 | 200.28 | 45.36 | 146.5 | 345.74 | 0 |
| Water, sewage and sanitation | 1180.78 | 1211.38 | 838.31 | 270.2 | 46.72 | 51.66 | 154.32 | 0 | 62.32 | 123.22 | 328.73 | 0 |
| Transport infrastructure & mobility | 240.17 | 75 | 126.27 | 119.03 | 158.35 | 205 | 496.52 | 0 | 329.61 | 402.77 | 304.27 | 0 |
| Power infrastructure | 10.4 | 0 | 0 | 322.8 | 319.29 | 66 | 108.16 | 0 | 20.06 | 54.97 | 0 | 6.66 |
| Solid waste management | 11.37 | 72.45 | 67.47 | 2.29 | 228.48 | 0 | 96.05 | 56.43 | 93.69 | 0 | 151.64 | 8.44 |
| City beautification, area-based development and green city | 46.86 | 33.74 | 57.4 | 125.98 | 20.71 | 85.4 | 35 | 0 | 325.52 | 0.4 | 341.87 | 30.42 |
| Housing and slum redevelopment | 0 | 0 | 0 | 0 | 0 | 39 | 2 | 0 | 11.38 | 3.4 | 0 | 0 |

(continued)

**Table 8.3** (continued)

| Type of project | Total Spendings[a] in various projects (in crores) | | | | | | | | | | | |
|---|---|---|---|---|---|---|---|---|---|---|---|---|
| | Punjab | | | Uttar Pradesh | | | | | | | | |
| | Amritsar | Ludhiana | Jalandhar | Agra | Kanpur | Aligarh | Bareilly | Saharanpur | Lucknow | Jhansi | Varanasi | Moradabad |
| Smart solutions | 15.41 | 6.28 | 6.04 | 9.92 | 115.68 | 87.63 | 236.21 | 32.53 | 92.25 | 15.2 | 308.22 | 0 |
| Other | 551.63 | 38.4 | 365.45 | 11.2 | 28.84 | 138.21 | 34.89 | 0 | 2.48 | 4.57 | 278.29 | 0 |

*Source* Compiled from various city's profiles at smartcities.gov.in (Accessed on June 04, 2023)

[a]Completed Projects and Projects under the work order stage have been considered

systems, wifi hotspots, smart city portals, air and water monitoring systems, etc. Major spending is done on underground wiring & strengthening of the transformer (150 Crores), waste processing and disposal system (100 Crores), intelligent LED street lights (98.11 Crores), etc. Various projects have also been completed such as road maintenance, bust charging stations, development of market zones, multi-level card parking, etc. In Aligarh city, key projects are launched and implemented. The city developed IT connectivity and digitalization for transport and mobility management at the cost of Rs. 54.17 Crores. The city has also improved the living conditions of the urban poor by upgrading nine slum areas of the city at the cost of Rs. 39 Crores. A common facility centre for lock and hardware clusters has been developed to boost the economy. Several other projects have also been undertaken such as an improved traffic management system, uninterrupted power supply and smart metering, CCTV surveillance, rainwater harvesting, energy-efficient lighting, etc.

In the city of Bareilly, special attention was paid to mobility and transport solutions where it developed smart mobility for area-based development investing the sum of Rs. 107.33 Crores. Other important works done under the smart city project are the installation of new LED lights, the construction of charging stations, construction of pumping and wastewater stations. An integrated command and control centre has also been developed under the mission. Further, the major projects completed under the SCM in Saharanpur include making land for disposal available for solid waste management. This project cost Rs. 30 Crores. Further, a sum of Rs. 18.93 Crores has been used for providing household bins for segregating waste and procurement of solid waste management equipment. Other completed projects in the city included establishing mobile clinics, smart e-libraries and stadium reconstruction. Several projects of the city are still in the work order stage. It includes the development of the Integrated Command and Control Centre, rejuvenation of parks, rejuvenation of water bodies, electrification, etc.

The capital city of Lucknow undertook several key development projects related to infrastructure and services. The most important project paid attention to was under the area development plan-development of the Gomati River Front with a total cost of Rs. 228.35 crores. Other important projects carried out in the city included a smart solid waste management system (Rs. 84.08 Crores), Lucknow smart city management portal (Rs 45.36 Crores), health ATM and smart health centres (Rs. 25 Crores), water supply strengthening (Rs. 43.42 Crores), etc.

The works undertook in Jhansi include the installation of a capacitor bank which cost Rs. 237.03 Crores. Projects under area-based development are the construction of a railway over a bridge at Shipari bazaar (Rs 132.41 Crores), electric lines (Rs. 43.82 Crores), construction of night shelters, pink toilet blocks, installation of the solar power plant, development of urban *haat* retail stores, etc. Several more important projects such as the construction of the Integrated Command and Control Centre, transformation of arterial roads into smart roads, etc.are still in the work order stage or the tender stage in the Jhansi smart city. Historic and religiously important city of Varanasi has also undertaken key projects under the mission. The city has recognized managing rapid growth in the path of achieving its smart city targets as per its strategies. In order to boost the economy and identity of the city, a cultural cum spiritual convention centre has been constructed by investing Rs. 192.25 Crores. Pan city repair of roads (Rs. 99.15 crores), renovation of Homi Bhaba Cancer Hospital (140 Crores), installation of advanced city surveillance system (159.74 Crores), and redevelopment of parking areas (96.62 Crores) are key projects undertaken in the city. Several other projects such as the beautification of riverfronts, IT solutions, redevelopment of the old Kashi area, redevelopment of parks etc. have also been undertaken for the development of the city.

In these select cities, various major projects have been undertaken for development, rehabilitation and renovation purposes. The projects are decided in the context of the development strategies of the cities concerned. It may be observed that there is a loose link between the plans and strategies and implementations of projects in each of the cities. These projects can be classified under a few themes: (i) the construction of Integrated Command and Control Centre (ICCC); (ii) the Master System Integrator (MSI) for the monitoring and providing the citizen's amenities in better ways, (iii) the development of transport infrastructure and mobility to make cities congestion free and providing better connectivity in the city; (iv) the improvement of water, sewage and sanitation including drainage modification, water utilization, toilet construction, sewage line repair and construction; (v) the power infrastructure development, which included the construction of new power lines, rewiring, construction of a smart power distribution system (more of them found in state of Uttar Pradesh); (vi) the solid waste management initiatives included providing ground for decomposing, providing modern equipment for waste collection etc.; (vii) the city beautification, area-based redevelopment and providing green space are other major initiative of the city; and (viii) housing and redevelopment (more of them found in the state of Uttar Pradesh).

In most of the cities of Uttar Pradesh, the utilisation of funds is found in varied categories, whereas in Punjab, major attention has been paid to water, sewage and sanitation improvement (Fig. 8.3). The major spending under SCM in Punjab is in sectors of water, sewage and sanitation. The spending for the same is 54.51% in Amritsar, 67.68% in Ludhiana and 53.13% in Jalandhar. For others, in Amritsar, the transport infrastructure accounted for 11.09%, whereas in Ludhiana, the cost incurred was 19.7% for the construction of ICCC and MSI. City plan of Jalandhar incurred 23.16% cost on sports facility development and skill development centres as well. Thus, cities under the SCM utilised the funds and resources in different ways for different purposes.

In the state of Uttar Pradesh, in Agra, the construction of ICCC and MSI utilised 32.89% of total funds in all projects. It is followed by improvement of power infrastructure (25.15%), water, sewage and sanitation improvement (21.05%) and beautification, area-based development and green space development (9.81%). In Kanpur city major spending has been towards power infrastructure which accounts for 26.43% of total spending. It is followed by the construction of ICCC and MSI (24%), solid waste management (18.91%) and transport infrastructure and mobility (13.11%).

In Aligarh city, major spending of 28.2% has been done on transport and mobility followed by smart city solutions (12.5%), city beautification (11.75%), construction

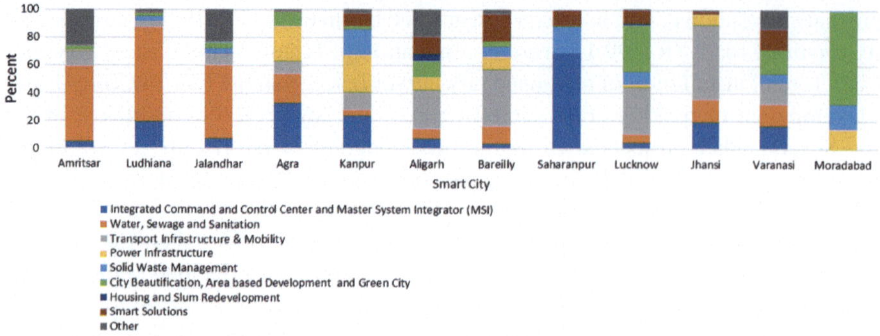


**Fig. 8.3** Spending on various projects in select smart cities, 2023. *Source* Smart Cities, MoHUA, New Delhi

of ICCC and MSI (7.45%), etc. In Bareilly city, transport and mobility projects cost around 41.11% of total cost followed by smart city solutions (19.56%) and water sewage and sanitation (12.75%). In smart city Lucknow, the major spending has been done on transport and mobility development and city beautification (33.54% and 33.13% respectively) followed by smart solutions which account for 9.39%. In Jhansi, transport infrastructure and mobility development incurred 53.63% of total spending followed by the construction of ICCC and MSI (19.51%) and water sewage and sanitation (16.41%). In the city of Varanasi, the funds have been almost equally utilised for different projects. It included 16.79% for the construction of ICCC and MSI, 16.61% for city beautification, 15.97% for water sewage and sanitation, etc. In Moradabad and Saharanpur, very few projects have been completed. This included the construction of ICCC and MSI in Moradabad with 69.2% of the spending and City beautification in Moradabad with 66.83% of the spending. Table 8.4 shows the types of projects approved for cities under the SCM in both states. A cursory look reveals that most of the projects are under the pan-city projects as compared to area-based development. However, Varanasi has a good number of projects under both project schemes. Since the Smart Cities Mission is a national flagship mission, most of projects undertaken in select cities of Punjab and Uttar Pradesh are common, equally prioritized and as core interventions in urban change. Though political differences remain in the running of these states, the SCM has been able to set a largely common urban interventions in the field of infrastructure, development and public facilities. Locational and spatial differences in the SCM projects is an inevitable result as each city's structure, design and composition differs clearly.

**Table 8.4** Types of projects approved under the SCM

| Aspects | | Pan-city projects | | Area based development | |
|---|---|---|---|---|---|
| Cities | | Number | Details | Number | Details |
| Punjab | Amritsar | 30 | SWM door-to-door collection system, procurement of truck mounted road sweeping machine, master system integrator for implementation and maintenance of smart solutions, etc. | 15 | Redevelopment of parks open spaces, facade improvement, upliftment beautification of area outside hathi gate, development of sports facility center at golbag etc. |
| | Jalandhar | 42 | ICCC, storm water sewer, construction of new 50 MLD sewerage treatment, infrastructure development (Civil) for construction & demolition (C&D) waste processing plant, etc. | 37 | Park development phase, railway station facade improvement, construction of CC street, industrial area park, etc. |
| | Ludhiana | 34 | ICCC, sewerage and septage system, fire fighting upgradation, STP rehabilitation, installation of roof top solar panels, smart road, etc. | 17 | Retrofit Sarabha nagar market, wayfinding and signages installation, water front development including landscaping beautification along Sidhwan canal, Façade lighting, etc. |
| Uttar Pradesh | Agra | 50 | Water supply, ICCC, storm water drainage system, sewage treatment plants, electric buses, etc. | 10 | Development of the Taj complex, buildings beautifications, developing mughal museum, developing Taj corridor, etc. |
| | Aligarh | 35 | Rain water harvesting, ICCC, CCTV surveillance, ensuring water supply, power line improvement, etc. | 18 | Redevelopment of bus stand and railway station, improvement in traffic management, rail over bridge construction, etc. |

(continued)

**Table 8.4** (continued)

| Aspects | | Pan-city projects | | Area based development | |
|---|---|---|---|---|---|
| Cities | | Number | Details | Number | Details |
| | Bareilly | 65 | ICCC, retrofitting of govt schools with smart class, solid waste storage and processing, smart mobility through infrastructural development, etc. | 22 | Strengthening of solid waste management system, slum redevelopment, rejuvenation of Akshar Vihar lake, etc. |
| | Jhansi | 23 | ICCC, transformation of arterial roads into smart roads, night shelter, conversion of 250-W sodium fitting, etc. | 11 | Construction of pink toilet blocks, development of open gym, development of urban haat-retail stores, interactive kiosks for tourist and information centre, Jhani fort-facility upgradation etc. |
| | Kanpur | 42 | Underground wiring & strengthening of transformer, road improvement and maintenance, ICCC, waste processing and disposal, smart city operation centre, etc. | 17 | Tree plantation tree guard, develop open gym, cuisine and handicraft festivals, upgradation of Nanarao Park, etc. |
| | Lucknow | 56 | Water supply strengthening, smart solution for existing parking, PNG network, smart city knowledge management centre, construction of Lucknow smart city office, ICCC, etc. | 24 | Establishing of open gym, beautifications of parks, development of heritage area in qaiserbagh Amirudaula library, restoration of Lal Baradari, etc. |

(continued)

**Table 8.4** (continued)

| Aspects | | Pan-city projects | | Area based development | |
|---|---|---|---|---|---|
| Cities | | Number | Details | Number | Details |
| | Moradabad | 13 | 640 kV solar roof top on government buildings, renovation and beautification of overhead tank, digital Pathshala, C & D waste recycle plant, LED-based smart street light, etc. | 2 | Redevelopment & beautification of parks, solar-based smart toilets with zero disposal |
| | Saharanpur | 21 | ICCC, sensor-based garbage bins, mobile clinic & telemedicine facility, solar power rooftop installations, LED street lights, etc. | 3 | Development of water bodies, rejuvenation of parks, construction of basket ball track |
| | Varanasi | 59 | pilot project for sweeping, collection and transportation of solid waste, city beautification, ICCC, augmentation of existing water supply system, ghats revitalization, construction of interceptor sewer and pumping, etc. | 52 | Development of 81 heritage sites, renovation and restoration of man Singh observatory, illumination or façade lighting, redevelopment of Harishchandra Ghat, etc. |

*Source* Compiled from Smart City Profiles (2023)
*Note* The projects details of Prayagraj (Allahabad) Smart City is not available

## 8.7 Scope of Smart Urban Planning: Conventional City to Smart City

The purpose of SCM was to transform cities into sustainable, citizen-centric, and technologically advanced urban centres in India. Smart cities have been seen as solutions to several challenges associated with cities globally. According to Jawaid and Khan (2020), "Smart development has been identified as a sustainable worldwide solution to the existing urban planning issues, whose principles aim to provide a better quality of life and advertise liveable communities". The *mission* in India aims to improve the quality of life for residents by leveraging ICT to enhance urban services and infrastructure. The SCM refrains from providing any template for smart cities but lays down six fundamental principles for the development of smart cities. These principles included keeping the community at the core for planning and implementation of planning purposes, more or less leading to enhance the outcomes with limited use of resources, strengthening cooperative and competitive federalism with healthy competition for the development, integrating innovation and sustainability in the city planning and utilising the technology as a means for the development of the city. The SCM also focussed on sectorial and financial convergence to achieve the goals of the mission. The selected cities in India receive funding and support to develop and implement smart solutions across various sectors such as transportation, energy, governance, waste management, water supply, and public safety. The key objectives of the mission have been the infrastructure development with a focus on creating and upgrading urban infrastructure to provide better connectivity, efficient transportation systems, reliable power supply, and sustainable waste management, technology integration and data-driven solutions to enhance service delivery, optimize resource utilization, and improve decision-making processes for urban governance. It also focussed on citizen engagement by involving residents in decision-making processes, encouraging feedback and suggestions, and enabling transparent and accountable governance. The sustainability and resilience goal of SCM encourages cities to adopt sustainable practices, promote renewable energy, reduce carbon emissions, and enhance disaster management capabilities to build resilient urban environments.

Thus, the SCM can help build efficiency in resource management through the establishment of ICCC and MSI which further help in the proper utilization of available resources and even manage human resources. The vision and implementation of the smart city initiatives in India have paid major attention to the smart solutions and interconnection of the people, administration and city systems. The integration of smart grids, efficient waste management, and intelligent transport networks can reduce energy consumption, minimize wastage, and promote sustainability. Kumar and Rattan (2020) argued, "in the 1990s, all the cities were conventional cities because at that time there was no concept of the Internet". This points to a transition in urban planning and development processes of our country. The smart city conception and its realization is a transition package wherein technology-in-system becomes a new normal. Smart is recognised as a solution to the kind of problems urban areas face. Studies also reveal that construction of smart buildings has the potential to reduce energy consumption by up to 40% and water consumption by up to 30% (Bajaj 2018). This will lead to an improvement in the quality of life in cities. The cities under the *mission* have identified the challenges being faced by city dwellers in their day-to-day lives. The cities have made provisions for solving these challenges, which reflects in the projects implemented in the cities. The environmental sustainability is at the core of smart development plans. The focus also is on reducing dependence on non-renewable power resources as many lights in these cities have been replaced with power-saving LEDs and solar lights. These cities emphasize renewable energy and incorporate green infrastructure. It will help to mitigate the environmental challenges. The smart city aimed at bridging the gap through active citizen participation in planning and policy. This innovative approach may foster transparency, inclusivity and accountability at urban local bodies. The cities under the SCM do face funding crunch and issues of financial sustainability arise. The high cost involved in implementing smart cities in India is substantially high in constructing infrastructure and technology solutions. Funding remains a significant constraint in smart city projects in India, with cities facing challenges in raising funds from diverse sources. According to Bajpai and Biberman (2021), "Funding shortfalls across the program clearly indicate that more was awarded to these area-based developments than was actually readily available, and that more could have been made available for "smarter," more cost-effective city-wide plans by deemphasizing area-based development, provided similar levels of funding would have been available for scaling these up". Safeguarding public data and ensuring privacy is also important in the smart city. Stakeholder capacity building is very much needed as stakeholders in such projects lack the understanding and ways of implementation of projects.

The mission bridged the infrastructure gaps by developing efficient transportation systems, upgrading utilities, and providing improved urban services. The establishment of ICCC played a crucial role in this as it enables the monitoring and delivery of services seamlessly. Intelligent traffic management systems and integrated public transportation networks, aim to reduce traffic congestion and improve mobility within cities. The incorporation of digital platforms for citizen engagement, e-governance, and transparent administrative processes is an unprecedented initiative. This can create a cycle of development by implementing the success strategies of existing smart cities in new cities and towns. Encouraging innovation and research in areas such as emerging technologies, data analytics, and urban planning to find innovative solutions for urban challenges can help in the proper implementation of development plans. With such strategies, conventional cities can be turned into smart cities. The takeaway from the ongoing projects are:

- Conducting a comprehensive assessment of the city's current state, identifying its strengths, weaknesses, and priority areas for improvement. This assessment should take into account the city's infrastructure, services, governance, and citizen needs.
- Defining a clear vision and strategy for the smart city transformation, outlining the desired outcomes, objectives, and focus areas. This should align with the unique characteristics and challenges of the city.
- Involvement of all relevant stakeholders, including government authorities, citizens, businesses, and academia, in the decision-making process, is a crucial step for smart city development. Seeking their input, feedback, and active participation to ensure the transformation aligns with their needs and aspirations.
- Developing a comprehensive and integrated master plan should prioritize the use of technology, sustainability, and citizen-centric solutions.
- Upgradation and development of urban infrastructure are essential to achieve to support smart city initiatives. Some of them include improving transportation networks, enhancing connectivity, deploying digital infrastructure, and promoting green and sustainable infrastructure solutions by identifying and deploying appropriate technologies to enable smart solutions in different domains.
- In order to achieve the target of smart governance, the implementation of data platforms and systems that enable the collection, processing, and utilization of data to make informed decisions, optimize resource allocation, and enhance service delivery is required to achieve the goal of smart city.

- Citizens can be encouraged to provide feedback, suggestions, and active involvement in monitoring and evaluating smart city projects.
- Providing technical and digital systems training and capacity building of relevant stakeholders involved in the smart city transformation can help sustain the smartness of the cities.

## 8.8 Concluding Remarks

The paper demonstrates that the SCM has come as a timely response to the need to resolve the ways urban problems and challenges are handled and addressed in the context of rapid urbanization in the country. The *mission* intends to make the city smart and is not limited to technological solutions and its integration, but people-centric approach is given a direction. The states of Punjab and Uttar Pradesh do not provide a systematic comparative perspective, but the cities selected under the mission so a picture of what kinds of projects are chosen and being implemented differently. Punjab has a high level of urbanization that is concentrated in important cities, whereas urbanization in Uttar Pradesh is comparatively low and spread over many cities and towns. The cities under the mission prepared a plan setting up the Pan City initiative that was focussed on the comprehensive development of the city. In both states, smart cities focussed on improving the basic amenities under the Pan City initiative including the development of water pipelines, sewage lines and sanitation facilities more in Punjab rather than Uttar Pradesh. The cities in Uttar Pradesh primarily focussed on developing an Integrated Command and Control Centre for dispensing the services and controlling the transport and mobility in the cities. Key area-based development plans catered to the specific needs of the cities. The historic and cultural heritage-rich cities, which witness the influx of tourists and pilgrims such as Amritsar, Varanasi, Agra, Prayagraj etc., did consider projects to provide better facilities to tourists in ways such as assuring digital connect, beautification, smooth mobility, safety, etc. Whereas, industrial-based cities such as Ludhiana, Jalandhar, Kanpur, Aligarh, Saharanpur, and Moradabad have taken into account provisions for industrial clusters development and strengthening the robust industrial infrastructure under the area-based development plans. Mobility and transport have been one of the most important themes for the development across the cities. Though the smart city in these states has made important progress, there is a lot more to be done. The cities of Uttar Pradesh and Punjab have shown substantial progress in developing the cities as smart. Projects in cities under the SCM seems common in both the state despite political differences and bureaucracy at functioning. The SCM has been a powerful tool to bring select cities on common agenda of urban change and transformation. However, to ensure the successful implementation of smart city initiatives, it is crucial to address inherent challenges such as funding, infrastructure, data privacy, stakeholder collaboration, and inclusivity. However, the special purpose vehicle formed for the purpose of implementation and regulation of projects in each is a centralized structure and how they would add to achieve democratic urban planning and effective role of urban local bodies is yet to be researched.

## References

Arku G, Marais L (2021) Global south urbanisms and urban sustainability: challenges and the way forward. Front Sustain Cities 3:1–4. https://doi.org/10.3389/frsc.2021.692799

Bajaj R (2018) Smart cities in India: creating a smarter nation (White Paper). Sterlite Tech. https://stl.tech/wp-content/uploads/2023/03/smart-cities-in-india-creating-a-smarter-nation.pdf. Accessed 20 Aug 2023

Bajpai N, Biberman J (2021) India's smart city program: challenges and opportunities. CSD working paper series: towards a new Indian model of information and communications technology-led growth and development, ICT India working paper #62. Centre for Sustainable Development. Columbia University

Bhatikar T, Neha S (2021) Cities and regions in South Asia. In: Orum AM, Ruiz-Tagle J, Haddock SV (eds) Companion to urban and regional studies. John Wiley & Sons Ltd., New York, pp 3–20

Census of India (1971–2011) General population tables. Directorate of Census Operation, India

Chakravorty S (1996) Too little, in the wrong places?: mega city programme and efficiency and equity in Indian urbanisation. Econ Pol Wkly 31(35/37):2565–2572

Cohen M (2013) Austerity and the global crisis: lessons from Latin America. Soc Res 80(3):929–952

Eremia M, Toma L, Sanduleac M (2017) The smart city concept in the 21st century. Procedia Eng 181:12–19. https://doi.org/10.1016/j.proeng.2017.02.357

Gavsker KK (2020b) Urbanization, changing economic landscape and policy responses in India: a study with reference to empowered action group states. Int J Geogr Reg Plan 6(2):160–172

Gavsker KK (2020a) New media technologies and the urban condition: changing socio-spatial relations and emerging geographies. In: More B (ed) Emergent technologies-new media and urban life. Common Ground Research Networks, Illinois, pp 1–16

Government of India (2012) Twelfth five-year plan: faster, more inclusive, sustainable growth. Planning Commission, New Delhi

Greenpeace (2017) Airpocalypse: assessment of air pollution in Indian cities. Greenpeace, New Delhi

Harrison I, Donnelly A (2011) A theory of smart cities. In: Proceedings of the 55th annual meeting of the ISSS-2011, Hull, UK, 55(1). https://journals.isss.org/index.php/proceedings55th/article/view/1703. Accessed 20 Aug 2023

Hoelscher K (2016) The evolution of the smart cities agenda in India. Int Area Stud Rev 19(1):28–44. https://doi.org/10.1177/2233865916632089

Hollands RG (2008) Will the real smart city please stand up?: intelligent, progressive or entrepreneurial? City 12(3):303–320. https://doi.org/10.1080/13604810802479126

Jawaid MF, Khan AR (2020) The smart city mission in India and prospects of improvement in the urban environment. IOP Conf Ser: Mater Sci Eng 955(1):012001. https://doi.org/10.1088/1757-899X/955/1/012001

JNNURM (2005) Jawaharlal Nehru national urban renewal mission: overview. Ministry of urban employment and poverty alleviation, Ministry of Urban Development, Government of India. https://mohua.gov.in/upload/uploadfiles/files/1Mission%20Overview%20English(1).pdf. Accessed 20 Aug 2023

Kumar A, Rattan JS (2020) A journey from conventional cities to smart cities. In: Sara Shirowzhan S, Zhang K (eds) Smart cities and construction technologies. Intechopen. https://doi.org/10.5772/intechopen.91675

Mathur OP, Naqvi AH, Laroiya A et al (2021) State of the cities India. Institute of Social Science, New Delhi

Mckinsey Global Institute (2018) Smart cities in Southeast Asia, Mckinsey & Company in Southeast Asia, Jakarta
MoHUA (2019) Ministry of housing and urban affairs smart cities mission: city-wise fund release and utilization. https://pib.gov.in/PressReleasePage.aspx?PRID=1780910. Accessed 16 Sept 2023
Moir E, Moonem T, Clark G (2014) What are future cities? Origin meaning an uses. In: Compiled by cities for the foresight future of cities project and guture cities Catapault, office of Science
NITI Aayog (2021) Reforms in urban planning capacity in India. NITI Aayog, Government of India, New Delhi
OECD (2020) Smart cities and inclusive growth-building on the outcomes of the 1st OECD roundtable on smart cities and inclusive growth. https://www.oecd.org/cfe/cities/OECD_Policy_Paper_Smart_Cities_and_Inclusive_Growth.pdf
Public Policy Research Centre (2014) Performance evaluation of Jawaharlal Nehru national urban renewal mission (JNNURM) in BJP Ruled States. Public Policy Research Centre, New Delhi. http://pprc.in/upload/PPRC%20Report%20on%20JNNURM.pdf. Accessed 16 Sept 2023
Rai PT (2012) Townships for sustainable cities. Procedia-Soc Behav Sci 37:417–426
Smart City Mission (SCM) (2015) Smart city mission statement and guidelines. Ministry of Urban Development, Government of India
Söderström O, Paasche T, Klauser F (2014) Smart cities as corporate storytelling. City 18(3):307–320. https://doi.org/10.1080/13604813.2014.906716
The Economist Intelligence Unit (2016) Startup my city-smart and sustainable cities in Asia. The Economist Intelligence Unit. https://startupmycity.economist.com/wp-content/uploads/2017/01/EIU-Startup-My-City-Briefing-Paper.pdf. Accessed 20 Aug 2023
Toli AM, Murtagh N (2020) The concept of sustainability in smart city definitions. Front Built Environ 6(77):1–10
Town and Country Planning Department (2021) Urbanization in Uttar Pradesh. Town and Country Planning Department of Uttar Pradesh. https://uptownplanning.gov.in/page/en/urbanization-in-uttar-pradesh. Accessed 15 Sept 2023
United Nations (2014) Population facts. Department of Economic and Social Affairs Population Division, United Nations
Veselitskaya N, Karasev O, Beloshitskiy A (2023) Drivers and barriers for smart cities development. 14(1): 85–110. https://um.ase.ro/no141/5.pdf Accessed 25 Sept 2023
Weinstein L, Sami N, Shatkin G (2014) Contested developments. In: Shatkin G (ed) Contesting the Indian city: global visions and the politics of the local. Wiley Blackwell, Chichester, UK
World Cities Report (2016) Urbanization and development: emerging futures world cities report 2016. United Nations Human Settlements Programme (UN-Habitat)
WUP (2019) World urbanization prospects: the 2018 revision. Department of Economic and Social Affairs. United Nations, New York. https://population.un.org/wup/publications/Files/WUP2018-Report.pdf. Accessed 25 Sept 2023
Xie W, Deng D, Chong Z (2019) The spatial and heterogeneity impacts of population urbanization on fine particulate (PM2.5) in the Yangtze river economic belt, China. Int J Environ Res Public Health 16 (1058):1–17

# Part III
# Geo-Environmental Sustainability and Tourism

# Chapter 9
# Challenges for Achieving Sustainable Water Supply and Sanitation in Dhaka City Under Rapid Urbanization

**Shimul Roy, Md Sariful Islam, Tanmoy Roy Tusher, and Shanta Dutta**

**Abstract** Dhaka is one of the fastest-growing megacities in Asia, where the urban population is increasing noticeably. Currently, the megacity is home to more than 21 million people, with the average annual growth rate of the population is 4.2% annually. By 2030, the city's population is expected to reach about 28.1 million. The growing urban population creates an enormous load on the city's existing infrastructures and amenities. This chapter summarizes the urbanization trend in Dhaka city, the status of water supply and sanitation services, existing infrastructural facilities, and significant challenges and prospects towards ensuring sustainable water supply and sanitation in light of the United Nation's sustainable development goal (SDG). The study shows that the rapid urbanization process in Dhaka city creates tremendous pressure on the city's infrastructure and essential services, including water supply and sanitation. Moreover, due to resource constraints, inadequate infrastructures, and lack of coordination in the service sector, it is challenging to ensure adequate water supply and sanitation services. However, considerable progress has been observed mainly in the city's water supply sector, including several projects and plans that have already commenced to comply with the SDG in the water supply and sanitation sector.

**Keywords** Dhaka · Urbanization sanitation · Water supply

S. Roy (✉) · T. R. Tusher
Department of Environmental Science and Resource Management, Mawlana Bhashani Science and Technology University, Tangail, Bangladesh
e-mail: royshimul@mbstu.ac.bd

M. S. Islam
Media Lab, Massachusetts Institute Technology, Cambridge, MA, USA

S. Dutta
Department of Civil and Environmental Engineering, The Hong Kong Polytechnic University, Hung Hom, Kowloon, Hong Kong, China

 
S. Anand et al. (eds.), *Emerging Geosustainability Transformations in India*, Advances in Geographical and Environmental Sciences,
https://doi.org/10.1007/978-981-97-7993-2_9

## 9.1 Introduction

Urbanization is the most common and visible phenomenon in the present world, mainly in developing countries. The United Nations reported that rapid urbanization is mostly found in developing countries where more than half of the population lived in urban areas between 1970 and 1994 (United Nations 1995). It is projected that by 2030, around 60% of the people in developing countries will live in urban areas (Cohen 2006). Sekkat (2017) reported that among the 30 largest urban agglomerations, 21 were in developing countries in 2010, and it could reach 24 out of 30 in 2025. Rapid urbanization in developing countries has seriously transcended most cities' capacity to deliver adequate services for city dwellers (Cohen 2006). Rana (2011) asserted that the major cities in developing countries are confronting and will continue to face the numerous problems and challenges of uncontrolled urbanization that may cause massive distress to city dwellers.

Bangladesh is one of the most densely populated countries worldwide where the rate of urban population growth is increasing rapidly. Rapid urbanization in the major cities of Bangladesh, particularly in Dhaka, is deteriorating the environmental condition and becoming a threat to environmental sustainability. Rana (2011) reported that this condition might be more severe due to uncontrollable rural–urban migration, lack of good governance, and unequal distribution of resources.

As a developing country, Bangladesh faces tremendous challenges of immense urban population growth, mainly due to rural–urban migration, lack of employment opportunities, rural landlessness, effects of natural disasters, and so forth (Khan and Siddique 2000; Islam et al. 2014; Rana 2011). As a result, the environmental conditions in most of the urban areas of the country are degrading due to many factors, including the discharge of human and industrial wastes into the surrounding surface water bodies, groundwater contamination from insufficient sewerage facilities, poor solid waste management and disposal systems, etc. Moreover, the urban population's inconsistent growth creates enormous pressure on the country's infrastructures and services such as water supply, sanitation, solid waste collection and management, etc. Khan and Siddique (2000) reported that these problems are most severe in Dhaka city, where a comparatively higher population growth rate is observed and where ~45% of households have income below the poverty line. In Bangladesh, the development of adequate urban infrastructures, water supply, and sanitation services is hindered due to a large number of urban populations and limited financial resources.

Based on the findings from available relevant literature, this study investigates the problems associated with rapid urbanization and water supply and sanitation services in Dhaka city. However, the study intended to explore some specific research questions: (i) What are the factors influencing rapid urbanization in Dhaka city? (ii) What is the current status of water supply and sanitation services? (iii) What are the government and city corporation's roles towards sustainable water supply and sanitation services?

## 9.2 Urbanization in the World and Bangladesh: an Overview

Globally, the urban population is increasing rapidly over time, with a much higher contribution from the less developed regions than the more developed regions. During 1990–2015, the world's urban population increased by ~3% annually, while the growth in the urban population in the less developed regions and more developed regions of the world was 4.2% and 0.7%, respectively (Table 9.1). The United Nations (2019) projected that in 2030, the world's total population could reach more than 8.5 billion, with a significant contribution by the less developed regions (i.e., ~85%). It is also estimated that the world's urban population will increase significantly between 1990–2030 (~2.3 times), with a significant share from the less developed countries (i.e., ~80%). Compared to 1990, the share of urban population to the world's total population is expected to increase noticeably in 2030 (i.e., 43% to ~61%) (Table 9.1). Cohen (2006) asserted that currently, around half of the world's population is in urban areas resulting from continued urbanization past couple of decades. It is also expected that by 2030, ~61% of the world's population will live in urban areas.

In Bangladesh, the population is increasing at an unprecedented rate compared to other developing countries. Currently, the country has a population of more than 163 million (World Bank 2021), with an average annual growth rate of 1.42% (Table 9.2). Most of the population increase occurs in urban areas, which results in an increased urbanization level. For instance, the level of urbanization in Bangladesh was very low before 1951 (i.e., 2.54–3.66%), which increased gradually in 1961 (i.e., > 5%), and a rapid increase had been observed from 1974–2001 (i.e., ~9–23%), as presented in Table 9.2 (BBS 1981; BBS 1991; BBS 2001). It is observed that most of the increases in urban population occur in the country's major cities, which are creating enormous pressure on the city's existing infrastructures and services and degrading the environmental quality and standards. For example, "the four major cities of Bangladesh (e.g., Dhaka, Chittagong, Khulna, and Rajshahi) hold nearly 60% of

**Table 9.1** Total and urban populations by the development group (1990–2030)

| Development group | Population (billions)[a] | | | | | | | |
|---|---|---|---|---|---|---|---|---|
| | 1990 | 1995 | 2000 | 2005 | 2010 | 2015 | 2020[b] | 2030[b] |
| *Total population* | | | | | | | | |
| World | 5.33 | 5.74 | 6.14 | 6.54 | 6.96 | 7.38 | 7.79 | 8.55 |
| More developed regions | 1.15 | 1.17 | 1.19 | 1.21 | 1.23 | 1.26 | 1.27 | 1.29 |
| Less developed regions | 4.18 | 4.57 | 4.96 | 5.33 | 5.72 | 6.12 | 6.52 | 7.26 |
| *Urban population* | | | | | | | | |
| World | 2.29 | 2.58 | 2.87 | 3.22 | 3.59 | 3.98 | 4.38 | 5.17 |
| More developed regions | 0.83 | 0.86 | 0.88 | 0.92 | 0.95 | 0.98 | 1.00 | 1.05 |
| Less developed regions | 1.46 | 1.72 | 1.98 | 2.30 | 2.64 | 3.00 | 3.38 | 4.12 |

*Notes* [a]United Nations (2019); [b]Projected Populations

**Table 9.2** Growth of urban population in Bangladesh (1901–2019)

| Year | Total national population (million) | Annual growth rate of national population (%) | Total urban population (million) | % of urban population[b] |
|---|---|---|---|---|
| 1901 | 28.2 | 0.70 | 2.43 | – |
| 1911 | 31.65 | 0.94 | 0.80 | 2.54 |
| 1922 | 33.25 | 0.60 | 0.87 | 2.61 |
| 1931 | 35.60 | 0.74 | 1.07 | 3.01 |
| 1941 | 41.99 | 1.70 | 1.54 | 3.66 |
| 1951 | 44.17 | 0.50 | 1.83 | 4.34 |
| 1961 | 55.22 | 2.26 | 2.64 | 5.14 |
| 1974 | 76.37 | 2.48 | 6.00 | 8.89 |
| 1981 | 89.91 | 2.32 | 13.56 | 15.54 |
| 1991 | 111.45 | 2.17 | 22.45 | 20.15 |
| 2001 | 123.85 | 1.54 | 28.60 | 23.01 |
| 2011 | 144.04 | 1.47 | 33.56 | 23.30 |
| 2019[a] | 163.05 | 1.04 | 60.99 | 37.41 |

*Notes* BBS (1981, 1991, 2001, and 2011); **a.** World Bank (2021); **b.** Indicates the level of urbanization in Bangladesh

the total urban population, while Dhaka alone holds 35%" (BBS 2001). However, Table 9.2 shows that the rapid growth in the urban population in Bangladesh occurred after 1974.

## 9.3 Urbanization in Dhaka City

Dhaka, the capital of Bangladesh, is one of the most densely populated megacities around the world. Currently, the megacity is home to more than 21 million people, with an average annual growth rate of 4.2% annually (World Population Review 2021). A rapid urbanization process and growth in the urban population have been observed in Dhaka city for the past couple of decades (Islam et al. 2014). It is observed that the current density of the population is much higher than in the last few decades. For instance, the current population density in Dhaka city is reported 8,229 persons per sq. km. (Sharmin et al. 2015) while it was 4,795 persons per sq km in 1991 (Department of Environment, Waste Concern and ITN-BUET 2004).

It is observed that Dhaka city's contribution to the national urban population increased significantly from 1981–2001. In 1981, its contribution to the national urban share was 25%, which increased considerably to 31% in 1991 and 34% in 2001 (BBS 1997). According to the United Nations (1998), the population growth rate in Dhaka city during 1974–2000 was 6.9%, which was higher than any other city in the world at that period. The rapid population growth in Dhaka city results in increased poverty and social vulnerability, lack of infrastructural facilities and essential services, environmental degradation, and inefficient urban management capacities (Hossain 2006). However, it is predicted that by the year 2030, the population of the city will reach 27.3 million (Zaman 2019), triggered by rapid economic development. This increased population in the city could severely hinder the development processes or activities like water supply, sanitation, housing, solid waste collection, disposal, and management systems.

### *9.3.1 Factors Behind Rapid Urbanization and Associated Impacts*

Apart from natural growth in the urban population, several factors are involved in the rapid urban population growth in Bangladesh, particularly in Dhaka. These include rural–urban migration, natural hazards, uneven development, local politics, poverty, unemployment, necessary service facilities like water supply and sanitation, housing facilities, health services, and so forth. Among the different factors, migration seems to be the most significant factor responsible for the rapid urbanization in Bangladesh's major cities. Generally, rural people and those who live in the areas adjacent to the cities tend to migrate to urban areas due to better employment opportunities, improved lifestyles, and essential services. Therefore, the contribution of frequent migration to the increased urban population growth in Bangladesh is much higher. Islam (1999) stated that during 1974–1981, migration contributed 40% to the urban growth in Bangladesh, which was nearly the same from 1981–1991, and for large cities like Dhaka could be up to 70%. However, the country's major cities migration rate, including Dhaka, increases due to the push–pull factors. According to Rana (2011), the pull factors include employment opportunities, better lifestyles, social security, and better income opportunities, while the push factors include unemployment, rural poverty, and natural hazards. It should be mentioned that because of push–pull factors, Bangladesh confronted a tremendous urban population growth rate of 10.3% during 1974–1981 (BBS 2001).

The rapid urbanization in Dhaka city has severe impacts on existing infrastructures, services, resources, and environmental quality directly or indirectly. Several studies reported deteriorating urban environmental conditions in Dhaka city due to the direct or indirect impacts of rapid growth in the urban population in the city (Akbar et al. 2007; Hossain 2006; Islam et al. 2014; Du et al. 2010; Roy et al. 2016, 2019; Rana 2011). These include changes in landscapes, inadequate water supply, poor sanitation, solid waste management, drainage congestion, water pollution, degrading ambient air quality, loss of wetlands, loss of urban greeneries, urban heat island impacts, and degraded physical and social environmental quality.

## 9.4 Status of Water Supply and Sanitation Services

Although safe and adequate water supply and sanitation are fundamental human needs and rights, inadequate water supply and sanitation facilities are prevalent in most developing countries where Bangladesh is not an exception. Generally, in Bangladesh, the primary source of water supply is groundwater. Khan and Siddique (2000) asserted that about 95% of the domestic and industrial water uses are extracted from the groundwater source in the urban areas of Bangladesh. In Dhaka City, 78% of water is extracted from groundwater sources, and the remaining 22% is from surface water sources (DWASA 2019). Due to groundwater's over-extraction, the water table is lowering rapidly (2–3 m annually). It also accelerates the extraction costs and results in severe water shortage during the summer when the water demand becomes high.

However, in Dhaka city, the Dhaka Water and Sewerage Authority (DWASA) provides water supply and sanitation services. Due to the rapid growth in the urban population, the demand for water supply has increased significantly over time (Fig. 9.1). The study shows a significant gap in water demand and supply scenario during 1963–2018. During this period, the water deficit ranges from 20 to ~500 million liters in Dhaka city. However, it became much lower after 2010, as the amount of water production capacity of the DWASA has increased significantly. For instance, during 1963–2010, the average annual water production was ~1,132 million liters, while from 2011–2018, it reached more than 2,386 million liters (Fig. 9.1).

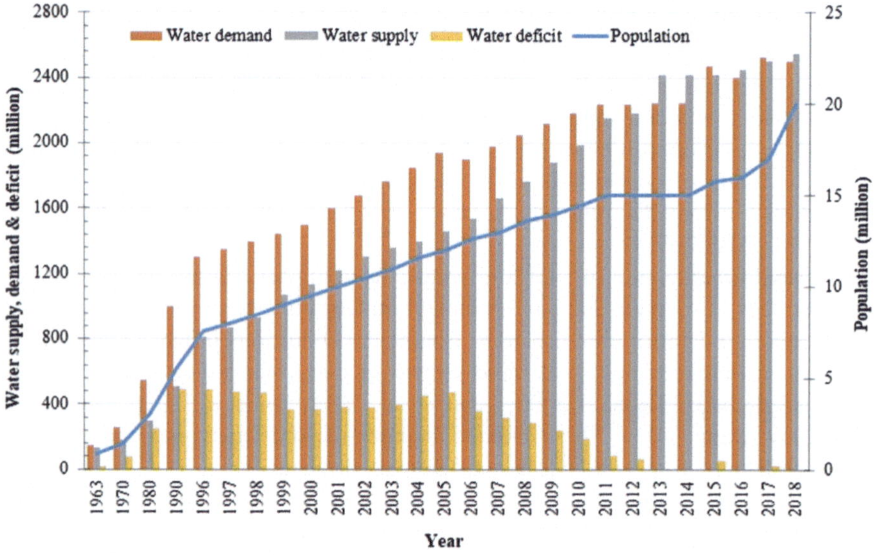


**Fig. 9.1** Water demand, supply, and deficit scenario with the growing population in Dhaka city. *Source* Adapted from (DWASA 2019, Yeazdani 2016, Haque 2019)

Ensuring equal and sustainable access to water for the rapidly growing Dhaka city population is a rising concern. According to Siddiquee and Ahamed (2020), meeting the growing water demands of the increased urban population is challenging for the DEWASA, as there is a significant gap between supply and demand. Streatfield and Karar (2008) reported that based on current demands, around 2.1 billion liters of water are needed daily, while the DWASA can provide about 1.45 billion liters per day, mainly from groundwater extraction. The rest of the water is supplied from the nearby *Shitalakshya* River, which is severely polluted due to the discharge of untreated or poorly treated industrial effluents. As a consequence, people are becoming infected by using this low-quality water. Moreover, it is essential to mention that vast numbers of slum populations reside in Dhaka, and their proportion is increasing rapidly. These people have limited access to the water supply and sanitation facilities (Islam et al. 2014). According to Streatfield and Karar (2008), instead of getting the recommended amount of water (200 L/day) for all purposes, many slum people cope with less than 10 L per day. Although the DWASA is responsible for supplying the pipe water in Dhaka city, most people who live in the squatter settlements or informal settlements are not adequately supplied with water or are deprived of it. According to CBSG (2010), only 30% of the slum residents receive adequate water services provided by DWASA. Previous studies also reported unsatisfactory water services to the urban poor in Dhaka city (e.g., Yeazdani 2016; Roy and Dutta 2017). However, this study identified that to meet the water demands, the poor people living in the squatter settlements or informal settlements have to rely on other water sources like tube wells, nearby ponds or rivers, dug wells, rainwater, etc. Some of them also rely on water vendors to meet their basic needs of water.

The sanitation services in the urban areas of Bangladesh are unsatisfactory to the urban dwellers. The waterborne sewerage system covers only some parts of Dhaka (DWASA 2019; Khan and Siddique 2000). However, in the City's vast area, the use of unsanitary latrines, and direct disposal of sewage in open fields is a common scenario. For instance, it is estimated that only 40% of the city's population can access sanitary waste disposal facilities. Streatfield and Karar (2008) mentioned that only two big cities of Bangladesh (i.e., Chittagong and Dhaka) have a piped water-based sewerage system, which is very old and poorly maintained. The sewerage system serves only 30% of the city dwellers of these areas, another 30% uses septic tanks, and the rest of the population has no primary sanitation facility.

The urban population in the squatter settlements in Dhaka city is increasing due to rapid migration. These poor people do not set up latrines due to limited space and financial constraints. Moreover, government bodies are becoming unable to provide adequate sanitation services to them. As a result, many households must share one latrine in an unsustainable manner. Although the DWASA is responsible for providing sanitation services to the urban dwellers in Dhaka city, these services of the DWASA are considered inadequate and not satisfactory to urban dwellers, especially to the city's urban poor. For instance, the sewer and sewage treatment system of DWASA serves about 15–20% of the city's population, ~25% have septic tanks on site, 15% of city dwellers use sanitary pit latrines, ~5% use bucket latrines, and another 35–40% rely on unsanitary systems (Hossain 2006).

### 9.4.1 Status of Existing Infrastructural Facilities

As a fast-growing megacity with a huge urban population, Dhaka city faces tremendous challenges and problems in almost all infrastructure sectors such as water supply, sanitation, solid waste disposal and management, and sewerage system. Siddiquee and Ahamed (2020) asserted that due to rapid urbanization, this situation is aggravated by the proliferation of informal, unplanned, and underserved settlements. In Bangladesh, water and sanitation services are provided by different public authorities such as the Water Supply and Sewerage Authority (WASA), the Department of Public Health and Engineering (DPHE), and the Local Government and Engineering Department (LGED). However, in Dhaka, the water supply and sanitation services are served by the DWASA. However, the city corporation's role is considered crucial, as it is the city corporation's responsibility to join the households in the water supply systems. But the matter of fact is that the services provided by these authorities are not satisfactory to the urban dwellers still now, mainly because of poor governance and management practices, coordination gaps with other relevant organizations or service providers, corruption, and lack of financial support. Previous studies (e.g., Roy and Dutta 2017, Rahman and Malik 2003, Rana 2011) reported a lack of coordination among the local or central government, private sectors, and non-government organizations (NGOs) in terms of water supply and sanitation services. This coordination gap indicates the institutional failure to ensure adequate services to the urban dwellers in Dhaka city.

The existing infrastructural facilities in water supply and sanitation in Dhaka city are considered inadequate to meet the urban population's growing demands. Islam et al. (2014) reported that a significant proportion of Dhaka city's inhabitants are dissatisfied with the existing water supply and sanitation facilities (i.e., 67% and 48%, respectively). Although significant progress has been observed in recent times, particularly in the water supply sector, the sanitation sector is still less focused (Tables 9.3 and 9.4). As shown in Table 9.3, the current water supply services have improved significantly in Dhaka city, which fulfilled the water demand by >94% (average) during 2010–2015, mainly through installing deepwater tube wells. In contrast, before 2010, this figure was 51–75%. A similar finding (i.e., 71%) has been reported by Rana (2011). However, in reality, it is challenging to fulfill the water demand by more than 90%, as adequate water supply could be hindered by various factors. These include water loss in the supply system through leakage in water pipelines, illegal water connections or networks by some DWASA's authority, and system loss or mismanagement through corruption. As a result, a portion of city dwellers (e.g., people living in slums or squatter settlements) are truly deprived of adequate water services.

As the main water supply option in Dhaka city is the pipe water supply, the large numbers of the urban poor living in the squatter settlements have minimal access to it. They rely on untreated or poorly treated surface water as an alternative option and become infected by various waterborne diseases. Hossain (2006) reported that currently, 60% of Dhaka city's residents have access to municipal piped water, and

**Table 9.3** Growth in water supply service in Dhaka (1990–2015)

| Year | Water demand (ML) | Water supply (ML) | % fulfilled of demand | No. of deep tube-wells |
|---|---|---|---|---|
| 1990 | 1,000 | 510 | 51.0 | 216 |
| 1996 | 1,300 | 810 | 62.3 | 216 |
| 2000 | 1,500 | 1,130 | 75.3 | 308 |
| 2005 | 1,940 | 1,460 | 75.3 | 418 |
| 2010 | 2,180 | 1,990 | 91.3 | 560 |
| 2015 | 2,475 | 2,420 | 97.8 | 702 |

*Notes* Adapted from DWASA (2019); ML-million litres

**Table 9.4** Infrastructural facilities in the water supply and sanitation sector in Dhaka

| Items | 2010–11 | 2011–12 | 2012–13 | 2013–14 | 2015–16 | 2016–17 | 2017–18 | 2018–19 |
|---|---|---|---|---|---|---|---|---|
| Deep tube-well | 586 | 615 | 644 | 672 | 760 | 795 | 827 | 887 |
| Water treatment plant | 4 | 4 | 4 | 4 | 4 | 4 | 4 | 4 |
| Water production (ML/day) | 2,150 | 2,180 | 2,420 | 2,024 | 2,430 | 2,450 | 2,500 | 2,550 |
| Water line (km) | 2,800 | 3,040 | 3,040 | 3,040 | 3,500 | 3,600 | 3,720 | 3,750 |
| Water connection | 3,02,132 | 3,10,314 | 3,25,717 | 3,40,756 | 3,61,938 | 3,71,766 | 3,79,686 | 3,90,642 |
| Sewer line (km) | 882 | 883 | 885 | 885 | 930 | 930 | 934 | 934 |
| Sewage treatment plant | 1 | 1 | 1 | 1 | 1 | 1 | 1 | 1 |

*Notes* DWASA (2014); DWASA (2019)

15% have indirect supply, while the remaining population depends on private well water and surface water. Although the city residents collect rainwater in recent times to meet the water demands, it becomes difficult for them to manage the rainwater properly due to congested and unplanned housing.

The sections discussed above show that the urban population in Dhaka city is not satisfied with the existing water supply and sanitation services. Furthermore, the current infrastructural facilities seem reasonably insufficient to meet the growing urban population's demands. Although considerable progress has been observed in the infrastructural facilities in the last couple of years, it is still challenging for the government to ensure sustainable water supply and sanitation in Dhaka city. If the current urban population growth trend continues in Dhaka city, the situation is expected to be more complicated and could become a barrier to achieving sustainable water supply and sanitation in the coming decade. Therefore, based on the current circumstances, it can be undoubtedly stated that uncontrolled urbanization might not be the blessing or spell for achieving sustainable water supply and sanitation in Bangladesh, particularly for the densely populated Dhaka city.

### *9.4.2 Step Toward Sustainable Water Supply and Sanitation Practices in Light of SDGs*

Sustainable water supply and sanitation are pivotal for healthy urban life. It is also one of the biggest challenges to meet the need for proper water supply and sanitation in the developing world's fast-growing cities. It is projected that most of the future urban growth will occur in small and medium-sized cities in the developing world (Fragkias et al. 2013). Meeting the projected future population's needs means developing new systems that are safe, affordable, and sustainable for future generations. Moreover, providing sustainable water supply and sanitation to all has many challenges to overcome. By 2017, 71% of the total global population used safe drinking water, and 45% of the total population managed to use safe sanitation services (United Nations 2020). According to the Centre for Affordable Water and Sanitation Technology (CAWST), there are five significant challenges to achieving sustained, universal access to safe water and sanitation: (i) including the scale of need for safe water and sanitation, (ii) the spatial variability of water and sanitation problems, (iii) how to sustain essential wash service for the long run, (iv) how to reach to the people who are in most need, and (v) the integration of water, sanitation, and hygiene for health. In developing countries, local governance also plays a vital role as a potential barrier to the implementation of SDG 6 (Herrera 2019).

To overcome the challenges, investment in sustainable water supply and sanitation is needed globally. Hutton et al. (2007) stated that investment in proper water and sanitation services is cost-effective across all world regions. Based on the cost–benefit analysis, they found that per US$ 1 investment in water and sanitation services produces economic benefits up to US$ 46. A sustainable sanitation system with resource recovery can produce products like fertilizers, irrigation water, or even biogas with market value. A proper and healthy water supply and sanitation systems are beneficial economically and reduce the global disease burden. Prüss-Üstün et al. (2008) found that water, sanitation, and hygiene can reduce the global disease burden by 9.1% and can prevent 6.3% of the total deaths related to it.

With more than 160 million population, Bangladesh has made remarkable progress toward providing water supply and sanitation during the last few decades. With a goal for safe and sustainable water supply and sanitation for all, the National Strategy for Water Supply and Sanitation 2014 was developed in Bangladesh. This strategy's main objective was to provide a uniform guideline to all the stakeholders, including government institutions, the private sector, and NGOs (LGRD 2014).

According to the World Bank (2018), access to water and sanitation has been gained from 1990–2015 by 20% and 29%, respectively, in Bangladesh. A joint survey by the Bangladesh Bureau of Statistics (BBS) and UNICEF Bangladesh also confirmed that Bangladesh had progressed significantly toward achieving SDG goals. This survey's key findings reported that almost 98.5% of people have access to improved drinking water sources while about 84.6% of Bangladesh households have access to improved sanitation (BBS 2019).

As a step towards complying with the SDG, different initiatives have been taken, including increased surface water collection and treatment facilities, rain-water harvesting projects, increased sewerage treatment plants, etc. For instance, several large water treatment plant (WTP) projects have already been undertaken to raise surface water reliance by up to 70% (DWASA 2019). Currently, the *Said-abad* Water Treatment Plant (Phase-III) is under implementation and will supply 450 million liters of water daily to the city dwellers. Additionally, two large WTPs such as *Gandharbpur* and *Padma* WTP (Phase-I) have already been undertaken to meet the growing water demands and to fulfill the target of using surface water. Collectively, these WTPs could supply 950 million liters of water daily, collected from two large rivers (i.e., *Padma* and *Meghna*). In terms of rainwater collection, significant progress has been observed in Dhaka city. For instance, in the northern part of Dhaka (*Uttara*), 169 buildings are under construction by the *Rajdhani Unnayan Kartripakkha (Rajuk)* that will be equipped with rainwater harvesting (RWH) systems. Furthermore, By 2021, the government intends to build 12,000 and 60,000 apartments in the *Jhilmil* and *Purbachal*, respectively, with RWH facilities (Abdullah 2019).

Apart from the water supply sector, a sewerage Master Plan has already been commenced by the government and city authority to provide a better and well-planned sewerage system in the city, including at least four sewage treatment plants. Among the planned sewage treatment plants, the 'Dasherkandhi Sewage Treatment Plant' is already under implementation, and another three will be implemented in *Uttara, Mirpur,* and *Rayerbazar* in Dhaka city in the future (DWASA 2019). These activities indicate considerable progress in the water supply and sanitation sector, a step towards achieving the UN's SDG. However, more needs to be done to attain the target by 2030. The progress is promising but requires effective actions from the government side to overcome the current problems and challenges involved in the service sector to ensure better water supply and sanitation access to comply with the SDG. Figure 9.2 shows the current practices and required actions towards the sustainable water supply and sanitation services in Dhaka city in the future.

## 9.5 Reflective Discussion

Urbanization is a growing phenomenon for most developing countries, and Bangladesh is not an exception. The country experienced higher population growth during the past few decades, and most of the population growth occurred in the urban areas of Bangladesh like *Dhaka, Chittagong,* and *Khulna.* The study revealed that

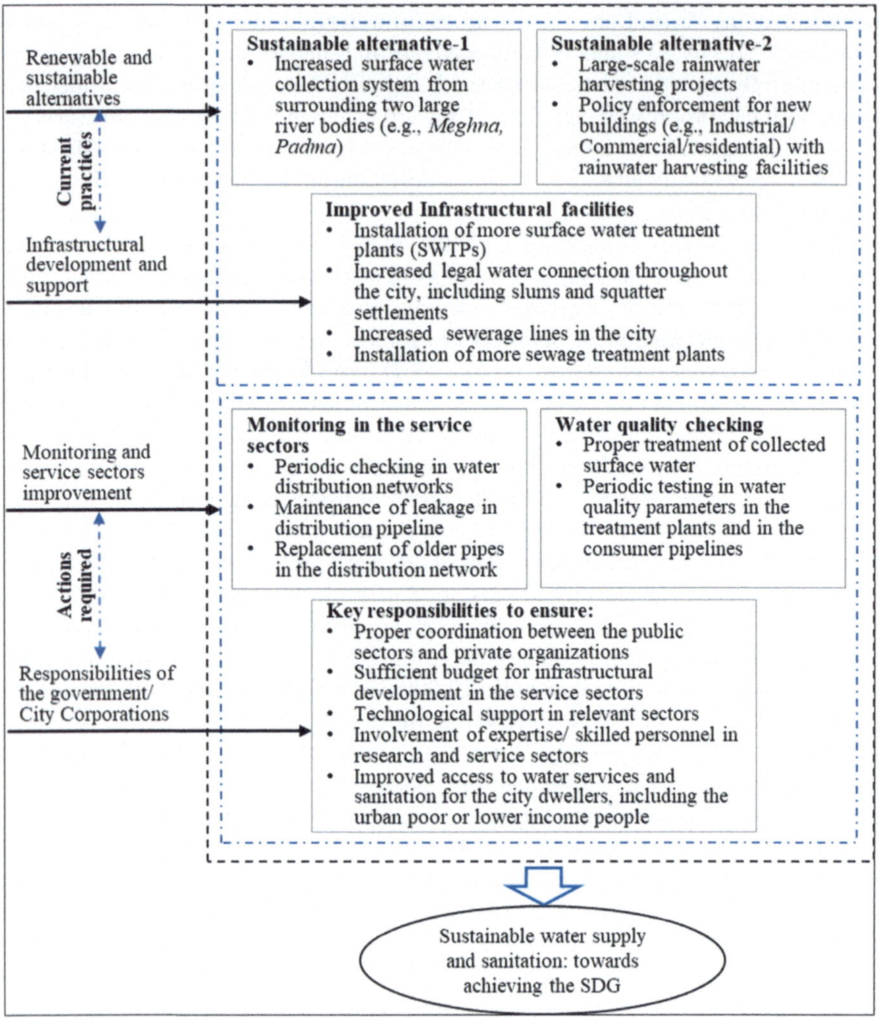


**Fig. 9.2** Prospects for sustainable water supply and sanitation in Dhaka city

in the recent decade, Dhaka city experienced huge urban population growth triggered by rural–urban migration, which negatively impacted all development sectors, including infrastructure, water supply, sanitation, public health, waste disposal, and management. The government authorities and city corporations are struggling to handle the existing situation. Compared to the rapid growth in the urban population, the infrastructural facilities and basic services are inconsistent in most urban areas. Studies show that combined with limited financial resources, the urban population's size places constraints on adequate urban infrastructural development sectors, particularly water supply and sanitation (Khan and Siddique 2000; Roy and Dutta 2017).

Besides, urban corruption is another factor that puts constraints on the development of these sectors. For instance, corruption is frequently observed during the water supply to the squatter settlements or informal settlements in Dhaka city. Akbar et al. (2007) noted that some public water suppliers (e.g. DWASA) workers do not want to supply the informal settlements with water adequately because this would reduce their extra income through bribes.

Environmental degradation caused by the rapid increase in urban population seems to be severe in Bangladesh's urban cities. For example, a vast urban population is living in the slums of Dhaka that have minimal access to water supply and sanitation facilities and, therefore, polluting the urban environment. Khan and Siddique (2000) and Islam et al. (2014) mentioned that in Bangladesh, due to inadequate sewerage systems and insufficient sanitary-waste disposal services, untreated or poorly treated domestic sewage is polluting the water and increasing the mortality rate through waterborne infections.

As mentioned earlier, the primary source of water supply in Bangladesh's urban areas is groundwater, and excessive groundwater extraction reduces the water table, accelerating the costs. Thus, sustainable water supply is becoming a challenging task for the water supply and sewerage authority (i.e., DWASA). So, there need for large-scale surface water treatment plants as the alternative option to increase the water supply and reduce dependency on groundwater extraction. According to Rana (2011), it is evident that before the installation of the 'Sayedabad water treatment plant,' ~97% of the supplied water was from the DWASA's deep tube wells, and the remaining was from the surface water. However, the surface water collection and treatment have increased considerably as a step towards sustainable water supply in the city to comply with the SDGs. According to the DWASA annual report for 2018–2019, 22% of supplied water is currently from surface water sources due to more water treatment plants (DWASA 2019). Roy and Dutta (2017) stated that there are no alternatives for achieving sustainable water supply in fast-growing Dhaka city instead of large-scale uses of available surface waters with proper treatments through surface water treatment plants.

Additionally, to reduce the pressure on groundwater extraction and to satisfy the water demands of the growing urban population, rainwater collection would become a sustainable option in Dhaka city, as the city is characterized by heavy rainfall during the rainy season. If it could be adequately stored, it would reduce the water scarcity in the city, particularly in the dry season when water demands become high and the water table becomes lower to extract groundwater. Geographically, Bangladesh has the great potential of receiving a significant volume of rainwater. Yeasmin and Rahman (2013) reported that the average annual rainfall in Dhaka is over 2100 mm, which is ~3 times the world average, and it can meet the water demands by 15%. As a step towards reducing the water crisis, in 2002, the DWASA attempted to collect rainwater through the 'DWASA's Rooftop Rainwater Harvesting Project' and collected 11.5 million liters of rainwater reducing the water demand significantly (Haq 2006; Roy and Dutta 2017). To meet the increasing water demands and ensure a sustainable water supply, large-scale rainwater harvesting projects should be initiated in Dhaka by the concerned authorities. Moreover, existing building design regulations (i.e., buildings with rainwater harvesting systems) should be implemented effectively. Abdullah (2019) reported that although there are regulations and strategies for rainwater harvesting policies in the country, no implantation of these policies has been observed that could have dealt with the water crisis. Therefore, policy interventions in the service sectors and proper implementation of existing regulations are required to comply with the SDG in the water and sanitation sector in Bangladesh.

However, more research and global knowledge sharing are needed to highlight the benefits of achieving sustainable water supply and sanitation systems. Sustainable water supply and sanitation will not only eradicate the problem of clean water and sanitation (i.e., SDG 6) but also contribute to achieving other goals like no poverty (SDG 1), zero hunger (SDG 2), good health and well-being (SDG 3), sustainable cities and communities (SDG 11), etc. For instance, SDG 6 (i.e., clean water and sanitation) was found to have a substantial share of synergies with SDG 3 (good health and well-being) in the countries where about 2.7 billion people live (Pradhan et al. 2017; Mainali et al. 2018).

## 9.6 Summary

Bangladesh faces numerous challenges in different development sectors due to rapid urbanization and uncontrolled growth in the urban population. The uncontrolled growth in the urban population poses a threat to achieving sustainable water supply and sanitation in rapidly growing cities, including Dhaka. It is also considered a significant constraint towards attaining the Sustainable Development Goals (SDGs). In Dhaka, different government organizations (e.g., DWASA, city corporations) are mainly responsible for providing basic services, including water supply and sanitation, to the urban population. However, the basic services seem inadequate and unsatisfactory compared to the urban population growth rate, as uncontrolled urbanization exerts enormous pressure on existing infrastructure and resources. Moreover, lack of policy enforcement, poor governance and poor management practices are the key constraints to providing sustainable water supply and sanitation services. Although the government takes different initiatives (e.g., increased collection, treatment, and uses of surface water, rainwater harvesting, improved sewerage system, etc.) to comply with the United Nation's SDGs, plans or policies concerning sustainable water supply and sanitation should be adequately implemented by the concerned authorities.

## References

Abdullah M (2019) News. 'Rainwater harvesting: plans without implementation' (published on 19 September). https://www.dhakatribune.com/bangladesh/2019/09/19/rainwater-harvesting-plans-without-implementation. Accessed 12 Feb 2021

Akbar D, Horen BV, Minnery J, Smith P (2007) Assessing the performance of urban water supply systems in providing potable water for the urban poor: the case of Dhaka, Bangladesh. Int Dev Plan Rev 29(3):299–319
BBS [Bangladesh Bureau of Statistics] (2019) Progotir pathey, Bangladesh multiple indicator cluster survey 2019: key findings. Bangladesh Bureau of Statistics, Dhaka
BBS [Bangladesh Bureau of Statistics] (1981) Bangladesh population census. The People's Republic of Bangladesh
BBS [Bangladesh Bureau of Statistics] (1991) Bangladesh population census. The People's Republic of Bangladesh
BBS [Bangladesh Bureau of Statistics] (1997) Bangladesh population census 1991: urban area report. In: Statistics division, ministry of planning, and government of Bangladesh, Dhaka
BBS [Bangladesh Bureau of Statistics] (2001) Bangladesh population census. The People's Republic of Bangladesh
BBS [Bangladesh Bureau of Statistics] (2011) Bangladesh population census. The People's Republic of Bangladesh
CBSG [Capacity Building Service Group] (2010) Ensuring services to slum dwellers. Dhaka WASA Organization for Low Income and Slum Community Water Service Delivery. Water and Sanitation for Urban Poor (WSUP). London, UK, pp 1–49
Cohen B (2006) Urbanization in developing countries: current trends, future projections, and key challenges for sustainability. Technol Soc 28(1–2):63–80
Department of Environment, Waste Concern and ITN-BUET (2004) SAARC workshop on solid waste management. Country Paper, Bangladesh
Du N, Ottens H, Sliuzas R (2010) Spatial impact of urban expansion on surface water bodies: a case study of Wuhan, China. Landsc Urban Plan 94(3–4):175–185
DWASA (2014) Annual report 2013–2014. Dhaka Water Supply and Sewerage Authority. http://dwasa.portal.gov.bd/sites/default/files/files/dwasa.portal.gov.bd/page/b5e42944_f9c0_430e_add7_b0888ba9e0d2/2021-01-17-16-11-7ecdd2b0df417d49962f65ab247a2e98.pdf. Accessed 10 Feb 2021
DWASA (2019) Annual report 2018–2019. Dhaka Water Supply and Sewerage Authority (DWASA). http://dwasa.portal.gov.bd/sites/default/files/files/dwasa.portal.gov.bd/page/b5e42944_f9c0_430e_add7_b0888ba9e0d2/2021-01-17-16-30-16fd0be7f0983f62f789db46a615ae71.pdf. Accessed 10 Feb 2021
Fragkias M, Güneralp B, Seto KC, Goodness J (2013) A synthesis of global urbanization projections. In: Elmqvist T, Fragkias M, Goodness J et al (eds) Urbanization, biodiversity and ecosystem services: challenges and opportunities. Springer, Dordrecht, pp 409–435. https://doi.org/10.1007/978-94-007-7088-1_21
Haq KA (2006) Water management in Dhaka. Water Resour Dev 22(2):291–311
Haque LCME (2019) Assessment of future water demand of Dhaka city. MIST J Sci Technol 7(1):19–28
Herrera V (2019) Reconciling global aspirations and local realities: challenges facing the sustainable development goals for water and sanitation. World Dev 118:106–117
Hossain S (2006) Social characteristics of a megacity: a case of Dhaka city, Bangladesh. In: Proc TASA 2006 Conf, Perth, Australia, pp 4–7
Hutton G, Haller L, Bartram J (2007) Global cost-benefit analysis of water supply and sanitation interventions. J Water Health 5(4):481–502
Islam N (1999) Urbanization, migration and development in Bangladesh: recent trends and emerging issues. CPD-UNFPA Paper 1:1–26
Islam MS, Rana MMP, Ahmed R (2014) Environmental perception during rapid population growth and urbanization: a case study of Dhaka city. Environ Dev Sustain 16(2):443–453
Khan HR, Siddique QI (2000) Urban water management problems in developing countries with particular reference to Bangladesh. Water Resour Dev 16(1):21–33

LGRD [Local Government Division, Dhaka] (2014) The national strategy for water supply and sanitation-2014. Ministry of Local Government, Rural Development and Cooperatives, Bangladesh

Mainali B, Luukkanen J, Silveira S, Kaivo-Oja J (2018) Evaluating synergies and trade-offs among sustainable development goals (SDGs): explorative analyses of development paths in South Asia and sub-Saharan Africa. Sustain 10(3):815

Pradhan P, Costa L, Rybski D, Lucht W, Kropp JP (2017) A systematic study of sustainable development goal (SDG) interactions. Earth's Futur 5(11):1169–1179

Prüss-Ustün A, Bos R, Gore F, Bartram J (2008) Safer water, better health. World Health Organization, Geneva, Switzerland

Rahman G, Mallik B (2003) Analysis of disparity in water and sanitation services: a comparative study of Dhaka, Chittagong, Khulna, and Rajshahi cities in Bangladesh. Water and Sanitation for Cities. BIP and CUS, Dhaka

Rana MMP (2011) Urbanization and sustainability: challenges and strategies for sustainable urban development in Bangladesh. Environ Dev Sustain 13(1):237–256

Roy S, Dutta S (2017) Opportunities and challenges for promoting sustainable water services to the urban poor in Dhaka city. J Environ Sci Nat Resour 10(2):1–10

Roy S, Dutta S, Hoque MM (2016) Urban forestry and urban greening for sustainable urban development: a case of Dhaka north city corporation area (Zone-1). J Bangladesh Agric Univ 14(2):167–176

Roy S, Akhter M, Sumi F, Hoque MM, Dutta S (2019) Factors influencing urban heat island formation and its effects on public health in Bangladesh. IOSR J Environ Sci, Toxicol Food Technol (IOSR-JESTFT) 13(7):77–87

Sekkat K (2017) Urban concentration and poverty in developing countries. Growth Chang 48(3):435–458

Sharmin S, Glass K, Viennet E, Harley D (2015) Interaction of mean temperature and daily fluctuation influences dengue incidence in Dhaka, Bangladesh. Plos Negl Trop Dis 9(7):e0003901

Siddiquee MSH, Ahamed R (2020) Exploring water consumption in Dhaka city using instrumental variables regression approaches. Environ Process 7(4):1255–1275

Streatfield PK, Karar ZA (2008) Population challenges for Bangladesh in the coming decades. J Health Popul Nutr 26(3):261–272

United Nations (1995) World urbanization prospects: 1994 revision. United Nations, New York

United Nations (1998) World urbanization prospects: 1996 revision. United Nations, New York

United Nations (2019) Department of economic and social affairs, population division. World Population Prospects, Online Edition

United Nations (2020) SDG goal 6: ensure availability and sustainable management of water and sanitation for all. https://sdgs.un.org/goals/goal6. Accessed 3 Jan 2021

World Bank (2021) Population data, Bangladesh. The World Bank Group. https://data.worldbank.org/. Accessed 2 Feb 2021

World Population Review (2021) Dhaka population 2021. https://worldpopulationreview.com/world-cities/dhaka-population. Accessed 3 Feb 2021

Yeasmin S, Rahman KF (2013) Potential rainwater harvesting in Dhaka city: an empirical study. ASA Univ Rev 7(1):143–150

Yeazdani SMG (2016) State of drinking water and its management aspects in Dhaka city. J Nepal Geol Soc 50(1):59–64

Zaman M (2019) News. In: Future of Dhaka as a mega city: lessons from Shanghai-Pudong experience. https://www.thefinancialexpress.com.bd/views/future-of-dhaka-as-a-mega-city-lessons-from-shanghai-pudong-experience-1557328522. Accessed 3 Feb 2021

# Chapter 10
# Hydrological Analysis and Climatic Variability of Water Balance: A Study of Geo-Sustainability in Delhi Metropolitan Region

**Shweta Rani, Mahima, and Shabnoor Bano**

**Abstract** Water is the prime determinant resource ensuring the sustainability of urban regions. Since historical times, this prime resource has been freely available both as a surface and underground resource. With the rapid pace of urbanization and industrialization coupled with the growing human population and unwanted interventions, especially in big metropolitan cities, water resources have largely been tapped and over-extracted leading to its depleting nature. The burgeoning climate change adds to this grim scenario. The global rise in temperature results in more evaporation from water bodies and transpiration from green plants, thus, disturbing the natural and healthy hydrological cycle. It results in extreme weather events like floods and droughts. The growing megacity of Delhi is no exception to this. Unscientific, irrational, and unconscious use of water during monsoon and off-monsoon periods often leads to problems of water abundance leading to flood scenarios and scarcity. The study is based on secondary data collected from the Census of India, Master plans of Delhi, Central, State, and Ground Water Boards, and the Indian Meteorological Department. The study intends to estimate and calculate water balance based on the Thornthwaite-Mather water balance model for Delhi and to propose innovative measures to manage water during water deficit and water surplus months to ensure durability, self-reliance, and sustainability of water in the long run. This study is based on the Thornthwaite-Mather water balance model as proposed and applied by Nugroho in calculating the water balance in the Tambakbayan watershed of Yogyakarta in Indonesia. Water Surplus and Water Deficit months are identified and depicted spatially. The paper suggests innovative practices like rainwater

S. Rani (✉)
Department of Geography, Dyal Singh College, University of Delhi, New Delhi, India
e-mail: shwetarani@dsc.du.ac.in

Mahima · S. Bano
Department of Geography, Delhi School of Economics, University of Delhi, Delhi, India

 
S. Anand et al. (eds.), *Emerging Geosustainability Transformations in India*, Advances in Geographical and Environmental Sciences,
https://doi.org/10.1007/978-981-97-7993-2_10

harvesting, construction of green buildings, storage and recovery of aquifers, installation of shade cloth, etc. The findings will help in the proper estimation and identification of water surplus and water deficit months resulting in rational allocation and utilization of water, thus ensuring hydrological sustainability of the area.

**Keywords** Climate · Delhi metropolitan region · Geosustainability · Hydrology · Water balance · Water management

## 10.1 Introduction

Hydrology and climate are two interrelated domains of inquiry. While hydrology is the study of water and its movement and circulation within the earth's system, climate relates to the existing weather patterns with its component prevailing over a space for a longer time. In the present growing era, both these domains of study have influenced each other. For example, the climate-change-induced impacts on the hydrological cycle include variability in precipitation, high evapotranspiration, and larger surface run-off, thus disturbing the cycle and water balance regime in any area posing a threat to its hydrological geo-sustainability.

Water is the second most essential element for survival, next only to air. Therefore, its ready and fresh availability should be ensured, today, tomorrow, and always. Surface water resources like rivers, and lakes were the only sources of water for all living beings in earlier times. However, the increase in water demand compelled human beings to use groundwater and various other forms of precipitation. Currently, water resources are under more pressure on account of the rising population, changes in their consumption pattern, and global climate change. The World Resources Institute in their analysis and mapping of the Aqueduct Water Risk Atlas (WRI 2013) established that India is under severe and sustained water stress, with a global rank of 13th among all "water-stressed" countries". Many areas are experiencing sustained water scarcity and water risks, with the groundwater and surface water storages being severely exploited. Between 2002 and 2016, groundwater depleted at 10 to 25 mm per year. It is further predicted that by 2030, India's water demand is forecasted to be twice the available supply of water and the resultant deficit shall cause a 6% loss in the country's GDP (NITI Ayog 2019). Based on estimations, India's water demand will grow by 24% by 2025 and 74% by 2050, with implications for food security, health, and access to resources (Amarasinghe et al. 2008).

Rapid urbanization and population growth with no added water sources and inefficient water governance are making the situation even more tiring for urban India. The results of climate change are only adding to it; rising temperatures are causing periodic drying up of surface water sources like lakes, and rivers. Extreme weather conditions like droughts and floods are affecting livelihood and economic activities, and thus, paralyzing the whole ecosystem. Big cities and towns in India such as Mumbai, Jaipur, Bhatinda, Lucknow, Nagpur, and Chennai are grappling with the problem of acute water shortages and water scarcity amid climate change. The metropolitan city of Delhi is no exception to this. These rapidly urbanizing cities in India are termed 'water-stressed cities' (Fig. 10.1). Nayaketal (2015) found that a shift to direct seeded rice, along with improvement in irrigation efficiencies in the Sirhind command area in India would improve the sustainability of groundwater use. Reducing the area planted with rice by 25% almost restores the system to sustainable groundwater use and reduces the cost per hectare for groundwater irrigation. Sharma et al. (2019) in their study on Chandigarh city in India evaluated the existing water balance and opined that planning and implementation of water use is undertaken in silos with little or no interaction between and across sectors. This led to frequent water scarcity and water pollution. About 30% of people in India live in cities that are expected to double in population by 2050. With a growing economy and changing lifestyles, the pressure on already strained water resources is increasing. The government has also shown an interest in Integrated Urban Water Management (IUWM) as a new framework and approach for the nation.

Chatterjee and Roy (2021) study on water balance in some of the selected cities in India discussed urban water resilience. They explored the water supply and demand scenarios in four water-stressed cities namely, Chennai, Coimbatore, Bengaluru, and Delhi based on primary and secondary data. They cited the main reasons for water scarcity in the megacities to be population growth, rapid urbanization increasing demand with no net additions to water supply scenarios, and application of water-intensive infrastructure in urban areas. The deterioration in quality and quantity largely impacts the health, safety, and sanitation needs of its people. The need of the hour in the era of growing urbanization and industrialization is to optimize available water resources through a circular water balance framework regulated through an Integrated Urban Water Management (IUWM) process. This will act as a domestic game-changer and will also serve as a model for other developing nations.

Machiwal et al (2017) in their study on estimating water balance in the Kachchh region in India found that suitable measures need to be adopted to check evaporation and seepage losses from the reservoirs in arid regions in India. They established that evaporation and percolation are mainly responsible for the loss of rainwater within 85 days which is stored in reservoirs and only 21% of stored water remains as a result which could be utilized for supplemental irrigation. This will lead to improved and sustained agricultural productivity.

An integrated approach to effective management of water resources is, therefore, much needed for a rapidly urbanizing India. The 2020–21 Union Budget allocated Rs. 11,500 crores for measures to enhance water sustainability in India and launched plans to provide piped water supply to 2.86 crore households by 2024 to fight India's water woes (Financial Express 2020). The target is to focus on water conservation measures, such as rainwater harvesting; renovation, restoration, and rejuvenation of tanks and lakes; reusing and recycling the wastewater; construction of recharge structures; desalination; watershed development; and intensive afforestation to manage the water resources efficiently (Rustogi and Singh 2017). As far as water security is concerned, an analysis to determine what needs to be done starts with the study of water balance. Water balance is the ratio between water inflow and outflow estimated for varying spatial and temporal scales, i.e., for the planet as a whole, for oceans,

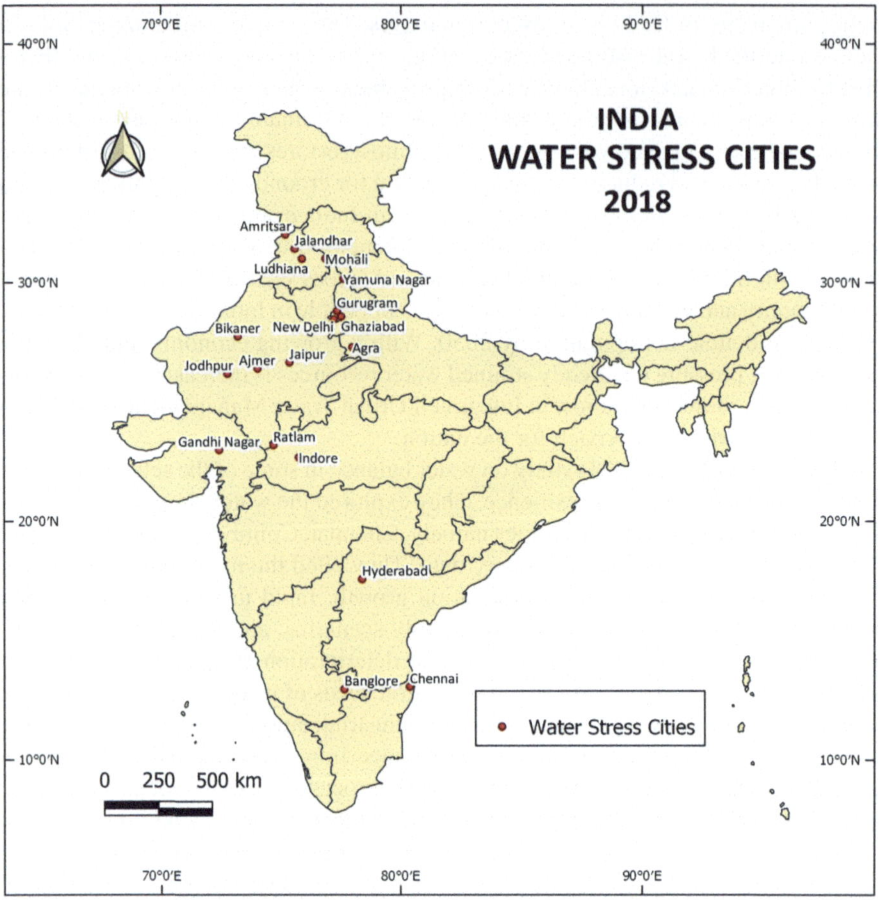


**Fig. 10.1** Groundwater-stressed cities in India (2018). *Source* Ministry of Statistics and Programmes Implantation (2018) derived from Chatterjee and Roy (2021)

continents, countries, natural-economic regions, and river basins, for a long-term period, or specific years and seasons. Its knowledge is necessary to evaluate the possible methods to minimize loss and to maximize gain and utilization of water which is so often the limiting factor. The study of water balance is valuable to the efficient management of water and soil and is of paramount importance to growing cities where the availability of water is a major concern.

Hence, the Delhi metropolitan region has been selected for the calculation of water balance and to analyze the status of water security and its sustainability. The National Capital Territory (NCT) of Delhi, which is the largest metropolis in north India, has undergone explosive demographic expansion witnessing increasing pressure on sources of water, i.e., in various river basins in north India as well as the internal groundwater resources. The city, which lies at the niche to manage and meet its water demand, is likely to face increasing water stress in the near future. The water balance equation propounded by Thornthwaite-Mather has been applied to study the monthly patterns of precipitation, temperature, and soil moisture in Delhi metropolitan regions providing the hydrological regime in the city in 2022 (Nugroho 2018).

## 10.2 Statement of the Problem

The city of Delhi is identified as one of the fastest-growing rates of urbanization. At the global level, the city is among the 10 largest cities and is recognized as the second most water-stressed city lying at high risk of severe water scarcity. With an elongated and extended network of water supply including 14,935 km long pipelines supported by 117 underground reservoirs, the fast-growing city of Delhi is supplying water to 20 million population (Economic Survey 2019). The available water resource in the city is facing the scenario of tremendous pressure and acute shortages owing to rapidly increasing population, rampant urbanization, and high incidences of industrialization. Consequently, the total potentiality of groundwater in Delhi has recorded an overdraft and decline of 130 million cubic meters (MCM) during the last 20 years from 428.07 MCM in 1983 to 292 MCM in 2003 as assessed by the Central Groundwater Board of India (MPD 2021). In addition to this, the various water authorities concerned with water reserves, their distribution, and supply observed and reported a decline in available groundwater level in 2018 which is of the order from 0.5 to 2 m per year over the past two decades. As a result of this declining trend in groundwater reserves without sound and rational replenishment activities, the national capital is under constant and increasing threat of water scarcity with 90% of the areas in Delhi classified as a semi-critical or critical water zone (CGWB 2016).

Thus, such a high trend in urbanization coupled with fresh migration waves negatively impacts both the urban infrastructure and the environment. Problems of deterioration in water supply, groundwater depletion, and stressed freshwater availability are the major concerns. Mencio et al. (2010) opines that the concept of sustainability is at the core of the water management model that the WFD puts forward. Sustainable water management is a multidimensional approach to the issue of interdependency between the natural, social, and economic variables that play a part in different water uses (WFD 2000).

## 10.3 Geographical Profile of the Study Area

The NCT of Delhi is situated in northern India. Geographically, it is located at 28°36′36″ N latitudes and 77°23′ 48″ E longitudes covering 1,483 sq km. Delhi, the capital city of India is bounded by the Indo-Gangetic alluvial plains in the North and East, by the Thar desert in the West, and by the Aravalli hill ranges in the South. Politically, it is surrounded by the state of Haryana on all sides except Uttar Pradesh in the east (Census of India 2001). The river Yamuna cuts the total area into two major parts (Fig. 10.2). The terrain of Delhi is flat. River Yamuna traverses through Delhi towards its east and the Aravalli range towards its south and west. The dominating flat plains in between are made up of fertile alluvium deposition recently formed. The Delhi Ridge and its four sections, the northern, the central, the south-central, and the southern, constitute the farthest extension of the Aravalli range. Administratively, Delhi has 9 districts, 27 Subdistricts, 113 towns including 3 statutory towns, and 110 census towns in the urban area and 112 rural villages (DCHB 2011).

The climate of Delhi interestingly represents an overlap between the monsoon-influenced humid subtropical climate (represented by sub-type Cwa) and semi-arid (BSh) as per Köppen climate classification. There exist large-scale seasonal variations

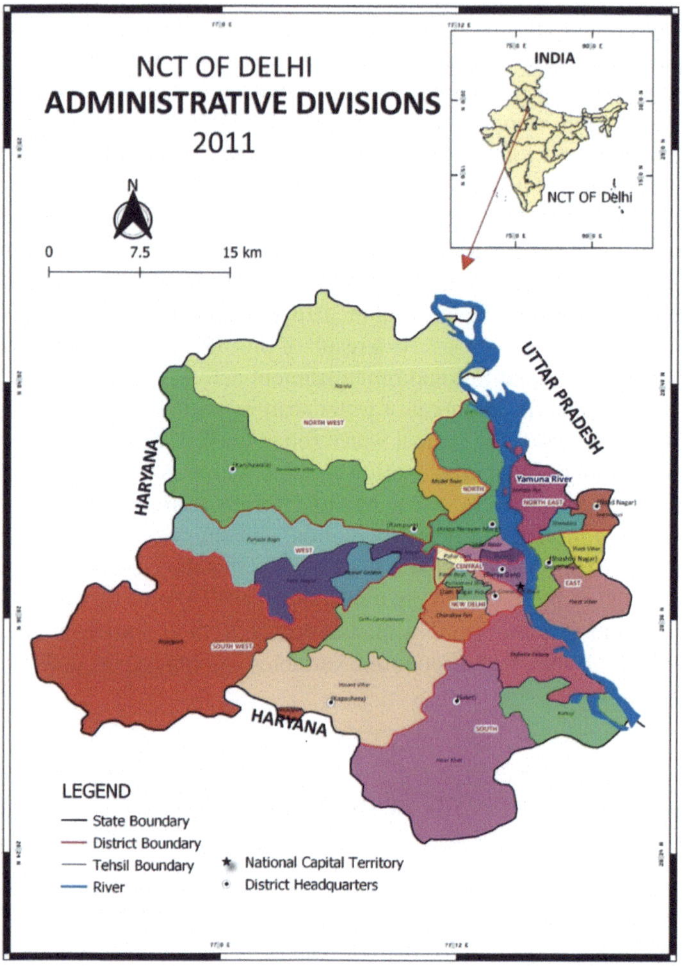


**Fig. 10.2** The study area, NCT-Delhi. *Source* Authors

in temperatures and precipitation. Summer begins in early April and assumes a peak in early June, with average temperatures being near 38 °C or even reaching a high of 45 °C due to occasional heat waves and blowing of local winds like 'loo'. The monsoon period starts in late June and ends in mid-September with the post-monsoon season commencing in late October. The average annual rainfall is 798 mm in Delhi and 75 per cent of it occurs in the three months of July, August, and September. This leads to contrasting problems of floods and droughts. Heavy rainfall in the watershed of the Yamuna aggravates flood scenarios in the city. The city also reels under drought in the drier months with high surface temperatures. Winter starts in November, and peaks in January retreats by the first week of March. Winters are generally cold with average temperatures reaching between 4 to 5 °C. In winter, Delhi experiences cold waves drastically lowering the temperature due to its closer proximity to the Himalayas. The capital city is known for its dense fog and haze in the winter season obstructing the rail, road, and air traffic with reduced visibility.

In 1950, the population of Delhi was 1.3 million. Since, then there has been a conspicuous rise in the population owing to several luring factors, i.e., a center of employment generation and higher education, the seat of national democracy, etc. The estimated population of Delhi as per Provisional Population Data, Census 2021 is 31.2 million with a head count of 31,181,376. These population estimates and projections are a result of the latest revision of the UN World Urbanization Prospects. These estimates represent the urban agglomeration of Delhi. This mega city of Delhi undoubtedly is the epitome of the cosmopolitan culture and diversity of India with the mutual co-existence of diverse religions, languages, and customs in splendid harmony.

## 10.4 Objectives of the Study

The major objectives of the present study are: (i) to identify the trends in population growth in the Delhi metropolitan region, (ii) to explore the gaps between demand and supply of water and to depict the water surplus and water deficit months in the study area, (iii) to analyze the future water supply scenarios in the study area, (iv) to explore the management plans regulating the smooth and healthy flow of water in urban metropolitan areas, and (v) to suggest the innovative solutions, measures, and techniques manage water deficit and surplus area in NCT-Delhi.

## 10.5 Materials and Methods

### *10.5.1 Sources of Data Collection*

The study is based on secondary data sources collected from the District Census Handbook of the Census of India (DCHB 1991, 2011) and the Primary Census abstract (1991, 2001, 2011) for state-level data. Data on water resources and water storage reservoirs are available from the Master Plan of Delhi (MPD 2021), Economic Survey (2019), Groundwater Year Book (2022–2023), and Ministry of Statistics and Programmes Implantation (2018). Hydro-meteorological data like temperature (t°C), Precipitation (P in mm), and theoretical/average sunshine hours (N) are obtained from the Indian Meteorological Department (IMD 2022). Also, public domain information on hydrology, hydrogeology, and urban water infrastructure is derived from the State and CGWB (2016).

### *10.5.2 Understanding Water Balance*

A water balance is an assessment of the major components of a hydrologic system and includes the interaction between surface and groundwater systems. It provides a general idea of the magnitude of replenishment and discharge components within a system. Figure 10.3 provides a synoptic view of the water balance in a basin of catchment aided by the components of precipitation, evaporation, and run-off.

Some of the significant characteristic features of water balance are discussed below:

- A water balance can be established for any area of the earth's surface by calculating the total precipitation as an input and the total of various outputs.
- The law of water balance states that the inflows to any water system or area are equal to its outflow and change in storage during a time interval.
- Floods and droughts are the two extreme situations in which inflow (precipitation) and outflow (evapotranspiration) both are very high. A balance between these two contrasting conditions can be achieved by the calculation of water balance.

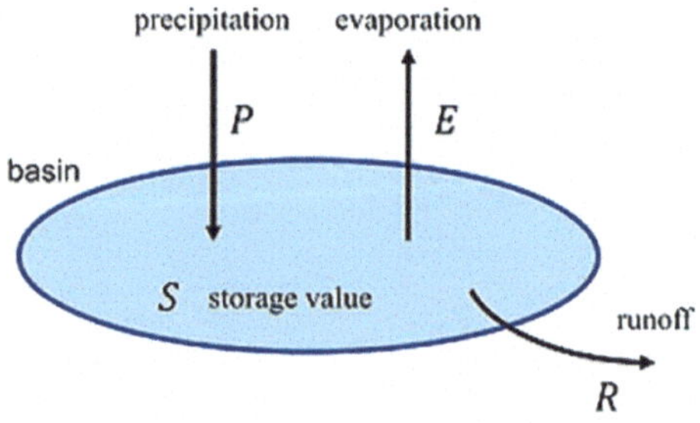


**Fig. 10.3** Water balance in a basin or catchment. *Source* Wikipedia (2022)

- The field of hydrology (the study of water) implies an input–output approach or systems approach.
- In hydrology, a water balance equation can be used to describe the flow of water in and out of a system where a system can be one of several hydrological or water domains, such as a column of soil, a drainage basin, an irrigation area or a city, lagoon, a River, and a dam.
- For the sustainment of human beings, striking a balance between the input of water and the output of water is significantly important.

### *10.5.3 Selected Indicators of Water Balance*

A water balance can be established for any area of the earth's surface by calculating the total precipitation (P), temperature (t °C), and average sunshine hours. Water balance attempts to describe the various ways in which water supply can be expanded. A study of water balance can be used to help manage water supply and predict where there may be water shortages. Altogether 4 indicators are selected for the calculation of water balance (Table 10.1). The data for these 4 indicators are collected for the 12 months of the year 2020 to show seasonal variations in the NCT of Delhi.

#### 10.5.3.1 Precipitation (P)

It is regarded as a prime indicator in calculating water balance. Precipitation is any liquid or frozen water that forms in the atmosphere and falls back to the earth. It comes in many forms like rain, sleet, and snow. Along with evaporation and condensation, precipitation forms a major part of the global water cycle. July receives the highest rainfall and November has the least rainfall.

#### 10.5.3.2 Temperature (T)

Temperature is a measure of how hot or cold something is. The month of May has the highest degree of temperature whereas January has the lowest degree of temperature.

**Table 10.1** Selected indicators

| S. No | Indicators | Unit of measurement |
|---|---|---|
| 1 | Temperature (T) | °C |
| 2 | Precipitation (P) | mm |
| 3 | Average/theoretical sunshine (N) | Hours |
| 4 | No. of days (d) | Number |

*Source* Authors

#### 10.5.3.3 Average Sunshine (N)

This indicator represents the average number of hours in a day when the sun is shining brightly. This is different from daylight hours, which is the time from sunrise to sunset April, May, and June receive the highest sunshine hours and January receives the lowest sunshine hours.

#### 10.5.3.4 Days (D)

It represents the total number of days in each month for a specific year. Two important criteria for the calculation of water balance are precipitation (input) and temperature. Mostly input is precipitation and the output is evaporation and transpiration. Input and output are required to assess whether the water balance is healthy or not. Evaporation can occur at any time and anywhere may it be day or night. But transpiration occurs from green parts of plants in sunlight. It occurs through the green chlorophyll of green plants. The two types of evapotranspiration are Potential Evapotranspiration (PET) and Actual Evapotranspiration (AET). PET is a measure of the ability of the atmosphere to remove water from the surfaces through the process of evaporation and transpiration. It is a hypothetical/predictive/assumption/theoretical measure of how much evaporation would have occurred if there were sufficient water available with no control over the water supply. AET is the quantity of water that is removed from a surface due to the process of evaporation and transpiration. It indicates the actual amount of water that is lost.

Figure 10.4 depicts the distributional pattern of rainfall in Delhi for 3 months in 2022. Based on the intensity of rainfall received, the city of Delhi has been broadly categorized as deficient, normal, and excess, and accordingly, spatial mapping has been done. These categories represent the percentage departure of realized rainfall from the normal rainfall. The magnitude of potential and actual evapotranspiration has also been assessed for each district to get a pattern of climatic regimes prevailing in Delhi.

### *10.5.4 Computation of Water Balance (Thornwaite Method)*

Mathematically, water balance is calculated using Eq. 10.1:

$$\mathbf{P} = \mathbf{R} + \mathbf{ET} + \Delta\mathbf{SM} \tag{10.1}$$

where

P = Precipitation, R = Stream Flow, ET = Evapotranspiration, ΔSM = Change in Soil Moisture.

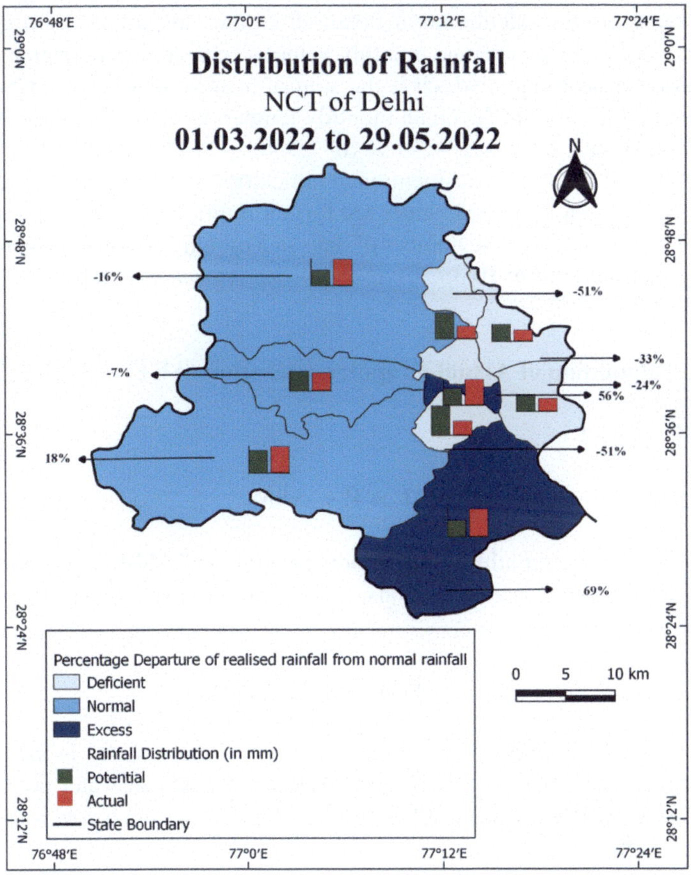


**Fig. 10.4** Distribution of rainfall and associated parameters in NCT Delhi. *Source* IMD (2022)

#### 10.5.4.1 Calculation of Potential Evapotranspiration (PET)

Water balance is generally known as the balance between the deficit and excess water, which is, assessed through Potential Evapotranspiration and Actual Evapotranspiration. Thus, to find out the level of the water balance of NCT Delhi, PET and AET need to be calculated.

The formula for PET is

$$\mathbf{PET(non-corrected) = 16(10 \times T)^a/I} \quad (10.2)$$

$$\mathbf{PET(corrected) = PET(non-corrected) \times N/12 \times d/30} \quad (10.3)$$

Formulae represent the calculation of Potential Evapotranspiration (non-corrected) (Eq. 10.2); where T is the average monthly value of temperature in degrees Celsius, '*I*' is the Annual heat index, which is the sum of heat indices ($\sum i = T/5^{1.514}$) for the 12 months (t represents the mean monthly temperature) and '*a*' is the empirical constant value ($675 \times 10^{-9} \times I^3 - 771 \times 10^{-7} \times I^2 + 1792 \times 10^{-5} \times I + 0.49239$). On computation using the above-mentioned indicators, non-corrected values of PET are obtained. The non-corrected values are further multiplied with monthly theoretical sunshine hours (N) and the number of days in a month (d) to get corrected values of PET as illustrated in Eq. 10.3.

#### 10.5.4.2 Calculation of Actual Evapotranspiration (AET)

The formula of AET is

$$\mathbf{AET = P + \Delta SM} \tag{10.4}$$

where P is the average monthly value of precipitation and ΔSM is the change in the storage of Soil moisture that varies considerably from place to place. Soil Moisture (SM) is calculated as

$$\mathbf{SM = AWC \times e^{(AWPL/AWC)}} \tag{10.5}$$

where *AWC* is the Annual water holding capacity of the soil, '*e*' is the empirical constant with a value being 2.71828 and AWPL is the annual water potential loss (Eq. 10.6). Annual Water Potential Loss is calculated using the below formulae:

$$\mathbf{AWPL = Preceding\ value\ of\ (P - PET) + Succeeding\ value\ of\ AWPL} \tag{10.6}$$

The monthly values of soil moisture (SM) obtained provide information about the quantity of water(in mm) available in storage at the end of each month (Eq. 10.5) and **Δ**SM is the change in storage capacity of soil moisture during the month (as illustrated below in the formulae Eq. 10.7); where a negative value indicates removal of soil moisture; a positive value indicates recharge of soil moisture.

$$\Delta \mathbf{SM = Preceding\ value\ of\ SM - Succeeding\ value\ of\ SM} \tag{10.7}$$

If P > PET, then AET = PET and if P < PET, then AET will be limited to the amount of precipitation (P), plus any moisture remaining in storage. Soil moisture deficit occurs when P < PET resulting in depletion of storage of soil moisture. In the present given example, SM = PET−AET, and indicates the amount of water that would have to be supplied to the soil to maintain normal plant growth. Soil moisture surplus (runoff), occurs when P > PET and the soil moisture is fully recharged. In this case, any water received over the amount needed by plants becomes runoff, and SM = P–PET.

## 10.6 Results and Discussion

### *10.6.1 Population Growth in the Delhi Metropolitan Region*

The city of Delhi is a massive metropolitan area in the country's north attracting the bulk of migrants owing to its strategic position as the national capital, an employment and educational hub, and a seat of political importance. The population of Delhi was 0.4 million in 1901 which rose to 0.5 million in 1921 (Fig. 10.5). It further rose to 1 million in 1941 and to 6.2 million in 1981. Under the influence of The New Economic Policy, Delhi registered a population of 13.8 million in 2001 which further rose to 16.7 million in 2011. As per the provisional Population Totals, the estimated population of Delhi will be 20.5 million in 2021.

At the global level, Delhi occupies the second position in terms of population next only to Tokyo. As per the United Nations projection, it is likely to surpass all the megacities in the world reaching a record-high population of 37 million by 2030* (World Urbanization Prospects, 2018). With such an unprecedented pace of growth, coupled with increasing trends in migration patterns, this city needs to be able to integrate its elegant past as well as the modern developments into an organic whole keeping pace with the future generations.

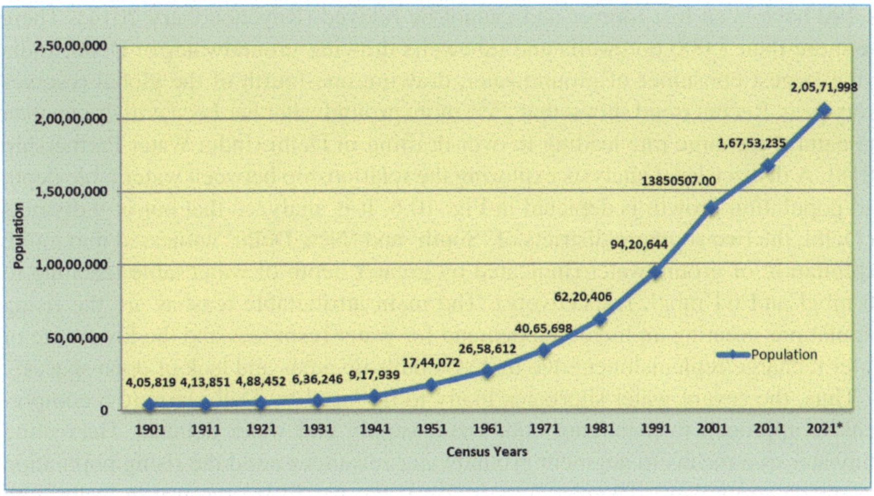


**Fig. 10.5** Trends in population growth in Delhi metropolitan region (1901–2021*). *Source* Census of India (2011) and Dhar (2022)

### 10.6.2 Availability and Supply of Water

Delhi recorded an average annual rainfall of 611 mm (Climate India 2022). However, the recharge of groundwater gets restricted due to non-permeable surfaces in the city which are the subsequent result of urbanization and water is not able to percolate and runoff to sewers. Out of the total water supplied by Delhi Jal Board, 60% comes from the river Yamuna, 34% from Ganga, and the remaining water is derived from groundwater resources. However, Yamuna River which is the major provider of Delhi's drinking water, is running heavily with pollutants covered with thick scum, froth, and toxic waste after a nearly clear-water Yamuna enters Delhi through Wazirabad. Moreover, almost 40% of Delhi Jal Board's water supply is deemed to be non-revenue water (NRW), marking a black hole with no revenue in return. Moreover, Yamuna River the major provider of the city's drinking water runs heavy with pollutants and is thick with toxic waste after a nearly clear-water Yamuna enters Delhi. The total water treatment and supply capacity of Delhi Water Authorities was 66 MGD (Million Gallons per Day) in 1956 which was raised to 240 MGD in 1979, further 437 MGD in 1990, 650 MGD in 2002, 906 MGD in 2014, and to a high of 916 MGD in 2020. The Delhi Jal Board (DJB) supplies around 935 gallons of water per day (MGD) against the demand of 1,140 MGD (MPD 2021).

A survey released by the Delhi Parks and Gardens Society (DPGS) has shown that out of the 611 water bodies in the capital, 274 have already dried up and as many as 190 have been lost forever and cannot be revived (Roychoudhary 2014). There are more than 5,000 borewells and tubewells drawing groundwater in Delhi. India is the largest consumer of groundwater, drawing one-fourth of the global reserves every year. Recent trend shows that 25% more groundwater has been withdrawn than the natural recharge rate leading to over-drafting in Delhi (India Water Partnership 2018). A district-level analysis exploring the relationship between water table depth and population growth is depicted in Fig. 10.6. It is analyzed that out of 9 districts in Delhi, the two southern districts of 'South' and 'New Delhi' witnessed maximum exploitation of groundwater (indicated by greater depth of water table reaching to 66 mbgl and 64 mbgl, respectively). The main attributable reasons are the rising populations creating an increased demand for water resources and the lower rate of water recharge/replenishment due to dense built-up areas and lack of open spaces.

Thus, the severe water shortages likely to be faced by Delhi require a comprehensive approach to managing both water supply and water demand. Harvesting rainwater as a means to augment groundwater resources amid the rising population was made compulsory 19 years ago, in 2001, but has little presence in Delhi with no solid data about its effective implementation (Nagpal 2019). While the government has placed a priority on meeting domestic water needs, the importance of the commercial, industrial, and agricultural sectors necessitates a considered approach to improve water security.

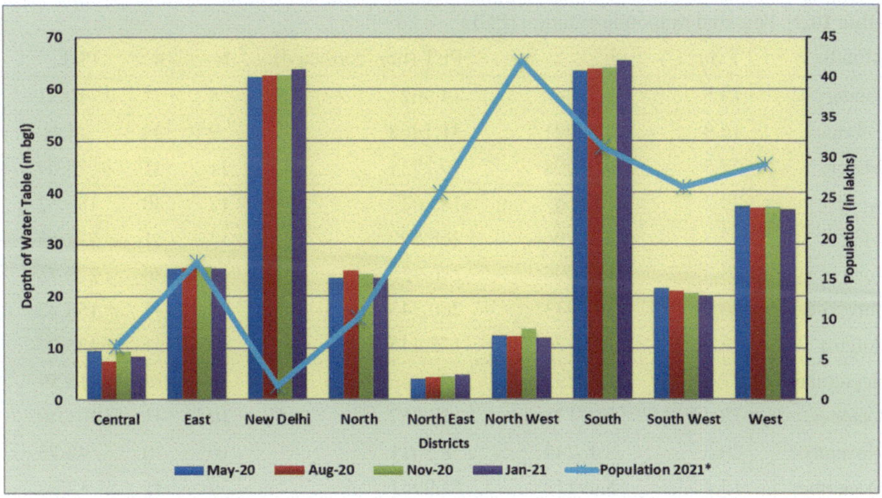


**Fig. 10.6** Maximum depth of water table and population in districts of Delhi. *Source* Central Groundwater Board (2022) and Dhar (2022)

### *10.6.3 Identification of Water Surplus and Water Deficit Months*

Water balance and its study have become increasingly significant in hydrology and hydrologic studies since the reserve of fresh water is limited. Water is added to the earth's surface through precipitation in its various forms and is regarded as the input system to the hydrological cycle. The water received is in a constant state of circulation on the earth's surface and is ultimately lost in the atmosphere through the process of evaporation and transpiration. This process is regarded as the output system, thus, completing the hydrological cycle. A balance should always be maintained between the input (storing) and output (spending) of water on the earth's surface through water balance. Imbulana and Manoharan (2020) in their study on evaluating water balance options for climate resilience highlighted that climate change coupled with population growth and agricultural expansion, increases the irrigation water deficits in the Cascade tank systems. Further, they explained that demand-side interventions in water management are more effective than supply-side management in reducing the demand–supply gap. Climate scientists predict that this shift will exceed the infiltration capacity of the soil resulting in frequent floods during monsoon and drought-like conditions during off-monsoon periods. Thus, the study provided a framework to design an optimal strategy to enhance climate resilience in smallholder irrigation systems to improve agricultural production and food security with the use of hydrological and water balance assessments.

In line with the preceding discussion, an effort is made to study the hydrological regime and estimation of water balance for 12 months for the Delhi metropolitan region for the year 2022. Tables 10.2 and 10.3 depict the monthly

**Table 10.2** Potential evapotranspiration (PET)

| Months | T (in °C) | i | PET (non-corrected) | N | d | PET |
|---|---|---|---|---|---|---|
| January | 13.5 | 4.49868 | 14.962 | 8 | 31 | 10.31 |
| February | 16.9 | 6.32091 | 31.1468 | 9 | 28 | 21.80 |
| March | 22.5 | 9.74908 | 79.2734 | 11 | 31 | 75.09 |
| April | 29.2 | 14.466 | 185.62 | 12 | 30 | 185.62 |
| May | 32.7 | 17.1706 | 268.599 | 12 | 31 | 277.55 |
| June | 33 | 17.4096 | 276.726 | 12 | 30 | 276.73 |
| July | 29.9 | 14.9943 | 200.543 | 10 | 31 | 155.42 |
| August | 28.7 | 14.0926 | 175.445 | 9 | 31 | 135.97 |
| September | 27.8 | 13.429 | 158.116 | 9 | 30 | 118.59 |
| October | 25.4 | 11.7133 | 117.757 | 10 | 31 | 101.40 |
| November | 20.5 | 8.46749 | 58.5011 | 10 | 30 | 48.75 |
| December | 15.4 | 5.49118 | 22.9961 | 9 | 31 | 17.82 |
| Total | | I = 137.803 | | | | |

*Source* Computed by Authors

**Table 10.3** Water balance in Delhi metropolitan region (2022)

| Months | P (in mm) | P-PET | AWPL | SM | ΔSM | AET | PET-AET |
|---|---|---|---|---|---|---|---|
| January | 23 | 12.70 | 0 | 160 | 101.71 | 124.71 | 0 |
| February | 33 | 11.20 | 0 | 160 | 0 | 33 | 0 |
| March | 20 | −55.09 | −55.09 | 113.40 | −46.61 | −26.61 | 101.70 |
| April | 14 | −171.62 | −226.71 | 38.80 | −74.60 | −60.60 | 246.22 |
| May | 20 | −257.56 | −484.26 | 7.76 | −31.04 | −11.04 | 288.59 |
| June | 74 | −202.73 | −686.99 | 2.18 | −5.58 | 68.43 | 208.30 |
| July | 208 | 52.58 | 0 | 160 | 157.82 | 365.82 | 0 |
| August | 183 | 47.03 | 0 | 160 | 0 | 183 | 0 |
| September | 99 | −19.59 | −19.59 | 141.57 | −18.44 | 80.56 | 38.02 |
| October | 13 | −88.40 | −107.99 | 81.47 | −60.10 | −47.09 | 148.50 |
| November | 5 | −43.75 | −151.74 | 61.98 | −19.49 | −14.49 | 63.24 |
| December | 8 | −9.82 | −161.56 | 58.29 | −3.69 | 4.31 | 13.51 |
| (AWC = 160 mm) | | | | | | | |

*Source* Computed by Authors

calculation for precipitation, potential evapotranspiration, soil moisture and actual evapotranspiration and computation of water balance.

The monthly trends in precipitation, potential evapotranspiration, and actual evapotranspiration for 12 different months in Delhi in 2022 are depicted in Fig. 10.7A.

The month-wise intensity of water is represented on the vertical Y-axis and different months are depicted on the horizontal X-axis. The month-wise distribution of precipitation, PET, and AET is presented to aid the process of computation of water balance for the study area.

Water surplus and water deficit months are identified through the computation of AET and PET (Fig. 10.7B). When PET > AET it indicates water deficit months and when precipitation exceeds the rate of potential evapotranspiration, water surplus months are identified. From Fig. 10.8, it can be analyzed that the 4 months of *January, February, July, and August* emerged as surplus months since enough water is available in Delhi; among which the month of August receives the highest rainfall. However, the remaining 8 months, i.e., *March, April, May, June, September, October, November, and December* are identified as deficit months. During these eight months, the city of Delhi faces deficit or severe shortages of water, with December receiving the lowest rainfall. The zone of water surplus is represented with light blue shading, while dark

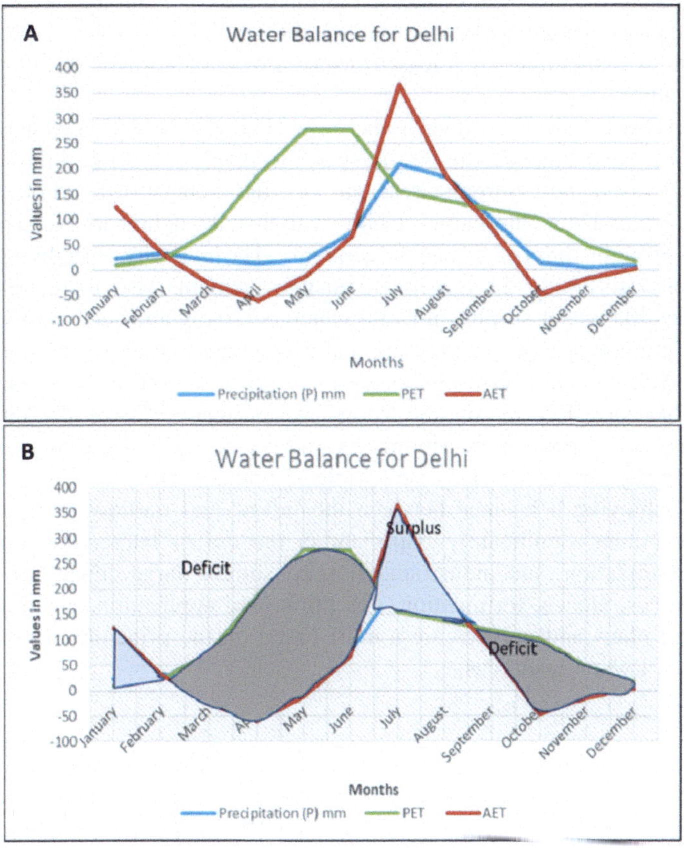


**Fig. 10.7** **A** Monthly trends for precipitation, PET and AET, **B** water surplus and water deficit months in Delhi. *Source* Prepared by authors based on calculations

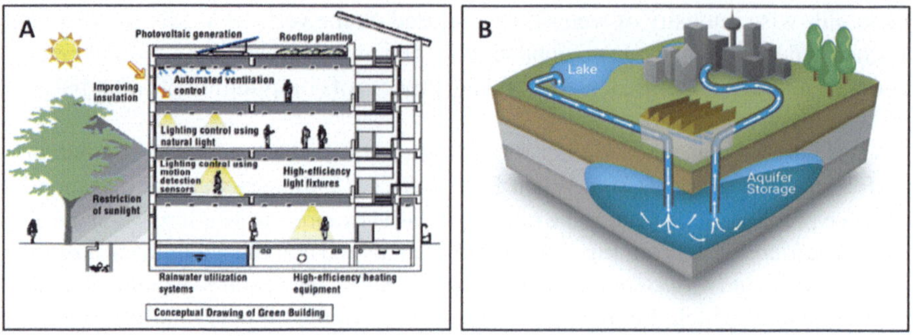


**Fig. 10.8** **A** Synoptic view of green water building. *Source* Zafar (2022); **B** Aquifer storage. *Source* Westward (2023)

grey areas represent the zone of water deficit in the Delhi metropolitan region in 2022. Mismanagement of water can be tackled through the calculation of water balance. It is only possible when people are aware of water deficit and water surplus months and accordingly, guidelines are released ensuring ways of storing water during water surplus months.

Any imbalance arrived at either in the form of surplus or deficit of water during estimation of water balance can be disastrous to mankind posing a threat to their survival and sustenance. Surplus areas can face the menace of flood due to heavy surface run-off and huge discharges. Contrary to this, the deficit areas can reel under continuous drought, thus paralyzing the economy. The megacity of Delhi is grappling with a water crisis amidst the year due to mismanagement. Also, there are instances of conditions like water surplus in Delhi which is largely attributable to variability in climatic parameters like temperature and precipitation and extremes of climatic conditions. Climate change results in untimely precipitation in Delhi which leads to a flood-like situation. In addition to this, the presence of excessive concrete or cemented areas and poor proper drainage systems also add to the growing havoc of floods especially during the monsoon periods (Krishnan 2019).

Hence, achieving a balance between the inflow and outflow of water within a system becomes significantly important in the era of burgeoning population, rampant urbanization, and industrialization consuming large chunks of available water resources. Hence, formulation, execution, and application of rational water management plans and policies are a must based on the principle of rational and justified use of water resources.

## 10.7 Suggestions and Innovative Solutions: A Way to Sustainable Development

Safe and uninterrupted water supply along with hygienic sanitation is very important and crucial for achieving the goal of 'Good Health and Well Being for All' (SDG-3) and maintaining socioeconomic sustainability of urban areas especially big towns and metropolises like Mega Cities. Rustogi and Singh (2017) in their study found that Najafgarh Lake situated in South-West Delhi occupying more than 300sqkm of area was a biodiversity hotspot. Presently, it stands as a topographical depression brimming with overgrown grass and garbage. The lake's disappearance has stolen a chunk of Delhi's culture and its use as a dumping ground has raised health concerns for the local population. Findings state that rejuvenating the Najafgarh Lake through rainwater, bio-intensive farming, creation of ponds, and implementation of community water management schemes with maximum people's participation, can ameliorate the water shortages.

It was expressed that the megacity of Delhi is reeling under thirst whereby the tanker mafia and the political governments both in the state and at the center play the blame game with each other (Shali and Ojha 2019). In line with the preceding discussion on the growing water-stressed scenario in the Delhi metropolitan region, rational and effective plans need to be formulated and implemented both at the state and central levels in collaboration. Some of the innovative solutions and suggestions that can be incorporated to maintain the balance of input and output of water in the Delhi metropolitan region by reducing wastage and countering extreme climate conditions (droughts and floods) are discussed below.

### *10.7.1 Emphasis on the Fundamental of 3R's*

One of the foremost and basic steps towards maintaining the geo-sustainability of water in metropolitan cities and towns is understanding and applying the fundamentals of the 3Rs in urban lifestyle. These three R's denote Reduce, Reuse, and Recycle. Among all the three 'Reduce' is deemed most effective.' The concept of 'Reduce' implies using fewer resources in the first place. By reducing the use of water, one can bypass reuse and recycle altogether. The misuse of water can be avoided in many ways like; checking the taplines leakages, using short showers, turning off the tap after use, etc. The second 'R' implies the 'Reuse' of water. The process of water reuse intends to reclaim water from a multitude of checkpoints; to be treated for sustainable purposes like uninterrupted supplies, recharge of groundwater, recreational and restoration purposes, etc. Moreover, these 'add-ons' to the available supply of water ensure security and sustainability. Recycling of wastewater is the third fundamental 'R' and is achieved through planned water reuse, for example, agricultural and landscape irrigation, industrial process, municipal water supply, process water for power plants, refineries, mills, and factories, indoor uses such as toilet flushing, dust control or surface cleaning of roads, construction sites and environmental restoration.

### *10.7.2 Management of Wastewater*

Production and generation of wastewater is a major environmental problem facing the metropolitan cities in India with Delhi being no exception to this. There exists a huge discrepancy between wastewater generation and its treatment, i.e., Delhi generates wastewater of 720 mgpd and only 473 mgpd gets treated. This excess of wastewater generated pollutes other water sources and it is quite visible in Delhi, especially along the banks of river Yamuna. The wastewater can be treated and reused for growing vegetables and fruits. Also, it can be stored for reuse in the peripheral areas of cities while planning urban development. Wastewater Treatment Plant should be encouraged that utilize sunlight and local algae species to remove nutrients and other contaminants from wastewater while generating large quantities of electricity. As a solution to wastewater management, Delhi received its first sewage treatment plant that will recycle sewage into drinkable water using advanced, tested technology (PTI 2015). The treatment plant is powered by solar energy, and that makes it highly eco-friendly (Madan 2015).

One such good practice to be witnessed in Delhi is the Delhi Jal Board (DJB) collaborating with NGO, SANA (Social Awareness, Newer Alternatives), commissioned the project called 'Sujala Dhara' or toilet to tap water. The major attraction of the plant is that it can 5,000 L of sewage to provide 4,000 L of drinkable water every hour. Water is treated through a bio-filter that consists of seven layers of organic and inorganic material including plants, earthworms, sand, and pebbles. Further, the water is passed through a nano-filter membrane with a size of 0.001 microns. The plant operates 24/7 and can efficiently produce 25 million litres of drinking water annually.

### *10.7.3 Water-Saving Aerator and Green Taps*

Recent innovative water-saving solutions for washbasins, sink taps, faucets, etc. are the 'water-saving aerators and green taps' (Water Conservation Techniques 2019). The purpose of these aerators is to compensate for the water pressure and dispense water at a defined flow rate, i.e., 2 to 8 L per minute. Water Saving Aerators and Green Taps easily convert taps into water-saving taps, thus, reducing both water wastage and savings on water bills. It has been established that applications of such application to such water-saving taps can lead to annual savings of up to 22,500 L/Year/tap (Conservation Mart 2011).

### 10.7.4 Rainwater Harvesting

Rainwater Harvesting ensures collecting the rainwater as and when it falls for future use. This collected water can be used as a deposit to be used when scarcity of water arises. It employs a simple technique to collect rainwater in the storage tank and to direct this stored water to groundwater aquifers to be used later during dry periods. Rooftop rainwater harvesting is the most widely used method to store water. Vashisth (2019) proposed some of the simple methods for rainwater harvesting especially in big cities and towns like Delhi with a burgeoning population and grappling with water scarcity in the major part of the year such as surface water collection systems, rooftop systems, dams, underground tanks, rain saucer, water collection reservoirs, barrage, and rain barrels. Rain-harvested water can be suitable put to multiple purposes like gardening, watering livestock, laundry, and flushing toilets.

Sharma and Midha (2021) explored the potentiality of Rainwater Harvesting as an alternative source of water supply for Delhi one of the waters stressed cities in the world. They applied the mass curve method to understand the feasibility of rainwater harvesting and found that an average roof of 60m$^2$ in Delhi will collect 3,64,800 of water approximately in a year for an average family size of 5 people. This method can act as a natural solution to recharge groundwater. Installing a rainwater harvesting system in urban areas reduces the amount of stormwater pollution by slowing the process of runoff and storing the water for future use and thus can add to the goal of sustainable use of water.

### 10.7.5 Green Water Building

Green building is an integrated whole systems approach for the design, construction, and operation of buildings that incorporates elements of water and energy efficiency, conservation of materials and resources, and indoor air quality. The concept of Green Water Building can largely be applied to metropolitan cities like Delhi where the construction of urban infrastructure including buildings becomes a necessity to accommodate the growing population. These green buildings are largely based on the principle of sustainability employing more efficient water-use devices such as low-flow fixtures, water-efficient washing machines, dry toilets, and a decline in water-guzzling air coolers (Fig. 10.8A).

### 10.7.6 Water-Efficient Dishwasher and Washing Machine

Using household appliances more effectively is a great way to save water, especially in populating and growing urban areas and big metropolitan cities like Delhi where urban demand is increasing by leaps and bounds. Energy Star provides a helpful guide for water-efficient dishwashers or washing machines. Such water and energy-efficient home appliances use 30% less water and 25% less energy, hence are termed water-efficient appliances. In the United States in appliances, there is EPA's Water Sense stamp of approval (EPA 2021). Any appliance with this EPA stamp indicates 20% less water usage than the average appliance.

### 10.7.7 Flood Management

A large portion of fresh water in the Delhi metropolitan area becomes devastating and useless due to heavy floods caused by large surface run-off during monsoon periods. Drainage areas of the Yamuna River have been utilized for flood management and flood security as a step toward water management (Vijay 2015). By construction of embankments and canals, a large part of the land can be conserved besides minimizing flood losses. The promotion of intensive afforestation activities in Delhi can serve the dual purpose of providing security from floods and interception processes by green plants recharging groundwater.

### 10.7.8 Rational Use of Groundwater

At the global level, almost 25% of the needs of water supply are met by groundwater resources. However, in metropolitan cities coupled with rapid population growth and urbanization, the demand for groundwater goes on increasing in proportion to its available quantity. Due to rapid exploitation and huge withdrawal, the available groundwater resources are shrinking posing a threat to its sustainability and mankind. The available groundwater resources in Delhi have been assessed to be 292 million cubic meters by the Central Ground Water Board (CWGB). In contrast to this, the withdrawal rates are 7% above it to meet the growing demands. 7 out of 9 districts in Delhi are categorized as overexploited districts in terms of groundwater availability. Hence, efforts should be made both at the public and private levels to pave the way for the recharge of groundwater by enhancing the infiltration capacity of the soil through afforestation programs, availability of drainage areas, a check on the concretization process that binds the gaps and pores hindering the permeability of the soil. Also, there should be stricter laws imposed where exploitation should be only in proportion to its recharging capacity. Moreover, there should be conscientious exploitation of groundwater by urban dwellers by adopting a sustainable lifestyle. Encouraging the use of less water-intensive crops and promotion of less water-utilizing industries should be put into practice.

### 10.7.9 Aquifer Storage

Aquifer storage refers to the process of replenishment of groundwater in an aquifer for future use. Aquifer storage and recovery is a method of aquifer recharge (Fig. 10.8B). This method involves direct injection of surface water supplies such as potable water, reclaimed water (i.e., rainwater), or river water into an aquifer for later recovery and use especially in big cities and towns like Delhi where the rainwater cannot percolate the soil or is not capable of percolating due to non-availability of open areas and rapid concretization to meet city's demand of urban infrastructure hindering the porosity of soil and obstructing direct infiltration. The process of Aquifer storage and recovery can prevent saltwater intrusion into freshwater aquifers and control land subsidence and the stored water in underground tanks is used for drinking purposes, industrial needs, and ecosystem restoration projects.

### 10.7.10 Evaporation Reduction Methods

#### 10.7.10.1 Physical Methods

The physical method of evaporation reduction includes bubble plumes and shade balls. These are discussed below:

**Bubble Plume**: It is the method of injecting air bubbles into water reservoirs. The vertical distribution of water indicates the presence of a zone of thermocline that separates the heated and less dense water above it at the surface called epilimnion and the cold and denser layer below called hypolimnion. A process of stratification occurs in the thermocline that acts as a barrier to normal convection circulation and prevents the hypolimnetic water from being mixed with the epilimnion water. On carefully injecting a bubble plume in a cold deep layer, Artificial destratification occurs which drastically reduces the rate of evaporation by uniforming temperature gradients over the depth of the storage.

**Shade Balls**: The use of shade balls is another physical method applied to reduce the rate of evaporation from the water surface (Fig. 10.9A). A study was conducted in the Nehru Pond at Namakkal district in the southern state of Tamil Nadu in India using 4-inch black balls from high-density polyethylene and carbon block to cover the reservoir. It was found that this method reduced the evaporation rates in the reservoir by 43 to 45%.

#### 10.7.10.2 Shade Cloth

Shade cloths are an effective way to prevent evaporation from the water surfaces by obstructing the wind action and blocking the incoming solar radiation via suspended structures installed over water surfaces with supporting poles and steel cables

**Fig. 10.9** Evaporation reduction methods **A** shade balls, and **B** shade cloth. *Source* Derived from Youssef and Khodzinskaya (2019)

(Fig. 10.9B). This method of water conservation is largely used for small reservoirs with an area of less than 10 hectares and has wider prospects in India. The recent advancement in Malaysia presents a new shade cloth knitting machine producing wider rolls, reducing the number of installed cables, thus, reducing the construction cost.

#### 10.7.10.3 Chemical Methods

Chemical mono-layers are insoluble compounds and when applied to water these compounds form an invisible film that can be used to cover a reservoir. The molecules of monolayers act as a physical barrier to water molecules escaping into the atmosphere and thus reduce the rate of evaporation. Moreover, these charge molecules tend to reduce the surface tension of water thus lowering the surface area available for evaporation. These chemical methods of water conservation have been in research and development since the 1920s.

#### 10.7.10.4 Method of Biological Covers

Biological covers including floating aquatic plants such as water lilies, small and great duckweed, and watermeal, etc. can help to reduce the rates of evaporation of water by preventing the connection between air and the boundary layer of water. Studies in Thailand have revealed that duckweed can reduce evaporation by up to 10%. However sound research should be conducted on the efficiency of such aquatic plants as water conservation methods since these aquatic plants can affect water quality and the nature flora dynamics.

### *10.7.11 Water Conservation by Municipal Bodies*

The municipal bodies in the urban areas should be instrumental in managing the individual demand and supply of water as well as conserving water. There should be strict municipal laws providing for the collection of rainwater from rooftops and to implement it. At the same time, mass awareness is very important in water conservation. Every individual should develop an attitude of water conservation and prevent every drop of water from being wasted. Municipal bodies should maintain the depth of traditional water sources by controlling their cleanliness and water should be conserved in different forms.

### *10.7.12 Need for Water Conservation*

Maintaining a balance between different components environment forms the basis of water conservation. Climatic and environmental change results in a shortage in the quantity of fresh water. To this grim scenario, the 'World Water Commission' was constituted in 1999 at Hague to propose methods and solutions for water conservation and creating an environment to solve the riddle of water crisis by 2025 and 2050. In line with the fact that the prime resource of water is going to be scarce in the near future, necessary and mandate plans and actions should include, prevention of leakages, accountability of water use, proper maintenance of the system, replacement of old and rusted pipes, reuse of treated sewage and their utilization for air conditioning, industrial cooling, and other non-potable uses.

## 10.8 Conclusions

The analysis of hydrological regime and climatic conditions for metropolitan cities like Delhi is highly significant for geo-sustainability planning through a thorough knowledge and estimation of water balance along with managing the scarce resource of water. Therefore, this study aimed to explore the gaps between the demand and supply of water to depict the water surplus and water deficit months in the study area and to suggest measures and sustainable solutions for water management to achieve hydrological sustainability in the region. The above analysis clearly explained that the 4 months of January, February, July, and August are the surplus months since enough water is available in Delhi; among which the month of August receives the highest rainfall. The remaining 8 months, i.e., March, April, May, June, September, October, November, and December are identified as deficit months where Delhi faces deficit or severe shortages of water, with December receiving the lowest rainfall. The identification of patterns of water surplus and water deficit months can help the policy-makers look into the issues and planning for urban sustainability in the region. Therefore, the hydrological sector requires special attention from the policy-makers for its sustainability. The suggested ways to achieve geo-sustainability through analysis of water balance include measures like reducing the usage of water, wastewater management, rational use of groundwater, aquifer storage, emphasis on green water building and rainwater harvesting, application of water saving aerators and green taps concerning the Delhi metropolitan region. It is anticipated that if sufficient measures are taken while making decisions about the area, the geo-hydrological sustainability of the area can be ensured.

## References

Amarasinghe UA, Shah T, Anand BK (2008) India's water supply and demand from 2025–2050: business-as-usual scenario and issues. In: IWMI conference proceedings, international water management institute, no 235165. pp 23−61

Available via. https://www.governancenow.com/views/columns/delhi-in-deep-waters-cag-report-flood Accessed 11 Oct 2023

Available via. https://www.epa.gov/watersense/watersense-products Accessed 21 Aug 2023

Census of India (2001) Demographic profile NCT of Delhi. https://delhiplanning.delhi.gov.in/sites/default/files/Planning/economic-survey-2003-04/ch3.pdf. Accessed 8 Apr 2023

CGWB [Central Groundwater Board] (2016)Aquifer mapping and groundwater management plan of NCT Delhi. State Unit Office, New Delhi

Chatterjee B, Roy A (2021) Creating urban water resilience in India: a water balance study of Chennai, Bengaluru, Coimbatore, and Delhi, Observer Research Foundation, New Delhi

Climate India (2022) An assessment of extreme weather events centre for science and environment, New Delhi. https://www.cseindia.org. Accessed 28 Dec 2022

Conservation Mart (2011) Water saving aerators: a mindless way to save money. https://blog.conservationmart.com/water-saving-aerators-a-mindless-way-to-save-money. Accessed 10 Nov 2023

DCHB [District Census Handbook] (1991) Census of India, NCT of Delhi

DCHB [District Census Handbook] (2011) Census of India, NCT of Delhi. https://censusindia.gov.in/2011census/dchb/0700_Part_B_DCHB_NCT%20OF%20DELHI.pdf. Accessed 8 Apr 2023

DCHB [District Census Handbook] (2011) Directorate of economics and statistics. Government of NCT, Delhi

Dhar M (2022) Projection of district-level annual population by quinquennial age-group and sex from 2012 to 2031 in India. IIPS, Mumbai, India. https://www.iipsindia.ac.in/sites/default/files/FULL_REPORT_WITH_FINAL_TABLES.PDF. Accessed 8 July 2023

Economic Survey (2019) Delhi government. New Delhi. https://delhiplanning.delhi.gov.in/sites/default/files/Planning/final_economy_survey_english.pdf. Accessed 22 Oct 2023

EPA (2021) Water sense products. United States Environmental Protection Agency

Financial Express (2020) Budget 2020: government allocates Rs 11,500 crore for Jal Jeevan Mission for FY21. Press Trust of India, New Delhi. https://www.financialexpress.com/budget/budget-2020-govt-allocates-rs-11500-crore-for-jaljeevan-mission-for-fy21/1846403/. Accessed 11 Oct 2023

Ground Water Year Book (2022–23) Government of India. Central Board, Delhi. https://cgwb.gov.in/cgwbpnm/public/uploads/documents/17032373871370440403file.pdf. Accessed 4 Jul 2023

http://cgwb.gov.in/AQM/NAQUIM_REPORT/Delhi/old/Naquim%20Report%20Delhi%20.pdf. Accessed 06 Oct 2023

Imbulana N, Manoharan S (2020) Hydrological and water balance studies to evaluate options for climate resilience in smallholder irrigation systems in Sri Lanka. Water Policy 22(6):1024–1046

IMD [India Meteorological Department] (2022) District-wise rainfall distribution New Delhi. https://mausam.imd.gov.in. Accessed 11 April 2023

India Water Partnership (2017–18) Marching ahead-towards better water resources management. Haryana. https://cwp-india.org Accessed 15 April 2022

Krishnan S (2019) Delhi in deep water. Governance Now

Machiwal D, Dayal D, Kumar S (2017) Estimating water balance of small reservoirs in arid regions: a case study from Kachchh, India. AgricRes 6(1):57–65. https://doi.org/10.1007/s40003-016-0243-5

Madan (2015) Delhi's first solar-powered plant will recycle 'sewage into drinking water. Planet Custodian. https://www.planetcustodian.com/delhis-first-solar-powered-plant-will-recycle-sewage-into-drinking-water/4286. Accessed 5 Aug 2023

Mencio A, Folch A, Mas-Pla J (2010) Analyzing hydrological sustainability through water balance. Environ Manage 45(5):1175–1190

MPD [Master Plan for Delhi] (2021) Delhi Development Authority. Government of Delhi, New Delhi. http://52.172.182.107/BPAMSClient/seConfigFiles/Downloads/MPD2021.pdf. Accessed 11 Jul 2023

Nagpal A (2019) Delhiis running out of water, and it is everybody's problem. The Wire, New Delhi. https://thewire.in/urban/delhi-water-crisis-problem. Accessed 28 Dec 2023

Nayak PC, Wardlaw R, Kharya AK (2015) Water balance approach to study the effect of climate change on groundwater storage for Sirhind command area in India. Int J River Basin Manag 1–19. https://doi.org/10.1080/15715124.2015.1012206

NITIAayog (2019) Composite water management index. In association with ministry of Jal Shakti and ministry of rural development, New Delhi. https://social.niti.gov.in/uploads/sample/water_index_report2.pdf. Accessed 28 Dec 2023

Nugroho AR (2018) Thornthwaite-Mather water balance analysis in Tambakbayan watershed Yogyakarta. Indonesia.

PTI [Press Trust of India] (2015) DJB and NGO to recycle raw sewage into portable drinking water. https://zeenews.india.com/news/eco-news/djb-and-ngo-to-recycle-raw-sewage-into-portable-drinking-water_1626897.html. Accessed 5 Dec 2023

Roychoudhary S (2014) Delhi's water bodies face threat of extinction. India Today. https://www.indiatoday.in/india/story/delhi-water-bodies-extinct-mcd-delhi-parks-and-lakes-society 183229-2014-03-01. Accessed 21 Aug 2023

Rustogi P, Singh SK (2017) Revival and rejuvenation strategy of water bodies in a metropolitan city: a case study of Najafgarhlake in Delhi, India. Int J Adv Res 5(2):189–195. https://doi.org/10.21474/IJAR01/3131

Shali P, Ojha S (2019) Delhi's water bodies face threat of extinction. India Today. https://www.indiatoday.in/mailtoday/story/delhi-city-andtanker-mafia-rules-as-politicians-play-blame-game-1544667-2019-06-08. Accessed 30 Nov 2024

Sharma NS, Kanwar VS, Kandra HS (2019) Water balance evaluation of chandigarh region, India. Int J Innov Technol Explor Eng 8(11):4072–4083

Sharma R Midha N (2021) Potential water balance using rainwater: an analysis of Delhi, megacity in India. J Environ Sci Pollut Res 7(1):447–450

Vashisth A (2019) Rainwater harvesting (RWH) techniques for urban and rural areas

Vijay CR (2015) Urban flood management in Delhi. https://www.business-standard.com/article/current-affairs/urban-flood-management-in-delhi-s-changing-climate-115121601352_1.html. Accessed 11 Oct 2023

Water Conservation Techniques (2019) http://gosmartbricks.com/water-conservation-techniques-10-products-that-help-reduce-water-use-in-buildings. Accessed 10 Nov 2023

Westward (2023) Aquifer storage and recovery: a bank of water for future withdrawal. https://www.westwardenv.com/news/aquifer-storage-amp-recovery-asr-a-bank-of-water-for-future-withdrawal. Accessed 20 Dec 2023

WFD [Water Framework Directive] (2000) The European Water Framework Directive. https://www.eea.europa.eu/policy-documents/water-framework-directive-wfd-2000. Accessed 16 Feb 2024

Wikipedia (2022) Water balance. https://en.wikipedia.org/wiki/Water_balance#:~:text=The%20law%20of%20water%20balance,and%20out%20of%20a%20system.). Accessed 11 March 2022

WRI [World Resources Institute] (2013) Adequate water risk atlas. https://www.wri.org/data/aqueduct-water-risk-atlas. Accessed 6 Oct 2023

Youssef Y, Khodzinskaya A (2019) A review of evaporation reduction methods from water surfaces. In: E3S web of conferences, vol 97, p 05044. https://doi.org/10.1051/e3sconf/20199705044

Zafar S (2022) Features of green building. EcoMENA. https://www.ecomena.org/salient-features-of-a-green-building/. Accessed 1 Nov 2023

# Chapter 11
# Exploring Ground Water Quality Assessment in Jaipur City, Rajasthan

**Rashmi Singh, Harish Kumar, Usha Rani, and Ashwini Kumar**

**Abstract** Assessing the quality of water is essential to assuring the long-term availability of clean drinking water, which is essential for both human health and economic growth. This chapter looks at Jaipur's groundwater quality while acknowledging the problems caused by urbanization, agriculture, and population increase. The current study assesses whether groundwater is suitable for human consumption by examining physicochemical, biological, and microbiological factors. Urbanization and human activity are putting more and more strain on Jaipur's water supplies, which are located in Rajasthan's semi-arid area. This study focuses on a number of important variables including conductivity, pH, alkalinity, hardness, nitrate, fluoride and TDS. It is based on the data from gathered secondary Sources such as Depat. Of Groundwater in Rajasthan. The findings show that these characteristics are present in variable amounts at various city stations. Higher levels of chloride, calcium, magnesium and total dissolved solids raise water quality concerns. The results ranged from outstanding to extremely low quality across various locations in the study area. The water quality index is an important method for evaluating the overall quality of the water. Furthermore, the study finds correlations between various parameters related to water quality, providing insights into possible sources of pollution, and influencing human activities related to groundwater quality. This study's results highlighted the necessity of implementing efficient water management plans, such as enhanced monitoring and treatment protocols, in order to protect public health and assure the long-term use of groundwater resources in Jaipur City.

R. Singh
Department of Geography, Miranda House, University of Delhi, Delhi, India
e-mail: rashmi.singh@mirandahouse.ac.in

H. Kumar (✉)
Department of Geography, Delhi School of Economics, University of Delhi, Delhi, India
e-mail: hkharishkumar.162@gmail.com

U. Rani
Department of Geography, Shivaji College, University of Delhi, Delhi, India
e-mail: ushar489@gmail.com

A. Kumar
Department of Geography, C.H.L. Government College, Chhara, Jhajjar, Haryana, India



S. Anand et al. (eds.), *Emerging Geosustainability Transformations in India*, Advances in Geographical and Environmental Sciences,
https://doi.org/10.1007/978-981-97-7993-2_11

**Keywords** Groundwater · Sustainability · Correlation matrix · Water quality index · Public health

## 11.1 Introduction

Life depends on water, and everyone needs access to a sufficient (appropriate, safe, and accessible) supply (WHO 2008). Enhancing the availability of clean drinking water might yield substantial health advantages. According to the Guidelines, drinking safe drinking water does not pose a serious risk to one's health over the course of a lifetime, even for those who may have varying sensitivity levels at various stages of life. Infants and youngsters, the elderly, and those with disabilities are the groups most vulnerable to waterborne illnesses, particularly while living in unhygienic conditions. To ensure that drinking water is as safe as possible, every effort should be made (Garg et al. 2010). Individuals who are typically susceptible to contracting a waterborne infection would need to take extra precautions. Water quality has a direct bearing on human health as many chemical components of water have been identified to be hazardous. Water should be free from such chemical constituents, which impair health. Hydrochemistry is the field that deals with the chemical, physical, and bacteriological aspects of water to assess whether or not it is suitable for certain uses.

Water is the most valuable resource that humans need to survive on Earth and is one of nature's greatest gifts (Abebe 2013). Humans, animals, and plants all depend on water for their survival and growth. It preserves bodily fluids and controls body temperature. Sweat is one of the ways it helps the body get rid of waste. Water is necessary for many things, including drinking, cooking, bathing, cleaning, drying, and ablution. It is also necessary for agriculture, power production, transportation, and the disposal of garbage. Since water is essential to life, its quality is a critical concern for all of humanity. The world is referred to as a "Blue Planet" because water covers more than two-thirds of its surface, giving the planet a blue appearance from space (Yadav et al. 2010). Water is naturally composed of dissolved materials, particulate matter that isn't dissolved, and living things. These elements are essential to the maintenance of essential biogeochemical cycles, making them essential components of high-quality water. Normally substances commonly bring about water quality issues that are hazardous to human well-being (Van Halem et al. 2009). Because they control the microclimate of each urban center, lakes and tanks are recognized as ecological barometers of the health of the city (Benjamin et al. 1996). The groundwater table and the groundwater nature of the close by springs are extraordinarily impacted by the surface water quality in inland water bodies because of the immediate connection between surface and groundwater. As a result, both natural processes (such as precipitation inputs, erosion, and weathering of crustal materials) and anthropogenic impacts (such as urban, industrial, and agricultural activities) have an impact on the quality of groundwater and surface water (Papatheodorau et al., 2006). The vulnerable ecology has to preserve a condition of natural balance

**Table 11.1** Water standards recommended by ICMR and BIS

| S. no. | Indicators/parameters | Standard value |
|---|---|---|
| 1 | pH value | 6.5–8.5 |
| 2 | Magnesium | 30 (mg/L) |
| 3 | Calcium | 75 (mg/L) |
| 4 | Sulphate | 150 (mg/L) |
| 5 | Chloride | 250 (mg/L) |
| 6 | Nitrate | 45(mg/L) |
| 7 | Electrical conductivity (ECμS/cmat 25 °C) | 300 |

*Source* Central Pollution Control Board (2020)

with its current surroundings, especially in light of pollution and human expansion. However, urban sewage runoff, agricultural practices, and population growth have accelerated the eutrophication process, resulting in a tenfold increase in nutrient imports to the level of natural occurrence in recent decades (Chaudhary et al. 2010). In 1975 and 1993, respectively., the Indian Council of Medical Research (ICMR) and the Bureau of Indian Standards (BIS) set the standard for the quality of water in India (Table 11.1).

Therefore, the biological, microbiological and physiochemical indicators that represent the status of the abiotic ecosystem are. The Jaipur district's groundwater quality has been the study's primary emphasis. The number of people on the planet, industrialization, fertilizer use in agriculture, and numerous other man-made activities all contribute to rising water contamination. The parameters of the water are another important factor that has an impact on the growth of organisms in water bodies including temperature, turbidity, nutrients, hardness, alkalinity, etc. (Smitha 2013).

## 11.2 The Study Area

Jaipur district spans from 26°25′ to 27°51′ north latitude and 74°55′–76°15′ east longitude, making up 3.23% of the state's total area of 11,117 km$^2$ (Fig. 11.1). It is well-positioned among the World Heritage Monuments, nestled at the base of the Aravali range, with plains to the west and south and hills to the north and east. Jaipur, which is currently the capital of the state of Rajasthan, was formerly the capital of the Kachawaha kingdom. The city's roadside buildings and parapets were painted pink during the Prince of Wales' visit in 1876, and the city came to be known as "The Pink City of India" (Census of India 2011).

The city is 390 m above mean sea level and is around 110 km in width from north to south and 180 km in total length from east to west. Situated in the eastern region of the state, it is encircled by the Aravali hills on both the eastern and northern fronts. The city's eastern neighboring districts are Alwar, Dausa, and Sawai Madhopur; its

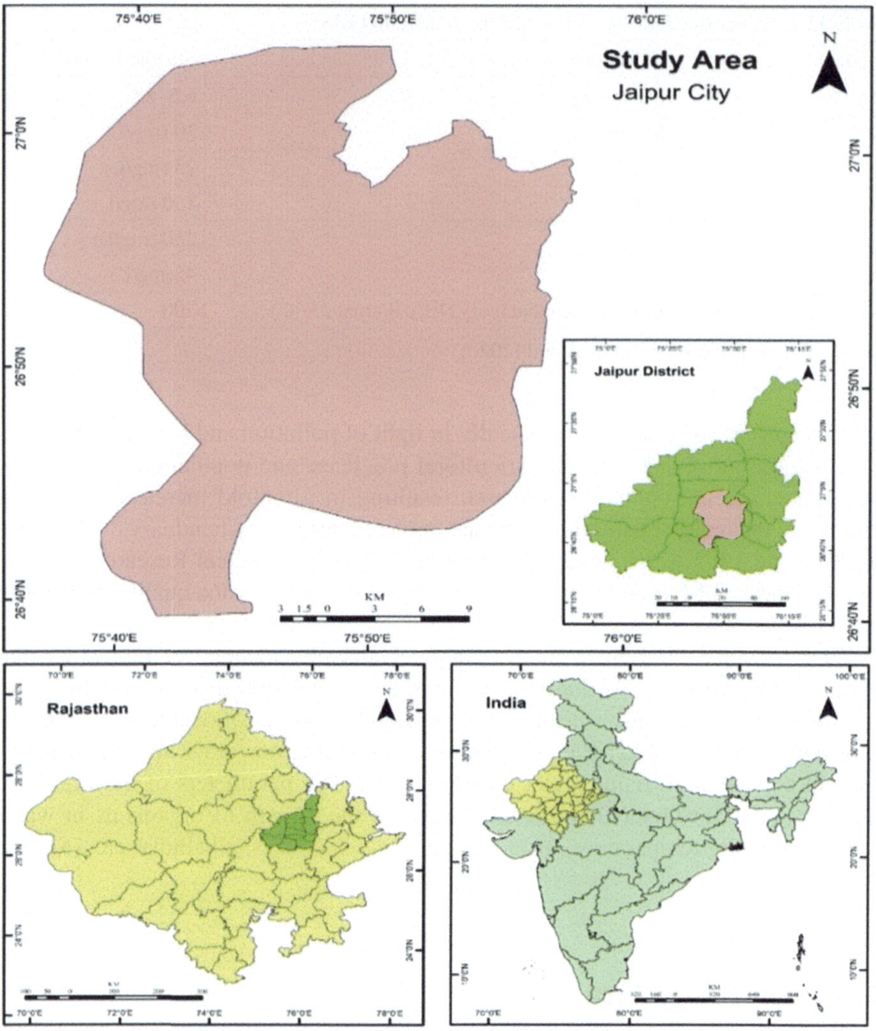


**Fig. 11.1** Location map of the study area. *Source* Census of India (2011)

western neighbors are Nagaur and Ajmer, located in north Sikar and south Tonk. (Census of India 2011).

## 11.3 Material and Methods

The executive engineer of the Department of Groundwater in Jaipur provided the secondary data that formed the basis of the study. A variety of water quality parameters, including color, odor, TDS, Ca+2 ions, Chloride, and Nitrates etc., were assessed on groundwater samples that were chosen for studies that concern BIS, ICMR, and WHO standards. It is quite challenging to regularly monitor every water quality parameter. As a result, numerous approaches have been taken to evaluate the water's quality. Due to its capacity to fully define water quality information, the water quality index method is regarded as the best and most frequently utilized method for defining the level of water quality. It provides a framework for expressing a numerical expression determined cumulatively, according to Li et al. (2010). In order to assess water quality and provide compatibility of physic-chemical water parameters, statistical correlation techniques, and WQI have been used. The World Health Organization and Indian Standard, among other national and international bodies, recommend standard values for the various physicochemical characteristics of underground water samples. These values are compared to the observed values of these parameters.

### *11.3.1 Water Quality Index*

A wide range of water quality indicators have been established and are now being utilized to assess the specific area's water quality. These indexes enable to extract the essential information about the properties of the samples and reduce a vast data set into a much more manageable and useful form the weighted Index and Nemerow's Pollution Index (Sulthonuddin et al. 2020) have been used in this work to determine the physicochemical factors producing pollution and to evaluate the current state of the water quality, respectively.

### *11.3.2 Correlation Method (Water Quality Index)*

The Water Quality Index (WQI) is a well-known and significant tool for expressing water quality. It provides relevant information about the water quality status and unit of measurement to policymakers, stakeholders, and researchers. Thus, in the present situation, it has emerged as a crucial technique for managing and evaluating the quality of surface and groundwater (Chauhan and Singh 2010). Physical, chemical, and biological parameters are frequently used to determine the water's quality. At the point when a review includes an enormous number of tests, the evaluation of water quality turns out to be very difficult in light of the fact that each example is assessed in view of the groupings of various parameters. A Water Quality Index converts a lot of data about water quality into a format that is easy to understand

and useful to the general public. Based on multiple water quality factors, it uses a single number to represent the area's overall water quality. In a nutshell, WQI is a useful mathematical tool that eliminates the need for subjective assessments of water quality and the biases of some sector specialists in this area. It can also be used to express the area's water quality level in a single word (Husain et al. 2014).

## *11.3.3 Water Quality Index Calculation Methods*

Brown et al. (1972) used the weighted arithmetic index method to calculate the water quality index in the following steps (Kumar et al. 2015).

### 11.3.3.1 Step I- Sub-Index of Quality Rating (Qn) Calculation

Comparing the parameter's relative value in the contaminated water to its standard permissible value, the quality rating, or sub-index (qn), for the Nth parameter in a collection of n water quality parameters is derived. The worth is acquired utilizing this articulation.

$$\mathrm{Qn} = 100\,[(\mathrm{Vn} - \mathrm{Vi})/(\mathrm{Sn} - \mathrm{Vi})]$$

where

- Qn is the Quality rating for the Nth water quality parameters represented by Qn.
- Vn is the calculated (estimated) value of Nth parameters at a particular sample of satisfactions.
- Sn is the allowed standard value for the Nth parameter.
- Vi is the enchanting value of the nth parameter in pure water.
- Except for pH = 7.0 and Dissolved oxygen of 14.6 mg/L.
- All Idyllic values (Vi) are taken as 0 for drinking water.

### 11.3.3.2 Calculation Formula for pH Quality Rating

Clean and Natural water should have a pH value is 7.0, however, contaminated water might have a greatest pH esteem is 8.5. As a result, the pH quality rating is calculated using the following relation:

$$\mathrm{qp} = 100\left[(\mathrm{VpH} - 7.0)/(8.5 - 7.0)\right]$$

where

VpH is the pH value that was observed during the study.

**Formula for Calculation of Quality Rating for Dissolved Oxygen**

While the standard allowed level for dissolved oxygen in drinking water is 5 mg/L, the ideal value (VDO) for dissolved oxygen is 14.6 mg/L. The following equation is used to determine the quality rating:

$$\mathrm{qDO} = 100[(\mathrm{VDO} - 14.6)/5 - 14.60]$$

where VDO is the measured value of dissolved oxygen.

#### 11.3.3.3 Step II—Calculation of Unit Weight (*Wn*)

The suggested values for each water quality measure have an inverse relationship with the determined unit weight (Wn) for the fundamental water quality boundaries.

$$Wn = K/Sn$$

where

*Wn* = the weight unit for the nth parameter Standard value for the nth parameter is *Sn*. *K* is the proportionality constant.

#### 11.3.3.4 Step III—Formula for Calculation of Water Quality Index

Water quality list is obtained by utilizing the accompanying condition or equation.

$$WQI = \frac{\sum_{qn} Wn}{\sum Wn}.$$

**Nemerow's Pollution Index (NPI)**

A simplified form of Raw's Pollution Index (RPI), Nemerow's Pollution Index (NPI) was introduced (Saeedi et al., 2010). It is obtained as

$$\mathrm{NPI} = \mathrm{CCC} - \mathrm{LCC}$$

A given parameter's relative pollution is represented by each NPI number, which should be less than or equal to one. An NPI value of more than 1.0 indicates the presence of contaminants in water, according to Saeedi et al. (2010).

## 11.4 Results and Discussion

### *11.4.1 pH*

The pH value allows for the quantitative measurement of an aqueous or other liquid solution's acidity or basicity. pH can be used to measure the acidity and alkalinity of water. Pure water, or "normal pH," is about 7 at 25 °C (Bates 1973). Pure water has a higher pH than acidic water. Drinking water that is too alkaline or too acidic is not advisable. Higher pH values have been shown to reduce the growth of scale in water heating equipment and the germicidal properties of chloride, despite the fact that pH has no direct negative effects on human health (Table 11.2). The highest pH value recorded in the Jaipur district is 8.4 at Jaipur, while the lowest is 7.5 at the station Chomu.

### *11.4.2 Magnesium*

Those who are not used to it have a laxative effect from magnesium hardness, especially when linked to the sulfate ion (Khurshid and Basheer 1998). For most adult aging populations, magnesium intakes of <150 mg/L per day are considered safe. Magnesium in very excessive doses may be harmful to the body. The average magnesium value found across the city is 43.24 mg/L. At Phagi, the magnesium content is the greatest at 87.96 mg/L, while at Mauzamabad, it is the lowest at 5.83 mg/L.

### *11.4.3 Calcium*

Different kinds of rocks affect this drinking water parameter. Water pipes that have a modest amount of calcium in them corrode less. One of the most neglected ions in freshwater is calcium, which is essential for the formation of bones, shells, and plant precipitation. At the Jaipur station, the maximum calcium concentration found in the water in the city of Jaipur was 336 mg/L. Despite the fact that the lowest level, 22 mg/L, was recorded at station Mauzamabad during the same season, many species absorb calcium. (Table 11.2).

### *11.4.4 Chloride*

By measuring the amount of chloride in the water, one can determine whether it is pure or not. The high levels of chloride, according to Tresh et al. (1944), indicate the

**Table 11.2** Physicochemical characteristics of groundwater

| S. no. | Village | pH | Alkalinity | TH | CaH | $Ca^{+2}$ions | MgH | $Mg^{+2}$ions | $Cl^-$ | $NO_3$ | F | TDS | EC |
|---|---|---|---|---|---|---|---|---|---|---|---|---|---|
| 1 | Jaipur | 7.6 | 435 | 732 | 336 | 134.4 | 396 | 96.22 | 315 | 131 | 0.93 | 1740 | 2486 |
| 2 | Jamwa Ramgarh | 7.7 | 368 | 188 | 78 | 31.2 | 110 | 26.73 | 176 | 20 | 0.42 | 920 | 1314 |
| 3 | Sanganer | 7.7 | 562 | 115 | 46 | 18.4 | 69 | 16.76 | 80 | 12 | 1.3 | 1034 | 1477 |
| 4 | Bassi | 7.7 | 426 | 163 | 70 | 28 | 93 | 22.59 | 60 | 70 | 1.13 | 760 | 1085 |
| 5 | Chaksu | 7.4 | 243 | 264 | 121 | 48.4 | 143 | 34.74 | 33 | 29 | 0.71 | 792 | 1131 |
| 6 | Phulera | 7.6 | 707 | 112 | 45 | 18 | 67 | 16.28 | 65 | 19 | 4.67 | 1200 | 1714 |
| 7 | Mauzamabad | 8.2 | 582 | 46 | 22 | 8.8 | 24 | 5.83 | 20 | 22 | 1.4 | 1055 | 1507 |
| 8 | Phagi | 7.7 | 409 | 568 | 206 | 82.4 | 362 | 87.96 | 484 | 8 | 0.9 | 2644 | 3777 |
| | | 8.2 | 458 | 284 | 121 | 48.4 | 163 | 39.6 | 140 | 22 | 2.9 | 1680 | 2400 |
| 9 | Katputli | 7.9 | 411 | 529 | 237 | 94.8 | 292 | 70.95 | 400 | 56 | 0.71 | 2216 | 3165 |
| | | 8.4 | 305 | 115 | 43 | 17.2 | 72 | 17.49 | 31 | 15 | 1.44 | 778 | 1111 |
| 10 | Shahpur | 7.9 | 748 | 360 | 151 | 60.4 | 209 | 50.78 | 278 | 22 | 0.37 | 2100 | 3000 |
| | | 8.3 | 462 | 161 | 67 | 26.8 | 94 | 22.84 | 176 | 18 | 1.8 | 1470 | 2100 |
| 11 | Viratnagar | 7.5 | 651 | 516 | 172 | 68.8 | 344 | 83.59 | 137 | 28 | 2.12 | 1696 | 2422 |
| | | 8.4 | 396 | 105 | 40 | 16 | 65 | 15.79 | 123 | 12 | 1.99 | 1298 | 1855 |
| 12 | Chomu | 7.5 | 586 | 192 | 78 | 31.2 | 114 | 27.7 | 333 | 78 | 2.05 | 2146 | 3065 |
| 13 | Amber | 8.4 | 258 | 158 | 64 | 25.6 | 94 | 22.84 | 202 | 86 | 1.14 | 1590 | 2271 |
| | | 7.8 | 333 | 156 | 67 | 26.8 | 89 | 21.62 | 163 | 44 | 0.79 | 1191 | 1701 |

*Source* Central Ground Water Board, Jaipur, Rajasthan (2020)

presence of organic debris, most likely from animals. The average chloride concentration in the district of Jaipur is 67.799 mg/L. Phagi's standard upper limit of 484 mg/L is lower than the maximum chloride level of 163 mg/L. The lowest concentration observed was 31 mg/L. The greatest value of chloride is Kathputli, which is inside the allowed limit.

**Nitrate**

In the atmosphere of the Earth, nitrogen is one of the most important gases for all living things. The nitrogen cycle is the primary mechanism that controls how nitrogen is distributed in nature. Nitrogen can undergo a wide variety of intricate chemical and biological transformations in the atmosphere or soil. Before being returned to the soil or the air, nitrogen is broken down into living and non-living components through the nitrogen cycle. The Jaipur stations had the highest concentration of nitrate, at 131 mg/l, indicating that sewage pollution is most likely the cause of the nitrogen-rich water that enters the lake.

### *11.4.5 Total Dissolved Solids*

TDS can be used to detect water that contains organic compounds and inorganic salts in trace amounts. Customers may find high levels of TDS in water unpleasant due to the bitter taste and severe scaling in household appliances and water pipes. A potential problem that requires additional investigation is suggested by a higher TDS concentration. For the purpose of determining the total dissolved solids, the gravimetric method was used. This demonstrates that human influence may be linked to agricultural activities that contribute to the local, regional, and temporal variability of runoff, as all of the sites exceed the 500 mg/L threshold set by the BIS, ICMR, and WHO.

### *11.4.6 Electrical Conductivity (EC µS/cm at 25 °C)*

Electrical conductivity indicates the ionic load and pollutants in water. Groundwater often has a high electrical conductivity. The hydroponic solution's strength may be evaluated with the use of the EC. The maximum EC level that can be found in drinking water is 300 S/cm, as stated in the ICMR Standard. EC values in the research region ranged from 610 to 1739 µS/cm, above the ICMR suggested level of 300 µS/cm. The maximum electrical conductivity measurement was 1739 µS/cm at the S15 station, while the lowest was 610 µS/cm. The average EC value in the Jaipur district is 1005 µS/cm.

### 11.4.7 Alkalinity

The presence of certain chemicals, such as hydroxides, carbonates, and bicarbonates, determines water's alkalinity. Because it has an impact on the treatment procedures, measuring it is crucial. To determine the total alkalinity (TA), water samples were titrated against a reference acid using methyl orange and phenolphthalein as indicators. By adding a phenolphthalein indicator, hydrogen ions were transformed into water and carbonate ions ($CO^{2-}$) into bicarbonate ions ($HCO^{3-}$). The bicarbonate ions were then neutralized by methyl red indicator into water and carbon dioxide ($CO_2$). The most extreme amount of alkalinity was seen at Shahpur (748 mg/L).

### 11.4.8 Water Quality Index (WQI)

A single value that is derived from a variety of water quality indicators is used to represent the total water quality at a given location and time. The WQI is the most widely used and comprehended of all index types. The WQI values include a number of significant physicochemical properties of drinking water (Table 11.3). The Water Quality Index (WQI) has a range of 0 to 100. The water is unsafe for consumption if the number is greater than 100. Excellent water quality was observed between 0–25. Very bad quality was reported between 76 and 100 (Fig. 11.2).

### 11.4.9 Calculated Value of Water Quality Index

Seven parameters were used in the calculation of the WQI. These characteristics include pH, magnesium, calcium, sulfate, chloride, nitrate, and electrical conductivity. These parameter samples were utilized to investigate the WQI of the Jaipur district. The following table contains the computed value of one of the sample stations in the research region (Table 11.3).

Due to high levels of TDS, TH, EC, and $NO_3$, the quality of the water in Jaipur, Bassi, and Muazamabad declined, indicating extremely low water quality with a low WQI rating. Viratnagar and Chomu have high WQI scores as a result of the dense population and the dumping of untreated or insufficiently treated sewage. Rapid nitrate contamination in Viratnagar and Chomu may have been brought on by permeable landforms, a low groundwater table, and an inadequate sewage infrastructure. The WQI values in Jotwara, Phagu, and Kathputli have not changed much (Table 11.4). Because of the growth of paved areas and the shrinking of open spaces brought about by building development, the natural recharge of groundwater in these places is decreasing.

**Table 11.3** Status of water quality in Jaipur city based on water quality index

| S. no. | Parameters | Observed value | Standard value (Sn) | Qualify rating | Unit weight (Wn) | Qn * Wn |
|---|---|---|---|---|---|---|
| 1 | Potential of hydrogen (pH) | 7.3 | 6.5–8.5 | 20 | 0.2941 | 5.882 |
| 2 | Magnesium (Mg) | 19.456 | 30 | 64.8533 | 0.0834 | 5.4087 |
| 3 | Calcium (Ca) | 80.16 | 75 | 106.88 | 0.0334 | 3.5697 |
| 4 | Sulphate ($SO_4$) | 22 | 150 | 14.6667 | 0.0167 | 0.2449 |
| 5 | Chloride (CI) | 21.276 | 250 | 8.5104 | 0.01 | 0.0851 |
| 6 | Nitrate ($NO_3$) | 39 | 45 | 86.6667 | 0.5004 | 43.367 |
| 7 | Electrical conductivity (ECμS/cm at 25 °C) | 780 | 300 | 260 | 0.0208 | 5.408 |
| | | | | ΣQn = 561.5771 | ΣWn = 0.9588 | $\Sigma Qn * W_n$ = 63.7125 |
| Water Quality Index (WQI) = ΣQn Wn/ΣWn WQI = 63.7215/0.9588 = 66.4596 | | | | | | |

*Source* Calculate by the authors

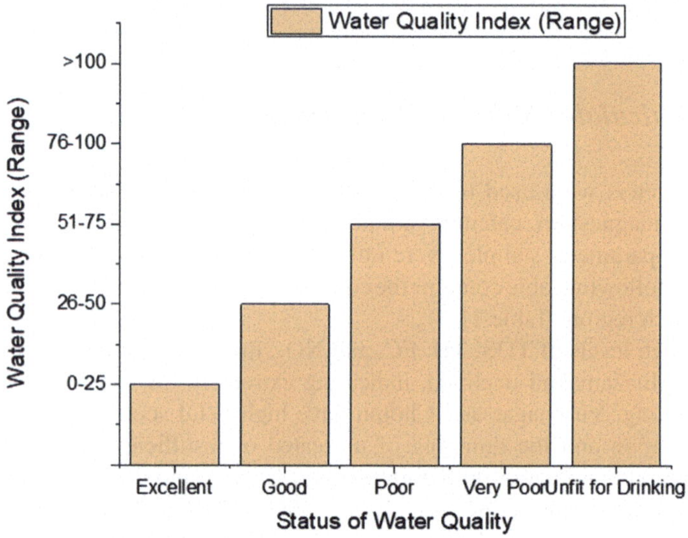


**Fig. 11.2** Status of water quality based on WQI. *Source* WHO (2004)

**Table 11.4** Status of water quality in Jaipur district

| S. no. | Sampling station | Source of drinking water | WQI | Results |
|---|---|---|---|---|
| 1 | Jaipur | Hand pump | 221.58 | Very poor |
| 2 | Jamwa Ramgarh | Tubewell | 74.90 | Good |
| 3 | Sanganer | Handpump | 100.67 | Good |
| 4 | Bassi | Hand pump | 286.00 | Very poor |
| 5 | Chaksu | Hand pump | 122.52 | Poor |
| 6 | Phulera | Hand pump | 120.25 | Poor |
| 7 | Mauzamabad | Hand pump | 362.74 | Unlit for drinking purposes |
| 8 | Phagi | Well | 99.29 | Good |
| 9 | Katputli | Hand pump | 78.05 | Good |
| 10 | Shahpur | Hand pump | 160.9 | Poor |
| 11 | Viratnagar | Hand pump | 239.86 | Very poor |
| 12 | Chomu | Hand pump | 396.67 | Unlit for drinking purposes |
| 13 | Amber | Hand pump | 204.1 | Very poor |

*Source* Calculated by the Authors, 2020

### *11.4.10 Correlation Matrix*

TDS and EC have a very strong positive association in the current study area (r = 0.995). This demonstrates that as the value of TDS rises or falls, the values of EC also appear up or down. TDS has a strong positive correlation with F, TH, and Cl. Cl, NO3, and K exhibit a strong positive correlation with TH. This shows that TH in the water sample may be caused by the presence of Ca $(NO_3)_2$ and $Mg(NO_3)_2$, $CaCl_2$, $MgCl_2$, $CaF_2$, and $MgF_2$. The results demonstrated the positive correlations between Cl & F, Cl & $NO_3$, Cl & K, K & F, and Na & K. The findings show that, overall, there is a positive correlation between 32 and 41 unions, but that, in the remaining unions (23 unions and 15 combinations), there is a negative correlation. The majority of the factors have a negative correlation with turbidity (Table 11.5). There is a negative association between temperature and Na.

Sustainability is essential to the long-term viability of water sources in the current scenario. Implementing sustainable water management strategies, such as efficient water use, pollution prevention, and ecosystem preservation, is crucial to the quantity and quality of groundwater. Building a sustainable water future for Jaipur and its people also involves encouraging community involvement, spreading knowledge about water conservation, and putting environmental protection regulations into action. Stakeholders may solve present issues and protect water resources for future generations by incorporating sustainability concepts into water management plans.

**Table 11.5** Water quality correlation matrix of Jaipur district

| | | pH | TDS | EC | TH | Cl | No3 | F | Na | K |
|---|---|---|---|---|---|---|---|---|---|---|
| pH | Correlation | 1 | 0.079 | 0.088 | 0.427 | 0.318 | 0.434 | 0.235 | 0.116 | 0.602 |
| | Sig. (2-tailed) | | 0.827 | 0.808 | 0.218 | 0.371 | 0.210 | 0.513 | 0.750 | 0.065 |
| TDS | Correlation | 0.079 | 1 | 0.995[b] | 0.749[a] | 0.915[b] | 0.629 | 0.780[b] | 0.530 | 0.353 |
| | Sig. (2-tailed) | 0.827 | | 0.000 | 0.013 | 0.000 | 0.051 | 0.008 | 0.115 | 0.317 |
| EC | Correlation | 0.088 | 0.995[b] | 1 | 0.737[a] | 0.901[b] | 0.614 | 0.746* | 0.515 | 0.325 |
| | Sig. (2-tailed) | 0.808 | 0.000 | | 0.015 | 0.000 | 0.059 | 0.013 | 0.128 | 0.359 |
| TH | Correlation | 0.427 | 0.749[a] | 0.737[a] | 1 | 0.906[b] | 0.957[b] | 0.640[a] | 0.217 | 0.796[b] |
| | Sig. (2-tailed) | 0.218 | 0.013 | 0.015 | | 0.000 | 0.000 | 0.046 | 0.547 | 0.006 |
| Cl | Correlation | 0.318 | 0.915[b] | 0.901[b] | 0.906[b] | 1 | 0.818[b] | 0.753[a] | 0.515 | 0.592 |
| | Sig. (2-tailed) | 0.371 | 0.000 | 0.000 | 0.000 | | 0.004 | 0.012 | 0.128 | 0.072 |
| No3 | Correlation | 0.434 | 0.629 | 0.614 | 0.957[b] | 0.818[b] | 1 | 0.501 | 0.169 | 0.833[b] |
| | Sig.(2-tailed) | 0.210 | 0.051 | 0.059 | 0.000 | 0.004 | | 0.140 | 0.640 | 0.003 |
| F | Correlation | 0.235 | 0.780[b] | 0.746[a] | 0.640[a] | 0.753[a] | 0.501 | 1 | 0.419 | 0.573 |
| | Sig. (2-tailed) | 0.513 | 0.008 | 0.013 | 0.046 | 0.012 | 0.140 | | 0.228 | 0.083 |
| Na | Correlation | 0.116 | 0.530 | 0.515 | 0.217 | 0.515 | 0.169 | 0.419 | 1 | 0.009 |
| | Sig. (2-taileda) | 0.750 | 0.115 | 0.128 | 0.547 | 0.128 | 0.640 | 0.228 | | 0.980 |
| K | Correlation | 0.602 | 0.353 | 0.325 | 0.796[b] | 0.592 | 0.833[b] | 0.573 | 0.009 | 1 |
| | Sig. (2-tailed) | 0.065 | 0.317 | 0.359 | 0.006 | 0.072 | 0.003 | 0.083 | 0.980 | |

[a]Correlation is significant at the 0.05 level (2-tailed)
[b]Correlation is significant at the 0.01 level (2-tailed)

## 11.5 Conclusions

The methodological WQI approach makes it easier to compare the drinking water quality of many sample locations. More than 80% of test locations had water that was deemed unsafe to drink, while 20% had less than the recommended level of pollution. It shows that fewer than 20% of water samples had excellent enough quality to drink. The collective WQI scores of the seven stations indicate that the water quality is unfit for human consumption and necessitates treatment before consumption. Total alkalinity, electrical conductivity, chloride, fluorides, nitrates, and total dissolved solids are the primary pollutants found in this study. However, some example areas likewise uncover generally speaking hardness, calcium, and magnesium as the essential contaminations notwithstanding different boundaries. However, it is also possible to observe that, in some samples, all parameters, with the exception of pH, exhibit characteristics indicative of pollution by examining the NPI values for both seasons.

## References

Abebe BG (2013) Urban water supply in Addis Ababa: problems and prospects. Approved PhD thesis, Department of Geography, Banaras Hindu University

Bates R (1973) Determination of pH: theory and practices. Wiley, New York, USA

Benjamin R, Chakrapani BK, Naarthana AV, Ramchandra TV (1996) Fish mortality in Bangalore lakes, India. Electric Green J 1(6):73–79

Brown RM, McCleiland NJ, Deininger RA, O'Connor MF (1972) A water quality index—crossing the psychological barrier. In: Proceedings of the international conference on water pollution research, Jerusalem, 18–24 June 1972, pp 787–797

Census of India (2011) Primary census abstract. Office of the Registrar General and Census Commissioner, Ministry of Home Affairs, New Delhi

Chauhan A, Singh S (2010) Evaluation of Ganga water for drinking purpose by water quality index at Rishikesh, Uttarakhand, India. Report Opin 2(9):53–61

Choudhary P, Routh J, Chakrapani GJ (2010) Organic geochemical record of increased productivity in Lake Naukuchiyatal, Kumaun Himalayas, India. Environ Earth Sci 60(4):108–113

CPCB [Central Pollution Control Board] (2020) Annual report-2022–21, Delhi. https://cpcb.nic.in/annual-report.php. Accessed 12 Feb 2023

Garg RK, Rao RJ, Uchchariya D, Shukla G, Saksena DN (2010) Seasonal variations in water quality and major threats to Ramsagar reservoir, India. Afr J Environ Sci Technol 4(2):61–76

Husain A, Hussain J, Kumar S (2014) Fluoride distribution in groundwater and survey of dental fluorosis in villages of Didwana tehsil of Nagaur district of central Rajasthan, India. India Water Week 2012: Water, Energy and Food Security

Khurshid S, Basheer A (1998) Pollution assessment and water quality status in parts of Cochin. Indian J Environ Prot 18(4):246–249

Kumar K, Kumaran S, Magesh NS (2015) Hydro-geochemistry and application of water quality index (WQI) for groundwater quality assessment, Anna Nagar, part of Chennai City, Tamil Nadu, India. Appl Water Sci 5(4):335–343

Li P, Qian H, Wujian H (2010) Ground water quality assessment based on improved water quality index in Pengyang county, Ningxia, Northwest China. E–J Chem 7(SI):S209–S216

Saeedi M, Abessi O, Sharifi F, Meraji H (2010) Development of ground water quality index. Environ Monitor Assess 163:327–335

Smitha S (2013) Physio-chemical analysis of the freshwater at River Kapila, Nanjangudu Industrial Area, Mysore, India. Int Res J Environ Sci 2:59–65

Sulthonuddin I, Hartono D, Said C (2020) Using Nemerow's pollution index method for water quality assessment of Cimanuk river in West Java. In: Proceedings of the proceedings of the 1st international conference on environmental science and sustainable development, ICESSD 2019, 22–23 October 2019, Jakarta, Indonesia. https://doi.org/10.4108/eai.22-10-2019.2291494

Tresh JC, Suckling EV, Beale JF (1944) Chemical and zooplankton studies of lentic habitats in northeastern New South Wales. Austr J Mar Freshw Res 21:11–33

Van Halem D, Bakker SA, Amy GL, Van Dijk JC (2009) Arsenic in drinking water: not just a problem for Bangladesh. Drink Water Eng Sci Discuss 2(51):64

WHO [World Health Organization] (2004) Guidelines for drinking-water quality. 3rd Edition, Incorporating First and Second Addenda. https://www.who.int/publications/i/item/9789241547611. Accessed 12 Feb 2023

WHO [World Health Organization] (2008) Progress on Drinking water and sanitation: special focus on sanitation. https://www.who.int/publications/i/item/9789280643138. Accessed 12 Feb 2023

Yadav AK, Khan P, Sharma SK (2010) Water quality index assessment of groundwater in Todarai Singh tehsil of Rajasthan state, India: a greener approach. E-J Chem 7(S1):S428–S432

# Chapter 12
# Morphometric Analysis of the Koyna Watershed Using Geospatial Techniques for Groundwater Prospects

**Namdev V. Telore and Yuvraj U. Kadam**

**Abstract** Morphometric analysis was carried out utilizing geographic methods to look at the drainage structure and features of the Koyna watershed using topographical maps and satellite data. This 1107 km$^2$ study centers on one of the Krishna River basin's sub-basins in the Satara district of Maharashtra. Using techniques from Horton, Schumn, Miller, Strahler, and others, morphometric properties such as stream order, bifurcation ratio, stream length, drainage density, circularity ratio, relief ratio, stream frequency, elongation ratio, and compactness constant were calculated. During the field visit, GPS was used to mark the necessary places. The Koyna basin has a dendritic to sub-dendritic drainage structure and is categorized as a sixth-order drainage basin. The drainage density of the basin is coarse, indicating a good potential for groundwater. The majority of the stream, or 50.55% of the total length of 3505.8 km, is made up of first-order streams, which get shorter as the stream order gets higher. From first-order to sixth-order streams, the stream length ratio, which ranges from 0.97 to 3.66, indicates rising groundwater potential. Across stream orders, the mean stream length ranges from 0.35 to 0.63 km. The elongated basin form with a high peak flow of short duration is indicated by the estimated form factor value of 0.192. When combined with the drainage density, the basin's 6.65 streams per km$^2$ stream frequency indicates an acceptable stream frequency. Overall, the analysis shows that the region has excellent groundwater prospects, which are reinforced by a network of streams that are evenly spaced out and contribute to the provision of groundwater.

**Keywords** Morphometric analysis · Groundwater prospects · Geospatial techniques · Koyna watershed

---

N. V. Telore (✉)
Department of Geography, Raja Shripatrao Bhagawantrao Mahavidyalaya, Satara 415510, Maharashtra, India
e-mail: nvtelore@gmail.com

Y. U. Kadam
Shivaji University, Kolhapur 416004, Maharashtra, India



S. Anand et al. (eds.), *Emerging Geosustainability Transformations in India*, Advances in Geographical and Environmental Sciences,
https://doi.org/10.1007/978-981-97-7993-2_12

## 12.1 Introduction

Delineating catchment areas, updating them, and examining their morphometric properties have all been demonstrated to be facilitated by the application of geospatial techniques. Any hydrological inquiry, including the evaluation of groundwater potential and groundwater management, must include a drainage basin analysis. The physiographic features of the catchment area, including their size, shape, slope, density, and tributary length, can be linked to several significant hydrologic phenomena (Rastogi and Sharma 1976).

In addition to conventional data, geospatial data can be used to delineate ridgelines, characterize areas, evaluate priorities, identify problems, assess potentials and management needs, identify areas that are prone to erosion, select locations for watershed management structures (Dutta et al. 2002; Eze and Efiong 2010; Inamdar et al. 2016).

Many authors have used remote sensing-based geographic approaches to perform morphometric analyses of Water basins to better plan water supply schemes and manage irrigation potential. This study describes the drainage features of the Koyna watershed in the Satara district, Maharashtra state, India, using a geospatial-based morphometric approach. The present study help to analyze the basin's hydrological characteristics.

Geospatial approaches are highly valuable for Morphometric analysis of basins because they are low-labor, low-cost, and time-efficient instruments. Geospatial technologies are being used by several researchers to analyze the morphometrics of basins. Using geospatial approaches for statistical shape analysis, Thomas et al. (2012) studied two tropical mountain Water basins in the southern Western Ghats and concluded that drainage networks represent tectonic framework, climate, geology, and the interplay between drainage characteristics.

Kamala and Samynathan (2018) used SRTM data in their geospatial Morphometric analysis of the Amaravathi River Basin, using ArcGIS software to extract important information. Their research revealed the drainage system's consequent nature, which is impacted by the land face slope. Specifically, lower-order streams primarily flow sequentially towards higher-order main streams.

Jasmin and Mallikarjuna (2012) carried out morphometric analysis using geospatial techniques for the study of groundwater potential. The work is significant since it emphasizes how critical these morphometric studies are to comprehending the basin's characteristics and determining significant ramifications for Tamil Nadu's aquifer prospects.

Das and Mukherjee (2005) discovered a consistent correlation between stream sequence and cumulative stream length, concentrating on fifteen different watersheds. Data on the bifurcation ratio and low drainage density imply that physical disruptions have little impact. The area is primarily covered with dense vegetation, with resistive, permeable bedrock beneath it. Higher groundwater levels are indicated by correlations in areas with low infiltration numbers, stream frequencies, and drainage densities.

Ansari et al. (2012) Eight watersheds with a high potential for artificial recharge but low infiltration quantities are identified by a thorough statistical shape analysis. Important information for the region's management of water resources is provided by this study. When it comes to the delineation, updating, and Morphometric analysis of catchment areas, geospatial techniques prove to be useful instruments. Using these methods, this study focuses on three sub-basins in the Agra district's Fatehabad area. The River pattern, which is mainly dendritic, indicates that the material is uniformly underlain.

Ansari et al. (2012) carried out a study on three sub-basins in the Agra district's Fatehabad area for indenting potential artificial recharge structure through morphometric analysis using geospatial data he found dendritic drainage patterns in this area due to uniform geological conditions.

In his study Circularity ratio values indicate long basins with minimal outflow, High underlying material permeability is indicated by the area's low drainage density and high vegetation density. Due to its greatest infiltration rate and low runoff, the surface water resources are more in that area.

Using SRTM data and GIS techniques, Javed et al. (2009) analyzed the morphological characteristics of the Wadi Aurnah watershed in the western Arabian Peninsula. They identified zones appropriate for artificial recharge structures and discovered the basin's potential for groundwater recharge. Topographical maps and geographic information systems (GIS) were used by Sreedevi et al. (2005) to analyze the morphometric characteristics of the Pageru River drainage in Andhra Pradesh. The runoff features of the basin and located good groundwater potential zones are learned.

Nooka Ratnam et al. (2005) evaluated the morphometric parameters and river patterns of the watershed using Geographic Information Systems (GIS) and satellite imagery techniques, and based on their findings, they identified locations with high groundwater potential. They discovered that drainage density, bifurcation ratio, and stream size determine the area with high groundwater potential.

The analysis indicates that sub-basin-II looks to have limited permeability, while sub-basin-III exhibits porosity and permeable subsoil conditions. This exhaustive study provides helpful data for managing the water resources in the area. Somashekar and Ravikumar (2011) developed theme strata for water drainage, soil, hydrological soil group, and LU and LC using IRS-1D and LISS-III satellite data (FCC). Utilizing the Soil Conservation Service's (SCS) curve number model, runoff estimation for sub-watersheds 1–4 yielded measurements of 1960, 2066, 1870, and 1810 mm, correspondingly.

The watershed was assessed for the Morphometric analysis in terms of three categories: aerial (shape factor, elongation ratio, circulatory ratio, drainage density), relief (basin relief, relief ratio, relative relief, ruggedness number), and linear (stream order, length, length ratio, bifurcation ratio, overland flow). Based on its estimated drainage density of 1.23 km/km$^2$, the examined region is classified as a very coarse-textured watershed. The Deccan Traps' Yerla River basin, which is located in a rain shadow area, is described by Telore (2020) as having an elongated shape, a dendritic pattern, being less dangerous, and being in a phase of transition of development in terms of morphometric criteria.

Using a Geographic Information System (GIS), Waikar and Nilawar (2014) carried out a morphometric study within the Charthana drainage area in Maharashtra, India. The GIS-based analysis classifies it as a fourth-order basin due to its sub-dendritic to dendritic River pattern, homogeneous texture, and lack of structural control.

The basin contains 148 streams, and the length of the stream diminishes as the order increases. With an estimated drainage density of 1.45 km/km$^2$, the watershed is thought to have a rough face and medium to good groundwater potential. The study also demonstrated how morphometric features can reveal information about the basin's climate. In order to determine the links between hydrological processes and morphometric parameters such as basin order, relief, size, form, and drainage density, drainage basin analysis is therefore crucial.

## 12.2 Material and Methods

### *12.2.1 The Study Area*

The research area is located in the western region of Satara district, Maharashtra, and encompasses 1107 km$^2$ within the Koyna River basin. It is distinguished geologically by basalt deposits from peninsular India's Deccan Traps. Both moist deciduous and tropical evergreen woods make up the forest cover. With 7365 streams, the Koyna watershed shows a dendritic to sub-dendritic drainage structure. This study region also includes the Koyna Dam. The Koyna River valley has a subtropical monsoon climate. Up to Koyna Dam, the river's course's catchment area is 891.78 km$^2$ (Swain et al. 2022) and receives well over 5000 mm of rainfall on average. The area extends from 73° 48″ E to 73° 57″ E and from 17° 15″ N to 17° 54″ N. The main tributaries of the Koyna River are Kera, Wang, Morna, and Mahind (Fig. 12.1).

### *12.2.2 Data Source and Methodology*

The morphology of the watershed was analyzed using the ALOS PALSAR DEM, which has a spatial resolution of 12.5 m; Landsat satellite imagery, which has a spatial resolution of 30 m; and the Land Survey of India topographical maps No. 47 G/10, 47 G/11, 47 G/14, and 47 G/15, which have a scale of 1:50,000. A stream order has been created using the Arc Hydro tool, Strahler's (1953) system, and ArcGIS-10.3 software by measuring the size and length of streams and the watershed. For the study, the investigation focused on the relief, aerial, and linear characteristics.

The linear aspect is further divided by the ratio of bifurcation (Rb), stream length (L), and stream order (Nu). The seven categories that comprise the aerial aspect are Drainage Density (Dd), Stream Frequency (Fs), Elongation Ratio (Re), Form Factor

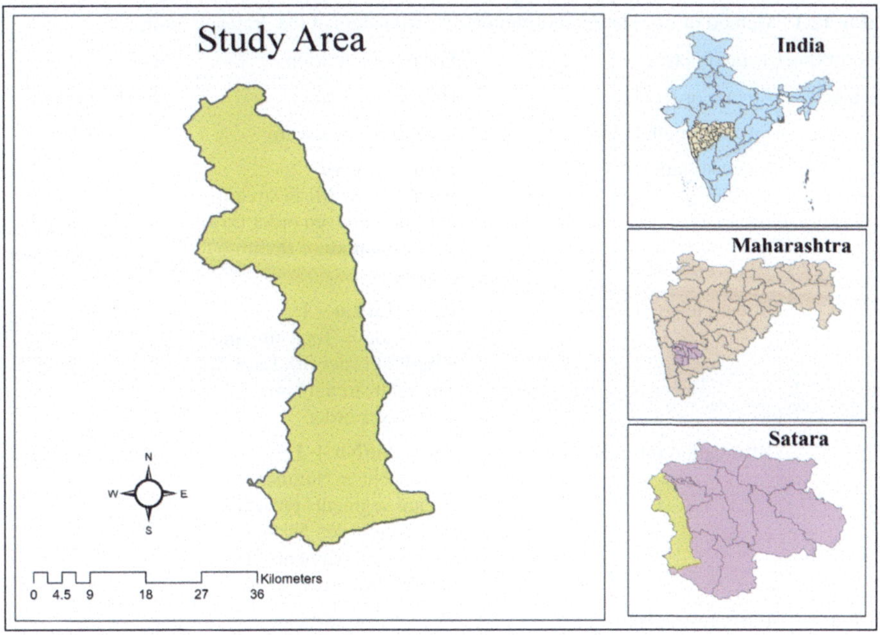


**Fig. 12.1** Location of the study area

(Rf), Circularity Ratio (Rc), and Length Overland Flow (Lo). Three categories are further distinguished within the Relief Aspect: Average Slope (Sa), Total Relief (M), and Relief Ratio (Rr). The formula for determining the morphometric parameters is displayed in Table 12.1. The geographical projection used for estimating the systems of rivers and basins is WGS-1984 or UTM zone 43N. The Arc GIS 10.3 program is used to create the longitudinal profile of the research area.

## 12.3 Results and Discussion

### *12.3.1 Morphometric Analysis of Basin*

The various techniques for measuring morphometric properties are covered in the following sections. The numbers of these parameters are further determined by applying the techniques that various researchers have recommended for the research area and described in their individual descriptions. Using various morphometry methodologies, Horton (1945) and Rai et al. (2017b) examined major drainage basin morphometry metrics such as stream number, stream length, river basin area, stream order, drainage density, and stream frequency.

**Table 12.1** Method of calculating morphometric parameters of the drainage basin

| Morphometric parameters | | Formula/definition | Source |
|---|---|---|---|
| Linear | Stream order (U) | Hierarchical order | Strahler (1964) |
| | Stream length (LU) | Length of the stream | Hortan (1945) |
| | Mean stream length (Lsm) | **Lsm = Lu/Nu**<br>where, Lu = Mean stream length of a given order (km), Nu = Number of stream segments | |
| | Stream length ratio (RL) | **RL = Lu/Lu – 1**<br>where, Lu = Total stream length of order (u), Lu − 1 = The total stream length of its next lower order | |
| | Bifurcation ratio (Rb) | **Rb = Nu/Nu + 1**<br>where, Nu = Number of stream segments present in the given order, Nu + 1 = Number of segments of the next higher order | Schumn (1956) |
| Relief | Basin relief (Bh) | The vertical distance between the lowest and highest points of the basin | |
| | Relief ratio(Rh) | **Rh = Bh/Lb**<br>where, Bh = Basin relief, Lb = Basin length | |
| | Ruggedness number(Rn) | **Rn = Bh x Dd**<br>where, Bh = Basin relief, Dd = Drainage density | |
| Aerial | Drainage Density (Dd) | **Dd = L/A**<br>Where, L = Total length of the stream, A = Area of basin | Hortan (1945) |
| | Stream frequency (Fs) | **Fs = Ln/A**<br>where, Ln = Total number of streams, A = Area of basin | |
| | Texture ratio (T) | **T = N1/P**<br>where, N1 = Total number of first-order streams, P = Perimeter of basin | |
| | Form factor (Rf) | **Rf = A/(Lb)**<br>Where, A = Area of basin, Lb = Basin length | |
| | Circulatory ratio (Rc) | **Rc = 4$\pi$A/P**<br>where A = Area of basin, $\pi$ = 3.14, P = Perimeter of basin | Miller (1953) |

(continued)

**Table 12.1** (continued)

| Morphometric parameters | | Formula/definition | Source |
|---|---|---|---|
| | Elongation ratio (Re) | **Re = √(Au/π)/Lb**<br>where A = Area of basin, π = 3.14, Lb = Basin length | Schumn (1956) |
| | Length of overland flow (Lg) | **Lg = 1/2Dd**<br>where Dd = Drainage density | Hortan (1945) |
| | Constant channel maintenance (C) | **C = 1/Dd**<br>where Dd = Drainage density | |

*Source* Compiled by authors

#### 12.3.1.1 Linear Aspects

The linear components of the basin's morphometric research are stream length, stream length ratio, mean stream length, bifurcation ratio, and stream order. The next section discusses these.

Stream Order (U)

The following writers use four different ordering strategies for accessible streams: Horton (1945), Gravelius (1914), Scheidegger (1970), and Strahler (1952). Strahler's (1952) technique, which is essentially a little variation of Horton's system (1945), which is embraced due to its simplicity. A small emerging stream is known as first-order; second-order is formed when two first-order streams come together; the third-order stream is formed when two two-order streams come together and so no. If two networks with different orders connect, the higher order is kept. The trunk stream is the highest-order stream segment. It is found that the tributaries of the Koyna River are sixth-order. Of the 7365 streams discovered, the first order is represented by 3723, the second order by 1835, the third order by 904, the fourth order by 496, the fifth order by 206, and the sixth order by 201 (Fig. 12.2). The streams in the basin imply a uniform texture and no physical control over the system (Telore 2020). A detailed analysis of the characteristics of the stream patterns is necessary to comprehend the basins (Strahler 2002).

Stream Number

The "stream number," which is inversely correlated to the stream order, is the total number of stream segments in various orders. Romshoo et al. (2012) provided that lower permeability and infiltration are indicated by a higher order of magnitude. The maximum frequency for first-order streams is found, as Table 12.2 and Fig. 12.2

**Fig. 12.2** Drainage system and stream ordering

**Table 12.2** Results of the linear aspect of the study area

| Stream order | No. of stream | Stream length meters | Stream length (km) | Percentage stream length | Mean stream length (Lsm) | Stream length ratio | Bifurcation ratio |
|---|---|---|---|---|---|---|---|
| 1 | 3723 | 2,334,867.06 | 2334.87 | 66.60 | 0.63 | 1.80 | 2.03 |
| 2 | 1835 | 638,395.1576 | 638.40 | 18.21 | 0.35 | 1.19 | 2.03 |
| 3 | 904 | 263,593.6107 | 263.59 | 7.52 | 0.29 | 1.12 | 1.82 |
| 4 | 496 | 129,624.804 | 129.62 | 3.70 | 0.26 | 0.78 | 2.41 |
| 5 | 206 | 68,756.73577 | 68.76 | 1.96 | 0.33 | 0.95 | 1.02 |
| 6 | 201 | 70,605.33401 | 70.61 | 2.01 | 0.35 | – | – |

*Source* Calculated by authors

demonstrate. A decrease in the frequency of streams is also observed with the increase in stream order.

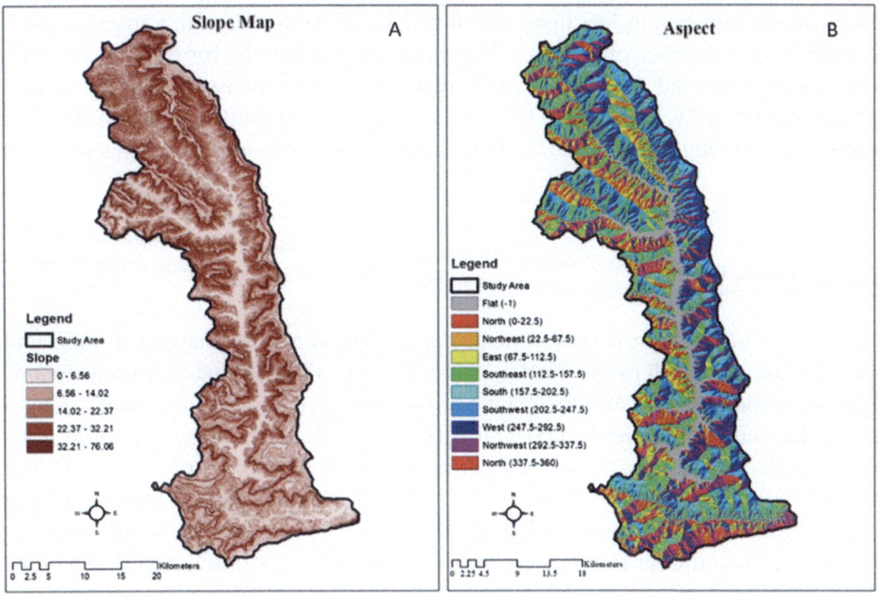


**Fig. 12.3** Slope and aspect of the study area

Stream Length ($L_u$)

We have used Horton's (1945) law to calculate the stream's length. Stream length forms one of the basin's most significant hydrological characteristics since it reveals characteristics of face runoff. Areas with relatively tiny streams tend to have higher slopes and finer textures. Longer streams often have flatter grades. First-order streams normally have maximum segment lengths, but higher-order streams frequently have shorter segments (Table 12.2). Using ArcGIS 10.3 software, the number of streams within a watershed and their length from mouth to drainage divide are measured (Srinivasa Vittala 2004).

The total distances of the first, second, third, fourth, fifth, and sixth order streams are displayed in Table 12.2 (Singh and Singh 1997) and are 2334.87, 638.40, 263.59, 129.59, 68.76, and 70.61 km, respectively. Rai et al. (2017a) suggested that the total amount of streams flowing from a high elevation, lithological variety, and relatively steep slopes form an inverse geometric pattern with the number of segments of stream in the same ordering (Fig. 12.3). Horton's (1945) law of stream number is confirmed by stream order observation.

Mean Stream Length ($L_{sm}$)

Strahler (1964) and Devne et al. (2019) stated that mean stream length is one characteristic that sets the system of drainage and the surfaces that it is connected to apart.

The mean stream length has been calculated by dividing the entire stream length of order by the number of streams. The mean stream lengths for the first, second, third, fourth, fifth, and sixth orders in the research region are displayed in Table 12.2 and are, respectively, 1.59, 2.87, 3.43, 3.83, and 2.99. Up until the fourth order, the average stream length increases; for both the fifth and sixth orders, it then somewhat decreases.

Stream Length Ratio ($R_L$)

The obvious connection between mean stream length and discharge can be found using the stream length ratio (Horton 1945). The basin's RL values between streams of various orders reveal variations in the basin's terrain and slope. The stream length ratio in the study area varies from 0.97 to 3.66 (Table 12.2). This observation suggests that groundwater prospects often rise from first-order to sixth-order streams. The stream length ratio is a helpful indicator of potential groundwater availability; greater values, particularly as stream orders grow, suggest more favorable conditions for groundwater resources. This is also concluded by Fayera Gudu Tufa (2018) in his paper.

Bifurcation Ratio ($R_b$)

Schumn (1956) defined the bifurcation ratio as the ratio of the number of stream segments of a given order to the number of sections of the next higher order. Horton (1945) looked at the bifurcation ratio as a relief and dissections metric. Strahler (1957) reveals that the bifurcation ratio shows a narrow range of variance for different locations or different environmental circumstances, with the exception of areas where geology predominates. Evidence suggests that Rb is different for different orders. The research area's mean Rb varies from 1.02 to 2.03, but the basin's average Rb is 1.86 (Table 12.2). These values are generally observed in areas where geologic formations have little effect.

#### 12.3.1.2 Relief Aspect

Relief Ratio ($R_h$)

Schumm (1956) defined the relief ratio as the ratio of the horizontal distance along the longest dimension of the basin, which is parallel to the main drainage line, to the highest relief. When drainage area and catchment sizes shrink in a given drainage basin, the Rh usually rises (Gottschalk 1964). Schumm (1956) defined relief ratio, which measures, measures the overall steepness of a drainage basin as well as the intensity of the act of erosion taking place on the slope of the basin. The steep slope and hard basalt give the basin a Rh value of 0.93, indicating significant relief.

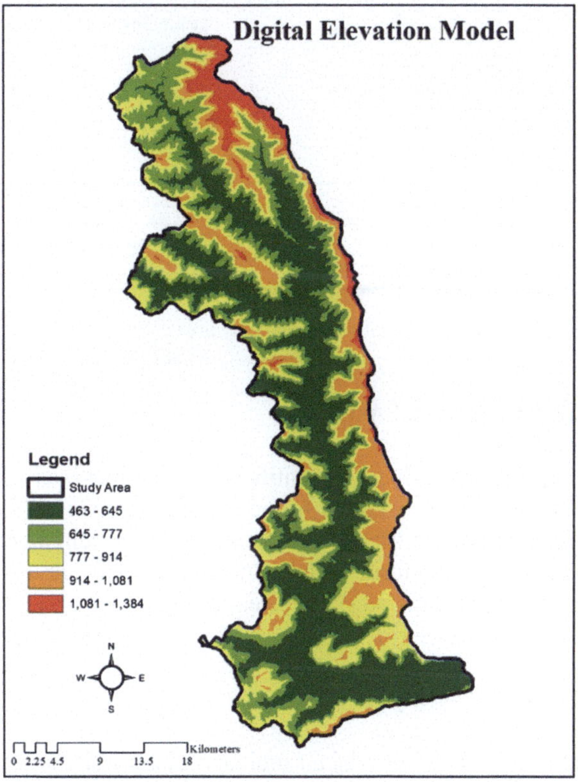


**Fig. 12.4** DEM of study area

Relative Relief ($R_{bh}$)

This word was first used by Melton (1957). It is acquired in the current research area by visually inspecting the DEM that was produced using data from the Shuttle Radar Topography Mission (SRTM). Based on elevation fluctuation from 463 and 1384 m, the land exhibits a mild to moderate slope (Fig. 12.4).

Ruggedness Number ($R_n$)

Ruggedness number (Rn) is the sum of the values of the density of drainage (Dd) and maximum basin relief (H), both given in the same unit of measurement. The toughness number is exceptionally high when both components are large and the slope is steep (Strahler 1956). The roughness number in the research area is high, 22.52 (Fig. 12.5).

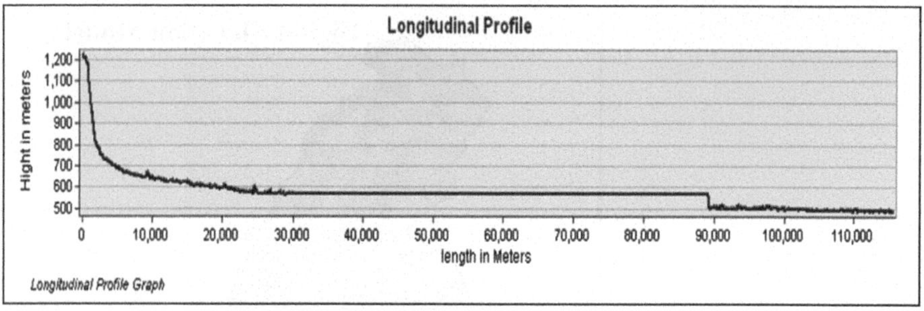


**Fig. 12.5** Longitudinal profile of study area

#### 12.3.1.3 Aerial Aspects

This section includes the entire area extending over a horizontal plane that provides a flow of river overland to the channel segment of the given order, in addition to all lower-order tributaries. Along with the length of river flow, it also contains the following: form factor, drainage density, drainage texture, elongation ratio, circularity ratio, and stream frequency.

Drainage Density ($D_d$)

Horton (1932) introduced drainage density (Dd) as an important indicator of the linear scale of landform characteristics in topography degraded by streams. It can be expressed as the sum of the lengths of all the channel segments inside the basin and its area. By calculating the drainage density, one may determine the median length of the river channel for the whole basin based on the closeness of channel spacing. Drainage density observations over a wide range of geology and climatic types have revealed that low drainage density is more common in places with high levels of highly permeable sub-face material, vegetation cover, and low relief.

In his investigation study, Carlston (1963) also came to the conclusion that aridity which is caused by a reduction in the number of plants covering the thickness of the soil layer is a result of elevated temperatures, increased evapotranspiration, and decreased rainfall. These factors control the differences in drainage densities and the capacity of the earth's face to infiltrate.

A high drainage density is seen in areas with minimal vegetation and weak or impermeable subsurface. High drainage density results in fine drainage texture, whereas low drainage density results in coarse drainage texture, according to Strahaler (1964). Higher drainage densities are associated with steeper landscape slopes (Chougale and Sapkale 2017; and Alemayehu et al. 2024). The moderate drainage density of 3.18 km/km$^2$ in the research region indicates a coarse density in the basin. This characteristic suggests that groundwater resources in the Koyna basin could be fruitful. The greater density of river systems and branches in the area contributes to the enhanced groundwater recharge and availability. Because of its modest drainage density, the basin has a relatively permeable subsoil and vegetative cover, according to Nag (1998).

### Stream Frequency ($F_s$)

In a unit region, the expression for stream frequency is the total number of stream segments for all orders (Hortan 1945), which indicates that as drainage density increases in the watershed, so does the total number of streams. It also demonstrates a positive correlation with drainage density. 6.65 streams per km$^2$ are estimated to exist in the study region (Horton 1932). This figure illustrates the favorable correlation between the watershed's drainage density and a moderate stream frequency. Additionally, it emphasizes the region's exceptional groundwater resources. A moderate stream frequency and positive correlation suggest that the groundwater resources in the studied area benefit from a scattered and well-balanced network of streams.

### Drainage Texture Ratio (T)

Drainage texture ratio (T) refers to the total number of stream segments within a region (Horton 1945). Numerous natural elements, such as temperature, precipitation, vegetation, type of soil and rock, infiltration capacity, relief, and growth stage, all have an impact on it (Asode et al. 2016). The drainage texture ratio, which rises with the intra-basin area, either directly or indirectly reflects the drainage density. With a texture ratio of 9.33, the basin is classified as high in nature according to the most current study.

### Form Factor (Ff)

Boundaries is known as the circulatory ratio, or Re (Miller 1953). It is influenced by geological formations, climate, land use/cover, frequency, length of streams, and slope of the basin (Fig. 12.3). The basin is categorized as having great to average relief with a Re score of 0.22. Structural instabilities also appear to have a substantial impact on the drainage system. The circulatory ratio's high value suggests that the terrain is young.

### Circulatory Ratio ($R_c$)

Boundaries are known as the circulatory ratio, or Re (Miller 1953). It is influenced by geological formations, climate, land use/cover, frequency, length of streams, and slope of the basin (Fig. 12.3). The basin is categorized as having great to average relief with a Re score of 0.22. Structural instabilities also appear to have a substantial impact on the drainage system. The circulatory ratio's high value suggests that the terrain is young.

#### Elongation Ratio ($R_e$)

The area of a basin divided by the area of a circle with the same circumference as the basin's perimeter is known as the circulatory ratio, or Rc (Miller 1953). It is affected by the basin's slope (Fig. 12.3), climate, land use/cover, length and frequency of streams, and geological formations. The basin's Rc value, which is 0.22, indicates that it has high to moderate relief. Structural disruptions appear to have a significant impact on the drainage system as well. The young stage of topography is indicated by the circulatory ratio's high value. These values fall into three categories: less elongated (<0.7), oval (0.9–0.8), and round (>0.9). In the research area, the Re value is 0.25 indicates an elongated basin. Along tectonic lineaments, a smaller Elongation Ratio value denotes a higher elevation and greater headward erosion. Less groundwater is available at high elevations than at low elevations.

#### Length of Overland Flow ($L_g$)

As per Horton (1945) the length of land flow is the volume of water that travels across the face of the planet before collecting into a particular stream channel. Lg is one of the most important independent variables affecting the c and physiographic evolution of the catchment area. The length of overland flow is approximately half of the equivalent of drainage density. This component has an inverse relationship with the mean slope of the passageway and is mostly connected to the sheet flow length. The Lg value for the research area is 1.59. This implies that there is good groundwater potential in the study area.

#### Constant Channel Maintenance (C)

Schumm (1956) used the inverse of drainage density as a feature referred to as the stream management constant. This constant's magnitude increases with the scale of the landform unit since it has a length dimension (measured in square feet per foot) also stated by Singh et al. (2022). As per Khan et al. (2012) the number of square feet of watershed face required to sustain one linear foot of stream is particularly indicated by the constant C. The basin has a steep slope, strong face discharge, and low to moderate permeability, as indicated by its value C of 0.31.

## 12.4 Summary and Conclusions

With 7365 streams overall, the Koyna drainage basin is categorized as a sixth-order basin. The basin's Bifurcation Ratio (Rb) of 1.86 suggests that geological formations have little to no effect on the river pattern. With a relief ratio of 0.93, the area appears to have little relief. 3.18 km/km$^2$ of drainage density indicates a moderate level of drainage density and a good potential for groundwater. With a Form Factor (Ff) of 0.91, the basin is classified as elongated. Good groundwater potential is further advised by the Length of Overland Flow (Lg) of 1.59. With a slope of 0.31, which denotes a steep gradient, strong face runoff, and slight to average permeability, the constant C, which represents the volume of watershed face required to sustain one linear foot of stream, is present. The basin has high to moderate relief, as indicated by its Circularity Ratio (Rc) of 0.22. The drainage system is greatly impacted by structural disturbances. The findings suggest that, in contrast to the upper portion of the basin, the bottom portion of the basin has substantial groundwater potential.

**Acknowledgements** Yuvraj Kadam is grateful to SARTHI for awarding the Chhatrapati Shahu Maharaj Fellowship, Government of Maharashtra.

## References

Ansari ZR, Rao LAK, Yusuf A (2012) GIS-based Morphometric analysis of Yamuna drainage network in parts of Fatehabad area of Agra district, Uttar Pradesh. J Geol Soc India 79:505–514. https://doi.org/10.1007/s12594-012-0075-2

Asode AN, Sreenivasa A, Lakkundi TK (2016) Quantitative morphometric analysis in the hard rock Hirehalla sub-basin, Bellary and Davanagere districts, Karnataka, India using RS and GIS. Arab J Geosci 9

Alemayehu A, Tesfaye A (2024) The morphometric investigation of the Gelda watershed in the Lake Tana sub basin: implications for managing soil and water resources. Model Earth Syst Environ 10:4207–4222

Carlston CW (1963) Drainage density and stream flow. US Government Printing Office, Washington. https://doi.org/10.3133/pp422C

Chougale SS, Sapkale JB (2017) Effect of terrain on stream junctions and drainage density: a study of Kadvi river basin, Maharashtra. Indian J Sci Technol 10(28):1–6

Das AK, Mukherjee S (2005) Drainage morphometry using satellite data and GIS in Raigad district, Maharashtra. J Geol Soc India 65:577–586

Devne B, Hagos G, Ashenafi W (2019) Assessing the effects of climate variability on river basin morphology in Ethiopia. Hydrol Earth Syst Sci 23(1):123–145. https://doi.org/10.5194/hess-23-123-2019

Dutta D, Sharma JR, Adiga S (2002) Watershed characterization, prioritization, development planning and monitoring remote sensing approach. Technical Report Isronnrms-Tr-103-2002, ISRO, Bangalore

Eze EB, Efiong J (2010) Morphometric parameters of the Calabar river basin: implication for Hydrologic Processes. J Geogr Geol 2(1):18–26

Fayera GT (2018) Hydrological and geomorphological characteristics of river systems in the Ethiopian Highlands. Addis Ababa University Press

Gottschalk LC (1964) Reservoir sedimentation. In: Chow VT (ed) Handbook of applied hydrology. McGraw Hill Book Company, New York

Gravelius H (1914) Die geographische Gliederung der Erde. Geographisches Institut, Leipzig

Horton RE (1932) Drainage basin characteristics. Trans Am Geophys Union 13:350–361

Horton RE (1945) Erosional development of streams and their drainage density: a hydro-physical approach to quantitative geomorphology. Geol Soc Am Bull 56(3):275–370

Inamdar P, Singh TP, Metha K et al (2016) Assessment of irrigation and agriculture potential of the Krishna river basin using geospatial techniques. Indian J Sci Technol 9(44):1–9

Jasmin I, Mallikarjuna P (2012) Morphometric analysis of Araniar river basin using remote sensing and geographical information system in the assessment of groundwater potential. Arab J Geosci 6:3683–3692

Javed A, Khanday MY, Ahmed R (2009) Prioritization of sub-watersheds based on morphometric and land use XE "Land use" analysis using remote sensing and GIS techniques. J Indian Soc Remote Sens 37:261–274

Kamala M, Swaminathan M (2018) Morphometric analysis of drainage basin using GIS techniques a case study of Amaravathi river basin, Tamilnadu. Int J Recent Sci Res 97:28142–28147

Khan M et al (2012) Monitoring river morphology changes: erosion and sedimentation along the Jamuna River using remote sensing and GIS techniques. Geomorphology 175:1–15. https://doi.org/10.1016/j.geomorph.2012.03.008

Khan I, Bali R, Agarwal KK, Kumar D, Singh SK (2021) Morphometric analysis of Parvati Basin, NW Himalaya: a remote sensing and GIS based approach. J Geol Soc India 97:165–172

Melton MA (1957) An analysis of the relations among elements of climate, faceproperties, and geomorphology. Technical Report-11, Proj.NR389-042. Department of Geology, Columbia University, New York

Miller V (1953) A Quantitative geomorphic study of drainage basin characteristics in the Clinch mountain area, Virginia and Tennessee. Project NR 389-402, Technical Report 3, Columbia University, Department of Geology, ONR, New York

Nag SK (1998) Morphometric analysis using remote sensing techniques in the Chaka sub-basins, Purulia district, West Bengal. J Indian Soc Remote Sens 26:69–76. https://doi.org/10.1007/BF03007341

Nooka Ratnam K, Srivastava RK, Venkateswara Rao V, Amminedu E, Murthy KSR (2005) Check dam positioning by prioritization of micro-watersheds using SYI model and statistical shape analysis—Remote sensing and XE "GIS" GIS perspective. J Indian Soc Remote Sens 33(1):25–38

Rai PK, Mishra VN, Mohan K (2017a) A study of morphometric evaluation of the Son basin, India using geospatial approach. Remote Sens Appl: Soc Environ 7:9–20

Rai PK, Mohan K, Mishra S et al (2017b) A GIS-based approach in drainage Morphometric analysis of Kanhar river basin, India. Appl Water Sci 7(1):217–232

Rastogi RA, Sharma TC (1976) Quantitative analysis of drainage basin characteristics. J Soil Water Conserv India 26(1 and 4):18–25

Romshoo SA, Bhat SA, Rashid I (2012) Geoinformatics for assessing the morphometric control on hydrological response at watershed scale in the Upper Indus Basin. J Earth Syst Sci 21:659–686

Scheidegger AE (1970) Theoretical geomorphology. Springer-Verlag, Berlin, Heidelberg. https://doi.org/10.1007/978-3-662-01025-9

Schumm (1956) Evolution of drainage systems and slopes in badlands at Perth Amboy, New Jersey. Bull Geol Soc Am 67:597–646. https://doi.org/10.1130/0016-7606(1956)67[597:EODSAS]2.0.CO;2

Somashekar RK, Ravikumar P (2011) Runoff estimation and Morphometric analysis for Hesaraghatta watershed: a remote sensing and GIS approach. J Indian Soc Remote Sens 39(1):95–106

Srinivasa Vittala S, Govindaiah S, Honne Gowda H (2004) Morphometric analysis of sub watersheds in the Pavagada area of Tumkur district, South India using remote sensing and GIS techniques. J Indian Soc Remote Sens 32(4):351–362

Strahler AN (1950) Equilibrium theory of erosional slopes, approached by frequency distribution analysis. Am J Sci 248:673–696
Strahler AN (1952) Dynamic basis of geomorphology. GSA Bull 63(9):923–938. https://doi.org/10.1130/0016-7606(1952)63[923:DBOG]2.0.CO;2
Strahler AN (1953) Dynamic basis of geomorphology. GSA Bull 63(9):923–938. https://doi.org/10.1130/0016-7606(1952)63[923:DBOG]2.0.CO;2
Strahler AN (1956) Quantitative slope analysis. Bull Geol Soc Am 67:571–596
Strahler AN (1957) Quantitative slope analysis. Bull Geol Soc Am 67:571–596
Strahler AN (1964) Quantitative geomorphology of drainage basins and channel networks. In: Chow VT (ed), Handbook of applied hydrology. McGraw-Hill, New York. Section 4–11. Available at: http://geomorphometry.org/content/quantitative-geomorphology-drainage-basins and-channel-networks
Strahler A (1964) Quantitative geomorphology of catchment area and channel networks. In: Chow V (ed) Handbook of applied hydrology. McGraw Hill, New York, pp 439–476
Strahler AN (1969) Physical geography. 3rd ed. Wiley, New York
Strahler AN (2002) Physical geography: a landscape appreciation. 3rd ed. Wiley, New York
Telore NV (2020) Quantitative morphometric analysis of the Yerla river basin, Deccan trap region, India. In: Sahdev S, Singh R, Kumar M (eds) Geoecology of landscape dynamics: advances in geographical and environmental sciences. Springer, Singapore, pp 115–132. https://doi.org/10.1007/978-981-15-2097-6_9
Thomas J, Joseph S, Thrivikramji KP et al (2012) Morphometrical analysis of two tropical mountain River basins of contrasting environmental settings, the southern Western Ghats, India. Environ Earth Sci 66:2353–2366. https://doi.org/10.1007/s12665-011-1457-2
Singh S, Singh MB (1997) Morphometric analysis of Kanhar river basin. Natl Geogr J India 43(1):31–43. https://scholar.google.com/scholar?cluster=2825065573789179275&hl=en&oi=scholarr
Singh VG, Singh SK (2022) Analysis of geo-morphometric and topo hydrological indices using COP-DEM: a case study of Betwa River Basin, Central India. Geol Ecol Landsc 101–128
Sreedevi PD, Subrahmanyam K, Ahmed S (2005) The significance of Morphometric analysis for obtaining groundwater potential zones in a structurally controlled terrain. Environ Geol 47(3):412–420
Waikar ML, Nilawar AP (2014) Morphometric analysis of a drainage basin using geographical information system: a case study. Int J Multi Cur Res 2:179–184

# Chapter 13
# A Methodological Framework to Assess Land Suitability for Agriculture

**Rajendra B. Zolekar, Hemlata P. Patel, and Snehal N. Kasar**

**Abstract** Among all the resources, land is a vital resource, being an indispensable part of both physical and cultural environments. However, the relentless demand for land, coupled with improper utilization, has led to widespread land degradation and associated socio-economic challenges. To mitigate inappropriate human impacts on natural resources and to establish well-informed land management strategies, Land suitability analysis has emerged as a beacon of hope in addressing this complex land use challenge. It serves as a valuable method for uncovering the inherent capabilities and potential of land for various purposes, providing valuable insights for sustainable land management. The study explores the methodological framework, and applications of land suitability analysis in agriculture, demonstrating its potential to analyse geosustainability transformations. Land suitability analysis combines qualitative and quantitative data to assess various land characteristics and generate maps depicting suitability levels for different land uses. Modern technological advancements, such as remote sensing and spatial information technology have enhanced the precision of land suitability analysis, allowing for informed decision-making in agriculture and other sectors. A review of existing studies reveals that commonly utilized multi-criteria decision-making methods, such as the Analytical Hierarchy Process and Weighted Linear Combination methods are widely used to calculate criteria weights and assign scores to sub-criteria within land suitability analysis. An analysis of criteria usage in agricultural land suitability indicates that factors such as slope, soil depth, soil texture, soil moisture, land use/land cover, geology, rainfall, and temperature are among the most frequently considered criteria. These classifications and findings provide valuable insights for future modeling and research endeavors in this field.

R. B. Zolekar (✉) · S. N. Kasar
K.V.N. Naik Shikshan Prasarak Sanstha's Arts, Commerce and Science College Nashik, Canada Corner, Nashik, Maharashtra, India
e-mail: rajzolekar@rediffmail.com; raj4mezolekar@gmail.com

H. P. Patel
Department of Geography, Savitribai Phule Pune University Pune, Pune, India
e-mail: hemlatapatel@unipune.ac.in

 
S. Anand et al. (eds.), *Emerging Geosustainability Transformations in India*, Advances in Geographical and Environmental Sciences,
https://doi.org/10.1007/978-981-97-7993-2_13

**Keywords** Land suitability · Agriculture · Analytic hierarchy process · Pairwise comparison matrix

## 13.1 Introduction

Land encompasses a spectrum of physical resources like soils, minerals and vegetation and cultural attributes like settlements, industries and agriculture (FAO 1976; Bhagat 2013). This intricate blend of physical and cultural environments positions land as an indispensable part of human life. Whether viewed from a physical or cultural perspective, one cannot deny the global rush towards land owing to its inherent use and value. In the face of burgeoning populations and human activities, land elements are overused and exploited (Feizizadeh and Blaschke 2012). Inappropriate land use has triggered inefficient exploitation of natural resources, destruction of land resources, ecological distress, poverty, hunger, and other associated social problems.

Looking at the facts put forward by the Indian Council of Agricultural Research, about 120.40 million hectares (m ha) of land is affected by land degradation in India (Ahmad and Pandey 2018). The global scenario is equally distressing; with 250 million people directly affected by land degradation and 1 billion people being at risk (WMO 2005). As per the FAO report, about 852 million people in developing countries suffer from hunger and malnutrition, out of which a staggering 217 million are undernourished and 225 million confront food insecurity in India alone (FAO, WFP and IFAD 2012).

Paradoxically, despite India's endowed expanse of fertile terrains and abundant natural resources, the nation struggles to meet the demands of its increasing population. The pursuit of intensive agriculture, often touted as a solution, has been hampered by the absence of effective management strategies and a dearth of scientific understanding regarding land resources. This has further deteriorated this invaluable resource in the form of soil erosion, exhausted soil fertility, continuing deforestation, and land degradation; creating a paradox where agricultural advances coincide with land deterioration, and conversely, agricultural stagnation jeopardizes food security. To solve this complex riddle and get a solution for the land use problem, land evaluation (in terms of land suitability analysis) in support of rational Sustainable Land Management (SLM) offers a ray of hope amidst the intricate web of challenges.

## 13.2 Land Suitability Analysis: An Overview

Land Suitability Analysis (LSA) is an important step in discerning the environmental limits essential for sustainable land use planning. Its primary aim revolves around predicting the inherent capabilities of land to support a specific land use over prolonged periods without deterioration. Additionally, it can help to identify priority areas amenable to management interventions and policy measures aimed at land and soil restoration initiatives. Simply, LSA is a systematic procedure or tool used for the identification of the most suitable spatial arrangement for both present and future selected land use (Collins and Steiner 2001; Malczewski 2004). As LSA functions in harmony with specific requirements, preferences, or predictors associated with a defined activity (Malczewski 2004), it classifies land suitability into categories such as highly suitable, moderately suitable, marginally suitable, and not suitable by adopting a range of rational criteria (FAO 1976).

A judicious combination of qualitative and quantitative data on various criteria results in a holistic index of evaluation, enabling a thorough appraisal of land suitability. The plethora of land characteristics considered for land suitability covers physical attributes of land such as relief, slope, drainage, soil quality, depth, texture, rainfall, and humidity; social factors such as population and poverty; and economic indicators such as labor intensity, capital intensity, market accessibility, and transportation facilities. Various scholars (Arshada and Martin 2002; Chandio and Matori 2011; Chandio and Matori 2011; Mustafa et al. 2011; Feizizadeh and Blaschke 2012) have delved into these biophysical, social, and economic aspects to discern land suitability for diverse purposes. The output of an LSA results in a set of maps, each dedicated to a specific land use, depicting the level of suitability associated with each land parcel. The results of LSA assist planners and decision-makers in selecting the best alternative for land use that will be beneficial to humankind and the environment. A judiciously determined land use pattern aligns with the growing needs of the local populace (Collins and Steiner 2001). Thus, it serves as a methodological approach to determine optimal land utilization for diverse purposes such as agriculture, plantation, recreation, settlement, industry, and watershed management, among others.

For scholars across the globe, LSA has emerged as a versatile tool for a multitude of purposes, such as designing land use patterns (Eastman et al. 1993), fostering sustainable land management (Akainci and Ozalp 2013; Zolekar and Bhagat 2018), understanding inherent land capacities, identifying optimal sites for different land use (Hopkins 1977; Malczewski 2004; Bandyopadhyay et al. 2009), resolving spatial dilemmas to enhance land efficiency and productivity (Zolekar and Bhagat 2014), mitigating environmental conflicts (Cengiz and Akbulak 2009), conserving invaluable natural resources like soil, water, vegetation, minerals, and nutrients (Abdel Rahman and Arafat 2020), anticipating land limitations for crop production, suggesting alternative land use (Prakash 2003), and contributing to the formulation of land use policies (Dumanski 1997) and many others.

Among all its applications, agricultural LSA for crop production holds a pivotal position for ensuring sustainable agriculture and aligning with the Sustainable Development Goals (SDGs) of the United Nations aiming at attaining global food security. Food security has become a global concern with agricultural lands facing massive pressure from population growth, climate change, land degradation, and rapid urbanization. To address current and future food security by efficient use of land resources, researchers throughout the world are conducting assessments of the capability and suitability of disposable land through spatial-based analytical and optimization approaches, wherein they have realized the importance of LSA as one of the prerequisites to achieve optimum utilization of the available land resources (Kihoro et al. 2013).

Technological advancements in the fields of Geographic Information Systems (GIS), Remote Sensing (RS), Decision-Support Systems (DSS) and web-based applications have empowered more precise and sustainable decisions related to agriculture in determining suitable locations for farming and identifying crops best suited for specific areas. The decision-making process in LSA is influenced by a multitude of factors (Yu et al. 2011). This has led numerous researchers to use GIS-based Multi-Criteria Decision Making (MCDM) methodologies, wherein GIS serves as a tool for the input, storage, retrieval, analysis, and output of spatial data. The MCDM technique is adopted for handling multiple criteria simultaneously and is designed to generate solutions from a variety of alternatives (Jankowski 1995).

Over the past decade, several methodologies under the umbrella of Multi-Criteria Evaluation (MCE) have been proposed for LSA. It includes Weighted Linear Combination (WLC) (Carver 1991; Eastman1997), Analytical Hierarchy Process (AHP) (Saaty 1980; Store and Kangas 2001), Ordered Weighted Averaging (OWA) (Yager 1988; Malczewski et al. 2003; Malczewski and Rinner 2005; Boroushaki 2008), and Concordance Analysis (Carver 1991; Joerin et al. 2001). These methodologies are widely employed in conjunction with GIS techniques to identify potential sites for diverse purposes, such as agriculture, plantation, landscape evaluation, watershed management, settlement planning, industrial allocation, environmental impact assessment, and regional planning (Kalogirou 2002; Prakash 2003; Shalaby et al. 2006; Olayeye et al. 2008; Bandyopadhyay et al. 2009; Yu et al. 2011; Foshtomi et al. 2011; Samanta et al. 2011; Mustafa et al. 2011; Mahabadi et al. 2012; Halder 2013; Rabia et al. 2013). Moreover, these methodologies also extend into ecological applications, such as habitat suitability evaluation for various plant and animal species (Store and Kangas 2001).

## 13.3 The Framework of LSA

The foundation of LSA rests upon a detailed comprehension of land, its attributes, and the multifaceted role it plays within the context of SLM. This section begins with the core concept of land, land evaluation, and SLM, setting the stage for diving into the varied methodologies employed for LSA, which are discussed later.

### *13.3.1 Land*

The concept of land, as formulated within the framework of land evaluation, was first given by the Food and Agriculture Organization (FAO 1976). Land represents a reasonably stable or predictably cyclic portion of the Earth's surface, that includes diverse elements such as relief, soils, near-surface rocks, minerals, flowing water, groundwater, near-surface atmospheric components (temperature, rainfall, etc.), flora, fauna, and also manmade features like land use, settlements, industries, and agriculture (FAO 1976; Bhagat 2012). In short, land encapsulates an amalgamation of terrain, water, soil, nutrients, minerals, and vegetation, providing the basis for diverse land use.

Beyond its physical characteristics, land plays various important roles, as enumerated by FAO (1995) that can be classified into production functions, biotic environmental functions, climate regulative functions, hydrologic functions, storage functions, waste and pollution control functions, living space function, archive or heritage function, and connective space function. Owing to this myriad of roles, the land is often exploited beyond its capacity, leading to detrimental land degradation including deforestation, erosion, flooding, waterlogging, salinization, and yield reduction (Kertesz 2006).

The consequences of land degradation extend to cultivated lands, forests, and grasslands alike, with significant portions facing these challenges (Bai et al. 2008). This degradation significantly affects both the land itself and the broader landscape, raising the pressing need for SLM strategies, a challenge that researchers, planners, and governmental authorities are concerned with (Hurni 1997). Consequently, the pursuit by scholars to comprehend land suitability has driven diverse methodologies in land evaluation studies. The subsequent section will explore the theoretical framework of land evaluation, providing insights into its principles and methods.

### 13.3.2 Land Evaluation

Land evaluation may be defined as "the process of assessment of land performance when it is used for definite purposes" (FAO 1995), or as "all methods to elucidate or predict the use potential of land" (Van Lanen et al. 1992). Through this analysis, the suitability of land properties to meet the demands of intended land use is determined. Thus, land evaluation emerges as a strategic tool that examines the feasibility of land for different purposes (Halder 2013). By predicting both the expected benefits and constraints of productive land use, land evaluation provides crucial insights into environmental degradation as well (Rossiter 1996). As Rossiter (1996) noted, the crux of land evaluation lies in the fact that variations in land properties influence land use, and by mapping these variations, land behavior can be predicted with a certain degree of certainty under specific land use, thus aiding decision-makers including land users, planners, and agricultural support services.

The emergence of land evaluation found impetus in the early 1970s, when the Food and Agriculture Organization of the United Nations sponsored working groups, culminating in the publication of the "Framework for Land Evaluation" in 1976 (FAO 1976). Its initial stages embraced qualitative methods rooted in biophysical, economic, and social criteria. This era also marked the predominance of "pedocentric" approaches, that gave more emphasis on soil resources, perhaps because the FAO land evaluation methodology was developed by soil scientists with expertise in agricultural land suitability classification. While soil is a vital factor in land suitability, it is one of several natural, economic, and human resources, all are equally important in the theoretical framework. As land evaluation evolved, quantitative methodologies emerged that incorporated more criteria, detailed technical procedures, and computer-based models (Van Lanen et al. 1992).

Quantitative methods either involved assigning weights to different criteria influencing land suitability (as in Index Overlay or Weighted Linear Combination method), breaking down the evaluation process hierarchically (Analytical Hierarchy Process), or combining factors through weighted averages (as in Ordered Weighted Averaging Techniques). As multiple criteria were assessed simultaneously, researchers started making informed decisions about land suitability using Multi-Criterion Decision Analysis (MCDA). Elaalem et al. (2010) suggested a land evaluation model to create comprehensive suitability maps by integrating remote sensing and GIS technologies into land evaluation processes. A core challenge facing land evaluation is to incorporate uncertainty and expert knowledge, concurrently, within the evaluation process. Globally, several researchers (Mokarram et al. 2007; Keshavarzi et al. 2011; Triantafilis et al. 2001) have tackled this through techniques such as fuzzy logic, designed to represent imprecise information, and expert systems that emulate human expertise in decision-making.

The choice of technique for land evaluation relies on the generality of objectives, data scale and availability, knowledge of land and land use relationships, desired precision of the evaluation results, and the dynamic spatiotemporal characteristics of land utilization types.

### *13.3.3 Sustainable Land Management (SLM)*

The Rio Earth Summit held in 1992 augmented the momentum for the exploration of various aspects of sustainable development. Amid these discussions, SLM emerged as a global concern, attracting significant attention from both the scientific community and policy-makers. SLM is viewed as a holistic approach that encompasses technologies, policies, and activities integrating socioeconomic principles with environmental considerations (Symth and Dumanski 1993). The objective is to simultaneously maintain or enhance production, minimize production risks, protect natural resource potential, and prevent soil and water degradation. In short, SLM requires harmonizing biophysical conditions with economic and social demands.

In their exhaustive research, Dumanski (1997) and Lefroy et al. (2000) outlined five key objectives of SLM that shape its core foundation: (i) productivity: sustaining or enhancing production and services, (ii) security: reducing production risks, (iii) protection: safeguarding natural resource potential and preventing soil and water degradation, (iv) viability: ensuring economic feasibility, and (v) acceptability: gaining social approval.

In the preceding section, the importance of the FAO framework was highlighted in evaluating the suitability of land use alternatives. This framework, although valuable, cannot evaluate sustainability or the likelihood of achieving it. To address this, various researchers, institutes, organizations, and governments have attempted to develop frameworks for monitoring land degradation and achieving SLM objectives (Kalogirou 2002). Evaluating sustainability involves intricate indicators and thresholds, the application of which varies across different regions. Despite this, it's undeniable that the FAO framework serves as an intelligent starting point for constructing a more comprehensive framework to assess land sustainability (Smyth and Dumanski 1993).

Within the realm of agriculture, a global consensus has been achieved regarding key biophysical indicators of SLM, encompassing areas such as nutrient balance, yield trends, land use intensity and diversity, soil quality, agro-biodiversity, water quality, forest land quality, rangeland quality, and land contamination/pollution (Dumanski 1997). These indicators, which have been embraced internationally, have played a pivotal role for researchers, extension workers, local planners, and progressive farmers in assessing and monitoring potential land utilization. These indicators serve as valuable instruments to measure changes in land quality and offer early alerts about unfavourable trends and problem areas, within the domain of SLM.

## 13.4 Data Acquisition and Processing for LSA

In the process of conducting a multi-criteria assessment to determine land suitability, a wide array of factors is considered. These factors encompass diverse aspects such as geological composition, biophysical characteristics, relief patterns, atmospheric conditions, vegetation types, as well as economic and socio-cultural attributes. The specific criteria used for LSA can vary depending on the intended purpose, but in a broader context, most of these criteria come into play. For instance, when evaluating whether the land is suitable for agriculture, several geographical factors detailing the land's topography, specific qualities of its soil, prevailing climate conditions, and accessibility of land need to be considered. To gather data about the land's topography, one can rely on topographical maps and satellite images, which enable the creation of maps indicating slope and aspect variations. In-depth soil analysis, encompassing physical and chemical properties can be carried out through laboratory assessments of soil samples. The outcomes of these soil analyses can then be integrated with the topographic data, resulting in specialized datasets providing information about soil depth and moisture content.

Moreover, techniques related to image processing, like image enhancement and classification methods, provide valuable assistance for interpreting distinctive land characteristics. These image-based approaches contribute to a better understanding of the suitability of the land for various potential uses. The datasets gathered during this process serve as the foundation for subsequent statistical analyses that ultimately lead to the development of comprehensive maps illustrating the suitability of the land for specific purposes. Although the upcoming section will investigate data preparation and processing tailored, particularly, for agricultural applications, the core processing steps remain largely consistent across various domains of land use assessment.

### *13.4.1 Field Survey and Laboratory Analysis*

A pre-requisite for any LSA is validating the accuracy of its results by conducting field surveys and sample experiments. Field surveys offer a means to verify and validate remotely sensed data, and other existing secondary databases. Accurate assessment of complex terrains is accomplished through direct on-ground observations of slope, aspect, and landform characteristics. Field surveys involve collecting soil samples that are subjected to laboratory analysis for assessing distinct soil properties that may not be adequately captured by remote sensing alone. Laboratory analysis may also aid in micro-level evaluation of the significant roles played by organic matter, nutrients, and minerals in the growth and development of soils (Bandyopadhyay et al. 2009).

Soil samples analyzed using traditional chemical methods and/or spectroscopic techniques can yield useful information on the physical and chemical attributes of the soil. Researchers like Shalaby et al. (2006), Bandyopadhyay et al. (2009), Foshtomi et al. (2011), Mustafa et al. (2011) have conducted in-depth analyses of soil properties that includes physical parameters ranging from drainage, depth, texture and water holding capacity, to chemical traits of the soil like potential of Hydrogen (pH), Electronic Conductivity (EC), Organic Carbon (OC), Nitrogen (N), Phosphorus (P) and Potassium (K). In summary, field surveys, sample experiments, and laboratory analysis collectively enrich LSA by providing accurate and site-specific data. These components ensure that land assessments reflect real-world conditions, enhancing the credibility of the results.

### *13.4.2 Integration of Remote Sensing*

The integration of remotely sensed data, in conjunction with Geographic Information Systems (GIS), plays an important role in modern LSA, offering a multitude of advantages in understanding complex landscapes. LSA demands an extensive range of data covering diverse aspects such as physiography, soil properties, water availability, vegetation cover, climate variables, and socio-economic attributes. Traditional methods like fieldwork, laboratory analyses, and conventional maps, while valuable, are time-consuming, expensive, and often lack holistic information (Tahir et al. 2013). Conversely, remote sensing provides timely, accurate, and reliable information on physical, structural, and biophysical attributes of land surface (Elachi and Zyl 2006; Zheng and Moskal 2009). Numerous studies have proved the applicability of space technology in identifying suitable sites for SLM (Bhagat 2012).

Several researchers and planners have used datasets obtained through remote sensing at varying spatial, temporal and radiometric resolutions (Joshi 2011) to identify, classify, and estimate different land attributes (Shalaby et al. 2006; Govender et al. 2007; Adam et al. 2009). Imageries having high spatial resolution, such as QuickBird, IKONOS, and RapidEye, offer intricate landscape details and refined surface characteristics, thereby, enhancing our understanding of relationships that exist between different land elements (Hu et al. 2013). Noteworthy instances include the use of IKONOS data for individual tree identification in forestry applications (Wulder and White 2004) and SPOT-5 images for LULC mapping (Feizizadeh and Blaschke 2012).

Images with high spatial resolution often compromise temporal resolution, while those with low spatial resolution tend to offer better temporal resolution (Forkuor and Cofie 2011). Moreover, fine-resolution datasets find limited applications due to their higher costs and the time-consuming nature of their acquisition process (Hu et al. 2013). Hence, it is not uncommon to come across numerous studies that focus on utilizing low and moderate-resolution satellite data, like Landsat TM, ETM, ETM+, NOAA AVHRR, MODIS, and Terra ASTER, for LSA (Bojórquez-Tapia et al. 2001; Shalaby et al. 2006; Golmehr 2008; Bandyopadhyay et al. 2009; Wang et al. 2010; Mustafa et al. 2011; Feizizadeh and Blaschke 2012; Samanta et al. 2011; Bhagat 2012; Singh 2012). Among these, Landsat data's availability and spectral richness make it comparatively more conducive for land suitability assessment.

The rise of hyperspectral data marks an exciting frontier in LSA. These data sets, acquired from hyperspectral imagers, are capable of capturing a multitude of very narrow and contiguous spectral bands across different portions of the electromagnetic spectrum (Govender et al. 2007). Hyperspectral imagery can detect and delineate essential features such as soil properties, vegetation health, water content, and land degradation indicators. Several studies (Melgani and Bruzzone 2004; Govender et al. 2007; Vorovencii 2009) have used hyperspectral datasets to classify LULC to identify forestry, water, crops, urban areas, minerals, and human-made materials. Wang et al. (2009) employed hyperspectral remote sensing data for analyzing land degradation. Geologists have predominantly utilized hyperspectral imagery for mineral mapping (Leverington and Moon 2012) and detecting various soil properties, including soil texture, soil moisture, vegetation, crop residue, organic content, and salinity, among others (Bandyopadhyay et al. 2009). Although hyperspectral data presents promising opportunities, its application is still relatively limited due to costs, processing time, and image specificity.

It is important to bear in mind that although remote sensing offers precise and comprehensive information about landscapes, these methods cannot be effectively recommended or applied to land suitability studies without the integration of supporting ancillary data, such as information about soils, land use, agriculture, forestry, climatic conditions, water resources, and more (Tahir et al. 2013).

### 13.4.3 Image Enhancement Techniques

Image enhancement techniques are a vital component of digital image processing that focuses on improving the visual quality and interpretability of images. These techniques aid in enhancing the visual and digital analysis of land elements, which include identification, classification, and interpretation of the observed land features (Sawy et al. 2012). A diverse array of image enhancement techniques exists, each geared towards refining digital image quality without damage to the images. Notable techniques include contrast enhancement, hue, intensity, and saturation transformations, edge enhancement, and gray level slicing.

In the context of LSA, these techniques have been widely utilized. Several noteworthy methods include: (i) contrast enhancement used to differentiate land features, enhancing the identification and classification of elements (Jenson 2005; Shalaby et al. 2006); (ii)spectral indices, such as vegetation indices (Ratio Vegetation Index, Normalized Difference Vegetation Index, Soil Adjusted Vegetation Index); water indices (Normalized Difference Water Index, Modified Normalized Difference Water Index); and soil indices (Normalized Difference Soil Index, Soil Salinity Index, Soil Moisture Index); used to assess vegetation, water and soil conditions, respectively; (iii) false color composite images generated through image enhancement provides valuable insights into complex land features and characteristics (Shalaby et al.2006; Singh and Singh 2011), and (iv) principal component analysis is a powerful technique for reducing data dimensionality of multispectral images (Golmehr 2008; Caocao et al. 2008; Bhagat 2012). When strategically combined with other analytical methods, image enhancement techniques empower analysts to make rational decisions in SLM.

### 13.4.4 Image Classification

Image classification is the process of categorizing land areas by grouping pixels or regions within an image into different classes based on their spectral properties. The selection of either unsupervised or supervised classification algorithms is influenced by the characteristics of the data, its scale, research complexity, available computational resources, and the specific objectives of the study. In unsupervised classification, computer algorithms automatically cluster pixels into distinct groups based on their spectral attributes, without the use of pre-defined class labels. On the other hand, in supervised classification, spectral signatures of training areas act as reference points, allowing each pixel in the image to be categorized into land cover classes based on the similarity of its spectral features to those of the training areas (Dash and Nayak 2012).

Many studies have applied both supervised and unsupervised techniques in various combinations to analyze land characteristics (Bhagat 2012). However, there has been a greater preference for supervised classification due to its accuracy and interpretability in targeted studies. For example, researchers like Shalaby et al. (2006), Bandyopadhyay et al. (2009), and Panhalkar (2011) have used supervised classification methods to create LULC maps for different crops using Landsat ETM+ images. Additionally, Halder (2013) used supervised classification to delineate agricultural land with block boundaries as an area of interest tool.

In supervised classification methods, commonly used classifiers include the Maximum Likelihood Classifier (MLC), minimum-distance-to-means classifier, Parallelepiped classifier, Support Vector System (SVM), and Neural Network classifier, among others. These classifiers have demonstrated success across various geographical and environmental conditions, although their performance may vary spatially (Shalaby et al. 2006; Panhalkar 2011). Among these classifiers, MLC is widely used and preferred for precise land classification by several researchers (Rwetabula and De Smedt 2005; Shalaby et al.2006; Bandyopadhyay et al. 2009; Panhalkar 2011; Mansor et al. 2012; Vibhute et al. 2013) because it estimates both the variance and covariance of the spectral response category when classifying an unknown pixel.

Despite their advantages, supervised classification methods have limitations, including potential mismatches between information classes and spectral classes, variations in signature homogeneity and uniformity, incomplete representation of unique spectral classes in training sites, the influence of sampling techniques and size on classification accuracy, and the possibility of misclassifying very similar classes. Hence, it should be noted that although both supervised and unsupervised classification methods are commonly used, each has its strengths and limitations. Researchers and land analysts should carefully choose appropriate algorithms based on the specific characteristics of the study area and the objectives of land suitability assessment. Furthermore, advances in object-based classification offer promising avenues for improving the classification accuracy needed in LSA.

## 13.5 Application of LSA in Agriculture

To ensure the sustainable development of agriculture and food production, effective management of agricultural land is essential. LSA plays a crucial role in the process of land use planning (Yu et al. 2011) and is a fundamental step toward optimizing the use of available land resources (Kihoro et al. 2013). This analytical approach enables the identification of key factors limiting agricultural production and empowers decision-makers to design crop management strategies that enhance land productivity. The contemporary agricultural landscape aims at achieving higher productivity, profitability, and environmental well-being. Consequently, there is a shift towards selecting crop varieties that are best suited to specific areas. A detailed review of the criteria and techniques used in the earlier studies for assessing land suitability is presented in Table 13.1.

**Table 13.1** Data, criteria, and techniques used for land suitability analysis

| Author | Techniques | Criterion | Data | Suitability field |
|---|---|---|---|---|
| Wang (1994) | GIS-based-artificial neural networks | Soil depth, Slope, moisture, fertility, texture, aeration, salinity, temperature and accessibility | Thematic map | Agriculture |
| Bojórquez-Tapia et al. (2001) | GIS-based multivariate application | Land cover, vegetation, soil type, elevation, landforms, major roads, and urban areas | Landsat-TM and thematic map | Agriculture |
| Joerin et al. (2001) | Outranking multi-criteria analysis, Homogeneity index, Suitability index | Impacts on nature reserve, water table, landscape, air pollution, noise, climate, landslide, accessibility, distance to localities and public facilities | Thematic map | Agriculture |
| Kalogirou (2002) | Boolean classification method | Slope, soil mechanics, toxicities, rooting condition, flood and erosion hazard, water level and drainage | Thematic map | Crop |
| Shalaby et al. (2006) | Square root and storied method | Slope, drainage, soil depth LULC, texture, CaSO4, CaCo3, ECP, EC and organic matter | ETM+ and Thematic map | Crop (Perennial) |
| Perveen et al. (2007) | MCDM Approach | Slope, soil texture, soil moisture, soil consistency, drainage, pH, and soil organic content | Terra/ASTER | Crop suitability |
| Olayeye et al. (2008) | Index productivity | Soil depth, drainage, rainfall, temperature, slope, texture, humidity, EC, pH, N, P, K, and cation exchange capacity | Thematic map and field-base | Rice (Irrigated Low land) |
| Bandyopadhyay et al. (2009) | AHP | Soil type, LULC, organic matter, soil depth, and slope | IRS-1D LISS-III and satellite data | Agriculture |
| Cengiz and Akbulak (2009) | AHP | Soil depth, erosion hazard, land use capability class, elevation, slope, distance to source of water, distance to road, limiting soil factors | Thematic maps | Agriculture |

(continued)

**Table 13.1** (continued)

| Author | Techniques | Criterion | Data | Suitability field |
|---|---|---|---|---|
| Jafari and Zaredar (2010) | AHP | Slope, LULC, elevation, erosion, climate, soil hydrology, soil structure, soil depth, soil texture, vegetation density, vegetation type, rainfall, temperature, distance from surface water, and distance from population centers | Thematic maps | Rangeland management |
| Chandio et al. (2011) | AHP and WLC | Available land, land value, and population density | Thematic maps | Public parks |
| Chandio and Matori (2011) | PCM | LULC, topography, accessibility, and economic | Thematic maps | Hillside development |
| Foshtomi et al. (2011) | Square root and the Storied methods | Soil depth, texture, EC, OC, pH, N, P, K, and cation exchange capacity | Thematic maps | Tea |
| Mustafa et al. (2011) | MCDM Approach | Soil depth, texture, EC, OC, pH, N, P, K, ECP and CaCO3 | IRS-P6 LISS III satellite data | Crops |
| Feizizadeh and Blaschke (2012) | AHP | Topography, climate, soil properties, water resources | SPOT-5 satellite data, Thematic maps | Agriculture |
| Akinci et al. (2013) | AHP | Soil groups, soil depth, land use, erosion, slope, aspect, elevation, and soil parameters | Thematic maps and field base data | Agriculture |
| Garcia et al. (2014) | AHP | Accessibility, security, needs of the agricultural product warehouse, acceptance, and costs | Thematic maps | Agricultural product warehouses |

*Source* Compiled by authors

### *13.5.1 Criteria for Agricultural Land Suitable Analysis*

Determining the most suitable land use for a given area involves integrating data and a multitude of criteria from various fields and sources, including soil science, social science, meteorology, and management science. These criteria can be grouped into distinct categories, each with its own set of parameters.

#### 13.5.1.1 Soil Criteria

Physical factors such as soil depth, texture, moisture content, fertility, and chemical properties encompassing pH, Electrical Conductivity, the concentration of specific chemicals (CaCo3, N, P, K, etc.), salinity, and cation exchange capacity are frequently considered in LSA. Researchers like Wang (1994), Shalaby et al. (2006), Olayeye et al. (2008), and Zolekar (2018) have incorporated these criteria to assess the suitability of land for various crops.

#### 13.5.1.2 Land Characteristics

Physical attributes of land like slope, LULC, elevation, vegetation, and landforms are commonly used to evaluate land qualities and suitability for agriculture (Bojorquez-Tapia et al.2001).

#### 13.5.1.3 Environmental and Socio-Economic Criteria

In some cases, LSA extends beyond physical attributes to include environmental and socioeconomic factors. Joerin et al. (2001) considered parameters such as impacts on nature reserves, accessibility, and air quality to evaluate housing suitability. In addition, geographic factors such as proximity to roads and urban areas, as well as climatic conditions like temperature and precipitation, play a pivotal role in LSA (Feizizadeh and Blaschke 2012). These criteria are essential for understanding the broader context of land suitability.

### *13.5.2 Techniques Used in Multi-criteria Decision Analysis for LSA*

MCDA is a systematic and structured approach used in decision-making to evaluate and select the best option among several alternatives, considering multiple criteria or factors. This method proves especially useful when there is a need to balance conflicting objectives and make informed decisions by considering various quantitative and qualitative factors. These techniques merge qualitative and quantitative criteria into a single evaluation index to establish relationships for spatial decision-making (Elaalem 2012). Various data sources, including satellite images, land use maps, administrative zoning information, vegetation distribution, census data, and socio-economic information, have been employed in land suitability analysis. Multi-criteria evaluation techniques are valuable for integrating information related to different land elements in the decision-making process for SLM. Methods like AHP, WLC, OWA, and Concordance analysis are commonly used within the

MCDA framework (Saaty 1980; Malczewski 2000; Joerin et al. 2001; Mokarram 2007; Bandyopadhyay et al. 2009; Mustafa et al. 2011; Elaalem 2012; Rabia and Terribile 2013). Some of these methods are discussed in detail in the following section.

#### 13.5.2.1 Analytic Hierarchy Process (AHP)

AHP is a structured hierarchical framework for organizing and analyzing complex decisions based on Mathematics and Psychology (Saaty 1980). Originally developed by Prof. Thomas Saaty in the 1960s for solving spatial Multi-Criteria Decision-Making problems (Malczewski 2006; Attua and Fisher 2010; Chandio et al. 2011; Samanta et al. 2011; Yu et al. 2011), AHP is a popular method for assigning weights related to inconsistent judgments and provides a measure of the inconsistency of respondents' judgments (Duc 2006). AHP methods involve four steps: (i) establishing a set of criteria and sub-criteria, (ii) assigning ranks to the criteria and sub-criteria using expert opinions, literature, and farmers' input, (iii) creating a pairwise comparison matrix, and (iv) calculating weights using Expert Choice Software (ECS) and AHP in GIS environment. Due to its applicability, AHP has been widely adopted by researchers, particularly for agricultural LSA. For example, Perveen et al. (2007) integrated GIS with multi-criteria evaluation based on the AHP method to determine the physical suitability of rice crops. Mustafa et al. (2011) also employed AHP techniques for the LSA of crops in Kheragarh Tahsil (India). Numerous researchers in the field of LSA have used GIS-based AHP methods for various agricultural applications, including rangeland management, public parks, and plantation site selection.

#### 13.5.2.2 Pairwise Comparison Matrix (PCM)

PCM was developed by Saaty (1980) as part of the AHP procedure that aids in assigning relative weights to criteria by comparing their importance. The use of PCM helps decision-makers to account for the varying degrees of importance among criteria and factors. PCM typically uses a scale ranging from 1 to 9 to determine the weights of criteria, with 1 indicating equal importance, 3 moderate, 5 strong, 7 very strong, 9 absolute importance, and intermediate values of 2, 4, 6, and 8 reflecting varying degrees of importance (Saaty 1997). Thus, PCM is a valuable tool for achieving consistency in decision-making and is employed to derive composite suitability maps (Raghunath 2006; Zolekar and Bhagat 2015).

#### 13.5.2.3 Weighted Linear Combination (WLC)

Weighted Linear Combination (WLC) is a significant method widely employed in Multi-Criteria Evaluation (MCE) based on weighted averages (Drobne and Lisec 2009). This method is based on an overlay approach (Malczewski 2004) and is employed when the sum of the weights always equals one (Mahabadi 2012). WLC combines standardized scores for each criterion and the weight of the factor for that criterion. It involves four steps: (i) standardizing the suitability maps, (ii) assigning weights of relative importance to the suitability maps, (iii) combining the weights and standardized suitability maps, and (iv) obtaining an overall suitability score (Malczewski 2004). WLC is a flexible method suitable for site selection in various contexts, such as agriculture, plantation, watershed management, and more. It is easy to implement in a GIS environment and is often favored by decision-makers (Massam 1988). However, it is essential to understand the underlying assumptions of WLC, as its misuse can yield questionable results (Malczewski 2000).

#### 13.5.2.4 Concordance–Discordance Analysis

Elimination and Choice Translating Reality (ELECTRE) as a part of the multi-criteria decision analysis framework, helps to assess the concordance and discordance between alternatives based on multiple criteria. ELECTRE quantifies the degree of agreement (concordance) or disagreement (discordance) between alternatives, thus enhancing the precision of agricultural land suitability assessments. A pairwise comparison approach provides partial or ordinal rankings of alternatives. It helps in determining preferences between two alternatives but does not quantify the degree of preference. This method can be applied to raster or vector-based data, enabling the assessment of how one parameter outranks another in specified criteria (Carver 1991). However, Concordance–discordance analysis is computationally complex when many factors are involved. It is often effective when used in combination with other decision-support tools and expert knowledge to make informed decisions in LSA.

## 13.6 Conclusions

In summary, LSA represents a holistic approach that integrates physical, environmental, and socio-economic criteria to evaluate land's capability to support specific uses. In a developing country like India, where agriculture is still the backbone of livelihood for millions, effective land management remains a challenge. In such circumstances, LSA offers a pathway to optimize land use, improve crop selection, and enhance agricultural productivity. By tailoring land management strategies to specific land characteristics and environmental conditions, LSA contributes to sustainable agricultural practices that meet the growing demands of an expanding population. Moreover, it aids in mitigating the adverse effects of land degradation, thereby promoting environmental well-being. The framework of LSA, as outlined in this chapter, highlights the importance of considering diverse factors, such as soil properties, land characteristics, and environmental and socio-economic criteria.

By generating suitability maps, planners and decision-makers can make informed choices for achieving both higher productivity and environmental well-being.

The various techniques and methodologies discussed in this chapter, such as AHP, Pairwise Comparison Matrix (PCM), Weighted Linear Combination (WLC), and Concordance–Discordance Analysis, offer flexible approaches for conducting LSA. These methods empower stakeholders to prioritize land use, select crop varieties suitable for specific areas, and develop strategies for enhancing land productivity while minimizing environmental impacts. Among these techniques, AHP emerges as a particularly effective approach in agricultural LSA, offering a systematic and structured approach for land evaluation. By harnessing the capabilities of GIS and MCDA methodologies, LSA equips decision-makers with the tools to make informed choices about land use, crop selection, and sustainable agricultural practices. As technology continues to advance and our understanding of land resources deepens, LSA can be effectively utilized to ensure that land resources are used efficiently and responsibly, safeguarding the well-being of both current and future generations.

## References

AbdelRahman MAE, Sayed MA (2020) An approach of agricultural courses for soil conservation based on crop soil suitability using geomatics. Earth Syst Environ 4:273–285

Adam E, Mutanga O, Rugege D (2009) Multi-spectral and hyper-spectral remote sensing for identification and mapping of wetland vegetation: a review. Wetlands Ecol Manage. https://doi.org/10.1007/s11273-009-9169-z

Ahmad N, Pandey P (2018) Assessment and monitoring of land degradation using geospatial technology in Bathinda district, Punjab, India. Solid Earth 9:75–90

Akinci H, Ozalp AY, Turgut B (2013) Agriculture land use suitability analysis using GIS and AHP technique. Comput Electron Agric 97:71–82

Arshad MA, Martin S (2002) Identifying critical limits for soil quality indicators in agroecosystems. Agr Ecosyst Environ 88:153–160

Attua EM, Fisher JB (2010) Land suitability assessment for pineapple production in the Akwapim South District, Ghana. A GIS-Multi-Criteria Approach. Ghana J Geogr 2:48–83

Bai ZG, Dent DL, OlssonL, Schaepman ME (2008) Global assessment of land degradation and improvement. Identification by Remote Sensing, Report 2008/01, ISRIC, Wageningen

Bandyopadhyay S, Jaiswal RK, Hegde VS et al (2009) Assessment of land suitability potentials for agriculture using a remote sensing and GIS-based approach. Int J Remote Sens 30(4):879–895

Bhagat VS (2013) Use of remote sensing techniques for robust detection and estimations of soil organic carbon: a review. Recent Progr Space Technol 3:83–102

Bhagat VS (2012) Use of remote sensing techniques for robust digital change detection of land: a review. Recent Patents Space Technol 2:123–144

BoroushakiS MJ (2008) Implementing an extension of the analytical hierarchy process using ordered weighted averaging operators with fuzzy quantifiers in ArcGIS. Comput Geosci 34(4):399–410

Caocao C, Gaodi X, Lin Z et al (2008) Analysis of Jinghe watershed vegetation dynamics and evaluation on its relation with precipitation. Acta EcologicaSinica 28(3):925–938

Carver SJ (1991) Integrating multi-criteria evaluation with geographical information systems. Int J Geogr Inform Syst 5(3):321–339

Cengiz T, Akbulak C (2009) Application of analytical hierarchy process and geographic information systems in land-use suitability evaluation: a case study of Dumrek village. Int J Sust Dev World 16(4):286–294

Chandio IA, Matori AN (2011) GIS-based multi-criteria decision analysis of land suitability for hillside development. Int J Environ Sci Dev 2(6):468–473

ChandioI A, Matori AN, Lawal DU et al (2011) GIS-based land suitability analysis using AHP for public parks planning in Larkana City. Mod Appl Sci 5(4):177–189

Collins MG, SteinerFR RMJ (2001) Land-use suitability analysis in the United States: historical development and promising technological achievements. Environ Manage 28(5):611–621

Dash T, Nayak T (2012) Non-correlated character recognition using artificial neural network. In: Proceedings of national conference on dynamics and prospects of data mining: theory and practices, pp 79–83

DrobneS LA (2009) Multi-attribute decision analysis in GIS: Weighted linear combination and ordered weighted averaging. Informatica 33:459–474

Duc T (2006) Using GIS and AHP technique for land-use suitability analysis, International symposium on geo-informatics for spatial infrastructure development in earth and allied sciences, organized by Japan-Vietnam geo-informatics consortium institute for environment and resources, Vietnam. Polytechnic University of Hochiminh City, Vietnam

Dumanski J (1997) Criteria and indicators for land quality and sustainable land management. ITC J 3(4):216–222

Eastman JR (1997) Idrisi for Windows, Version 2.0: Tutorial Exercises, Graduate School of Geography-Clark University. Worcester MA

Eastman JR, Kyem PAK, Toledano J et al (1993) GIS and decision making. UNITAR, Geneva

Elaalem M (2012) Land suitability evaluation for sorghum based on Boolean and fuzzy-multi-criteria decision analysis methods. Int J Environ Sci Dev 3(4):357–361

Elaalem M, Comber A, Fisher P (2010) Land evaluation techniques comparing fuzzy AHP with TOPSIS methods. 13th AGILE International Conference on Geographic Information Science, Guimaraes, Portugal, pp 1–8

Elachi C, Zyl JV (2006) Introduction to the physics and techniques of remote sensing, 2nd ed. Wiley

FAO (1976) A framework for land evaluation, Soil Bulletin 32. Food and Agriculture Organization of the United Nations, Rome, Italy

FAO (1995) Planning for sustainable use of land resources: towards a new approach. FAO Land and Water 2, Rome

FAO, WFP and IFAD (2012) The state of food insecurity in the world 2012. Economic growth is necessary but not sufficient to accelerate the reduction of hunger and malnutrition. Rome, Italy

Feizizadeh B, Blaschke T (2012) Land suitability analysis for Tabriz County, Iran: a multi-criteria evaluation approach using GIS. J Environ Plan Manage 1:1–23

Forkuor G, Cofie O (2011) Dynamics of land-use and land-cover change in Freetown, Sierra Leone and its effects on urban and peri-urban agriculture- a remote sensing approach. Int J Remote Sens 32(4):1017–1037

Foshtomi MD, Norouzi M, Rezaei M et al (2011) Qualitative and economic land suitability evaluation for Tea in sloping area of Guilan, Iran. J Biol Environ Sci 5(15):135–146

Garcia JL, Alvarado A, Blanco J et al (2014) Multi-attribute evaluation and selection of sites for agricultural product warehouses based on an analytic hierarchy process. Comput Electron Agric 100:60–69

Golmehr E (2008) A remote sensing evaluation for agronomic land use mapping in Tehran Province, Iran. J Appl Sci Environ Manag 12(2):43–46

Govender M, Chetty K, Bulcock H (2007) A review of hyperspectral remote sensing and its application in vegetation and water resource studies. Water SA 33(2):145–152

Halder JC (2013) Land suitability assessment for crop cultivation by using remote sensing and GIS. J Geogr Geol 5(3):65–74

Hopkins L (1977) Methods for generating land suitability maps: a evaluation. J Am Inst Plan 34(1):19–29

HuQ WW, Xia T et al (2013) Exploring the use of Google Earth imagery and object-based methods in land use/cover mapping. Remote Sens 5:6026–6042

Hurni H (1997) Concepts of sustainable land management. ITC J 3(4):210–215

Jafari S, Zaredar N (2010) Land suitability analysis using multi-attribute decision-making approach. Int J Environ Sci Dev 1(5):441–445
Jankowski P (1995) Integrating geographical information systems and multiple criteria decision-making methods. Int J Geogr Inform Syst 9(3):251–273
Jenson JR (2005) Introductory digital image processing, 3rd ed. Prentice Hall, Upper Saddle River, NJ, USA, pp 266–269
Joerin F, Theriault M, Musy A (2001) Using GIS and outranking multi-criteria analysis for land-use suitability assessment. Int J Geogr Inf Sci 15(2):153–174
Joshi PC (2011) Performance evaluation of vegetation indices using remotely sensed data. Int J Geomat Geosci 2(1):231–240
Kalogirou S (2002) Expert systems and GIS: an application of land suitability evaluation. Comput Environ Urban Syst 26:89–112
Kertesz A (2006) The global problem of land degradation and desertification. Hung Geogr Bull 58(1):19–31
Keshavarzi A, Sarmadian F, Ahmadi A (2011) Spatially-based model of land suitability analysis using block kriging. Austr J Crop Sci 5(12):1533–1541
Kihoro J, Bosco NJ, Murage H (2013) Suitability analysis for rice growing sites using a multicriteria evaluation and GIS approach in great Mwea region, Kenya. Springer Plus 2:1–9
LefroyRB BH, Rais M (2000) Indicators for sustainable land management based on farmer surveys in Vietnam, Indonesia, and Thailand. Agr Ecosyst Environ 81:137–146
Leverington DW, Moon WM (2012) Landsat-TM-based discrimination of lithological units associated with the Purtuniq Ophiolite, Quebec, Canada. Remote Sens J 4:1208–1231
Mahabadi NY, Givi J, Khorasgani MN et al (2012) Land suitability evaluation for Alfalfa and Barley based on FAO and fuzzy multi-criteria approaches in Iranian arid region. Environ Desert 17:77–89
Malczewski J (2000) On the use of weighted linear combination method in GIS: common and best practice approaches. Trans GIS 4(1):5–22
Malczewski J, Chapman T, Flegel C et al (2003) GIS multi-criteria evaluation with ordered weighted averaging (OWA): case study of developing watershed management strategies. Environ Plan Press 35(10):1769–1784
Malczewski J, Rinner C (2005) Exploring multi-criteria decision strategies in GIS with linguistic quantifiers: a case study of residential quality evaluation. J Geogr Syst 7(2):249–268
Malczewski J (2004) GIS-based land suitability: a critical overview. Progr Plan 62:3–65
Malczewski J (2006) GIS-based multi-criteria decision analysis: a survey of the literature. Int J Geogr Inform Sci 20(7):703–726
Mansor SB, Pormanafi S, Mahmud AR (2012) Optimization of land use suitability for agriculture using integrated geospatial model and genetic algorithms. ISPRS Ann Photogram, Remote Sens Spat Inform Sci 1(2):230–234
Massam BH (1988) Multi-criteria decision-making (MCDM) techniques in planning. Prog Plan 30(1):1–84
Melgani F, Bruzzone L (2004) Classification of hyperspectral remote sensing images with support vector machines. Trans Geosci Remote Sens 42(8):1778–1790
Mokarram M, Rangzan K, Moezzi A (2007) Land suitability evaluation for Wheat cultivation by fuzzy theory approach as compared with parametric method. The International Arch Photogram, Remote Sens Spat Inform Sci 38(2):140–145
Mustafa AA, Singh M, Sahoo RN (2011) Land suitability analysis for different crops: a multi-criteria decision-making approach using remote sensing and GIS. Researcher 3(12):61–84
Olayeye AO, Akinbols GE, Marake VM (2008) Soil in suitability evaluation for irrigated lowland rice culture in Southwestern Nigeria: management implications for sustainability. Commun Soil Sci Plant Anal 39:2920–2938
Panhalkar S (2011) Land capability classification for integrated watershed development by applying remote sensing and GIS techniques. ARPN J Agric Biol Sci 6(4):46–55

Perveen F, Ryota N, Imtiaz U (2007) Cropland suitability analysis using a MCE and GIS approach. In: 5th international symposium on digital earth. The University of California, Berkeley, USA, pp 1–8

Prakash TN (2003) Land suitability analysis for agricultural crops: a fuzzy multi-criteria decision making approach. MSc thesis, The International Institute for Geo-information Science and Earth Observation (ITC), Enschede, The Netherlands, pp 6–13

Rabia AH, Figueredo H, Huong TL (2013) Land suitability analysis for policy making assistance: a GIS-based land suitability comparison between surface and drip irrigation systems. Int J Environ Sci Dev 4(1):1–6

Rabia AH, Terribile F (2013) Introducing a new parametric concept for land suitability assessment. Int J Environ Sci Dev 4(1):15–19

Raghunath M (2006) Application of remote sensing and GIS in urban land suitability modeling at parcel level using multi-criteria decision analysis. Indian Institute of Remote Sensing National Remote Sensing Agency Department of Space, Govt. of India Debra Dun, (Unpublished), pp 85–89

Rossiter DG (1996) A theoretical framework for land evaluation. Geoderma 72(3–4):165–190

Rwetabula J, De Smedt F (2005) Land use and land cover mapping of the Simiyu catchment (Tanzania) using remote sensing techniques. Remote Sens Photogram 27:1–16

Saaty TL (1980) The analytic hierarchy process: planning, priority setting, resource allocation. McGraw Hill International, New York, NY, USA

Saaty TL (1997) A scaling method for priorities in hierarchical structures. J Math Psychol 15:234–281

Samanta S, Pal B, Pal D (2011) Land suitability analysis for rice cultivation based on multi-criteria decision approach through GIS. Int J Sci Emerg Technol 2(1):12–20

Sawy S, Abdel-Hameed A, Sultan KA (2012) A GIS-based digital land resources framework, for optimal soil management in Barda and Awaje Basin, Syria. In: International conference on applied life sciences, pp 191–197

Shalaby A, Ouma YO, Tateishi R (2006) Land suitability assessment for perennial crops using remote sensing and geographic information systems: a case study in North-western Egypt. Arch Agron Soil Sci 52(3):243–261

Singh P, Singh S (2011) Landuse pattern analysis using remote sensing: a case study of Mau District, India. Arch Appl Sci Res 3(5):10–16

Singh S (2012) Land suitability evaluation and land use planning using remote sensing data and geographic information system techniques. Int J Geol, Earth Environ Sci 2(1):1–6

Smyth AJ, Dumanski J (1993) FESLM: an international framework for evaluating sustainable land management. World Soil Resources, 73. FAO, Rome

Store R, Kangas J (2001) Integrating spatial multi-criteria evaluation and expert knowledge for GIS-based habitat suitability modelling. Landsc Urban Plan 55(2):79–93

Tahir M, Imam E, Hussain T (2013) Evaluation of land use/land cover changes in Mekelle City. Ethiopia Using Remote Sens GIS, Comput Ecol Softw 3(1):9–16

Bojórquez- LA, Diaz-Mondragon S, Ezcurra E (2001) GIS-based approach for participatory decision making and land suitability assessment. Int J Geogr Inform Sci 15(2):129–151

Triantafilis J, Waed WT, Mcbratney AB (2001) Land suitability assessment in the Namoi valley of Australia, using continuous model. Amsterdam J Soil Res 39:273–290

Van Lanen HAJ, Van Diepen CA, Reinds GJ (1992) A comparison of qualitative and quantitative physical land evaluations using an assessment of the potential for sugar beet growth in the European Community. Soil Use Manag 8(2):80–89

VibhuteAD, Nagne AD, Gawali BW (2013) Comparative analysis of different supervised classification techniques for spatial land use/land cover pattern mapping using RS and GIS. Int J Sci Eng Res 4(7)

Vorovencii I (2009) The hyperspectral sensors used in satellite and aerial remote sensing. Bull Transilvania Univ Brasov 2(51):52–56

Wang F (1994) The use of artificial neural networks in a geographical information system for agricultural land-suitability assessment. Environ Plan 26(2):265–284

Wang J, Sammis TW, Gutschick VP (2010) Review of satellite remote sensing use in forest health studies. Open Geogr J 3:28–42

WMO (2005) Climate and land degradation. World Meteorological Organization (WMO) Publication No. 989. http://www.wmo.int

Wulder MA, White JC (2004) Comparison of airborne and satellite high spatial resolution data for the identification of individual trees with local maxima filtering. Int J Remote Sens 25(11):2225–2232

Yager RR (1988) On ordered weighted averaging aggregation operators in multi-criteria decision making. IEEE Trans Syst Man Cybern 18(1):183–190

Yu J, Chen Y, Wu J (2011) Cellular automata-based spatial multi-criteria land suitability simulation for irrigated agriculture. Int J Geogr Inf Sci 25(1):131–148

Zheng G, Moskal LM (2009) Retrieving leaf area index (LAI) using remote sensing. Theories Methods Sens 2719–2745

Zolekar RB, Bhagat VS (2014) Use of IRS P6 LISS-IV data for land suitability analysis for cashew plantation in a hilly zone. Asian J Geoinform 14(3):23–35

Zolekar RB, Bhagat VS (2015) Multi-criteria land suitability analysis for agriculture in the hilly zone: remote sensing and GIS approach. Comput Electron Agric 118-C:300–321

Zolekar RB, Bhagat VS (2018) Multi-criteria land suitability analysis for plantation in Upper Mula and Pravara basin: remote sensing and GIS approach. J Geogr Stud 2(1):12–20

Zolekar RB (2018) Integrative approach of RS and GIS in the characterization of land suitability for agriculture: a case study of Darna catchment. Arab J Geosci 11(24):780

# Chapter 14
# Debating Coastal Tourism and Sustainability

## Insights of Land Resource Management in Calangute and Anjuna Beach Destinations of Goa

**Nandkumar Sawant, Venkatesh PrabhuGoankar, Sagar Mali, and Anil Yedage**

**Abstract** Tourism is one of the fastest-expanding sectors that provide employment and a source of income while it adversely affects people and places. Coastal tourist destinations, especially tropical lands, have special attractions as they provide the right setting of sun, sand, and sea to the tourist. Some of the coastal tourist destinations are experiencing the impacts of the same. Therefore, there is a need to revisit and reassess the development of sustainable tourism. One of the prominent tourism-oriented states of India, Goa has experienced a voluminous influx of tourists over the decades and the twenty-first century has experienced exponential growth, far exceeding the residential population. Thus, the present research attempts to envisage the development of tourism and the dynamics of decadal changes and transformation of landuse and landcover features over nearly three decades, especially after globalization. The study also assesses Butler's, (1980) Tourism Area Life Cycle, in which a tourist destination goes through six predictable stages by taking the case study of Calangute and Anjuna Beach tourist destinations. The study applies geospatial technology to achieve its objectives. The study shows that these destinations have reached the final stage (stagnation) of the Tourism Area Life Cycle and areas close to the beach have been converted into tourist services and commercial utilities. The applied algorithms state that there has been a significant increase in the built-up area. Lack of intervention may lead to further commercialization and development of residential structures and commercial on the beaches, which will result in erosion and coastal regulation zone violation and raise concerns for the sustainability of the beach destination.

N. Sawant (✉) · V. PrabhuGoankar · A. Yedage
Parvatibai Chowgule College of Arts and Science, Margao, Goa, India
e-mail: nns001@chowgules.ac.in

S. Mali
Vidya Prabodhini College of Commerce, Education, Computer and Management, Vidyanagar, Alto Porvorim, Goa, India

 
S. Anand et al. (eds.), *Emerging Geosustainability Transformations in India*, Advances in Geographical and Environmental Sciences,
https://doi.org/10.1007/978-981-97-7993-2_14

**Keywords** Coastal destination · Land use · Tourism · Transformation

## 14.1 Introduction

Tourism represents as most rapidly growing industry, making a significant contribution to employment opportunities, capital flow and income generation across various countries of the world. In the context of coastal regions of developing nations, tourism has emerged as a key driver for economic advancement and development (UNEP and UNWTO 2005). However, the swift growth of tourism industry has been associated with environmental degradation and socio-cultural impacts, leading to alterations in land use and land cover (LULC) patterns (Neto 2003; Joshi et al. 2011; Rawat et al. 2013). From this perspective, empirical studies provide valuable insights for tourism planning in specific contexts, but most of the cited research works remain disconnected from frameworks related to sustainable development (Collins 1999). Models have been developed within the tourism domain, encompassing descriptive, explanatory, and predictive (Chorley and Haggett 1967). Unlike descriptive and explanatory models, which primarily elucidate the process and function, Predictive models are built upon causal relationships to enable future projections (Thirumoorthy and Wong 2015). Considering this, scholarly works have envisaged the key concepts in tourism studies, such as the tourism system (Leiper 1979), tourist typology (Plog 1974; Cohen 1972; Smith 1977), and Butler's Tourist Area Life Cycle (TALC) (Butler 1980), understanding the travel behaviours, motivations of tourists, and the local community's perception of tourism. However, the key aspects have been the ramifications, both positive and negative, drawing up a wider canvas of complex studies and future predictions.

Coastal areas, in particular, are ever-changing and sensitive to human activities can quickly lead to ecological vulnerability (Huang et al. 2009). One of the remarkable dynamics and visible aspects is the LULC change due to physical and anthropogenic activities. Land cover is a reflection of the ecological status of an area and, therefore, assessment of LULC is one of the foremost important parameters to be considered in planning land resource management and monitoring environmental changes (Jagadeesh et al. 2015) for sustainable development. Understanding the coastal zones and their importance to the tourism industry, the need to protect these resources is not only vital in economic development but also presents a growing dilemma for sustenance (Phillips and Jones2006).

Goa, a tiny state on the Western coast of India, is a globally known tourist destination famous for its sandy beaches, swinging palms and Indo-Portuguese historical legacy. The stretch of 110 km coastline of Goa is dotted with more than 40 beautiful beaches. Ever since tourism took root in Goa just after a decade of liberation in 1962, more and more tourists flock to Goa every year, making it a favourite holiday destination for Indians as well as for international tourists (Diniz et al. 2014). Tourism was introduced in Goa in the early 1970s, in the villages of Calangute, Anjuna, and Baga. These villages, therefore, transformed. Traditional jobs of farming and fishing were abandoned and new tourism-related entrepreneurship started (Sawant and Majumdar 2009). During the past two decades, the development of tourist infrastructure, the increase in population and construction activities have created tremendous pressure on land resources in Anjuna and Calangute. The present study is, therefore undertaken to evaluate the tourism life cycle and decadal changes in LULC changes in Anjuna and Calangute. The present study involved temporal analysis for its transformational changes at these destinations.

The core aim of the present research is to evaluate the tourist inflow in Goa and seek alternatives. Therefore the objectives are to apply Butler's Model to assess the tourism level in Goa with special reference to Calangute and Anjuna, as beach destinations, and to assess the impacts of the LULC on the coastal landscape.

## 14.2 Study Area

The coastal belt of Anjuna and Calangute were selected as study areas, as these are the most sought-after tourist destinations in Goa by the tourists. The study area is a part of Bardez Taluka, Goa, India (Fig. 14.1). Popularly known as Calangute-Anjuna-Baga beach destination, it is one of the famous beaches, with magnetic force attracting voluminous domestic and international tourists. The Calangute-Baga-Anjuna coastal stretch is a longitudinal beach, stretching almost 5 km. Anjuna Beach lies north of Baga Beach and is famous for its flea market popular among tourists. Calangute is the most popular holiday destination in Goa. The stretch of Anjuna coast is 5.48 km with an area of 13.85 sq km (1365 ha) while Calangute has a coastline of 4.89 km and an area of 11.68. km (1168 ha).

## 14.3 Data Source and Methodology

Apart from the assessment of TALC, the present study includes an analysis of changes and transformations in the LULC of Anjuna and Calangute in Bardez taluka of Goa India. This study is a culmination of both primary and secondary data sources. Ground truthing and field observation were carried out using GPS whereas satellite images and multi-temporal satellite data were used as a secondary source. Anderson Classification Scheme was used for the classification of LULC changes. To reduce the spatial resolution from 30 to 15 m, these imageries were geo-processed and clipped to get better results. Accordingly, the area was digitized and classified into 11 categories (scrubland, barren land, sandy beach, built-up areas, agriculture land, vegetation, coconut plantation, water bodies, wetlands, fallow land and public utilities). This was carried out using the Regional Plan of 1990 and 2001 and Google Earth images of 2004 and 2018.

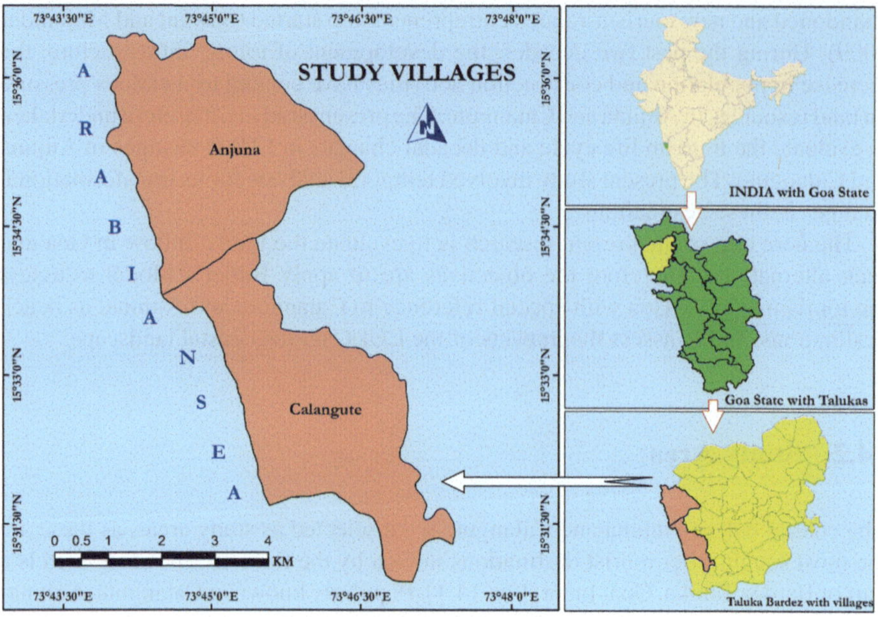


**Fig. 14.1** Location map of the study area

# 14.4 Results and Discussion

## *14.4.1 Butler's Model and the Tourism Level in Goa*

A popular model for examining the development of a specific tourist site is Butler's (1980) Tourism Area Life Cycle (TALC). According to the concept, a tourism region goes through six predictable stages of growth: discovery, involvement, development, consolidation, stagnation, and decline or rejuvenation (Fig. 14.2).

Butler's Tourism Area Life Cycle Model is a simplistic linear model that explains the six stages of tourism. The *Exploration Stage* of the model is indicative that a group of people explore the location, for its natural beauty or cultural richness. The natives also see an opportunity for initiating small business entrepreneurship within the framework of the assets that they possess. The promotion of destination is more so through word of mouth and gradually there is an inflow of tourists. In the second stage, i.e., the involvement stage, residents engage themselves in providing facilities (tourist services) to the tourists. As the tourism economy starts developing, local natives give up traditional occupations and involved in new business entrepreneurship i.e. hotels, restaurants, souvenir shops, local transport and other services.

As the destinations gain more popularity and the tourist flow increases, the tourist facilities provided by the locals become limited. Professional interventions through government or private corporations start setting up much-needed infrastructure.

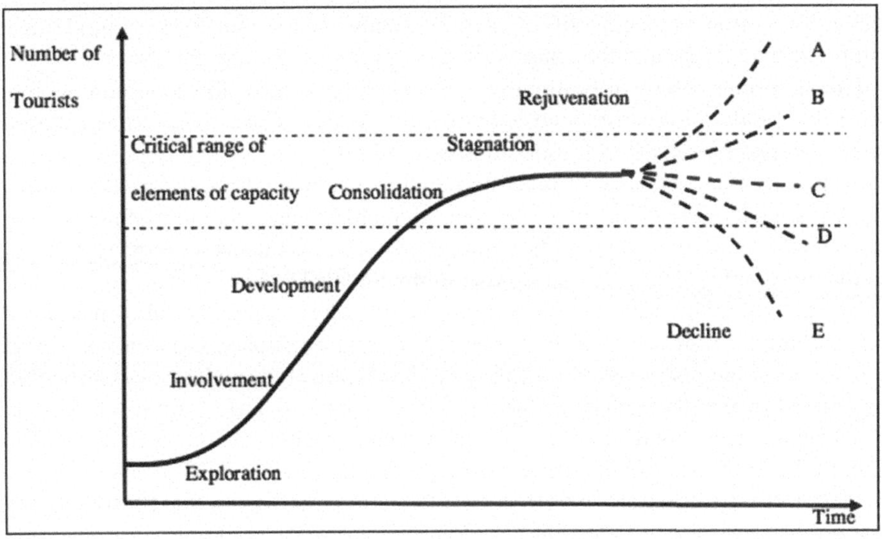


**Fig. 14.2** Butler's model. *Source* Butler (1980)

There is planned marketing of the destination to popularize the destination. Efforts are towards the development of tourism as an industry. Therefore, it is referred to as "Development Stage". This also marks the divide between the locals and the tourists, as tourists stay in hotels rather than in guest houses. In the subsequent years, there is exponential growth in tourist arrivals, exceeding the local population. Amenities provided by commercial enterprises now dominate the local tourism economy. In this consolidation stage, the locals feel they are losing stakeholder ship in tourism of their place or region (Berry 2000). The mounting tourist influx reaches its peak i.e. stagnation stage. The tourist inflow exceeds the carrying capacity and negative implications in the form of environmental, social and economic can be witnessed. Slowly, there is a shift of tourists from core destinations to peripheral destinations as there is overcrowding. Thus, the tourist destination now witnesses a decrease in tourist arrivals. The duration of stay of the tourists is shorter and tourist establishments find it difficult to sustain. This is the declining stage. Finally, for the sustenance of tourism, rejuvenation is required through new perspectives or enhancement. This rejuvenation can create a re-emergence of a tourist economy through investment from the State and businesses spurn new development (Sahli 2020).

In short, the tourism area life cycle (TALC) model illustrates the development process of a tourism destination through different recognizable stages. As the tourism activity expands and the number of visitors increases, environmental and social problems will rise. This will significantly reduce the magnetism of the destination which will ultimately lead to its decline (Almeida and Correia 2010). Butler recognizes that change is constant and that, ultimately, the initial reasons a location becomes a desirable tourist destination will no longer exist and the location will either need to seek rejuvenation or face decline. Therefore, embedded within Butler's model is a call for sustainability and the conservation of resources (Hwang 2017).

Goa, a former colony of Portugal, has changed in its social and economic sectors since the 1960s. Tourism-related activities in the state have, therefore, increased and developed rapidly. Beach tourism is currently the State's most popular form of tourism and offers a variety of other activities (Sawant and Majumdar 2009). On a 110 km stretch of coastline, there are more than 40 beaches, and more than 95% of visitors enjoy beach vacations. Therefore, tourism is also the one of most significant sectors for employment and contributors to the State's Domestic Product.

The pivotal moment in the tourism history of Goa is signified by the emergence of the "Hippie Trail" in the late 1960s. This event positioned Goa on the global tourism stage but also marked a significant development in modern tourism within the context of the developing world. From the context of TALC, It represented an initial phase of tourism growth and can be interpreted as a form of exploration (Fig. 14.2). As the number of tourist arrivals, both international and domestic, steadily increased, the government undertook development initiatives in the 1980s actively support the tourism sector, leading to enhancements in essential tourist services and infrastructure (Sawant 2022). Subsequently, the liberalization of the economy and the subsequent boom in information technology contributed to an upsurge in the popularity of the State, resulting in a rise in domestic tourist influx that persisted into the new millennium.

Before the commencement of globalization, the tourist inflow was less than a million, but spilled over into the new millennium and witnessed leap-frogging growth. The tourist arrivals in 2003 crossed 2 million and the State received 4.6 million visitors in 2010 (Table 14.1). The tourist arrivals doubled within five years. Goa reached in 'developing stage', and it was then predicted that in the long-term, the local tourism sector would have to compete with huge business houses engaged in tourism (Kamat 2010). This was the developing stage for tourism in the State (Fig 14.3).

In the next ten years, there is a consistent increase in tourist arrivals and the figures crossed 8 million. The resident population is 1.5 million, which means that tourist arrival is more than 5 times the Goa's Population. This exerts pressure on local resources, especially, in the coastal areas, as more than 95%of tourists prefer beach tourism, marking the phase of consolidation and early symptoms of stagnation. Tourism development has caused unfavourable impacts on the social environment of Goa. Resource exploitation, resource and environment management, economic and local sustainability, community engagement, socio-cultural sustainability and destination management influence the sustainability of tourist destinations in Goa (Venugopalan 2021).

**Table 14.1** Tourist arrivals in Goa, 1990–2019

| Year | 1990 | 1995 | 2000 | 2005 | 2010 | 2015 | 2016 | 2017 | 2018 | 2019 |
|---|---|---|---|---|---|---|---|---|---|---|
| Tourist Arrivals (millions) | 0.88 | 1.10 | 1.24 | 2.34 | 4.62 | 5.30 | 6.33 | 7.68 | 8.11 | 8.05 |

*Source* Department of Tourism, Government of Goa (2019)

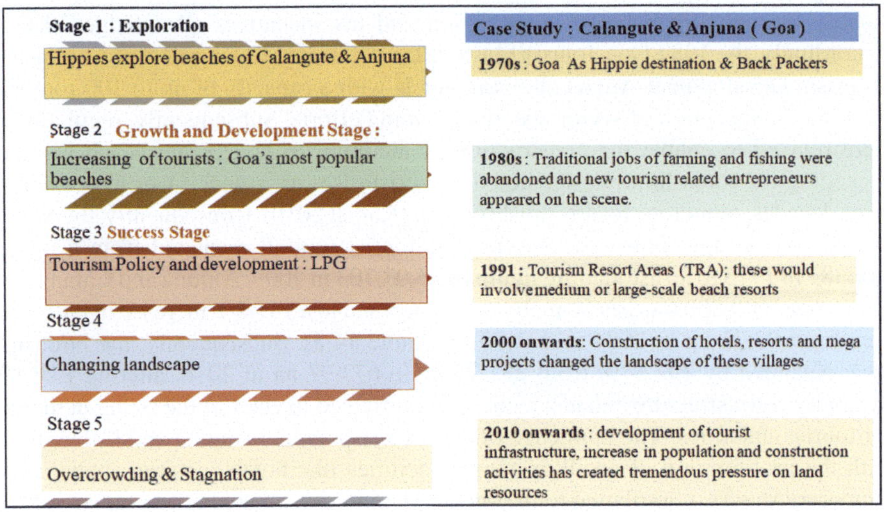


**Fig. 14.3** Calangute and Anjuna coastal tourist destination, assessing Butler's theory *Source Authors*

In the realm of beach tourism, the coastal beach tourism in Anjuna and Calangute was deliberaltylychosen for present research as this location widely recognized as a top tourist spot in Goa, that act as a magnetic force for both domestic and international visitors. Further, the tourist services and related activities have expanded significantly over the decades. The Calangute–Baga–Anjuna coastal stretch spans approximately 5 km, forming a linear coastline.

### *14.4.2 Tourism and Landuse/Land Cover Change*

The tourism sector serves as a primary driver of Land Use and Land Cover (LULC) modification. LULC plays a crucial role in comprehending the nexus between human actions and the environment that emphasizes the need to investigate the change (Dar et al. 2017). Evaluating the impacts of tourism progress involves appraising the visible repercussions in specific areas and assessing the substantial effects of tourism. In short, the ramifications of tourism expansions extend beyond just physical alterations but also encompass changes in multiple dimensions. (Duffield and Walker 1984). As the popularity of coastal tourism increases, subsequently population, workforce and tourist facilities increase thereby impacting the LULC, and fuelling spatial expansion and sprawl.

The early 1990s witnessed a significant transformation in India's economic strategies. The adoption of liberalization, privatization, and globalization policies led to a notable surge in tourism activities. The 1992 economic policy shift in India primarily focused on fostering tourism by enhancing infrastructure development, notably through beach resort construction and tax incentives (Routledge 2000). Accordingly, the 1987 Goa Tourist Master Plan delineated Calangute and Anjuna as 'Tourism Development Areas' allowing hotels with a capacity of up to 100 rooms, marking the inception of sector-specific planning efforts. Subsequently, regulations were relaxed to enable the construction of medium- to large-scale beach resorts exceeding the room limit, designated as 'Tourism Resort Areas'. Over the ensuing three decades, numerous resorts mushroomed (Kamat 2010). Consequently, the State experienced a rapid influx of visitors. The tourist footfall escalated from a mere 119,387 in 1973 to an astounding figure of 25,03,703 in 2009. Anjuna and Calangute attract over 600,000 tourists annually. This surge led to a 147.2% increase in built-up area from 93.75 ha in 2004 to 231.75 ha (Table 14.2). Subsequently, the built-up area expanded by 192.54% from 231.75 ha to 677.97 ha in 2018, altering LULC dynamics. Infrastructure enhancements were upscaled to cater to the rising demand for tourist amenities. The eminence of tourist hotspots like Anjuna and Calangute, with the proliferation of ancillary tourist facilities like hotels, eateries, pubs, and temporary shacks, contributed to the substantial increase in the built-up area.

In the year 1990, one-third of the geographical land was under scrub cover followed by agriculture and vegetation. These three class categories occupied more than 80% of the LULC of the study area (Table 14.2). In 2018, the major portion included the built-up area which increased from 623.17% from 93.75 ha in 1990 to 677.97 ha in 2018. This increase in built-up area was proportionate to the decrease in the area of the other categories. By 2018, the trend of waning was observed for areas under the category of agriculture, sand dunes, vegetation, scrubland and barren land (Fig. 14.4). Scrubland and coconut plantation areas also witnessed a considerable decline. Precisely, scrubland decreased by 45.63% from 786.89 ha to 427.8 ha while sandy beach cover decreased by 74.68% from 91.84 ha to 23.25 ha. Other categories that showed a decline were coconut plantations, barren land and agricultural land (Fig. 14.4).

The increase in built-up area and proportionate decrease in other categories can be attributed to multiple factors. The economy of India underwent significant policy shifts at the beginning of the 1990s. Liberalization, privatization and globalization policies brought a remarkable increase in tourism activities, especially in the study area. The phase of economic development within India sought liberalization of the economy and integration into the global market. As a significant component of India's new economic policy, the emphasis was on the promotion of tourism by creating infrastructure, especially the creation of beach resorts and providing tax concessions (Routledge 2000). The first sector-specific effort towards planning was the master plan for tourism development in Goa in 1987, by demarcating Calangute and Anjuna as 'Tourism Development Areas' where hotels with up to 100 rooms were to be allowed. Also, demarcating it as 'Tourism Resort Areas' facilitated the construction of medium or large-scale beach resorts with over 100 rooms. Thereafter, many resorts came up in recent decades. On average, Anjuna and Calangute, the beach tourist destinations, areas receive more than 6 lakh visitors. Owing to this, in 2004, the built-up area showed a 147.2% increase from 93.75 to 231.75 ha (Table 14.2). In 2018, the built-up area further soared by 192.54% from 231.75 to 677.97 ha. This has

**Table 14.2** Landuse/land cover for Calangute and Anjuna, 1990–2018

| Classes | 1990 | | 2004 | | 2018 | | 1991–2004 | 2004–2018 | 1990–2018 |
|---|---|---|---|---|---|---|---|---|---|
| | Area in hectares | Area in % | Area in hectares | Area in % | Area in hectares | Area in % | % increase / decrease | % increase / decrease | % increase / decrease |
| Scrub | 786.89 | 30.88 | 690.56 | 27.1 | 427.8 | 16.79 | −12.24 | −38.05 | −45.63 |
| Barren land | 128.59 | 5.04 | 159.25 | 6.25 | 125.22 | 4.91 | 23.84 | −21.37 | −2.62 |
| Sandy Beach | 91.84 | 3.62 | 72.18 | 2.83 | 23.25 | 0.91 | −21.41 | −67.79 | −74.68 |
| Built-up | 93.75 | 3.68 | 231.75 | 9.09 | 677.97 | 26.6 | 147.2 | 192.54 | 623.17 |
| Agriculture | 605.09 | 23.74 | 589.5 | 23.13 | 497.18 | 19.51 | −2.58 | −15.66 | −17.83 |
| Vegetation | 676.79 | 26.55 | 656.22 | 25.75 | 642.81 | 25.22 | −3.04 | −2.04 | −5.02 |
| Coconut plantation | 79.59 | 3.13 | 54.64 | 2.14 | 32.37 | 1.27 | −31.35 | −40.76 | −59.33 |
| Water bodies | 52 | 2.04 | 65.19 | 2.56 | 94.26 | 3.7 | 25.37 | 44.59 | 81.27 |
| Wetlands | 9.14 | 0.35 | 7.25 | 0.28 | 7.5 | 0.29 | −20.68 | 3.45 | −17.94 |
| Fallow land | 15.64 | 0.58 | 14.23 | 0.56 | 9.25 | 0.36 | −2.8 | −35 | −36.82 |
| Public Utilities | 10.14 | 0.39 | 7.69 | 0.3 | 10.87 | 0.43 | −24.16 | 41.35 | 7.2 |
| Total | 2548.46 ha | 100 | 2548.46 ha | 100 | 2548.46 ha | 100 | | | |

*Source* Landsat TM (1990), ETM + (2004) & OLI (2018), Computed by Authors

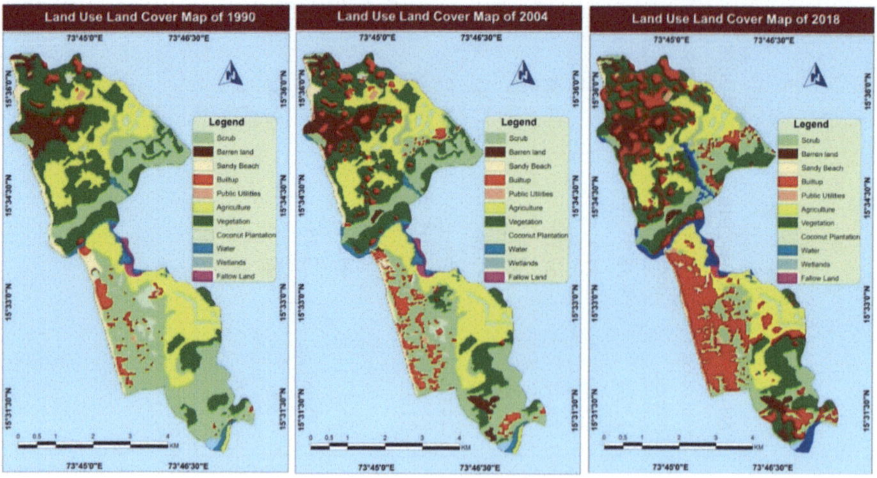


**Fig. 14.4** LULC changes in 1990, 2004 and 2018, *Source* Landsat TM (1990), ETM + (2004) & OLI (2018)

impacted the dynamics of land use and land cover. To meet the increasing demand for tourist services, infrastructural changes also upscaled. Tourist destinations like Anjuna and Calangute, which had a history of tourism from the early 1970s came to wider prominence also experienced an increase in tourism facilities like hotels, restaurants, pubs and temporary shacks, etc. which have contributed to the drastic increase of built-up area.

Change matrix analysis between 1990 and 2004 exhibited that six categories experienced a reduction in the area. Scrubland, sandy beach cover, agriculture, vegetation, coconut plantation and fallow land have experienced a negative change due to the reduction of land under these categories (Table 14.3). There was a loss of 72 ha of the total scrub land to cater to the expansion of the built-up area and 32 ha contributed to the expansion of barren land. Nearly, 9 ha of area under Sand Dunes came under water, probably due to inundation. Thus, the area under water also showed an increase. When the matrix of 2004 was compared to 2018, the number of categories that experienced negative change increased to seven (Table 14.4). Further, it has been observed that the scrub area was altered to barren, built-up, agriculture, and vegetation. Barren land has changed to scrub, built-up, vegetation and public utility for recreation. Sandy Beach area too, was reduced and contributed to the expansion of built-up areas and vegetation. The agricultural area was converted to built-up, vegetation, and recreational activity, and 19 ha had gone underwater. Thus, the huge expansion of the built-up area observed by 2018 was at the expense of a reduction in the area of scrubland, barren land, agricultural land, vegetation, coconut plantation, and encroachments on the sandy beaches. This could be due to the fact that at this time, there was a spurt in construction activities all over the state. Moreover, demand from tourism has led to an increase in built-up areas either for commercial and infrastructural use or for residential purposes. The need for expansion of built-up areas and recreational activities was the major demand of heavy load imposed by tourism. This prompted the construction of hotels, resorts, and mega projects to change the whole landscape of these villages (Kamat 2010).

In the transitional probability matrix, the columns show the LULC of 2004 while the rows show the LULC of 2018 (Fig. 14.5). The diagonal matrix of row and column indicates the percentage of the area which remains unchanged for each category, while the figures in the rows indicate the probability of transforming into other categories. The most significant observation from the probability matrix was the diminishing of specific areas namely, agricultural land, scrubland, sandy area, and vegetation, as compared to the drastic increase in built-up area.

The Sand Dunes are part of the Coastal Regulation Zone (CRZ) as they have ecological significance. Reduction in the sandy area by 10%is attributed to the dramatic growth and proliferation of innumerable edifices such as coastal hotels and resorts, residential flats, dwellings, small restaurants, beachside bars, roads and beach shacks along and across the coastal zone, most of which are located on sandy areas as dunes. (Mascarenhas 2002). An increase in the number of structures in the CRZ over a period of time was reported (Ramamohana and Suneetha 2013).

The above-mentioned tourist destinations recorded about a 50% increase in the number of buildings. Recently, the Goa Coastal Zone Management Authority (GCZMA) has issued demolition orders to two major resorts in Calangute and Anjuna for violation of CRZ/NDZ areas. Though the entire belt within 200 m of the High Tide Line (HTL) in CRZ-III areas, human settlements, shacks, and resorts have sprung up (Anonymous 2018). Since Goa's tourism activity is seasonal there are many temporary shacks and hotels erected during the peak tourist season and dismantled during the off-season. Therefore, the beach shack by itself being seasonal and temporary, does not induce any large-scale environmental damage. However, the major concern is the large number of shacks which is on the rise (Mascarenhas 2002).

The study was conducted by NSCSM (Netaji Subhas Computer Siksha Mission) for the coastal tourist destination of Goa in 2018. The assessment framework stipulates a minimum rating of 1 for tourist infrastructure, with the criteria for the quantity of shacks/huts/cottages and other provisional constructions per hectare being set at 100. Analysis of the aforementioned framework reveals that Arpora-Nagoa (529 shacks and provisional structures/ ha), Calangute (177), and Candolim (106) have surpassed the specified threshold of 100 (Fig. 14.4).This is indicative that the tourist destination exceeds the prescribed limit.

**Table 14.3** LULC class conversion from 1990 to 2004

| 1990 | | | | | | | | | | | | | |
|---|---|---|---|---|---|---|---|---|---|---|---|---|---|
| 2004 | Classes in hectares | Scrub | Barren land | Sandy beach | Built-up | Agriculture | Vegetation | Coconut plantation | Water | Wetlands | Fallow land | Public utilities | Grand total |
| | Scrub | 644 | 3 | 2 | 18 | 0 | 21 | 3 | 0 | 0 | 0 | 0 | −691 |
| | Barren land | 32 | 114 | 0 | 2 | 0 | 10 | 0 | 0 | 0 | 0 | 0 | +159 |
| | Sandy beach | 0 | 0 | 71 | 0 | 0 | 1 | 0 | 0 | 0 | 0 | 0 | −72 |
| | Built-up | 72 | 11 | 9 | 73 | 13 | 48 | 5 | 0 | 0 | 0 | 0 | +233 |
| | Agriculture | 0 | 0 | 0 | 0 | 588 | 1 | 0 | 0 | 0 | 0 | 0 | −590 |
| | Vegetation | 36 | 0 | 0 | 1 | 1 | 594 | 20 | 0 | 2 | 0 | 0 | −654 |
| | Coconut plantation | 0 | 0 | 0 | 0 | 2 | 0 | 52 | 0 | 0 | 0 | 0 | −54 |
| | Water | 4 | 0 | 9 | 0 | 1 | 0 | 0 | 52 | 0 | 0 | 0 | +65 |
| | Wetlands | 0 | 0 | 0 | 0 | 0 | 0 | 0 | 0 | 7 | 0 | 0 | −7 |
| | Fallow land | 0 | 0 | 0 | 0 | 0 | 0 | 0 | 0 | 0 | 14 | 0 | +14 |
| | Public utilities | 0 | 0 | 0 | 0 | 0 | 0 | 0 | 0 | 0 | 0 | 8 | +8 |
| | Grand total | 789 | 128 | 91 | 95 | 605 | 676 | 79 | 52 | 9 | 14 | 8 | 2548 |
| | | | Stable area | | | − | Negative change | | | + | Positive change | | |

*Source* Landsat TM (1990) & ETM + (2004); Computed by authors

**Table 14.4** LULC Class conversion from 2004 to 2018

| 2004 | | | | | | | | | | | | | |
|---|---|---|---|---|---|---|---|---|---|---|---|---|---|
| 2018 | Classes in hectares | Scrub | Barren land | Sandy Beach | Built-up | Agriculture | Vegetation | Coconut plantation | Water | Wetlands | Fallow land | Public utilities | Grand total |
| | Scrub | 364 | 2 | 0 | 4 | 5 | 47 | 8 | 0 | 0 | 0 | 0 | −430 |
| | Barren land | 5 | 120 | 0 | 0 | 0 | 1 | 0 | | 0 | 0 | 0 | −126 |
| | Sandy beach | 0 | 0 | 23 | 0 | 0 | 0 | 0 | 0 | | 0 | 0 | −23 |
| | Built-up | 214 | 14 | 31 | 217 | 61 | 123 | 12 | 0 | 0 | 3 | 0 | +676 |
| | Agriculture | 9 | 0 | 0 | 2 | 476 | 8 | 4 | 0 | 0 | 0 | 0 | −497 |
| | Vegetation | 91 | 22 | 17 | 10 | 28 | 474 | 0 | 0 | 0 | 1 | 0 | −643 |
| | Coconut plantation | 2 | 0 | 0 | 0 | 0 | 0 | 30 | 0 | 0 | 0 | 0 | −32 |
| | Water | 5 | 0 | 0 | 0 | 19 | 3 | 0 | 65 | 2 | 1 | 0 | +94 |
| | Wetlands | 0 | 0 | 0 | 0 | 0 | 0 | 0 | 0 | 5 | 0 | 0 | −5 |
| | Fallow land | 0 | 0 | 0 | 0 | 0 | 0 | 0 | 0 | 0 | 9 | 0 | −9 |
| | Public utilities | 0 | 2 | 0 | 0 | 1 | 0 | 0 | 0 | 0 | 0 | 8 | +11 |
| | Grand total | 691 | 159 | 72 | 233 | 590 | 654 | 54 | 65 | 7 | 14 | 8 | 2548 |
| | | | Stable area | | | – | Negative change | | | + | Positive change | | |

*Source* ETM + (2004) & OLI (2018); Computed by authors

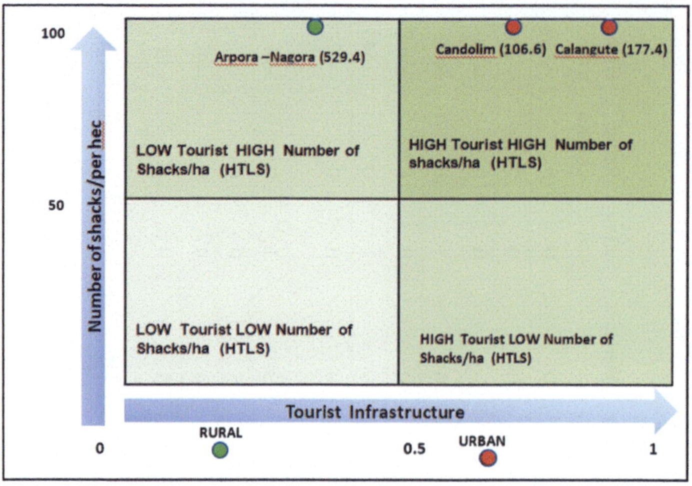


**Fig. 14.5** Transitional probability matrix: tourist infrastructure and number of shacks and other temporary structures/ha. *Source* NSCSM (2018)

## 14.5 Conclusion and Recommendations

In conclusion, it is undeniable that tourism serves as the primary source of revenue for Goa; however, there exists a necessity to reassess and reconsider the nature and structure of tourism. The long-standing presence of tourism in the coastal landscape of Goa has witnessed a significant transformation in the region, both positive and negative. The present study reveals the coastal hub of tourism, encompassing Calangute, Anjuna, and Baga, has reached its pinnacle in the context of built landscape thereby warranting the adoption of sustainable tourism practices to enrich and safeguard the preservation of the natural and societal aspects. To balance the economic revenue generation and preservation of the natural landscape, it is essential to formulate and implement a strategic tourism blueprint emphasizing the sustainable advancement of coastal tourism.

An integrated coastal zone management strategy is needed with holistic regional planning that fosters balanced natural ecosystems and human welfare. For this, existing positive aspects need to be studied and integrated. For inclusiveness in planning strategy, community participation along with authorities for grassroots planning needs to be fostered. This is crucial, especially, in the context of projected LULC in future years to come in the study area.

## References


Almeida A, Correia A (2010) Tourism development in Madeira: an analysis based on the life cycle approach. Tour Econ 16(2):427–441

Anonymous (2018) Goa tourism looks to increased footfalls in 2018–19. O Heraldo, 17 Mar 2018. https://www.heraldgoa.in/Goa/Goa-Tourism-looks-to-increased-footfalls-in-201819/128269. Accessed 23 Feb 2024

Berry EN (2000) An application of Butler's (1980) tourist area life cycle theory to the Cairns region, Australia, 1876–1998. PhD thesis, James Cook University, Queensland

Butler RW (1980) The concept of a tourism area cycle of evolution: implication for management resources. Can Geogr 24(1):5–6

ChorleyRJ HaggettP (ed) (1967) Models in geography. Methuen, London

Cohen E (1972) Towards a sociology of international tourism. Soc Res 39(1):164–182

Collins A (1999) Tourism development and natural capital. Ann Tour Res 26(1):98–109

Dar SN, Shah SA, Skinder S et al (2017) Geospatial approach analysis of land uses land/cover change in the tourist town of Leh (Ladakh). Int J Recent Sci Res 8(6):17676–17682

DinizR FSP, De Barros MM (2014) Local residents' perception of the psycho-social and economic impact of tourism in Goa. Int J Sci Res Publ 4(11):1–8

Duffield BS, Walker SE (1984) The assessment of tourism impacts. In: Clark BD, Gilad A, Bisset R, Tomlinson P (eds) Perspectives on environmental impact assessment. Springer, Dordrecht, pp 479–516. https://doi.org/10.1007/978-94-009-6381-8_31

Government of Goa (2019) Tourist arrivals, Tourism department, Goa. https://www.goatourism.gov.in/wp-content/uploads/2019/09/Tourist-Statistics-1.pdf. Accessed 23 Apr 2020

Huang J, Tu Z, Lin J (2009) Land-use dynamics and landscape pattern change in a coastal gulf region, southeast China. Int J Sust Dev World 16(1):61–66

Hwang L (2017) Butler's tourism area life cycle and its expansion to the creative economy. The SAGE International Encyclopedia of Travel and Tourism (ed) Sage Publications, Thousand Oaks, p 208

Jagadeesh CB, Naiklal HSS, Nagaraj S (2015) Dynamics of rapid urbanization of Bangalore and its impact on land use/land cover—a case study of Vrishabhavathi sub-watershed. Int Res J Eng Technol 02(03):2338–2345

Joshi RR, Warthe M, Dwivedi S et al (2011) Monitoring changes in land use land cover of Yamuna riverbed in Delhi: a multitemporal analysis. Int J Remote Sens 32(24):9547–9558

Kamat BS (2010) Assessment of Goa's tourism industry with respect to destination life cycle. South Asian J Tour Heritage 3(2):128–139

Leiper N (1979) The framework of tourism: towards a definition of tourism, tourist, and the tourist industry. Ann Tourism Res 6(4):390–407

Mascarenhas A (2002) Restoration of sand dunes along human-altered coast: scheme for Miramar Beach Goa. National Institute of Oceanography, Contribution No. 3720

Neto F (2003) A new approach to sustainable tourism development: moving beyond environmental protection. DESA Discussion Paper No. 29:212–222

NSCSM (2018) Carrying capacity of beaches of Goa for providing shacks and other temporary seasonal structures in private areas. Report submitted to Government of Goa, Ministry of Environment, Forest, Climate Change, Government of India, Chennai

Phillips M, Jones A (2006) Erosion and tourism infrastructure in the coastal zone: problems, consequences and management. Tour Manage 27(3):517–524

Plog SC (1974) Why destination areas rise and fall in popularity. Cornell Hotel Restaurant Admin Quart 14(4):55–58. https://doi.org/10.1177/001088047401400409

Ramamohana RP, Suneetha (2013) Coastal zone management using remote sensing and GIS: a case study of Goa. Int J Curr Res 5(3):718–725

Rawat JS, Biswas V, Kumar M (2013) Changes in land use/cover using geospatial techniques: a case study of Ramnagar town area, district Nainital, Uttarakhand, India. Egypt J Remote Sens Space Sci 16:111–117

Routledge P (2000) Consuming Goa: tourist site as dispensable space. Econ Pol Wkly 35(30):2647–2656

Sahli E (2020) Tourism destination development: an application of Butler's (1980) tourism area life cycle model to Hammamet, Tunisia. Bachelor Thesis, Rhine-Waal University of Applied Sciences, pp 3–6

Sawant NN (2022) Goa's landscape through maps. Dyanmangal Publication, Solapur

Sawant NN, Mazumdar D (2009) Beach tourism, globalisation and socio-economic transformation: a case study of Palolem beach community, Goa. In: Sawant NN, Shekhar S, Diddee J et al (eds) Globalisation: issue and challenges for India. Department of Geography, Parvatibai Chowgule College Margao-Goa, pp 221–234

Smith VL (ed) (1977) Hosts and guests: the anthropology of tourism. University of Pennsylvania Press, Philadelphia

Thirumoorthy T, Wong KM (2015) Tourism, Chapter 24. In: Idris A, Moghavvemi S, Musa G (eds) Selected theories in social science research. UM Press, Kuala Lumpur, pp 273–285

UNEP, WTO [United Nations Environment Programme, World Trade Organization] (2005) Making tourism more sustainable: a guide for policy makers. pp 11–12. https://wedocs.unep.org/20.500.11822/8741

Venugopalan T (2021) Sustainability of Goa tourism: perception of tourists. J Adv Sch Res Allied Educ 18(1):98–106

# Chapter 15
# An Impact of the COVID-19 Pandemic on the Sustainable Tourism Practices in Darjeeling District (West Bengal): A Focus on Hotels and Tourist Vehicles

**Virendra Nagarale, Piyush Telang, and Shweta Ghimireh**

**Abstract** The pandemic has brought significant changes to the global tourism sector, and Darjeeling, renowned for its scenic landscapes and tea plantations, has not been exempt from these effects. By analysing the impact of COVID-19 through a sustainability lens, the study explores its implications for the environment, local communities, and the long-term viability of the tourism sector. The research finds that the decline in tourist inflow during the pandemic has led to a reduced carbon footprint, as transportation-related emissions, including those from tourist vehicles, decreased due to fewer visitors. Moreover, the pandemic's impact on hotels and tourist vehicles has had profound effects on the local community. Many hotels and tour operators faced financial challenges, resulting in job losses and decreased income for those reliant on tourism. Analysing these effects highlights the need to develop strategies that enhance community resilience, such as diversifying livelihoods, supporting local entrepreneurship, and strengthening social safety nets. Assessing the economic sustainability of the tourism sector in Darjeeling by understanding the pandemic's impact on tourist inflow is crucial. The analysis of the study involves examining the dependency on external markets and identifying ways to stimulate domestic tourism, fostering a more resilient and self-sustaining tourism industry. The study concludes that the assessment of environmental impacts, community resilience, economic sustainability, destination management practices, and responsible tourist behaviour is essential for building a more sustainable and resilient tourism sector in Darjeeling. Lastly, the study notes the seasonality of tourist arrivals, with a lower number of visitors during the monsoon period due to factors such as landslides and heavy rainfall. Understanding these patterns is crucial for sustainable tourism planning and management in Darjeeling.

V. Nagarale (✉)
Department of Geography, SNDT Women's University, Pune Campus, Pune, Maharashtra, India
e-mail: drnagarale@gmail.com

P. Telang · S. Ghimireh
Department of Geography, SNDT Women's University, Pune Campus, Karve Road, Pune, Maharashtra, India

 
S. Anand et al. (eds.), *Emerging Geosustainability Transformations in India*, Advances in Geographical and Environmental Sciences,
https://doi.org/10.1007/978-981-97-7993-2_15

**Keywords** Sustainability · Tourism · Darjeeling · Tourist vehicles · COVID-19

## 15.1 Introduction

In recent times, the global landscape has been, significantly reshaped by an unprecedented event- the COVID-19 pandemic. This pandemic has not only posed severe threats to public health but has also triggered far-reaching socio-economic ramifications across various sectors. Among these, the tourism industry, a vital contributor to economic growth and cultural exchange, has been particularly impacted. This study delves into the intricate web of effects that the COVID-19 pandemic has woven around sustainable tourism practices in the picturesque Darjeeling district of West Bengal, India. Focusing on the crucial elements of hotels and tourist vehicles, this research aims to shed light on the intricate interplay between the pandemic and the sustainable tourism ecosystem in the region.

Darjeeling, nestled in the lap of the Eastern Himalayas, has long been celebrated for its breathtaking landscapes, vibrant culture, and rich heritage. The district's sustainable tourism practices have been fundamental not only for its economic vitality but also for its conservation of the unique environment and local traditions. However, the emergence of the COVID-19 pandemic has brought about a paradigm shift, challenging the very foundation of these practices. The hospitality sector, a cornerstone of Darjeeling's tourism landscape, has experienced unprecedented disruptions due to the pandemic-induced restrictions on travel, lockdowns, and safety concerns. Similarly, the operations of tourist vehicles, which facilitate the exploration of the region's beauty, have encountered hurdles such as reduced mobility and changing travel preferences. The pandemic's impact on these facets of tourism not only has immediate economic implications but also raises critical questions about the resilience and adaptability of sustainable tourism models.

This study seeks to explore the multifaceted repercussions of the COVID-19 pandemic on Darjeeling's sustainable tourism practices, with a specific focus on hotels and tourist vehicles. By examining the challenges faced, the strategies employed for survival, and the potential long-term transformations, this research endeavours to offer valuable insights into how the district can navigate the path toward a more resilient and sustainable post-pandemic tourism landscape. As we embark on this exploration, it is crucial to recognize that the insights garnered from this study could extend beyond Darjeeling, contributing to a broader understanding of how destinations worldwide can navigate the intricate nexus of global disruptions and sustainable tourism practices.

Tourism holds a pivotal role in India's economic landscape and is experiencing rapid expansion. According to the calculations of the World Travel and Tourism Council, in 2021, tourism contributed ₹13.2 lakh crore (US$170 billion) to India's GDP, accounting for 5.8% of the total, and employed 32.1 million individuals. Despite these figures falling short of pre-pandemic levels, the country's economy demonstrated substantial recovery in 2021 following the substantial downturn experienced in 2020. Projections indicate that the sector is poised to grow at an annual rate of 7.8%, reaching ₹33.8 lakh crore (US$420 billion) by 2031, equivalent to 7.2% of the GDP (WTTC 2022).

Among the nation's most profoundly impacted by the pandemic, India stands out. The steady escalation of infections and deaths compelled the Indian Government to implement an array of measures, including social distancing, community lockdowns, remote work arrangements, self-imposed or mandatory quarantines, and restrictions on travel, both domestically and internationally. These measures collectively brought tourism to an almost complete standstill for an extended duration. India, renowned for its distinctive natural beauty, rich cultural heritage, and warm hospitality, has historically been a favoured destination for both local and international tourists. The tourism sector plays a pivotal role in employing millions of individuals. In 2019–20, direct contributions of tourism to India's GDP and employment stood at 2.7 and 6.7%, respectively. When factoring in indirect contributions, these figures rose to 5.2 and 15.3%, respectively (Mehta 2021). However, the pandemic has not only imperilled the sector's growth trajectory but has also precipitated a substantial economic downturn that is expected to require considerable time to recover from.

## 15.2 Significance of the Study

This study is dedicated to examining the theoretical and empirical dimensions surrounding the impact of the COVID-19 outbreak on tourist vehicle inflow and hotels in the Darjeeling district of West Bengal. Through a comprehensive overview of theoretical frameworks and empirical findings, it seeks to illuminate the profound influence of the COVID-19 pandemic on the tourism landscape of the region. Within the Darjeeling district, tourism stands as a crucial pillar supporting both the local economy and social development. This study is designed with the primary objective of scrutinizing and assessing the factors that have triggered significant transformations in the lives of individuals engaged in the tourism sector. Central to its purpose is the endeavour to comprehend how the advent of COVID-19 has intricately altered the daily livelihoods of those dependent on the tourism industry. With a particular focus on hotels and tourist vehicles, this research is committed to comprehending and evaluating the far-reaching ramifications of the pandemic on the tourism industry of the Darjeeling district in West Bengal. It seeks to uncover the multifaceted influences arising from diverse factors such as occupation, migration status, educational qualifications, and the situational context during the pandemic. Through meticulous analysis of these demographic patterns, the study aims to arrive at well-founded conclusions concerning the COVID-19-induced disruptions in the tourism sector. Furthermore, the study aims to propose pragmatic preventive measures that can equip the people of Darjeeling to effectively manage similar challenges in the future. By formulating growth strategies that navigate the complexities of economies amidst epidemic crises, this research strives to foster resilience within the tourism sector, ensuring its sustenance despite the challenges posed by the COVID-19 pandemic.

## 15.3 Review of the Literature

In the realm of scholarly exploration concerning the impact of the COVID-19 pandemic, notable contributions have been made by scholars on the multifaceted challenges posed by the pandemic. A comprehensive investigation into the repercussions of the COVID-19 pandemic on India's tourism sector was carried out by Jaipuria et al. (2021). Employing an Artificial Neural Networks (ANN) predictive model, the study aimed to forecast foreign tourist arrivals. Additionally, the analysis assessed COVID-19's impact on tourism, including fluctuations in Foreign Exchange Earnings (FEE). The investigation encompassed an evaluation of tourist inflow patterns and socio-economic dynamics within the industry, focusing on drivers and hotel managers. While addressing socio-economic well-being, the study did not extensively examine employment issues or economic statuses during the pandemic. Ghosh (2022) explored a different tangent, scrutinizing the experiences of healthcare workers during the COVID-19 crisis. Investigating discrimination and operational adaptations, the study included insights from North Bengal Medical College. Utilizing theories like self-actualization and health communication, the research unveiled the complex behavioural changes induced by the pandemic. Primary data, including direct observations of healthcare workers, underpinned the study's emphasis on documenting nuanced shifts. Similarly, Faisal and Dhusia (2021) investigated the pandemic's repercussions on the tourism sector. Utilizing Convenience Sampling within a defined area, the study employed SPSS for analysis. Findings were presented through descriptive statistics. Faisal and Dhusia's (2021) study encompassed diverse respondents, including hotel managers and drivers. Purposive Sampling and reference management via Mendeley added depth to the research. Das and Chowdhury (2020) examined the impact of COVID-19 lockdowns in densely populated areas of North Bengal. Phone-based surveys were conducted, covering varied occupational backgrounds. The research spotlighted the economic hardships of daily wage earners and their vulnerability to work scarcity. In another vein, Mondal et al. (2021) encompassed a holistic investigation into West Bengal's sectors during the initial COVID-19 wave. Incorporating geospatial technology, the study mapped and analyzed pandemic effects, paving the way for informed planning and preparedness. Sandhya (2021) evaluated India's travel and tourism industry's resilience, drawing from past pandemics and COVID-19 research. The study incorporated practical data collection and robust analyses to illuminate sectoral challenges and present recommendations for post-pandemic revival. Das (2020) focused on the COVID-19 impact on West Bengal's tourism sector and analyzing tourist inflow trends. The research underscored the decline in both international and domestic tourists. Utilizing an online platform, surveys were disseminated across different districts in West Bengal known for their major tourist attractions. The collected data facilitated district-wise analysis through tabulation. In contrast, this dissertation delves into the Darjeeling district, spotlighting the pandemic's economic and social ramifications on individuals within the tourism sector. The investigation extensively employs field surveys in the Darjeeling District.

## 15.4 Methodology

The current study encompasses a comprehensive approach to data collection, incorporating both primary and secondary sources. The primary data was gathered through questionnaire surveys, employing purposive and random sampling techniques among hotel managers and drivers in the Darjeeling district. The sample size of 83 included both local residents and migrants seeking employment opportunities. The primary surveys were conducted during the post-COVID-19 period, spanning September to October. The questionnaire featured 22 questions, addressing areas such as general information, employment and economic details, social context, and the COVID-19 situation. To augment the primary data, in-depth interviews were conducted with respondents, enriching the dataset. During the field survey in the Darjeeling district, additional background information was gleaned to provide a holistic perspective.

Complementing the primary data, an array of secondary sources was harnessed, encompassing published articles, reports, research papers, news articles, newspapers, books, and pertinent documents (Fig. 15.1). The District Census Handbook (2011) of the Darjeeling district was a valuable resource for extracting demographic insights about the study area. To manage and analyze the collected data, an assortment of tools and techniques was employed. Digitalization of questionnaire responses was executed using platforms such as Google Forms and Google Docs. ArcMap facilitated the creation of various maps, while MS Excel played a pivotal role in data compilation, analysis, and presentation. Within MS Excel, diverse analytical techniques like calculating averages, percentages, tabulating information, and identifying and rectifying errors were applied. Additionally, MS Excel's functionalities were harnessed to generate graphs, charts, and tables, visually representing the findings of the study. This multifaceted toolkit ensured a comprehensive and detailed analysis of the collected data, fostering a nuanced understanding of the research area.

## 15.5 The Study Area: Darjeeling as a Tourist Centre

Situated in the northernmost region of West Bengal, India, Darjeeling District graces the eastern foothills of the majestic Himalayas. Encompassing 12 distinct Blocks (Fig. 15.2), this district spans a total area of 3149 $km^2$. Renowned for its picturesque hill station and the prized Darjeeling tea, the district finds its administrative centre in Darjeeling town. Geographically, Darjeeling District unveils a dual landscape, characterized by its hills and plains. These encompass the hill subdivisions of Darjeeling,

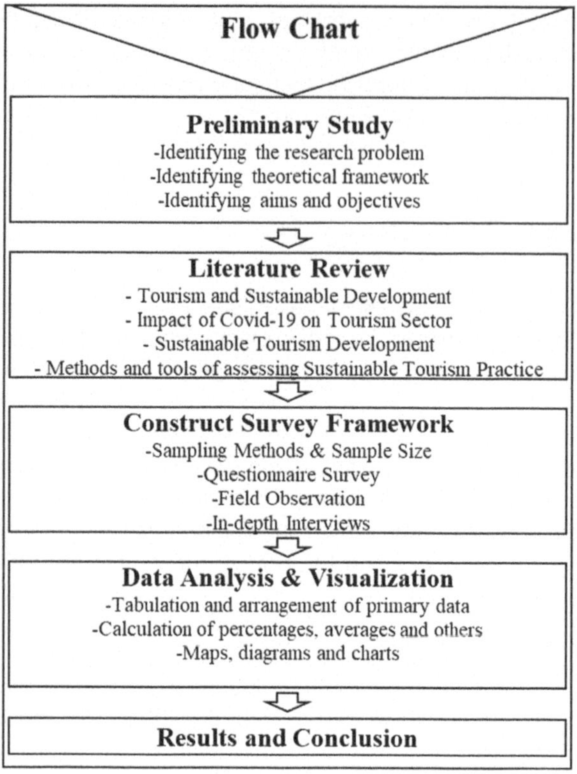


**Fig. 15.1** Methodological framework

Kurseong, and Mirik, as well as the Kalimpong district. The foothills of the Darjeeling Himalayas extend into the Siliguri subdivision, denoted as the Terai.

Bordered by Sikkim to the north, the district shares its southern boundary with Bihar's Kishanganj district. To the southeast, it is contiguous with Bangladesh's Panchagarh district. The eastern frontiers are marked by Kalimpong and Jalpaiguri districts, while the western periphery is shared with Nepal's easternmost Province No. 1. Stretching 18 miles (29 km) from north to south and 16 miles (26 km) from east to west, Darjeeling district occupies a unique geographic footprint. In the 2011 census, it emerged as West Bengal's second least populous district, following Dakshin Dinajpur.

The primary allure of the district revolves around the enchanting town of Darjeeling, often written as 'Darjiling,' renowned for its tranquil natural vistas and a vibrant blend of cultural and ethnic diversity that entices travellers from distant corners. Nestled between the Mechi and Teesta rivers, Darjeeling finds its idyllic spot in the heart of the Eastern Himalayas. Positioned within the Sadar subdivision, the town is perched at an average elevation of 2045 m (6709 ft) on the Darjeeling–Jalapahar range. This distinctive range, extending from south to north and originating

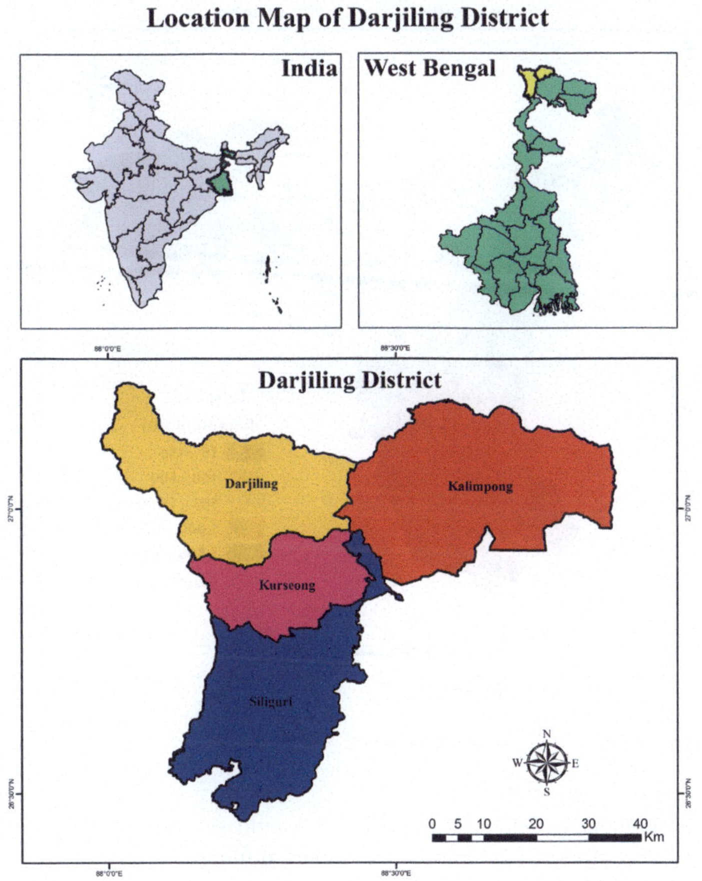


**Fig. 15.2** The Darjeeling district, study area. *Source Authors*

at Ghum, forms a Y-shape with its foundation established at Katapahar and Jalapahar. Two arms branch out beyond the Observatory Hill to the North (Fig. 15.3).

The northeastern arm rapidly descends, culminating in the Lebong spur, while the northwestern arm offers a gentler slope, passing through North Point before descending into the valley near the Tukvar Tea Estate (Fig. 15.4). Dominating the horizon is the awe-inspiring Kangchenjunga, the world's third-highest peak, soaring to an impressive 8,598 m (28,209 ft). Situated 74.4 km to the north, this majestic mountain stands as the prominent focal point of the panoramic vista.

The Darjeeling district boasts a temperate subtropical highland climate, shaped by its close proximity to the Bay of Bengal. This geographical positioning results in a consistently humid atmosphere, rendering Darjeeling with a perpetual spring-like climate for much of the year. While a brief monsoon season and a brisk winter interlude punctuate this pattern, the district aligns with the attributes of an oceanic

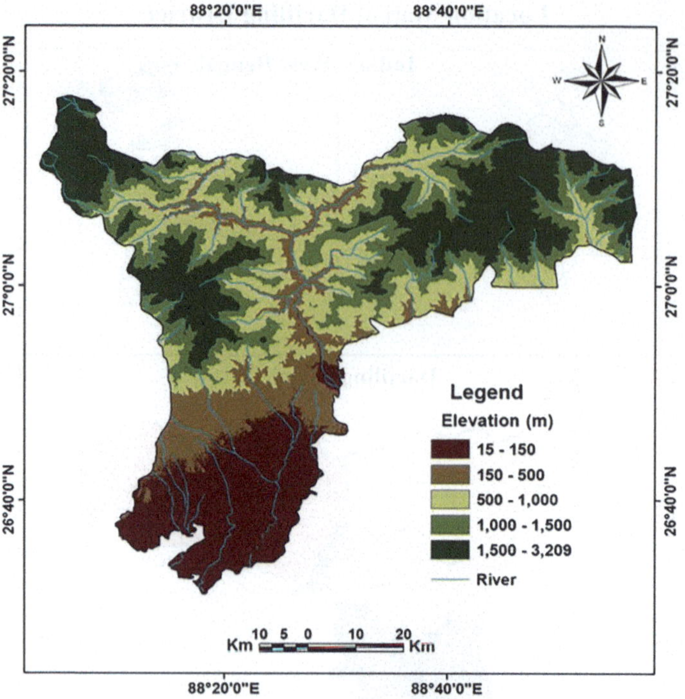


**Fig. 15.3** Physiography of the Darjeeling district. *Source* Authors

climate. Consequently, Darjeeling experiences mild summers and cooler winters, occasionally accompanied by snowfall at higher altitudes.

The soil composition within the Darjeeling district emerges as a product of both fluvial processes and the disintegration of lithological elements. The Kalimpong region predominantly features reddish soils, occasionally interspersed with darker variants attributed to the prevalence of phyllites and schists. The highland expanses spanning west to east are primarily characterized by mixed sandy loam and loamy soils, whereas the southern slopes of Mirik and Kurseong are characterized by clayey loam soils with a reddish hue. Sandy soils, on the other hand, predominate to the east of the Teesta River. Across these diverse soil types, acidity prevails (pH 3.5–5.0), with a slight increase in depth indicating a deficiency of surface bases and their accumulation in lower soil layers. The process of lateritic weathering significantly contributes to the transformation of the underlying substratum.

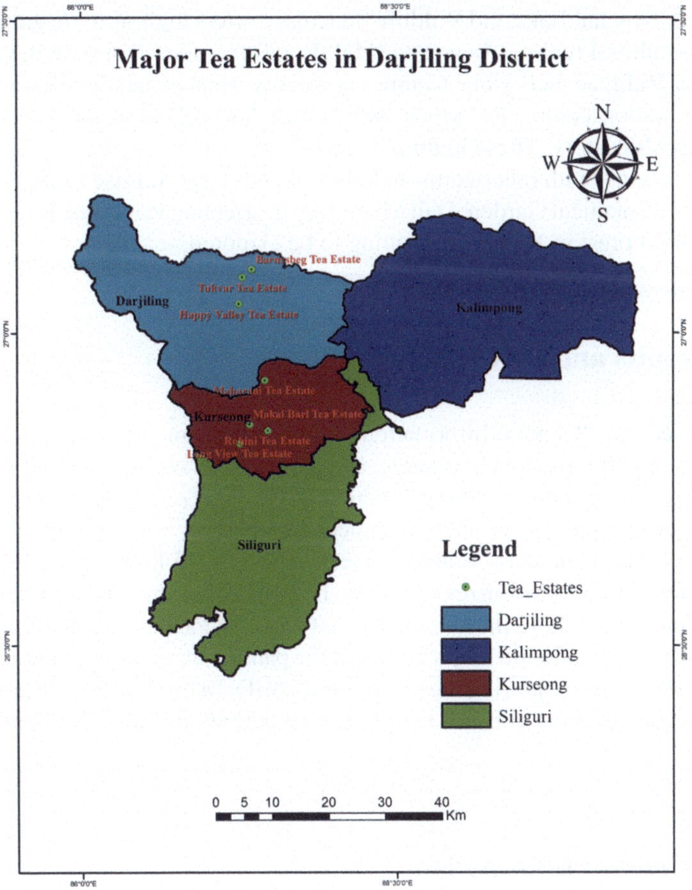


**Fig. 15.4** Major tea estates of the Darjeeling district. *Source* Field survey

## 15.6 Tourist Spots in Darjeeling

The district boasts an array of captivating tourist spots that cater to various interests. Among its historic and heritage attractions, Batasia Loop stands out with its picturesque spiral railway tracks and meticulously adorned garden, while the Rock Garden offers a man-made rocky haven adorned with waterfalls and bridges. The Himalayan Railway Station, a UNESCO world heritage site, enchants visitors with its toy train journey (Anonymous 2023). The Bengal National History Museum showcases indigenous flora and fauna. For natural and adventurous experiences, Tiger Hill provides mesmerizing sunrise views, while Padmaja Naidu Zoological Park harbours a diverse array of flora and fauna, including a significant snow leopard population. The Singalila National Park impresses with its altitude and panoramic views, while Senchal Lake and Wildlife Sanctuary offers high-altitude animal sightings. In the cultural realm, Chowrasta (Mall Road) provides shopping delights, and the Tibetan Refugee Self-Help Centre showcases Tibetan handicrafts and serves as a rehabilitation centre. Religious inclinations find solace at the Peace Pagoda and Ghoom Monastery. These highlights merely scratch the surface of Darjeeling's myriad attractions, with other gems including Jhandi Dara Sunrise Point, Surya Sen Park, Lloyd's Botanical Garden, Dali Monastery, Darjeeling Ropeway, Eagle's Craig, and Downhill Forest in Kurseong, waiting to be explored.

## 15.7 Results and Discussions

The initial section, "General Information" captures fundamental respondent details, encompassing particulars such as name, age, gender, and occupation. Subsequently, the "Employment-Economic Status" section delves into respondents' work-related and financial circumstances, including occupation type, work duration, monthly income, and supplementary sources of earnings. The following segment, titled "Social Information" scrutinizes elements like migration history and educational achievements. Concluding the survey, the "COVID-19 Pandemic Situation" section portrays the socio-economic landscape amid the pandemic, shedding light on respondents' conditions and exploring the impact of COVID-19 on their lives. Furthermore, it offers insights into the ebb and flow of tourist activity during this unprecedented period.

### *15.7.1 General Information*

In relation to the participants' age distribution, as evidenced by the field survey encompassing 83 individuals, the largest segment falls within the age range of 32–37 years (Fig. 15.5). Among the total respondents, 29 individuals (accounting for 34.94% of the total) belong to this specific age group, indicating a significant representation of middle-aged participants.

Within this cohort, there are 12 Drivers and 17 Hotel Managers, with a distribution of 8 females and 21 males, reflecting the highest figures for both occupation and gender groups. In contrast, the lowest level of participation was observed in the age bracket of 45–50, with a mere 5 individuals (equating to 6.02% of the total) taking part. This data underscores a notable inclination towards greater engagement in the tourism industry among the younger demographic, in contrast to the older segment.

Regarding the gender composition of the participants, within the group of 83 individuals, 15 respondents identified as female, while 68 identified as male (Fig. 15.6) This data illustrates that females constitute a modest proportion of 18.07% in the tourism industry, whereas males dominate at 81.93%. This gender disparity can be attributed to the industry's limitation, confining females primarily to the Hotel sector,

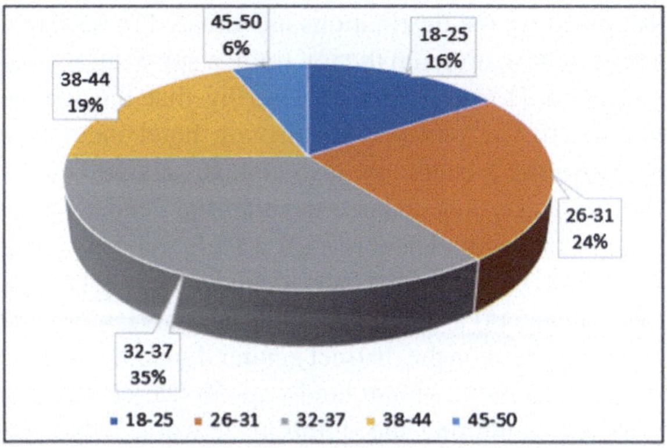


**Fig. 15.5** Age structure of the surveyed respondents. *Source* Primary survey

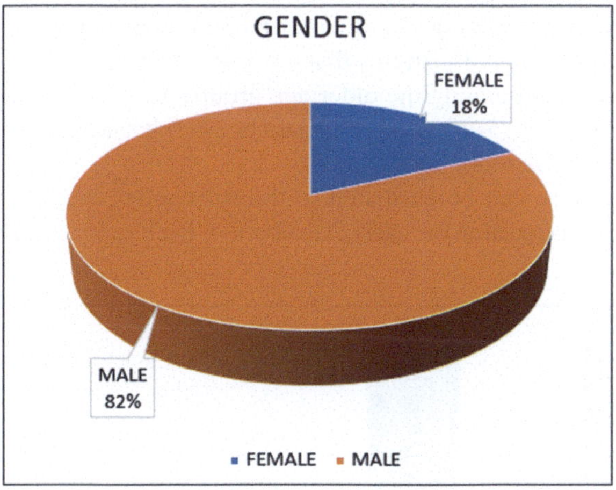


**Fig. 15.6** Gender composition of the surveyed respondents. *Source* Primary survey

and the absence of female drivers due to both challenging terrains and social factors. Meanwhile, among the male participants, 29 are Hotel Managers, while the remaining 39 are Drivers, further delineating the distribution of roles within the industry. As previously discussed, the demographic distribution and participation trends reveal insights into the industry's dynamics.

### *15.7.2 Employment and Economic Information*

Concerning occupational distribution, the survey reveals a dichotomy between two primary roles: Drivers and Hotel Managers. Among the 83 participants, 39 individuals are employed as drivers, while 44 are working as Hotel Managers. This bifurcation is a direct consequence of the consistent influx of tourists. Given that many prominent tourist destinations are situated in remote areas without local transportation options apart from the toy train, a substantial number of locals find employment as drivers. However, it's noteworthy that this profession is predominantly male-dominated. Conversely, the thriving hotel industry, which serves as the linchpin of Darjeeling's economy, also engages a significant portion of the local populace. The survey further underscores that the gender ratio among participants employed as hotel managers leans toward a male majority, with a ratio of approximately 2:1 (Fig. 15.7). Out of the 44 individuals in the hotel management role, 15 (34.09% of all hotel managers) are female, while 29 (65.91% of all hotel managers) are male. This data sheds light on the distinct gender dynamics within these professions, showcasing the nuanced employment landscape in Darjeeling's tourism industry.

When examining the duration of work, it becomes apparent that the category with the lowest representation among the 83 participants is those with 16–20 years of experience (Table 15.1). This group comprises only 5 individuals (6.02% of the total), consisting of 2 drivers and 3 hotel managers.

This data illustrates that the tourism industry in the Darjeeling district is not as prevalent among the older age groups. Conversely, the majority of participants fall within the 5–10 years of experience bracket, accounting for 37 out of 83 participants (44.58% of the total).

This can be attributed to the substantial increase in tourist inflow the industry experienced in the early 2015 period, leading to a surge in employment opportunities

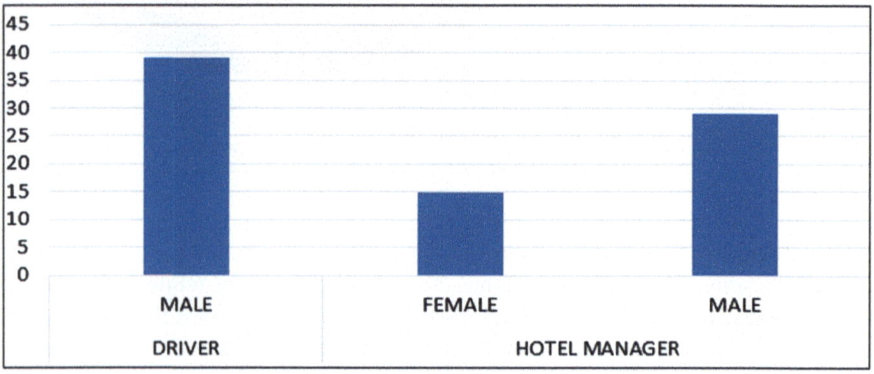


**Fig. 15.7** Occupational structure of the surveyed respondents. *Source* Primary survey

**Table 15.1** Duration of work of respondents

| Duration of work | No. of respondents |
|---|---|
| Less than 5 years | 25 |
| 5–10 years | 37 |
| 11–15 years | 16 |
| 16–20 years | 5 |

*Source* Primary survey

for younger individuals within this sector. The impact of the COVID-19 pandemic on the tourism industry in the Darjeeling district was substantial, resulting in a significant downturn in activities during the lockdown phase. This effect is evident in this study, wherein out of the total 83 participants, a striking 61 individuals (73.5% of the total) reported earning a monthly income of less than 5000 rupees (Fig. 15.8).

Notably, the highest income bracket observed was within the range of 7000–9000 rupees, a category that only 7 out of the 83 participants (8.43% of the total) fell into. This stark disparity in income levels can be attributed to the fact that those engaged in the local tourism industry heavily rely on their jobs as their primary source of sustenance, leaving them vulnerable to the industry's volatility and without alternative passive income sources.

Within the cohort of 83 participants, a substantial portion of 63 individuals (75.9% of the total) heavily rely on additional sources of income beyond their primary roles as drivers or hotel managers to sustain their families (Table 15.2).

These supplementary avenues include support from government schemes, pension or income from spouses, assistance from family members, engagement in other businesses, and various other resources. The survey results indicate that a majority of the respondents, accounting for 56.6%, do not engage in advertising their vehicles or hotels, while the remaining 43.4% utilize social media platforms such as MakeMyTrip, EaseMyTrip, and redBus for advertisement purposes (Fig. 15.9). It is noteworthy that a significant portion of those who opt for advertising are hotel

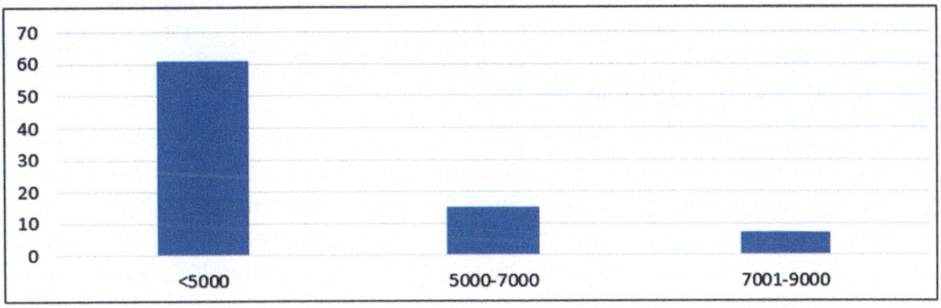


**Fig. 15.8** Income levels during COVID-19 of the surveyed respondents. *Source* Primary survey

**Table 15.2** Other income sources of the respondents

| Resources to support family members | No. of respondents | Percent |
|---|---|---|
| • Support from the government scheme | 2 | 3.2 |
| • Pension/income of spouse | 32 | 50.8 |
| • Support from family members | 27 | 42.9 |
| • Other business | 20 | 31.7 |
| • Other resources | 6 | 9.5 |

*Source* Primary survey

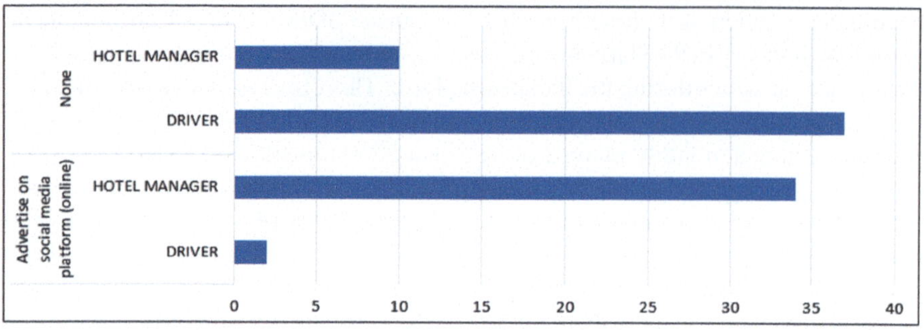


**Fig. 15.9** Number of respondents and the mode of advertisement. *Source*: Primary survey

managers. This discrepancy can be attributed to the fact that many drivers are affiliated with the Darjeeling Cab Association, adhering to their assigned junctions for clientele. Interestingly, none of the respondents employ hoarding boards as a means of advertisement.

### 15.7.3 Social Information

The survey reveals diverse migration patterns among the 83 participants. Around 50.6% (42 respondents) are native to the Darjeeling region, while 49.4% (41 respondents) migrated to the Darjeeling district for reasons such as marriage, education, employment, and calamity (Table 15.3).

The migrating respondents elaborated on their motivations, with Employment and Marriage emerging as the primary drivers. Many individuals seek better employment opportunities due to limited earnings in their current positions. Marriage prompts migration, particularly for women relocating to their marital households. Educational pursuits and escaping regional calamities are additional rationales for migration. Education significantly influences both social advancement and occupational competence. Among the 83 participants, the majority possess basic compulsory education, ranging from 8 to 10th standard. Interestingly, none of the participating drivers and only five out of the 44 hotel managers are graduates. This observation underscores

**Table 15.3** Migration status of the respondents

| Reasons for migration | No. of respondents |
|---|---|
| Marriage | 13 |
| Education | 7 |
| Employment | 34 |
| Calamity | 3 |

*Source* Primary survey

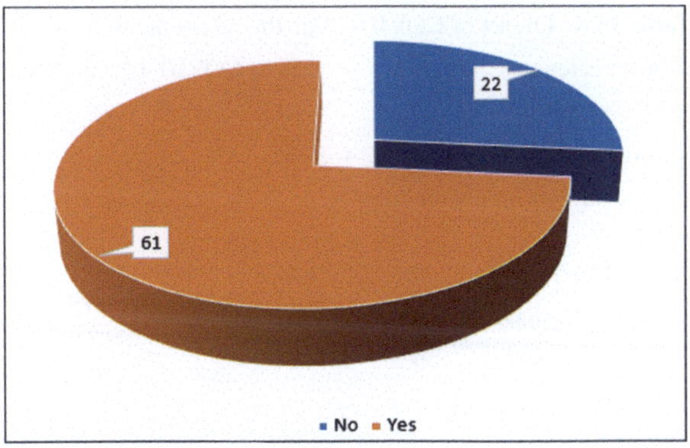


**Fig. 15.10** Transmission of COVID-19 among respondents. *Source* Primary survey

potential deficiencies in higher education accessibility and underscores prevailing literacy challenges.

## 15.7.4 *Situation During COVID-19 Pandemic*

Amid the COVID-19 pandemic, many individuals faced severe economic setbacks, leading to widespread job loss. The survey reveals that a substantial 79.5% of respondents were rendered jobless, while the remaining 20.5% were employed but with significantly reduced salaries.

### 15.7.4.1 Transmission of COVID-19 Among Respondents

The COVID-19 virus had a profound impact on the entire population, resulting in a sharp increase in the country's death toll. The survey highlights that out of the 83 respondents, 61 were infected by the COVID-19 virus, while the remaining 22 remained uninfected (Fig. 15.10).

### 15.7.4.2 Wave-Wise Impact of COVID-19

The survey underscores the dual impact of both waves of the pandemic on respondents. Among the 83 participants, a significant 60.2% were adversely affected by both waves, 31.3% were affected solely by the first wave (Table 15.4), and 8.4% were solely affected by the second wave. The majority experienced detrimental effects such as loss of savings, employment, income reduction, loss of family members, and changes in employer behaviour, which significantly impacted their economic stability.

**Table 15.4** Impact of COVID-19 on the respondents

| Impact category | First COVID-19 wave (%) | Second COVID-19 wave (%) |
|---|---|---|
| Loss of jobs | 64 | 77.1 |
| Decrease in income | 76 | 91.6 |
| Loss of life of family members/ relatives | 41 | 49.4 |
| Loss of savings | 77 | 92.8 |
| Change in behaviour of employer | 7 | 8.4 |
| Other reasons | 18 | 21.7 |

*Source* Primary survey

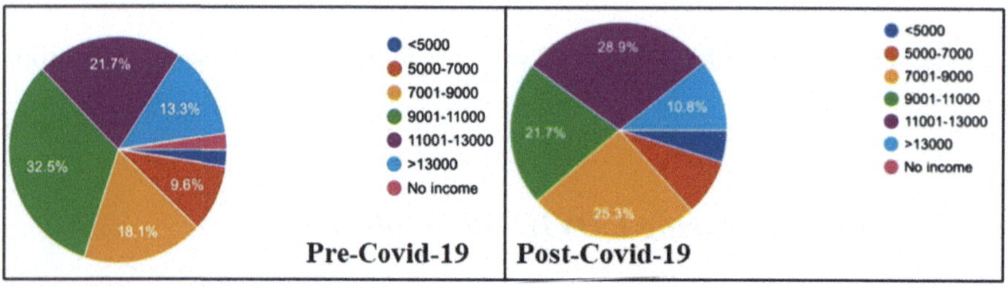


**Fig. 15.11** Respondents' income during pre- and post-COVID-19. *Source* Primary survey

#### 15.7.4.3 Income Status: Pre- and Post-COVID-19 Pandemic

Before the pandemic, the tourism industry in the Darjeeling district flourished, contributing to respondents' higher incomes as drivers and hotel managers (Fig. 15.11). However, the pandemic-induced lockdown brought tourist activities to a standstill, leading to a significant decrease in income. This study reveals the income disparity between pre and post-pandemic periods for drivers and hotel managers, attributing the decline to the absence of tourist arrivals.

#### 15.7.4.4 Price of Transportation Fare/Hotel After COVID-19 Pandemic

Following the income decline, 75 out of 83 respondents increased transportation fares or hotel room rates to compensate for pandemic-induced losses. In contrast, 8 respondents maintained their rates, leading to savings loss due to reduced income (Table 15.5).

**Table 15.5** Transportation and hotel prices after COVID-19

| Whether increased the price of transportation fare/hotel after the pandemic? | No. of respondents | % |
|---|---|---|
| YES | 75 | 90.4 |
| NO | 8 | 9.6 |

*Source* Primary survey

**Table 15.6** Precautionary measures for COVID-19 transmission

| Do the respondents take precautionary measures to control COVID-19 spread? | No. of number of respondents | % |
|---|---|---|
| YES | 80 | 96.4 |
| NO | 3 | 3.6 |

*Source* Primary survey

#### 15.7.4.5 Precautionary Measures for COVID-19 Transmission

Amid government recommendations for COVID-19 prevention, 80 out of 83 respondents adhered to precautionary measures in hotels and vehicles to mitigate virus transmission (Table 15.6). Only 3 respondents reported not taking any precautionary measures.

#### 15.7.4.6 Monthly Inflow of Tourists

Darjeeling's allure lies in its Himalayan landscapes and scenic beauty, attracting tourists throughout the year. Survey results indicate that tourists primarily flock to Darjeeling during the winter months from October to March. Notably, the influx peaks during the winter (Table 15.7), offering tourists the chance to witness snow-covered valleys. Conversely, the period between July and September sees fewer visitors due to monsoon-related landslides and heavy rainfall. Local tourists within West Bengal often visit during the colder months, while visitors from other Indian states constitute a significant portion of the total tourists. However, residents of Darjeeling district are less inclined to visit, given their familiarity with the region's climate and attractions.

## 15.8 Conclusions

This comprehensive study sheds light on the various facets of the tourism industry in the Darjeeling district and its intricate relationship with the challenges posed by the COVID-19 pandemic. The findings underscore the significance of adopting sustainable tourism practices to ensure the resilience and long-term viability of the tourism

**Table 15.7** Tourist inflow in Darjeeling

| Month | Count of tourist Inflow | Percentage of tourist inflow |
|---|---|---|
| Jan-March | 82 | 98.8 |
| April-June | 26 | 31.3 |
| July–September | 0 | 0 |
| Oct-December | 77 | 92.8 |

*Source* Primary survey

sector in the face of unprecedented disruptions. The study's exploration of demographic information, employment status, migration patterns, and economic conditions of respondents within the tourism sector vividly portrays the sector's reliance on a diverse workforce, including both drivers and hotel managers. The stark impact of the pandemic on employment, income, and overall economic stability highlights the vulnerability of tourism-dependent communities in times of crisis. The research also brings attention to the gender disparity within the tourism industry, particularly the underrepresentation of females, predominantly in the driver profession. This underscores the importance of inclusive strategies to promote gender equality and equal economic opportunities within the sector. Moreover, the study delves into the dynamic inflow of tourists in Darjeeling, highlighting seasonal trends and regional preferences. Sustainable tourism practices must be employed to strike a balance between attracting visitors, protecting the region's natural beauty, and ensuring the socio-economic well-being of the local population.

In the context of sustainable tourism practices, the study's findings advocate for diversification of income sources among tourism stakeholders. The emphasis on relying on additional resources beyond the primary occupation, coupled with innovative advertising strategies, indicates a step towards economic resilience. The survey's revelation of the disparity in educational qualifications among respondents points towards the need for skill development and educational initiatives to enhance employability and professional growth within the tourism sector. The study also underscores the importance of safeguarding public health and following precautionary measures, as the majority of respondents actively practiced them. This underscores the role of responsible tourism in ensuring the safety of both visitors and the local population. Considering the challenges posed by the pandemic, it is evident that sustainable tourism practices, encompassing economic diversification, gender inclusivity, education enhancement, and ecological conservation, are imperative for the tourism industry's long-term viability and contribution to the socio-economic development of Darjeeling district. By integrating these principles, the region can emerge stronger, more resilient, and better equipped to face future uncertainties while preserving its natural and cultural assets for generations to come.

## References


Anonymous (2023) Darjeeling toy train: an enchanting journey through time. Times of India, August 3

Das GK (2020) Tourism in trouble amid COVID-19 crisis in West Bengal. Frontier 12. Available via Frontier. https://frontierweekly.com/views/jul-20/12-7-20-Tourism%20in%20Trouble%20Amid%20COVID-19%20Crisis%20in%20West%20Bengal.html. Accessed 15 Dec 2023

Das S, Chowdhury S (2020) Impact of COVID-19 lockdown in North Bengal. VIKALP people's perspective for change. https://vikalp.ind.in/2020/06/impact-of-COVID-19-lockdown-in-north-bengal/. Accessed 15 Dec 2023

District Census Handbook (2011) Darjeeling district, West Bengal, Census of India

Faisal M, Dhusia DK (2021) Pandemic's (COVID-19) impact on tourism sector of India. Anais Brasileiros De Estudos Turisticos-Abet 11:14. https://doi.org/10.5281/zenodo.5770282

Ghosh N (2022) COVID-19 pandemic and its influence: case study in North Bengal medical college, Darjeeling district, West Bengal, India. J Pub Health Prim Care 3(6):86–89. https://doi.org/10.4103/jphpc.jphpc_24_20

Jaipuria S, Parida R, Ray P (2021) The impact of COVID-19 on tourism sector in India. Tour Recreat Res 46(2):245–260. https://doi.org/10.1080/02508281.2020.1846971

Mehta A (2021) India and the coronavirus pandemic: economic losses for households engaged in tourism and policies for recovery. National Council of Applied Economic Research, New Delhi

Mondal BK, Sahoo S, Paria P et al (2021) Multi-sectoral impact assessment during the 1st wave of COVID-19 pandemic in West Bengal (India) for sustainable planning and Management. Arab J Geosci 14:2448. https://doi.org/10.1007/s12517-021-08836-z

Sandhya H (2021) The survival of travel and tourism industry amidst the COVID 19 pandemic: challenges and opportunities of the Indian tourism sector. Atna J Tour Stud 16(1):41–66. https://doi.org/10.12727/ajts.25.3

WTTC [World Travel & Tourism Council] (2022) Travel and tourism impact-2022: global trends. https://wttc.org/Portals/0/Documents/Reports/2022/EIR2022-Global%20Trends.pdf. Accessed 15 Dec 2023

# Part IV
# Rural Sustainability and Livelihood Security

# Chapter 16
# Technology Penetration in Domestic Chores and Its Impact on the Sustainability: A Study of Aligarh District (Uttar Pradesh)

**Arti Sharma and Tejbir Singh Rana**

**Abstract** The present chapter establishes the link between technological advancement in cooking and its impact on women's quality of life for their sustainability. For centuries, cooking has been the prime work of women and the kitchen has a dominant place for women. Hence, finding how much technology has penetrated the kitchen is very interesting to study. Extensive primary data was collected from a total number of 400 households from urban and rural areas of the Aligarh district using a structured questionnaire. The study found that there is an alteration in the thinking of family members with each new addition of kitchen appliances in a household. With the advent of each new kitchen appliance, other family members also developed an interest in cooking food. But it also observed that the addition of new cooking appliances is given the least priority in the budget of middle- and lower-income households. In rural–urban areas, the rural area of the district still uses old-age technology, while urban areas are upgrading their technology faster than rural areas.

**Keywords** Sustainable development · Social norms · Technology · Aligarh district

## 16.1 Introduction

Since Independence, the Indian government has been continuously engaged in supporting and promoting science and technology through its various programs and policies. India was also one of the first countries to adopt and execute the Science Policy Resolution in 1958. The resolution states that in the modern age, the key to national prosperity and development lies in the effective combination of three factors:

A. Sharma (✉)
Department of Geography, Delhi School of Economics, University of Delhi, New Delhi, India
e-mail: artisharmaoe@gmail.com

T. S. Rana
Department of Geography, Shivaji College, University of Delhi, New Delhi, India

 
S. Anand et al. (eds.), *Emerging Geosustainability Transformations in India*, Advances in Geographical and Environmental Sciences,
https://doi.org/10.1007/978-981-97-7993-2_16

raw materials, technology and capital, In this the role of technology is widely recognized. The progress in the development of new scientific techniques can make up for a deficiency in natural resources, and reduce the capital demands (DST 2016).

Thus, there has been consistent sponsorship and political provisions for scientific investigation in India. This is also indicated by a huge science and technology institutional structure such as the Department of Science and Technology, Department of Information Technology, Department of Electronics, Indian Space Research Organization (ISRO), and several other institutes. Several programs created under various governments to promote and support research and development in various fields (TDB 2016). However, ironically penetration of science and technology still has not achieved desirable results in Indian society. Especially when it comes to technology penetration for female's comfort in life.

Maybe because India is an agricultural land and primarily patriarchal society. More than half of the population is directly or indirectly dependent on agriculture. In a patriarchal society, women own a very small proportion of the natural resources and often face discrimination compared with their male counterparts, i.e., obtaining land, education, employment, housing and holding on monetary resources (Satterthwaite et al. 1996).

In the Aligarh district of Uttar Pradesh wheat is the dominant crop. In wheat-growing areas of the Indian subcontinent, rural women face restrictions while men have greater opportunities. Consequently, women must struggle more to gain economic competency for survival (Agarwal 2010). Thus, this study tries to analyze the effects of technology on the very first need of human existence, i.e., food and how cooking has been done in a household. Cooking is considered as work of women. Food preparation for the family and society is the primary activity of women; hence this study is trying to find out the development opportunities for women and how their life became easy by the penetration of technology in their kitchen.

## 16.2 Literature Review

Inequality is an important aspect as it provides a detailed understanding of the politics of organizing women at the grassroots level. Further, it also helps to understand women's responses related to poverty, patriarchy, and their demand for survival (Agarwal 1989). Substantial gender gaps significantly affect women's voluntary participation at home and also in the community. This might be because of the gender gap in economic grants, sociocultural norms, people's perceptions, government rules, etc. These inequalities are also responsible for creating tendencies among women toward non-cooperation, or non-voluntary cooperation and non-voluntary non-cooperation at various levels (Agarwal 2007).

The devaluation of domestic work is not only a product of gender norms but also of caste and race. The ideologies of motherhood, and middle-class femininity for free domesticity have prevented the recognition of housework as work and paid labour, given its historical separation from the productive 'work' sites (Davis1981) and in light of women of colour and 'lower' castes consigned to perform it (Chakravarti 1993, 1995; Dickey 2000; Parrenas 2004; Ray 2000; Ray and Qayum 2009; Sangari 1993). According to Singh and Pattanaik (2020), unpaid domestic work is a backbone of productive activities and a big contributor to the well-being of households and the economy of the country. However, the preponderance of women in household work and keeping them out of financial activities put unpaid domestic work as invisible work, outside the production boundaries, and further outside the purview of economic policy.

The sustainable development goals (SDG) recognize and value unpaid care, domestic work through the public services, infrastructure, and social protection policies and the promotion of shared responsibility within the household and the family as nationally appropriate (Target 5.4) (UN 2023). However, much of women's unpaid work is unrecognized, unreported, and underestimated by the family, the market, and the state regardless of the benefits relished by these provisions. Furthermore, neoliberal macroeconomic policies have been formulated without recognizing, reducing and redistributing unpaid work in the economy. These new-liberal macroeconomic policies not only to increasing the penalty of women but also to harming economic development (Hirway 2005; Elson 2017; Folbre and Yoon 2008).

This has been found by Komatsu et al. (2011) indicating that households with modern technology, i.e., Solar Home Systems (SHS) successfully reduce their consumption of kerosene and dependency on rechargeable batteries, with the cost reductions accounting for 20–30% of monthly expenditures on Solar Home Systems. From all these literature reviews, it has been found that so many studies are recognizing the devaluation of women's unpaid domestic work. But very few studies regarding, how technological penetration can improve or reduce this unpaid domestic work. Hence, it is directly related to women's development and their sustainability in the world.

## 16.3 Research Questions

Based on these arguments research question has been framed for this study. In this study we have investigated that what are the effects of technological advancement on the rigidity of social norms for women. Do these norms become stronger, more liberal, or disappear altogether?

## 16.4 Data Sources and Research Methodology

Primary and secondary data sources have been utilized to achieve the objectives. The secondary data sources are; the Population Census of India, books, research papers, internet websites, etc. The primary data has been collected with the help of both quantitative and qualitative research techniques. In these methods, personal interviews, case studies, and observation methods through a participatory approach have been followed. For interviewing respondents, a detailed questionnaire has been prepared and this questionnaire has questions related to the use of home appliances. Case study and observation methods have been used for in-depth inquiry and it is supplementing the personal interviews and understanding of primary data at the stage of data analysis.

For the beginning of the field study, a pilot visit to the study area was done, marking the very first step in primary data collection. This pilot visit helped in knowing, the general structure of the society, the type of social groups dwelling in the district, who can be the key respondents during primary data collection, etc. A short report has been prepared based on information gathered from the pilot visit to the study area. Accordingly, the structure of the questionnaire and types of questions have been prepared and taken final shape for the interview. In the next stage, the finalization of the sample size for primary data collection has been done. The finalization of the sample size for primary data collection has been done based on the 2011 census data of the total population of Aligarh District. At a 5% confidence level, a total of 400 samples/persons has been fixed.

These 400 samples were further divided according to the proportion of the rural, urban, male, and female population in the total population of Aligarh district in the 2011 census. Samples have been distributed through stratified random sampling and cover both rural and urban areas of Aligarh district. A society consists of various economic classes, religions, rural–urban populations, genders, etc. In stratified random sampling, the population is divided into several groups/layers and sub-groups/layers that are individually more homogeneous than the total population of the study area. Subsequently, samples from each stratum according to the objectives of the study have been selected. Table 16.1 shows showing rural, urban, male, and female population and their percentage in the corresponding total population of the Aligarh district.

Based on the above proportion of the population, a total of 400 samples size is divided into strata of rural male, rural female, urban male, and urban female (Fig. 16.1). Out of 400 sampled households, 267 samples are from rural households and 133 samples are from urban households. Out of 267 rural households, 143 are rural male households (the population of rural males in Aligarh District in the 2011 census was 13,08,923) and 124 rural female households (the population of rural females in Aligarh District in the 2011 census is 11,47,775). Similarly, out of 133 urban households, 71 are urban male households (the population of urban males in

**Table 16.1** Total population and male–female population in Aligarh district

| Area | Total population | (%) | Male population | (%) | Female population | (%) |
|---|---|---|---|---|---|---|
| Rural | 24,56,698 | 66.87% | 13,08,923 | 67.06% | 11,47,775 | 66.66 |
| Urban | 12,17,191 | 33.13% | 6,43,073 | 32.94% | 5,74,118 | 33.34 |
| Total | 36,73,889 | 100% | 19,51,996 | 100% | 17,21,893 | 100 |

*Source* DCHB (2011)

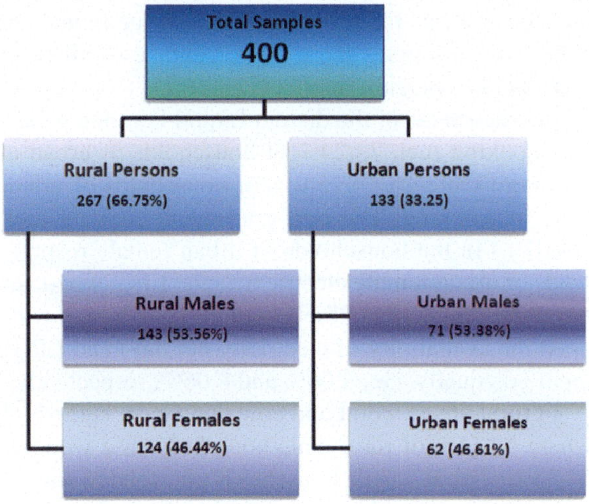


**Fig. 16.1** Flowchart representing the division of samples/sample design. *Source* Authors

Aligarh District in the 2011 census is 6,43,073) and 62 are urban female households (the population of rural females in Aligarh District in the 2011 census is 5,74,118).

Care had been made to interview the head of the family but when the head of the family was not available for an interview, preference had been given to another adult person in the household. From the flowchart, one can easily understand the overall base of division for surveyed households.

## 16.5 Results and Discussion

Available and popular kitchen appliances in the district have been studied for this purpose, for example, types of cooking fuel, mixer and grinder, refrigerator, and microwave oven. Each of these technological advancements has been discussed below.

### *16.5.1 Types of Cooking Fuel*

The types of cooking fuel are the first indicator of the monetary status of a household. Food is the first need of humans and every household whether poor or rich has invested a sizable portion of their income in food. Mostly in the district, as the economic condition improves, families prefer to invest in efficient cooking fuel for their households. Efficient cooking fuel can cook food hassle-free and fast and is an add-onto the quality of life. It has been noticed that female-headed households tend to invest more in kitchen appliances while male-headed households are more inclined towards the up-gradation of occupational and transportation technologies. Table 16.2 discusses the total percentage of all types of cooking fuel used in the household of Aligarh district (Fig. 16.2).

In urban areas of the district Liquid Petroleum Gas (LPG) is the highest consumable cooking fuel. 73.24% of households of urban male respondents use LPG for cooking (Table 16.2). 75.81%of households of urban female respondents also use LPG for cooking. The next convenient cooking fuel is wood and cow dung cake (14.07%) in the households of urban female respondents. However, in the households of urban female electric heaters, LPG, wood, and cow dung cake are found to be convenient cooking fuels both in combination with an electric heater (depending upon the availability of electricity per day) and LPG or wood & cow dung cake. It reported equally, i.e., 8.06% and 8.06%, respectively.

In rural areas, wood and cow dung cake is the highest consumable cooking fuel where 68.53% of rural male households use wood & cow dung cake for cooking. 54.83% of rural female households use wood & cow dung cake for cooking. The next maximum consumable cooking fuel is a combination of wood, cow dung cake, and LPG in the rural areas of the district. 20.88%of rural male respondents and 26.61%of rural female respondents reported that they use these 3 cooking fuels (LPG, wood, and cow dung cake) for cooking. These rural respondents said, they use wood and

**Table 16.2** Respondents profile using different types of cooking fuels

| Types of cooking fuels | Male | | | | Female | | | |
|---|---|---|---|---|---|---|---|---|
| | Urban male | | Rural male | | Urban female | | Rural female | |
| | Number | % | Number | % | Number | % | Number | % |
| Wood | [a]– | – | – | – | – | – | – | – |
| Cow-dung cake | 01 | 1.41 | – | – | – | – | – | – |
| Kerosene oil | – | – | – | – | 03 | 4.84 | – | – |
| Electric heater | 02 | 2.82 | – | – | 02 | 3.23 | 01 | 0.81 |
| LPG | 52 | 73.24 | 14 | 9.79 | 47 | 75.81 | 19 | 15.32 |
| LPG, Wood and Cow-dung cake | 03 | 4.23 | 30 | 20.98 | – | – | 33 | 26.61 |
| Electric heater and LPG | 02 | 2.82 | – | – | 05 | 8.06 | 01 | 0.81 |
| Wood and Cow-dung cake | 10 | 14.07 | 98 | 68.53 | 05 | 8.06 | 68 | 54.83 |
| LPG, Wood, Cow-dung cake, Electric heater | – | – | 01 | 0.7 | – | – | 01 | 0.81 |
| LPG and Cow-dung cake | 01 | 1.41 | – | – | – | – | – | – |
| Wood, Cow-dung cake and electric heater | – | – | – | – | – | – | 01 | 0.81 |
| **Total** | **71** | **100** | **143** | **100** | **62** | **100** | **124** | **100** |

*Source* Primary survey 2013–16

[a]No respondent falls in this category for the corresponding column

**Fig. 16.2** Different types of cooking fuels used in households of respondents. *Source* Authors

cow dung cake for cooking, and the use of LPG cylinders is restricted only for making tea, snacks, or any other light/small portion cooking recipes. The reason is burning wood and cow dung cakes take substantial time to emit heat. Thus, investment of time gives good output for cooking 2–3 recipes, breads, or a full meal for the family. For light cooking like tea, hot milk/water, *khichadi,* etc. liquid petroleum gas (LPG) is preferred by females of the family. Because heat generation by LPG is fast, highly efficient, and smoke-free.

The reasons for the maximum use of wood and cow-dung cakes as cooking fuel in rural areas are: first, the economic condition of the rural families is not good enough to pay for cooking fuel. Second, electricity supply is not regular in the rural areas of the district. This restricts all time use of the electric heater. Thirdly, some rural people say that the breads are easily digestible when cooked in the heat of burning wood and cow dung. Fourth, cheap and easy availability of cow-dung cakes in rural areas. As most of the households are engaged in animal husbandry. Consequently, cow dung cakes by default freely available and can be used as cooking fuel.

The scenario in the urban areas of the district is completely different. Urban areas experience a high density of population. The compact settlement gives no space for grass and other animal fodder to grow, resulting in no free supply of dry wood and cow dung cakes. Thus, LPG is the most popular cooking fuel in urban areas. The use of efficient and comfortable cooking fuel has an impact on quality of life. Wood and cow dung cakes are difficult to burn, and difficult to control heat as various stages of cooking, emit smoke. All these make the whole cooking process burdensome for a person. It can also cause weakening of eye-sight, burn, and blackening of cooking utensils and nearby walls of the kitchen.

### *16.5.2 Mixer and Grinder*

Mixer and Grinder are the next basic equipment used in the kitchen nowadays. Primary data collected during the field survey found that the use of a mixer and grinder is minimal in the rural areas of the district. In the primary data, only 11.19% of rural male respondents reported the use of a mixer & grinder in their kitchen. Only 15.32% of rural female respondents said utilization of mixer and grinder in their cooking. This is a very low percentage in rural areas. Table 16.3 shows primary data regarding the availability of mixers and grinders in the households of the district.

The households in urban areas reported more than 50% use of mixers and grinders in their kitchens. Figure 16.3 shows the availability of a mixer and grinder in the households of urban male, rural male, urban female, and rural female respondents. 57.75% of urban male respondents replied that they own a mixer grinder and only 42.25% said that the nonavailability of equipment in their kitchen. Urban female respondents have a better position than urban male respondents. 69.35% of urban female respondents replied yes to owning this technology in their kitchens and only 30.65% reported not owning a mixer and grinder.

Most rural females are illiterate. They find themselves inefficient in reading the user manual provided with the mixer. Hence, their handling of the mixer is not proper.

**Table 16.3** Respondents owning mixer and grinder

| | Male | | | | Female | | | |
|---|---|---|---|---|---|---|---|---|
| | Urban Male | | Rural Male | | Urban Female | | Rural Female | |
| | Number | % | Number | % | Number | % | Number | % |
| Yes | 41 | 57.75 | 16 | 11.19 | 43 | 69.35 | 19 | 15.32 |
| No | 30 | 42.25 | 127 | 88.81 | 19 | 30.65 | 105 | 84.68 |
| Total | 71 | 100 | 143 | 100 | 62 | 100 | 124 | 100 |
| | Total Urban | | Total Rural | | Total Male | | Total Female | |
| | Number | % | Number | % | Number | % | Number | % |
| Yes | 84 | 63.16 | 35 | 13.11 | 57 | 26.64 | 62 | 33.33 |
| No | 49 | 36.84 | 232 | 86.89 | 157 | 73.36 | 124 | 66.67 |

*Source* Primary survey, 2013–16

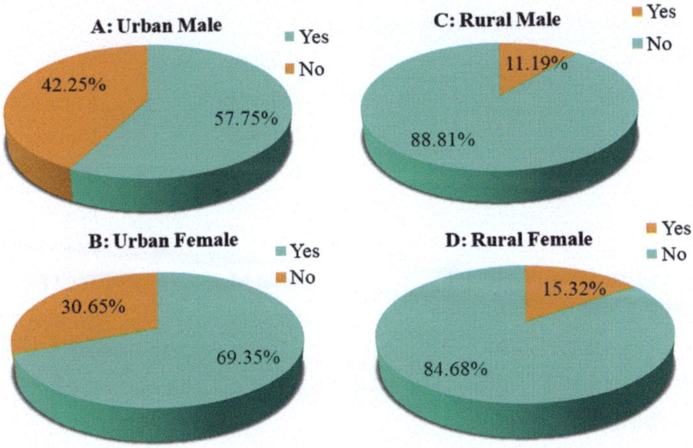


**Fig. 16.3** Respondents owning mixers and grinders. *Source* Primary survey 2013–16

Mostly rural females learn it from other females/males. Poverty and no regular supply of electricity are the other causes of the non-availability of this technology in rural areas.

## *16.5.3 Microwaves Oven*

A microwave oven is another kitchen equipment for fast cooking. But microwaves oven is not so popular kitchen appliance in the district. Most of the North Indian recipes do not require baking so the use of microwave ovens becomes very less for cooking of food. Hence, only affluent families in the district purchase it. Table 16.4 shows primary data for microwave ovens. In urban areas, few families have possession of it while in rural areas the use of microwave ovens is almost nil.

In the field survey, it was found only 8.45% of households of urban male respondents and 11.29% of households of urban female respondents owned microwave ovens (detailed data of primary survey given in Table 16.4. In rural areas, only 0.7% of households in rural male respondents owned microwave ovens. In rural female households, no one had been found with a microwave oven in their kitchen. Most households reported no use of microwave ovens. 91.55% of households of urban male respondents, 99.3% of households of rural male respondents, 88.71% of households of urban female respondents, and 100% of households of rural female respondents reported no use of microwave ovens in their kitchens (Fig. 16.4).

The impact of this technology could not be identified, as firstly it has no penetration in the rural households of the district. Secondly, the households in the urban area who own it do not use ovens very often. Thus, those urban respondents did not notice any substantial effects of microwave ovens on their lifestyle.

**Table 16.4** Respondents owning microwave ovens

| | Male | | | | Female | | | |
|---|---|---|---|---|---|---|---|---|
| | Urban Male | | Rural Male | | Urban Female | | Rural Female | |
| | Number | % | Number | % | Number | % | Number | % |
| Yes | 06 | 8.45 | 01 | 0.7 | 07 | 11.29 | 0 | 0 |
| No | 65 | 91.55 | 142 | 99.3 | 55 | 88.71 | 124 | 100 |
| Total | 71 | 100 | 143 | 100 | 62 | 100 | 124 | 100 |
| | Total Urban | | Total Rural | | Total Male | | Total Female | |
| | Number | % | Number | % | Number | % | Number | % |
| Yes | 13 | 9.77 | 01 | 0.37 | 07 | 3.27 | 07 | 3.76 |
| No | 120 | 90.23 | 266 | 99.63 | 207 | 96.73 | 179 | 96.23 |

*Source* Primary survey, 2013–16

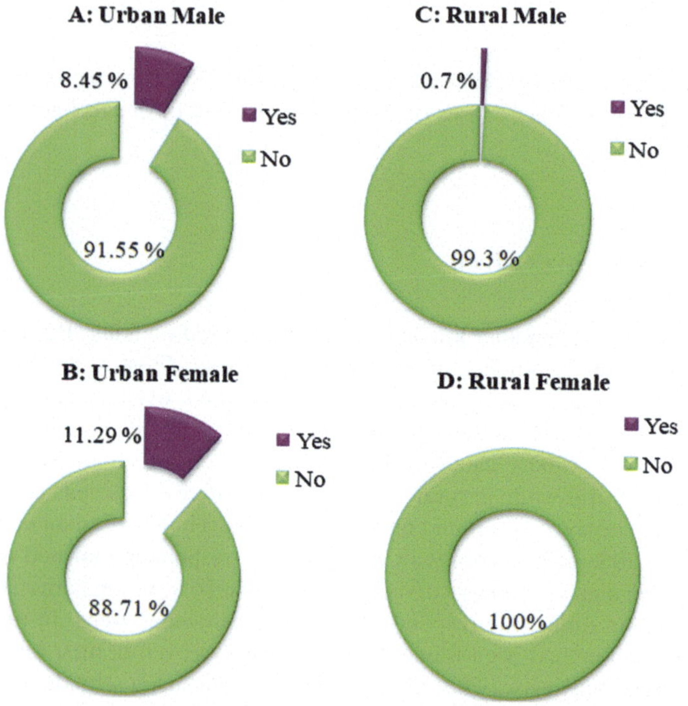


**Fig. 16.4** Respondents owning microwave ovens in households. *Source* Primary survey, 2013–16

### 16.5.4 Refrigerator

The refrigerator is the next technology related to the kitchen which makes life easier to perform cooking and related convenience. Again, the condition of rural households is not good in owning and using refrigerators. Primary data reflect that only 11.89%of households of rural male respondents own refrigerators. The condition of the households of rural female respondents is the same i.e., only 20.16% own refrigerators in their houses. 88.11% of rural male respondents and 79.84% of rural female respondents do not own refrigerators in their households (Table 16.5). Urban households are in better condition in owning refrigerators too. 60.56% of urban male respondents and 74.19% of urban female respondents accounted for owning a refrigerator. Only 25.81% of urban female respondents and 39.44% of urban male respondents did not own refrigerators. Table 16.5 elucidates the situation of urban and rural areas.

Respondents from the district were not able to demarcate the impacts of this technology on their lifestyle and social values. They were simply replying that they could get cold drinking water in summer and food could be stored while few urban males described the negative impact of refrigerators and mixer grinders on the health of female members (those females were housewives) because their manual labour for domestic work has been reduced. It further explains that the coming of these technologies made females reluctant to serve fresh food to their family members. Only one urban female listed many positive impacts of the refrigerator on her lifestyle (this respondent was working in a government job). She illustrated that she faces a shortage of time in the morning because of her office timing, and storing food in the refrigerator made her life easy and hassle-free and she can save time in rush hours. Figure 16.5 shows some pictures of refrigerators in respondent's houses.

**Table 16.5** Respondents owning refrigerators

| | Male | | | | Female | | | |
|---|---|---|---|---|---|---|---|---|
| | Urban Male | | Rural Male | | Urban Female | | Rural Female | |
| | Number | Percent | Number | Percent | Number | Percent | Number | Percent |
| Yes | 43 | 60.56 | 17 | 11.89 | 46 | 74.19 | 25 | 20.16 |
| No | 28 | 39.44 | 126 | 88.11 | 16 | 25.81 | 99 | 79.84 |
| Total | 71 | 100 | 143 | 100 | 62 | 100 | 124 | 100 |
| | Total Urban | | Total Rural | | Total Male | | Total Female | |
| | Number | % | Number | % | Number | % | Number | % |
| Yes | 89 | 66.92 | 42 | 15.73 | 60 | 28.04 | 71 | 38.17 |
| No | 44 | 33.08 | 225 | 84.27 | 154 | 71.96 | 115 | 61.83 |

*Source* Primary survey 2013–16

**Fig. 16.5** Types of refrigerators used in households. *Source* Authors

## 16.6 Conclusions

The adoption of technology is limited to urban areas only in the district. Rural areas of the district still have a long way to go in achieving technological advancement. The reason for this low adoption of technology in rural areas of Aligarh district is due to the low purchasing power. However, it has been noticed that technological advancement alters the thinking of people and social norms from one generation to another generation. For example, wood and cow-dung cakes as cooking fuel, no availability of mixer and grinding, microwaves ovens, refrigerators, etc. increases the labour for cooking in rural areas. Therefore, in rural areas cooking is the sole responsibility of females. While in urban areas, the availability of advanced technology in the kitchens, makes cooking easy and less time-consuming. Urban females even train other family members to operate kitchen appliances. This resulted in some flexibility in social norms related to cooking in urban areas. Advanced technologies also attract male members of the family to help with cooking. Hence, it can be concluded that urban areas are experiencing some flexibility in social norms with the coming of advanced technology in the kitchen. In contrast, rural areas are continuing with rigid social norms.

**Acknowledgements** The first author has sincere gratitude towards the University Grant Commission (UGC) and Department of Geography, Delhi School of Economics, University of Delhi for providing a research fellowship for conducting this research. Without this fellowship fund, this research could not culminate in research findings.

## References

Agarwal B (1989) Rural women, poverty and natural resources: sustenance, sustainability and struggle for change. Econ Pol Wkly 24(43):WS46–WS65

Agarwal B (2007) Gender inequality, cooperation and environmental sustainability. In: Baland JM, Bardhan P, Bowles S (eds) Inequality, cooperation, and environmental sustainability. Russell Sage Foundation, New York, pp 274–313

Agarwal B (2010) Rethinking agricultural production collectivities. Econ Pol Wkly 45(9):64–78

Chakravarti U (1993) Conceptualising brahmanical patriarchy in early India: gender, caste, class and state. Econ Pol Wkly 28(14):579–585

Chakravarti U (1995) Gender, caste and labour: ideological and material structure of widowhood. Econ Pol Wkly 30(36):2248–2256

Davis AY (1981) Women, race and class. Navayana Publishing, New Delhi

DCHB [District Census Handbook] (2011) Village and Town Directory-Aligarh. Series-10, Part XII-A. Census of India 2011 Uttar Pradesh. Directorate of Census Operations Uttar Pradesh

Dickey S (2000) Permeable homes: domestic service, household space, and the vulnerability of class boundaries in urban India. Am Ethnol 27(2):462–489

District Census Handbook Aligarh (2011) Village and Town-wise Primary Census Abstract (PCA). Series-10, Part XII-B. Census of India 2011 Uttar Pradesh. Directorate of Census Operations Uttar Pradesh

DST (2016) Department of Science and Technology. http://www.dst.gov.in/stsysindia/sprl958.htm. Accessed 25 June 2016

Elson D (2017) Recognize, reduce, and redistribute unpaid care work: how to close the gender gap. New Labor Forum 26(2):52–61

Folbre N, Yoon J (2008) Economic development and time devoted to direct unpaid care Activities. UNRISD Flagship Report on Poverty. United Nations Research Institute for Social Development, Geneva

Hirway I (2005) Unpaid work and economy: gender, poverty and millennium development goals. Integrating Unpaid Work into Development Policy Working Paper No. 838, Hudson

Komatsu S, Kaneko S, Ghosh PP (2011) Are micro-benefits negligible?: the implications of the rapid expansion of Solar Home Systems (SHS) in rural Bangladesh for sustainable development. Energy Policy 39(7):4022–4031

Parrenas RS (2004) Migrant Filipina domestic workers and the international division of reproductive labor. Gend Soc 14(4):560–580

Ray R (2000) Masculinity, femininity, and servitude: domestic workers in Calcutta in the late twentieth century. Fem Stud 26(3):691–718

Ray R, Qayum S (2009) Cultures of servitude: modernity, domesticity and class in India. Stanford University Press, Stanford

Sangari K (1993) The 'amenities of domestic life': questions on labour. Social Scientist 9(11):3–46

Satterthwaite D, Hart R, Levy C, Mitlin D, Ross D, Smit J, Stephens C (1996) The environment for children: understanding and acting on the environmental hazards that threaten children and their parents. Earthscan, London

Singh P, Pattanaik F (2020) Unfolding unpaid domestic work in India: women's constraints, choices, and career. Palgrave Commun 6:111. https://doi.org/10.1057/s41599-020-0488-2

TDB [Technology Development Board] (2016) Department of Science and Technology. http://www.tdb.gov.in. Accessed 25 June 2016

UN [United Nations] (2023) Achieve gender equality and empower all women and girls. https://sdgs.un.org/goals/goal5. Accessed 10 Aug 2023

# Chapter 17
# Rural Livelihood Generation and Sustainability Concerns: A Case Study of Bankura District, West Bengal

**Swati Sachdev, Aritra Chakraborty, Deep Narayan Pandey, Md Anisujjaman, Swati Thakur, and Barnika Saha**

**Abstract** Sustainable livelihood involves addressing issues of vulnerability, enhancing capabilities and assets in addition to providing basic means of sustenance over generations and in even periods of strife. In India, a majority of people still reside in villages; and rely on agriculture with limited rural diversification. They have a predominance of marginal farmers with poor earnings or wages and limited access to basic amenities mainly water, sanitation and health. For example, 70–75% of water does not meet the standard quality prescribed by the World Health Organization. Rural India is at a crossroads as environmental and pollution concerns are engulfing agriculture. Climate change can lower farmer income by 25%. All these have a more adverse impact on vulnerable groups and rural areas. Against this backdrop, the paper examines the nature of rural livelihood and the multi-faceted aspects of sustainability in general and especially in the Bankura district of West Bengal. The paper argues that sustainable livelihood entails providing rural areas with physical, economic, knowledge, societal, and electronic connectivity in the contemporary era. It uses indicators for the above parameters of connectivity and runs a Principal Component Analysis to examine the above hypothesis and identify

S. Sachdev (✉)
Department of Geography, Budge Budge College, affiliated to University of Calcutta, Kolkata, India
e-mail: swatisachdev@gmail.com

A. Chakraborty
Department of Geography, Presidency University, Kolkata, India

D. N. Pandey
Special Centre for Disaster Research, Jawaharlal Nehru University, New Delhi, India

M. Anisujjaman
Department of Geography, Sidho Kanho Birsha University, Purulia, West Bengal, India

S. Thakur
Department of Geography, M.M.H. College, Ghaziabad, India

B. Saha
In-House Trainer, Ufaber, Mumbai, India

 
S. Anand et al. (eds.), *Emerging Geosustainability Transformations in India*, Advances in Geographical and Environmental Sciences,
https://doi.org/10.1007/978-981-97-7993-2_17

the relative importance of different dimensions or factors. The paper also focuses on one arena of livelihood management to promote sustainable rural development, i.e., the Panchmura, Bankura district terracotta industry, which is part of folk culture and heritage. It has attained global recognition by registration under the 'Geographical Indications of West Bengal.' This along with local community participation has enhanced work security as well as a community more synchronized with the environment, making the village an example of how to strive towards sustainable rural livelihood.

**Keywords** Rural · Connectivity · Local · Sustainability

## 17.1 Introduction

The majority of Indians still reside in the villages. More than 65% of Indians live in rural areas. They still primarily rely on agriculture for their survival and are marginal farmers. Thus, they have neither adequate land holdings nor alternate employment avenues. They have poor earnings or wages and limited access to basic amenities mainly water, sanitation and health e.g. 70–75% of water does not meet the standard quality prescribed by the World Health Organization (WHO). Rural India is at a crossroads as environmental and pollution concerns are engulfing agriculture, e.g., climate change can lower farmer income by 25% (Government of India 2019). They also have poor access to social infrastructure throughout the year. In addition, too much reliance on agriculture, poor resource utilization and management practices have raised environmental concerns in rural areas as well in contemporary times.

In this context, sustainable livelihood becomes a primary concern. Sustainable Livelihood involves addressing issues of vulnerability and social exclusion, enhancing capabilities and assets in addition to providing basic means of sustenance over generations and in even periods of strife. The paper is set against this backdrop and seeks to examine the nature of rural livelihood and the multi-faceted aspects of sustainability through a case study of Bankura district in West Bengal.

## 17.2 Contextualization and Objectives

Chambers and Conway (1992) proposed the definition of sustainable rural livelihood as, "a livelihood comprises of capabilities, assets (stores, resources, claims and access) and activities required for a means of living; a livelihood is sustainable which can cope with and recover from stress and shocks, maintain or enhance its capabilities and assets, and provide sustainable livelihood opportunities for the next generation; and which contributes net benefits to other livelihoods at the local and global levels and in the short and long-term". Since the turn of the century, there have been several studies trying to focus on livelihood issues in rural areas. While

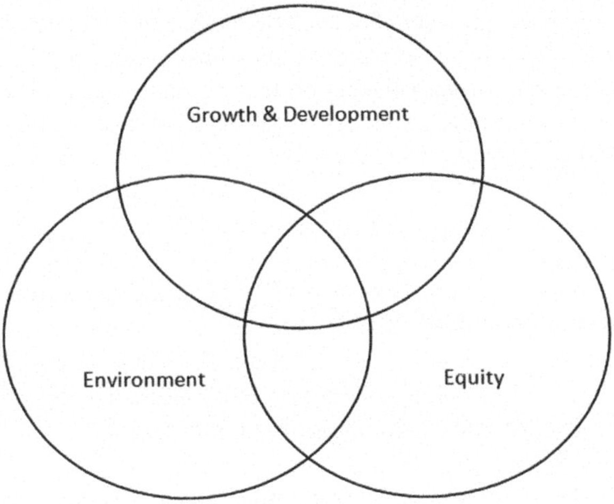


**Fig. 17.1** Growth-equity-environment: sustainability nexus. *Source* Adapted from Salite and Pipere (2006)

some studies reflect the need for access to basic amenities like drinking water, and sanitation, other studies focus on issues of employment—especially the Mahatma Gandhi National Rural Employment Guarantee Act-2005 (MGNREGA) scheme. Other fewer studies examine issues of rural education and health. However, there is a dearth of studies, which examine the issues in a holistic interconnected mode examining the issues of development-equity in tandem with the environment (Fig. 17.1). The lens of sustainable livelihood helps to traverse this gap and examine this intertwining and nexus between the various elements of development-livelihood and sustainability. The objective of this paper is to examine how rural sustainability should be measured and how to strive toward achieving sustainability through a case study.

## 17.3 Measuring Rural Sustainability and Conceptual Framework

Contemporary literature on sustainability has a bias toward examining urban sustainability issues. However, sustainability studies in rural areas have been gaining greater attention post-2020. It has been recognized that sustainability conceptions differ for rural and urban areas (Zulauf and Wagner 2021). Other studies have focussed on a comprehensive study of methods to attain rural sustainability and ascertain its drivers and concluded that the "Triple Bottom Line" of special, economic and environmental forces govern it (Nelson et al. 2023). Researchers have recognized the multifaceted nature of rural sustainability and highlighted the role of small businesses in attaining it as well (Frolova et al. 2020). In India, rural sustainability initially emphasized promoting sustainable agriculture primarily (Singh 2018), and only in recent years has the importance of sustainability been broadened to encompass other aspects. Other studies have emphasized the need to revitalize the rural economy through the identification of interconnections between rural and urban areas (Williams et al. 2021). Even the World Bank has developed a "Sustainable Rural Economy Program" for Eastern and Southern Africa in 2021 and UNDP has a program of "Rural Sustainable Development" initiated for Uzbekistan in 2022.

However, the concept of linkage of rural sustainability to livelihood was first popularised in academic deliberations in 1991 by Chambers and Conway in their seminal work on "Sustainable Rural Livelihood" which suggests that livelihood can be the integrating element between environment, social and economic aspects and result in sustainable rural development. They state that household livelihood wherein people's capabilities, the actual activities they undertake, the tangible (resources) and intangible assets (claims), and their earning form the basic components and flow of livelihood.

There have been many attempts to measure sustainability, the most popular being the Barometric Model of sustainability (IUCN 1996). This model (Fig. 17.2) proposes that ecosystem and human well-being must be seen in conjunction. Hence, the index of human and ecosystem well-being is taken and combined to arrive at an aggregate score. The intersection of these two scores along two scales represents progress toward sustainability. Studies based on these have been undertaken even in India to measure sustainability. Other attempts have also been made to examine the level of sustainability, e.g., through the Sustainable Livelihood Index (SLI); a tool first used to assess the preparedness of rural poor to receive entrepreneurial projects in Malaysia (Kamaruddin and Samsudin 2014). The SLI incorporated various dimensions and comprised five main conceptions of livelihood—asset, strategy, outcome, institutions and vulnerability.

Dr. A.P.J. Abdul Kalam, former President of India, in his Vision 2020, while elaborating on the PURA scheme (Providing Urban amenities in Rural Areas) mentioned that the development of a cluster of villages entails the following interlinked connectivity, namely physical, knowledge, electronic leading to economic and social connectivity (Kalam 2008).

This paper follows a combination of Chambers and Conway (1992) conception of livelihood and the above conception of connectivity in a single thread to examine the nature of rural livelihood sustainability (Fig. 17.3). The paper argues that the above five conceptions of connectivity will govern people's livelihood, capability and best utilization of resources thereby resulting in enhancing the dynamic nature and intergenerational sustainability of household livelihood embedded in local resources and community networks, which will help in enabling sustainable livelihood in rural areas as a single entity.

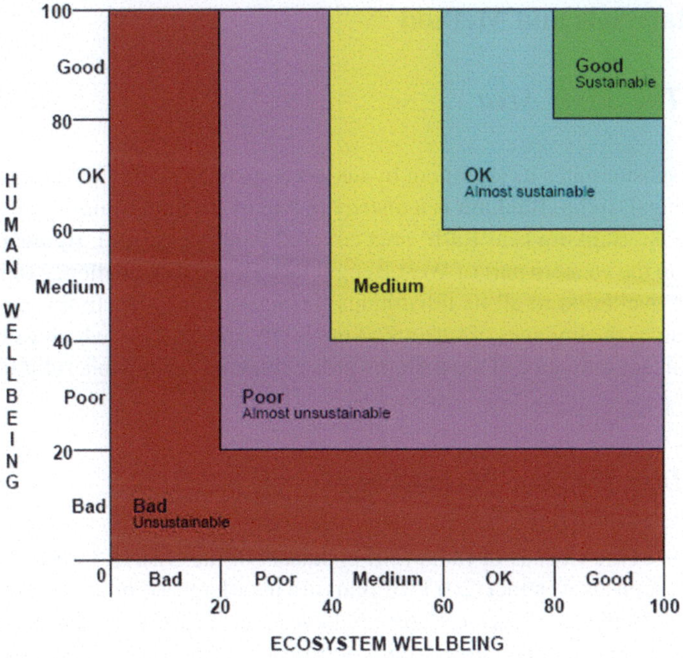


**Fig. 17.2** Barometer of sustainability. *Source* IUCN (1996)

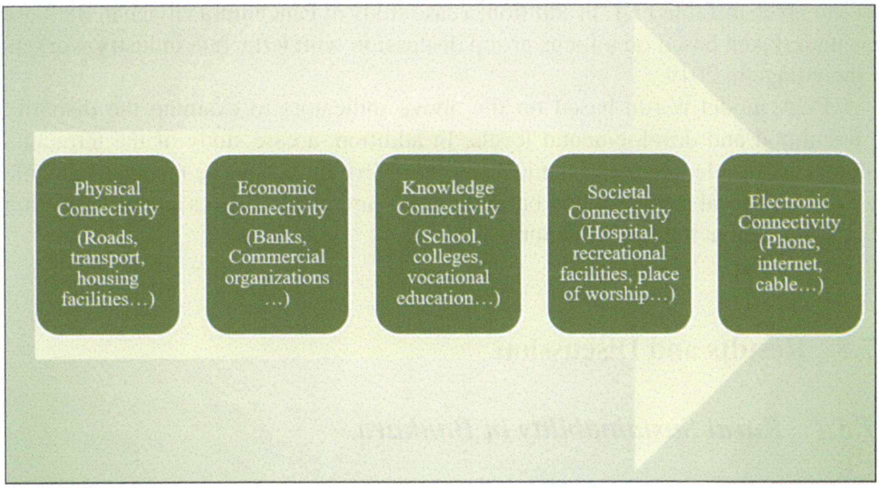


**Fig. 17.3** Rural connectivity development dimensions towards sustainability. *Source* Adapted and modified after Kalam (2008)

## 17.4 Materials and Method

### 17.4.1 The Study Area

The paper substantiates its argument by undertaking a case study of Bankura district of West Bengal, India. Bankura is a district in one of the agriculturally predominant states of India. Bankura is a 'Rarh' area covered with laterite soil. Being a drought-prone area in the western part of West Bengal faces the severe challenge of providing sustainable livelihood to all its inhabitants throughout the year. In this context, the paper examines the linkages of various connectivity and explores which would have a more significant influence in the path towards achieving sustainable rural livelihood.

### 17.4.2 Database and Methodology

The paper uses the Census of India (2011) dataset of the District Census Handbook and Primary Census Abstract (2011) of Bankura in order to examine sustainable rural livelihood in Bankura. It uses different indicators as measures of livelihood assets, capabilities, resources and connectivity and runs a Principal Component Analysis (PCA) to examine the various livelihood and developmental levels among different Community Development blocks in Bankura. The indicators considered for the exercise are given in Table 17.1. In addition, a case study of Panchmura village in Bankura was undertaken based on a focus group discussion with terracotta industry workers in the village in 2019.

A PCA model is run based on the above indicators to examine the disparity in livelihood and developmental levels. In addition, a case study of the terracotta industry was undertaken to examine how rural livelihood can be made sustainable by linking to local resources and building on community networks and conservation of environment at the same juncture.

## 17.5 Results and Discussion

### 17.5.1 Rural Sustainability in Bankura

The Bankura district is a drought-prone district and the Workforce Participation Rate (WFPR) as of 2011 was only 41%. Furthermore, as high as 65% of workers are still engaged in agriculture and almost 70% of their income is generated by agriculture. This excessive reliance on the agriculture of the district in spite of being a drought-prone area is a severe constraint for development.

**Table 17.1** Connectivity development dimensions and indicators considered for analysis

| Connectivity | | | | | Amenities and facilities | Environmental |
|---|---|---|---|---|---|---|
| Physical | Economic | Knowledge | Societal | Electronic | | |
| Surfaced roads | Workforce participation rate | Literacy rate | Awareness: sex ratio | Access to mobile | Safe drinking water | Having a sanitation facility within the premises |
| Ownership of bicycle | Availing banking services | Female literacy rate | Access to health facilities: hospital beds | | Access to electricity | Having linkage/outlet to drainage |
| | | Access to primary schools | | | Housing condition | |

*Source* Authors

To determine the factors that need to be targeted to improve the livelihood of people in Bankura, an examination of the various parameters of connectivity in all its dimensions was undertaken. To have sustainable rural development physical infrastructure in terms of accessibility is of prime importance. About 27% of Bankura only has surfaced roads although 72% own a bicycle. As regards Knowledge, an assessment of both infrastructure in terms of access to primary schools and literacy rate overall and female reveals a grim picture with only 10 primary schools per 10,000 population and 61% and 51% aggregate and female literacy rates, respectively. Electronic connectivity represented by ownership of mobile phones also represents that only 35% of households owned mobiles on average in Bankura. The societal aspect reveals a grim scenario with a sex ratio of only 956 females per 1000 males and only 8 hospital beds per 10,000 population. This has implications for economic connectivity which directly impinges on access to amenities and livelihood. Low accessibility, poor knowledge base, and poor access to educational and health infrastructure result in high social and economic disparities reflected in high sex ratio, poor WFPR (especially among females at 18%), and only around 52% of households having at least one member with a bank account on an average in the Bankura district.

Poor access to and use of economic and social assets leads to lower capabilities as well as poor conditions of livelihood. While 85% of households in Bankura have access to safe drinking water on average, only 41% have electricity, 36% have good housing conditions, and an even lower share have a 'pucca' roof. The consequence is that the surrounding environment and sanitation are also a cause of concern. Only 60% of households have sanitation facilities within or near their premises and only 13% of households have a proper outlet to drainage. The above aggregate picture is coupled with the fact that there are intra-district variations. Among the 22 CD Blocks in Bankura, four blocks namely Bankura I, Bankura II, Indus, and Sonamukhi are better off and Hirbandh and Ranibundh are more deplorable than the district average.

A PCA model run based on the above indicators revealed that the variation in developmental levels among CD Blocks in Bankura was strongly and significantly explained by the above model, indicating the importance of all the above parameters in developmental processes. The first four components explained more than 77% of the variation and the Kaiser–Meyer–Olkin Measure of Sampling Adequacy (KMO) also was valid at 0.6 (Table 17.2).

In addition, a separate model run for each connectivity development dimension e.g. economic, social, infrastructural, knowledge, electronic, amenities, and environmental revealed that only the education, economic and environmental determinants model significantly and strongly explained the variation (Table 17.3). Thus, rural livelihood is impacted more by the nexus of economic, knowledge and environmental aspects in the Bankura district. In this context, an exploration was undertaken to examine how in one particularly developed village of Bankura district, the economic scenario rooted in local knowledge and resource utilization and environmentally sustainable enterprise has enabled to ensure sustainable rural livelihood in that village.

**Table 17.2** Results of the PCA model run on 15 factors

| Total variance explained | | | |
|---|---|---|---|
| Component | Initial eigenvalues | | |
| | Total | % of variance | Cumulative% |
| 1 | 6.405 | 42.700 | 42.700 |
| 2 | 2.553 | 17.021 | 59.721 |
| 3 | 1.457 | 9.715 | 69.436 |
| 4 | 1.243 | 8.286 | 77.722 |
| Extraction Method: Principal Component Analysis | | | |

*Source* Authors

**Table 17.3** Results of the PCA model run on each connectivity development dimension

| Indicators | Percentage of variance explained by first component |
|---|---|
| Knowledge | 66 |
| Economic | 62 |
| Environment | 92 |

*Note* Significant results reported

### *17.5.2 Rural Sustainability Through Livelihood Management: Case Study of Terracotta, Bankura*

Analysis revealed that among the various blocks of Bankura, the location quotient for the household industry was high in a particular village of Panchmura in Taldangra Block (Fig. 17.4). Bankura being a drought-prone area, relied on agriculture which was causing distress scenario. Even as per the 2011 census, more than 91% of the district was rural. In the midst of this, Panchmura village relying on Terracotta artefacts and handicrafts was an outlier with a more sustainable livelihood scenario. Other studies have also highlighted the need for livelihood diversification and need for greater reliance on the rural non-farm sector (Lama 2014; Singh 2018).

The Panchmura village relies on the terracotta industry. Terracotta is a Latin word 'terracotta' meaning 'baked Earth.' Terracotta work implies baking clay after shaping it into specific designs to provide rigidity and sturdy shape. Locally available clay or '*entelmati'* forms the basic raw material for this household industry in this region, which is also supported by cooperatives. Even the colouring is done with natural colour, e.g., gad (yellowish clay), banak (reddish clay), and Sonamukhi plant flower powder. The predominant artefacts identified as terracotta of Panchmura for which it has now attained global recognition are Horse, Elephant, civilization-floral-faunal motifs, images of Sanskritaniyas and Folk Art (NCERT undated).

The basis of the industry in this village is an ancestral heritage in the region of terracotta clay craft since the Mauryan and Gupta Periods where terracotta sculptures were used to decorate temples. Even in the British period local rulers used terracotta

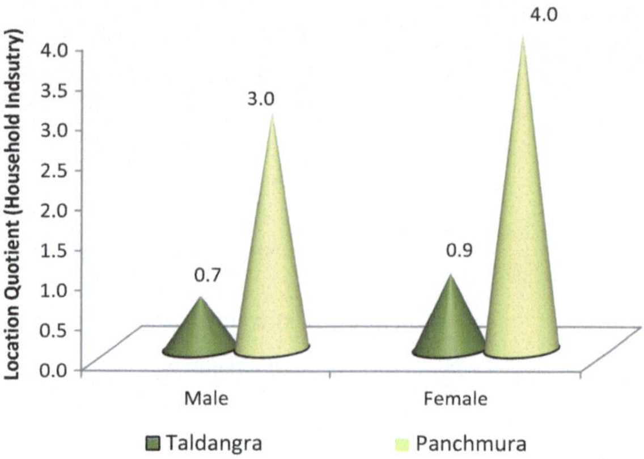


**Fig. 17.4** Location quotient for household industry workers. *Source* Calculation and diagram compiled by the author based on data available in the Directorate of Census Operations, Census of India, 2011

plaques and stucco patterns to build temples. Panchmura thus has a vibrant unique living tradition of terracotta. These skilled workers form the base for the development of household industry in this village and region.

The impact of having reliance on household industry rooted in local raw material and relying on social kinship and community network and capital through the formation of cooperatives ('Panchmura Mritshilpi Samabay Samity') has led to it being a prosperous and sustainable livelihood avenue. The fact that as high as 90% and 60% of men and women respectively in the village are literate and around 65% of both men and women have secondary or higher education has further strengthened the livelihood sustainability. The consequences on the standard of living and livelihood of local inhabitants are evident in the fact that WFPR in the village for both men and women is around 60%, well above the district average. They have better houses as more than 70% have semi-pucca or pucca houses and more than 50% have exclusive toilets and access to safe drinking water through tubewells or taps. Another study substantiates the higher standard of living in Panchmura wherein it reports (Fig. 17.5) that nearly 70% of households had more than one mobile and more than 60% of workers of terracotta had a monthly household income ₹5,000–₹19,999 and another 15% between ₹20,000 and ₹29,999 (Saha 2019).

It has now attained global recognition and is registered under the 'Geographical Indications of West Bengal', named 'Bankura Panchmura Terracotta Craft' on 28 March 2018 ('Bankura Horse') (Fig. 17.6). The government also awarded recognition to ancestral producers and the Panchmura Pottery Society, cooperative society was formed in 2005–2006 to help the artisans. Thus, in addition, besides improvement in the standard of living directly, these products are now offered for display even in museums in Russia and America. This has resulted in enhancing the rural economy through increased marketing avenues, boosting rural tourism and providing more stable jobs throughout the year thus providing enhanced sustainable employment avenues based on local resources. Because of the flourishing handicrafts in this region, even young adults of the next generation are engaging in the occupation.

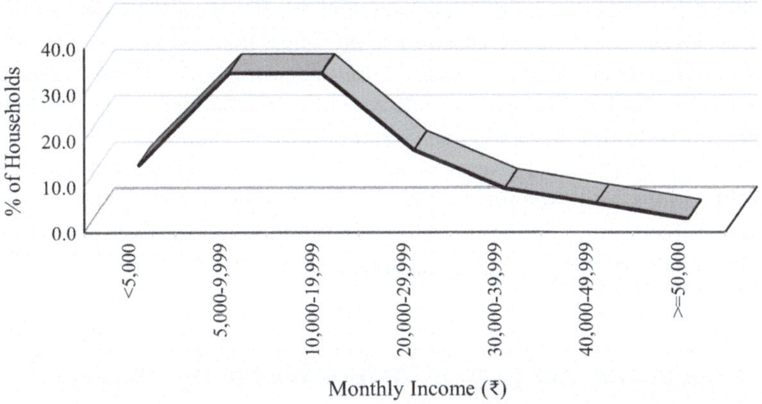


**Fig. 17.5** Monthly Income of households in terracotta work. *Source* Adapted from Saha (2019)

**Fig. 17.6** Terracotta work and 'Panchmura Terracotta Horse'. *Source* Saha 2019

This is not to say that there is no scope for improvement. The Panchmura still faces problems that needs to be addressed. Firstly, there is still gender segregation of occupation evidenced as women seem to be employed more as unpaid family labour rather than having their operations. There is also a dearth of women who undertake tasks of making mould and women are concentrated on more mechanical tasks of cleaning and colouring. Secondly, there are problems due to the presence of intermediaries which results in low rates as there is a lack of marketing skills among the artisans themselves. There is also a lack of diversity in products that led to problems in the marketing process. Thirdly, extraction and utilization of mud have also resulted in soil erosion where the clay is not excavated suitably through proper adoption of post-excavating management and conservational measures. Drought conditions are also resulting in scarcity of raw materials sometimes. Fourthly, the artisans with the cultural heritage and ancestral training in folk art lack the expertise in marketing and managing capital. Finally, even though households have a sustainable source of livelihood in Panchmura, the impact remains localized and has limited percolation effects unless it can be increased in scale. Thus, there is a need for training in marketing, diversification, and financial entrepreneurial skills. There is also a need for financial institutions that will provide capital to expand the scale of operation.

In spite of the above problems this model of development based on local household industry rooted in local resources and labour can be emulated in other regions with modifications as it exemplifies how local resources and traditional knowledge and skills coupled with community networks through cooperatives can generate a sustainable source of livelihood in rural areas.

## 17.6 Conclusion and Lessons–Pathways for Development

The above case study analysis supports this paper's findings that adequate knowledge, a source of sustenance and a stable economy do translate into an improved standard of living which tends to become sustainable if it is based on local resources and community networks and promoted through government policies. There is a need to further strengthen the livelihood scenario by providing access to training and modern financial institutions and instruments and marketing on one hand through already running government schemes like Grameen Kaushalya Vikas Yojanas and National Rural Livelihood Mission. However, the local specificities, environment and geographical milieu of the region must be considered when providing skill training for the household industry. In addition, to enhancing knowledge and skills, there has to be a complementary focus on local environment-related asset and livelihood creation such that the technologies and processes used create environment-friendly livelihoods, regenerate and conserve natural resource base, and at the same time have multiplier and spillover effects with other dimensions of development to ensure sustainability.

Finally, for future outlook and to tap the true potential of generating employment and boosting regional development through household industry instead of localized rural sustainability avenues, there is a need to acknowledge the various strands prevailing within the community and focus on involving women as well through the formation of SHGs to empower them and ensure they are not relegated to niche occupations and mechanical occupations and tasks generally considered an extension of their work at home (Sachdev 2016). According to 'The Centre for Environment Education, sustainable rural development entails not just focusing on income generation and better resource utilization for all segments of society but also integration of the environment with economic considerations' (CEE 2023). In addition, it also entails the integration of natural and cultural heritage. This conception is well evidenced in Panchmura, wherein Sustainable Livelihood Management has evolved as a result of global recognition and government support working in conformity with community participation through cooperatives based on the utilization of local clay resources and local ancestral skilled labour working in folk art rooted in cultural heritage and perpetuating cultural ethos using natural processes and in harmony with the environment to ensure synergized development.

## References


CEE [Centre for Environment Education] (2023) Sustainable rural development. https://www.ceeindia.org/sustainable-rural-development. Accessed on 25 Dec 2023

Census of India (2011) District census handbook: Bankura. Directorate of Census Operations, West Bengal

Chambers R, Conway G (1992) Sustainable rural livelihoods: practical concepts for the 21st century. IDS Discussion Paper 296. IDS, Brighton. https://www.ids.ac.uk/download.php?file=files/Dp296.pdf. Accessed on 22 Sep 2019

Directorate of Census Operations (2011) Census of India primary census abstract—Bankura. Directorate of Census Operations, West Bengal

Frolova SV, Litvinova IV, Salomatina MN, Bryazgunova NS (2020) Small business as a factor in sustainable rural development. In: BIO web of conferences, vol 17, pp 1–5. https://doi.org/10.1051/bioconf/20201700004

Government of India (2019) Economic survey 2018–2019. Ministry of Finance, New Delhi

IUCN (1996) Barometer of sustainability. Prepared for IUCN/IDRC project on Assessing progress toward sustainability. IUCN, British Columbia and IUCN, Switzerland. https://portals.iucn.org/library/sites/library/files/documents/Rep-1996-030.pdf. Accessed on 22 Sep 2019

Kalam APJ (2008) Speech by Sr. A.P.J. Adbul Kalam, former President of India during the inauguration of Symposium on Vision for professional India and corporate India by the year 2020. New Delhi, 2008. https://www.icsi.edu/media/webmodules/presentations/DrAPJSpeech.htm. Accessed on 25 Dec 2023

Kamaruddin R, Samsudin S (2014) The Sustainable Livelihoods Index: a tool to assess the ability and preparedness of the rural poor in receiving entrepreneurial project. J Soc Econ Res 1:108–117

Lama S (2014) Livelihood diversification and its determinants: evidence from the household survey in backward district of West Bengal. Galaxy Int Interdiscip Res J 2(3):68–80

NCERT (undated) Living craft traditions of India. New Delhi. http://ncert.nic.in/textbook/pdf/kehc102.pdf. Accessed 31 May 2019

Nelson KS, Nguyen TD, Francois JR, Ojha S (2023) Rural sustainability methods, drivers, and outcomes: a systematic review. Sustain Dev 31(3):1226–1249. https://doi.org/10.1002/sd.2471

Sachdev S (2016) Gender, work and space: home-based workers in garment industry in Kolkata. In: Raju S, Jatrana S (eds) Women workers in urban India. Cambridge University Press, Delhi, pp 195-227

Saha B (2019) An assessment of terracotta industry workers in West Bengal: characteristics and prospects, a comparative case study of Panchmura and Chattabaria. M.Sc. Dissertation Regional Planning, Vidyasagar College, University of Calcutta

Salite I, Pipere A (2006) Aspects of sustainable development from the perspective of teachers. J Teach Educ Train 6:15–32

Singh R (2018) Sustainable rural development in India: issues and challenges. Int J Res Soc Sci 8(3):330–341

Williams JM, Chu V, Lam WF, Law WWY (2021) Rural sustainability: challenges and opportunities. In: Revitalising rural communities. Springer Briefs on Case Studies of Sustainable Development. Springer, Singapore. https://doi.org/10.1007/978-981-16-5824-2_1

Zulauf K, Wagner R (2021) Urban and rural sustainability: divergent concepts and their consequences for marketing. Front Sustain 2:670866. https://doi.org/10.3389/frsus.2021.670866

# Chapter 18
# Evaluating the Role of Shuttlecock Manufacturing for Local Sustainable Economy in Jaduberia, West Bengal

**Asmita Mukherjee and Ranjita Banik**

**Abstract** Badminton is a popular international sport. The main component of this game is the shuttlecock. From the onset of independence, Bengal started its journey of shuttlecock making at Jaduberia in the Uluberia sub-division of the Howrah district with very few peasants. It is the only shuttlecock manufacturing hub in all of India where specialized craftsmen produce hand-made shuttlecocks. The main raw materials needed for making shuttlecock are duck feathers and hen feathers. However, the cost of raw materials is increasing and quality is decreasing day by day. The traditional craftsmen with their old manufacturing techniques are not able to stay in the competition with machine-made long life and cheaper shuttlecock which are China and Japanese-made. Above all, the Covid pandemic affected this industry intensively. Now, the industry is struggling to keep up its existence. The paper wants to focus on the present situation of the industry and the socioeconomic condition of the people associated with this business.

**Keywords** Shuttlecock · Manufacturing · Peasant · Craftsmen · Labour

## 18.1 Introduction

In the mid-nineteenth century badminton was played with feather shuttle cock and it was called lawn tennis. However, the start-up of a durable synthetic shuttlecock in the 1950s gave the game a wider appeal. We can remember our badminton superstars (Dey 2020). But we tend to forget the craftsmen of Jaduberia in the Uluberia sub-division of Howrah district, West Bengal which has been the hub of shuttlecock in

A. Mukherjee (✉)
Bijoy Krishna Girls College, Howrah, West Bengal, India
e-mail: asmitamukherjee81@gmail.com

R. Banik
Individual Researcher, Howrah, India
e-mail: ranjitabanik0017@gmail.com



S. Anand et al. (eds.), *Emerging Geosustainability Transformations in India*, Advances in Geographical and Environmental Sciences,
https://doi.org/10.1007/978-981-97-7993-2_18

India since its inception (Pathak 2017). This shuttlecock manufacturing industry is mainly a small-scale industry. The only location in India where shuttlecocks have been produced is Uluberia. In the Uluberia region, the manufacture of shuttlecocks has grown in popularity since 1930. Many individuals were employed directly or indirectly in the shuttlecock manufacturing industry as badminton gained popularity in India. Because the game was played in both high-level and low-level sports at the time, there was a great demand for Uluberia shuttlecocks. However, because Chinese and Japanese shuttlecocks have gained market share in India over the past few years, demand for Indian shuttlecocks has decreased. Indian shuttlecocks cannot compete with foreign shuttlecocks since they are handmade, and have poor longevity, whereas these international shuttlecocks are manufactured using machines, and have high longevity. Many people are working in this field, but neither the government nor any corporate organizations have given this industry the attention it deserves to grow. This primarily depends on the individual efforts of those involved in the shuttlecock production. But this industry's fame diminished day by day. The paper emphasizes the socioeconomic condition of those involved in this business as well as its current state.

The objectives of the study are (i) to trace the history of shuttlecock manufacturing industry at Jaduberia, (ii) to identify the production stages of shuttlecock making, (iii) to analyse the present socio-economic condition of the people associated with this industry, (iv) to evaluate the effects of Covid pandemic on this industry and people, and (v) to identify the problems and future prospects of this industry.

## 18.2 Material and Methods

### *18.2.1 Location of the Study Area*

The Howrah district is situated between 22°48′N and 22°12′N latitude and between 88°23′E and 87°23′E longitude in West Bengal state of India (Fig. 18.1). Uluberia Municipality is a part of Howrah district. Uluberia is situated at 22°41′N and 88°11′E. Ward No. 29 and 30 of Jaduberia are situated in this Uluberia Municipality (Fig. 18.2).

### *18.2.2 Data Source and Methodology*

The study is based on primary data to a large extent. However, secondary data has also been used wherever required. The Census data have been collected from a secondary source and a ward map from Uluberia Municipality. Ward numbers 29 and 30 are selected for door-to-door surveys with the help of local Toto drivers and people. Two types of questionnaires have been made, one for owners and another for labourers. The owners are classified based on their monthly income. The traditional

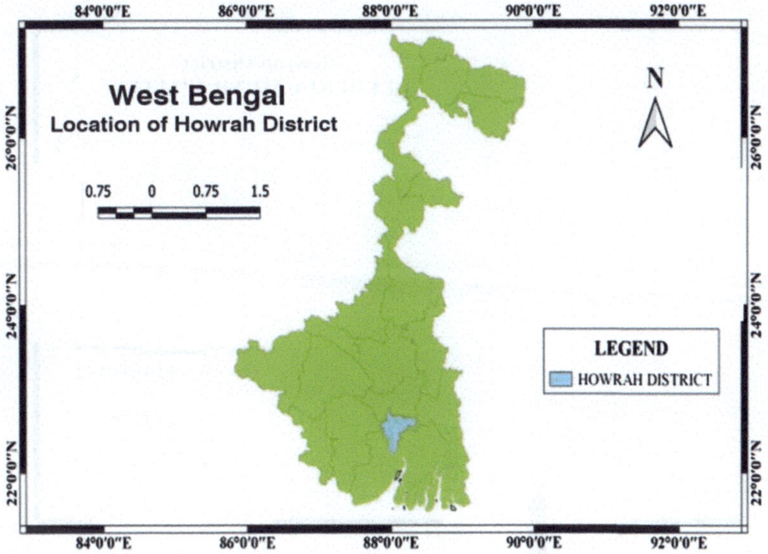


**Fig. 18.1** Location of Howrah district in West Bengal

manufacturing hub (Paid Goods and Services Tax-GST) and local manufacturing units (No paid GST) are also been marked. The socio-economic study has been done based on a door-to-door survey of the people associated with the shuttlecock-manufacturing hub. A total of 50 people have been surveyed out of which 16 are owners and 34 are labourers. As a part of the post-field activity, tabulation of data and representation have been made with the help of statistical techniques and QGIS software for preparing a map.

## 18.3 Results and Discussion

### *18.3.1 History of Shuttlecock Manufacturing Industry at Jaduberia*

Before independence, Bengal began producing shuttlecocks when J. Bose worked at a store in Calcutta selling sporting goods and managing imported shuttlecocks. That was way back in the 1920s. The vendors who provided shuttlecocks to clubs in Calcutta where the British officials practiced the sport, from where J. Bose had learned the craft. Then he demonstrated how to make shuttlecocks to the locals. They did not show much enthusiasm for this project. Despite being uneducated, Dhananjay Pandit stood out among the peasants. He is now credited for founding the cottage industry in this region. In 1984, Heera Pandit, daughter of Dhananjay Pandit,

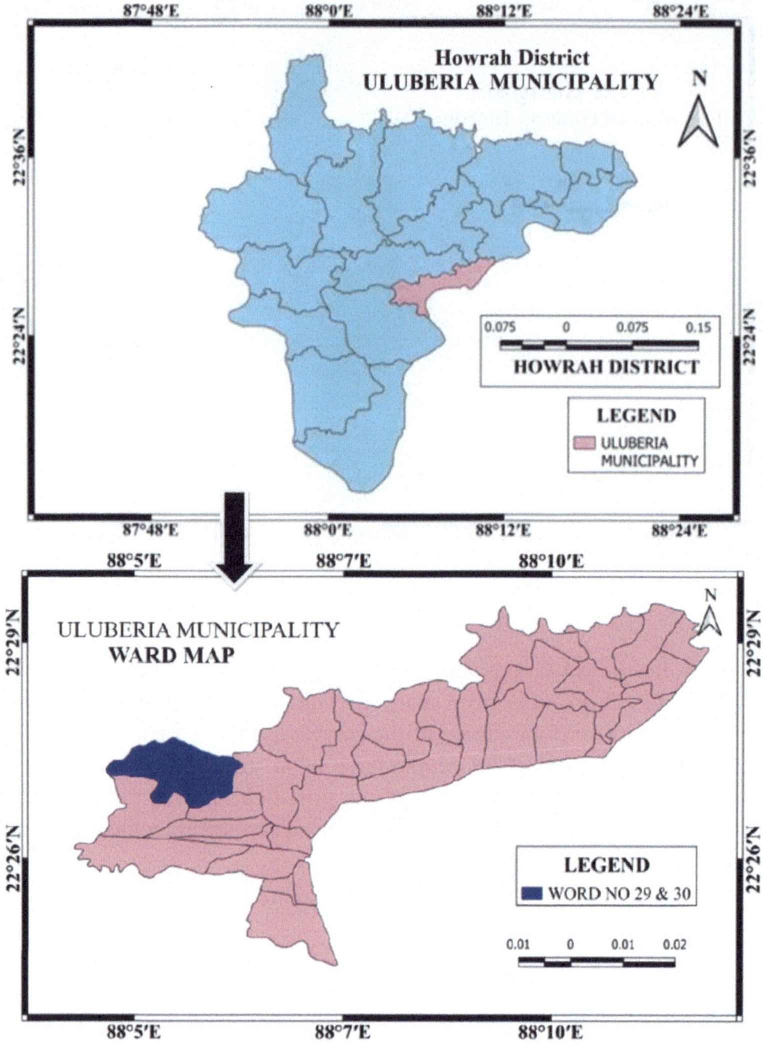


**Fig. 18.2** Location of the study area. *Source* Authors

controlled the renowned D.B. Pandit & Company, which is now a leading farm of shuttlecock. Other leading firms include S. Niyogi & Company.

### 18.3.2 Processes of Making Shuttlecock

Shuttlecock in Jaduberia is a handmade product. So it is labour-intensive industry. A few years back, more than 5000 people were employed in about 60–70 big and small units. But now the number of employees is decreasing day by day. On average 200–300 people have employed in about 30–50 traditional and local units.

After washing, the feathers are separated based on their quality. The production stages (Table 18.1 and Fig. 18.3) produce three types of shuttlecocks with duck feathers high quality (Rs. 40/piece), medium quality (Rs. 25/piece), low quality (Rs. 15/piece), and two types of shuttlecocks with hen feathers, high quality (Rs. 8/piece), low-quality (Rs. 5/piece) (Table 18.2 and Fig. 18.4).

The study includes a survey of 50 people associated with the shuttlecock manufacturing industry. Out of a total of 50 respondents, 16 are owners. They have both traditional manufacturing hubs and local manufacturing hubs. The rest of the 34 respondents are both in-factory and out-factory labourers. Based on their monthly income, owners are classified into three different groups (Fig. 18.5).

**Group 1**: Owners of this group have income above Rs. 50,000 per month. But the percentage of the owner is 12.50%, which is the lowest among the three groups. Their business is traditional near about 80 years. The owners of this group only make duck feather shuttlecocks. So it takes more labour cost. Shuttlecocks of different quality are made here for example, high-quality shuttlecocks are 40 rupees per piece and low-quality shuttlecocks are Rs. 10 per piece (Table 18.3).

**Group 2**: Owners of this group have income between Rs. 30000 and Rs. 50000 per month. The 50% of owners are in this group. The owners of this group used hen feathers and duck feathers both for making shuttlecocks.

**Group 3**: The owners of this group have an income lower than Rs. 30000 per month. Approx. 37% of owners are involved in this group (Table 18.4). The owners of this group only make hen feather shuttlecocks. The hen feather shuttlecock is mainly used for playing on the local ground. The longevity of this cock is very low and destroyed quickly. The per-piece price is very low. That is why the owners of these monthly profits are very low.

Here, it is seen that the shuttlecock made with duck feathers is more labour intensive than that made of hen feather shuttlecocks. Although the rate of production of shuttlecock with hen feathers is high the owners are not interested in making hen feather shuttle cock because hen feather shuttle cock is very low quality so demand is low (Fig. 18.5 right), labour wage is low so the artisan is not interested. If 1000 pieces of hen feathers are cut with a machine a labour wage is 50 rupees. On the other hand, if 1000 pieces of duck feathers are cut the labour wage is 100 rupees. In this industry, there are two types of labourers. In-factory labourers are males and out-factory labourers are females. A total of 34 respondents out of which are 21 male labourers and 13 female labourers. Based on their monthly income labour is classified into three different groups.

**Table 18.1** Production stages of shuttlecock manufacturing

| Production | Materials used | Nature of work | Types of labour | Mode of payment |
|---|---|---|---|---|
| Washing feathers | Detergent and ujala | Washed feathers and dipped Ujala's water to make them white | Male labour | Weekly |
| Cutting | Feather trimming machine | Feathers are trimmed to 3inches of the idle part | Male labour | Weekly |
| Shaping | Scissors | Making rounded-shaped feathers | Male labour | Weekly |
| Create holes in the bottom | Drill machine | Use to drill 16 holes into each bottom | Male labour | Weekly |
| Fixing | Glue (platina) | Feathers are fixed by hand in 16 holes with the help of glue | Male labour and out factory female worker | Per piece |
| Bounding | Yarn | The string of feathers bonded together with the help of yarn | Male labour and out factory female worker | Per piece |
| Attached sticker | Glue (Fevicol) | Sticker pasted onto the bottom | Male labour and out factory female worker | Per piece |
| Decoration | Ribbon | A ribbon is bonded to the bottom | In factory male labour and out factory female labour | Per piece |
| Packing | Barrel | Ten shuttlecocks of the same weight are packed | Male labour and out factory female worker | Per piece |
| Sealing | Heat paper | The barrel is wrapped with the heat paper | Male labour | Weekly |

*Source* Primary survey, 2022

The in-factory labourer whose monthly income is above Rs. 4500. All are skilled and engaged in this work for more than 35 years. They mainly work with duck feathers. The second class of in-factory labourers has their income between Rs. 3500/- and Rs. 4500/-. They mainly work with hen feathers and are semi-skilled labourers. Though the demand for hen feather shuttlecocks is higher than duck feather shuttlecocks the wage rate of the labourer is very low due to the low price per box.

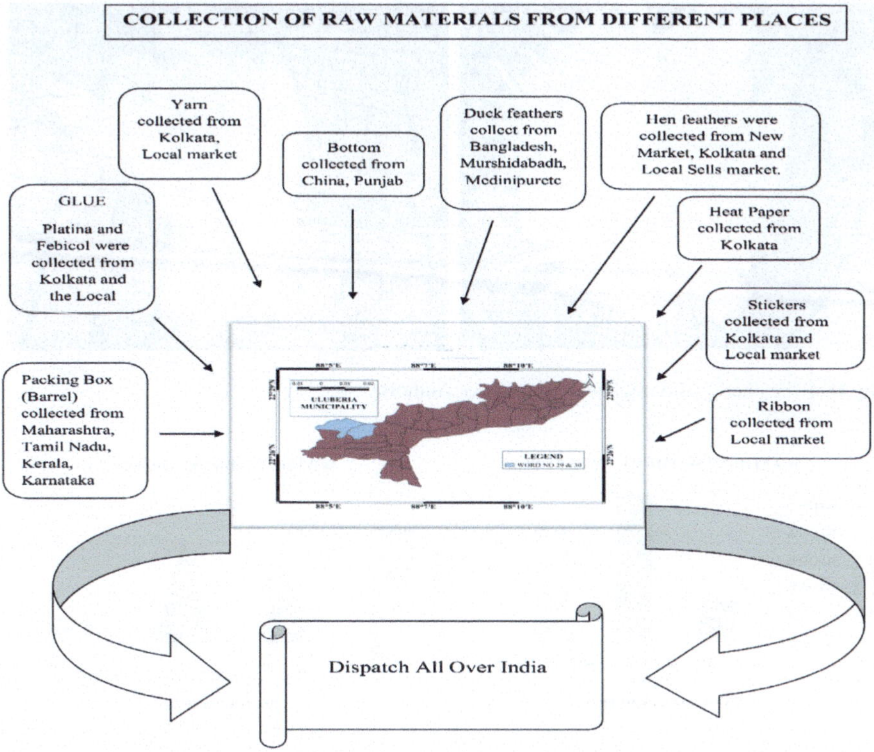


**Fig. 18.3** Collection of raw materials. *Source* Primary Survey, 2022

**Table 18.2** Prices of raw material for shuttlecock

| S. no. | Items | Price |
|---|---|---|
| 1 | Hen feather | 1000 Pieces = Rs. 50/- |
| 2 | Duck feather | 1000 Pieces = Rs. 300/- |
| 3 | Bottom | 1000 Pieces = Rs. 5000/- |
| 4 | Yarn | 1 KG = Rs. 800/- |
| 5 | Platina | 1 KG = Rs. 300/- |
| 6 | Fevicol | 5 Litres = Rs. 500/- |
| 7 | Heat paper | 1000 Pieces = Rs. 1400/- |
| 8 | Sticker | 1000 Pieces = Rs. 70/- |
| 9 | Ribbon | 1 Bandil = Rs. 45/- |
| 10 | Barrel | 1000 Pieces = Rs. 5500/- |

*Source* Primary survey 2022

**Fig. 18.4** Feather quality high and low. *Source* Authors

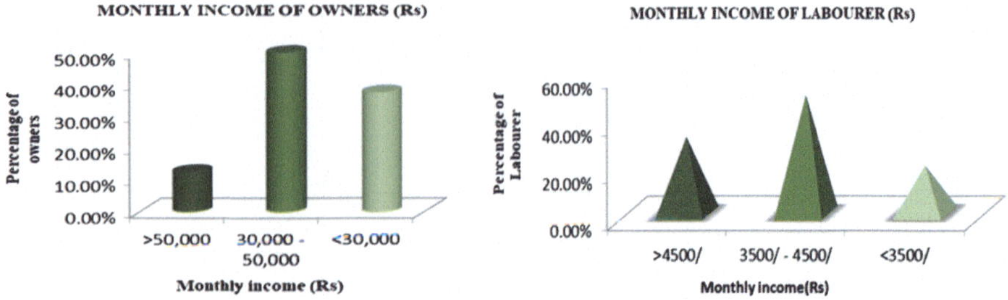


**Fig. 18.5** Monthly income of owners and labours. *Source* Primary survey (2022)

**Table 18.3** Group 1 (income above Rs. 50,000)

| Respondent | Number of labour (in-factory) | Number of labour (out factory) | Production (monthly) (boxes) | Selling prices (monthly) (Rs.) | Profit (monthly) (Rs.) | Investment (monthly) (Rs.) |
|---|---|---|---|---|---|---|
| A | 17 | 3 | 600 | 240,000 | 79,548 | 160,452 |
| B | 14 | 2 | 520 | 208,000 | 69,750 | 138,250 |

*Source* Primary survey (2022)

No medical and provident fund facilities have been given to the labourer. They only received Rs. 2000/- as a bonus during Durga Puja.

Now we have discussed in-factory labourers whose monthly income is between Rs. 3500 and Rs. 4500. It can be seen that most of the labourer are involved for about 20–25 years. Most of the labourers are involved in making hen feather shuttlecock so their monthly wage is very low. Many female members are also involved in this work. They are called out-factory labourers. They work from home. That is why their wage is very low. They are engaged in both duck feather and hen feather shuttlecock making.

**Table 18.4** Group 3 (income less than Rs. 30,000)

| Respondent | Number of labour (in factory) | Number of labour (out factory) | Production (monthly) (boxes) | Selling prices (monthly) (Rs.) | Profit (monthly) (Rs.) | Investment (monthly) (Rs.) |
|---|---|---|---|---|---|---|
| A | 7 | 1 | 1500 | 120,000 | 24,000 | 96,000 |
| B | 4 | 2 | 1300 | 104,000 | 20,800 | 83,200 |
| C | 8 | 2 | 1700 | 136,000 | 27,200 | 108,800 |
| D | 5 | 1 | 1200 | 96,000 | 19,200 | 76,800 |
| E | 9 | 2 | 1600 | 128,000 | 25,600 | 102,400 |
| F | 6 | - | 1100 | 88,000 | 17,600 | 70,400 |

*Source* Primary survey (2022)

### 18.3.3 Impact of COVID-19 on the Shuttlecock Manufacturing Industry

It is well-known that small-scale industries are the backbone of the Indian economy. Small-scale industries have arisen and making a major contribution to production, creativity, export, and inclusive economic development (Folbre 1986). The WHO declared Coronavirus a pandemic and due to this, the whole economy has to go into recession. Social distancing, restriction, health, and economic-driven demand shifts from COVID-19 are expected to shutter many small-scale industries (Baldwin and Mauro 2020; Bell and Lewis 2004).

The present study is the shuttlecock industry which is a small-scale industry. During the COVID-19 period, the shuttlecock industry is struggling and facing several problems. Due to COVID-19, extended lockdown had an impact on the socio-economic condition of this industry. The outbreak of COVID-19 brought labour layoffs and shut down of factories. Factories had to face financial fragility, the crisis of procurement of raw materials and labourers to work in the production and supply process disrupted production flows reduced demand for goods and services, and a decrease in the wage of labour.

### 18.3.4 Impact of COVID-19 on the Economy of This Industry

This business is gradually deteriorating due to various problems. Due to COVID-19, this business was on the verge of collapse. Production in any industry depends upon different factors like raw materials, the demand for goods and services, capital, and many other reasons. There was a significant difference in the market arrivals of raw materials before and during the COVID-19. Since the lockdown was enforced with strict restrictions on transport and other activities, the market demand has decreased (Dev and Sengupta 2020). Before COVID-19, the manufacturer brought 1,000 pieces

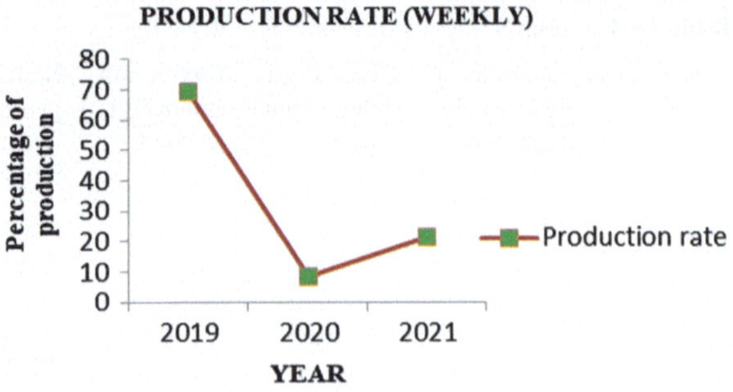


**Fig. 18.6** Production rate. *Source* Primary survey (2022)

of white ducks' feathers in the range of Rs. 800–900 but during the lockdown, the supply of feathers was so hampered that a huge crisis arose which escalated the cost. Manufacturers are buying 1,000 pieces of feathers for a whopping Rs. 1,900–Rs. 2,000. The price of feathers has doubled than normal but the qualities of feathers are not up to the mark.

During the pandemic, volumes meet about 30–40% of the demand due to the supply crisis of basic raw materials. So, the production level (weekly) was changed hugely (Fig. 18.6). The prices of raw materials were increased. However, manufacturers were unable to increase the prices of their products on account of competition. That's why many units were closed during the lockdown.

#### 18.3.4.1 Decrease Income

The Income of the manufacturers has decreased a lot during the Pandemic. According to the manufacturers, the price of the shuttlecock depends upon the quality per barrel before the Pandemic. However, during the pandemic, they had to sell their product at low prices because of very low demand. During the lockdown, they continued their manufacturing with the few raw materials already stored. As a result, the rate of production decreased automatically (Fig. 18.7).

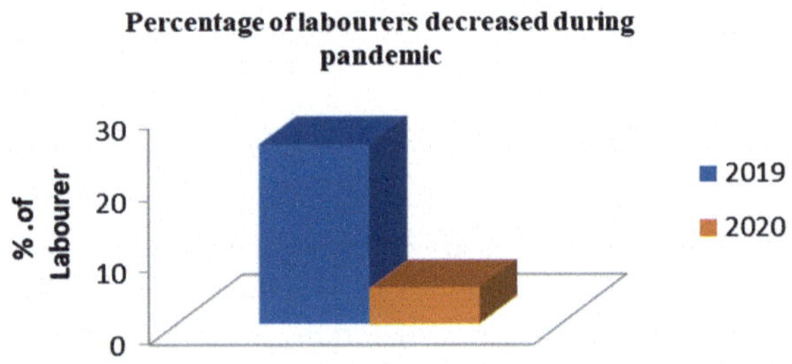


**Fig. 18.7** Decreased labour due to the pandemic. *Source* Primary survey (2022)

#### 18.3.4.2 Labourers Became Jobless

During the lockdown, the labourer in this industry immediately became jobless. Due to very low demand and irregular supply of raw materials, shuttlecock production decreased drastically. Which results layoff of labourer by industry owners. Labourers sometimes had to quit their jobs because of irregular wages or even no wages for months. Those who still worked during the lockdown period had to work with a salary decrease of 30–35%.

### *18.3.5 Impact of COVID-19 on the Social Life of the People Associated with This Industry*

During the pandemic period, the monthly income of the owners of each income group is greatly reduced. This intensely affects their daily life. The pandemic presents an unprecedented challenge to public health, food habits, the education system, and the overall daily life of people.

### *18.3.6 Occupation Shifting*

This pandemic had a devastating effect on the daily lives of labourers. The monthly wages of labourer shad decreased rapidly and many labourers became unemployed. During the pandemic, the labourer had to depend on ration shops and various charitable organizations for food and necessary items to maintain their livelihood (Chaddna et al. 2017; Dev 2020). They engaged themselves in the cultivation of their land, selling vegetables and fish, selling garments, envelop making, etc.

### *18.3.7 Impact on Family Members of the Labourers and Female Labourers*

The family members of the labourer were greatly affected by the lockdown situation. This time family income of the labourer cannot be able to fulfil their basic needs. To support their family, female members had to come forward and engage themselves in different work. In my study area, women are mainly out-factory workers, they have done part-time work and their household expenses are predominantly met by their income. Many female labourers became unemployed during the pandemic. They immediately shift their job. Females have engaged themselves as maidservant, babysitters, selling vegetables with husbands, working 100 days of government work, and selling homemade pappad, puffed rice and boric.

### *18.3.8 Impact on Child Education of the Labourer*

In my study area, COVID-19 had no significant impact on the education of the owners' children because they managed. But had a negative impact on the education of the labourers' children. Mostly uneducated, self-employed parents struggled to continue the education of their children by using new technology. As a result, a large no of students are forced to drop out of school.

## 18.4 Problems

Though small-scale industries in India are considered the line of life for the Indian economy and are the need of the hour for the financial and social progress of the country, the shuttlecock manufacturing industries are facing some genuine problems. The increasing price of raw materials, irregular and limited supply of imported raw materials, and decreasing quality of duck feathers are raw material-related problems (Ghosal 2012). Shuttlecock making is an art. People have not got proper training in shuttlecock making. Labour wages are not at all attractive, so the young generation does not have any interest in this industry. Many in-factory and out-factory labourer changed their occupation during the Covid pandemic. The shuttlecocks manufactured in Jaduberia have only been used in practice matches and domestic tournaments. During the lockdown, several tournaments were cancelled, demand was reduced, and production decreased. Production was also hampered due to increasing prices of raw materials. Japanese Yonex Shuttlecocks are made of superior feathers and last longer. China's plastic products are fast gaining acceptance. They both are fully machine-made and superior in quality. However, the inferior quality handmade Bengal shuttlecock cannot be able to stay in competition with Japan and China. Its demand in the market is decreasing day by day (Zhikuan and Zhang 2022).

Above all the increased rate of GST charges (from 8 to 12%) gives impetus for declining of this industry. But some rays of hope are also there for the prospects of this industry. The Indian Postal Department has taken steps to bring Jaduberia's shuttlecock manufacturing industry to the world through "Permanent Pictorial Cancellation" (Ananda Bazar Patrika 2021). The Uluberia Chamber of Commerce applied for a GI (Geographical Indication) tag for the Uluberia domestic shuttlecock. The government of West Bengal is planning to tie up with Haringhata Farm, Nadia, and Vietnamese duck poultry farm for a steady supply of white feathers (the main raw material) for shuttlecock manufacturing.

## 18.5 Conclusions

The shuttlecock manufacturing industry is a unique industry at Jaduberia because quality shuttlecocks are supplied from this place all over India. This is a traditional cottage industry at Jaduberia in Howrah district of West Bengal. Though this is the only shuttlecock manufacturing center in India, currently it has been facing an inherent crisis at various levels. High rate of raw material price, irregularity in quality raw material supply, traditional man-made production process, lack of interest of the young generation to be involved in this industry and above all the effect of lockdown during the pandemic are tremendously affecting the industry and the socio-economic condition of the people associated with it. But the rays of hope lie in the initiatives taken by both central and state governments to rejuvenate the past glory of this industry. The government has taken many initiatives like "Permanent Pictorial Cancellation", The Uluberia Chamber of Commerce applied for a GI (Geographical Indication) tag for the development of the Uluberia shuttlecock. We hope that these initiatives will be able to elevate the present stagnant condition of this industry by overcoming its present problems and will gain popularity all over the world.

## References

Baldwin RD, Mauro BW (2020) Economics in the time of COVID-19. CEPR Press, pp 109–112

Bell C, Lewis M (2004) The economic implications of epidemics old and new, vol 5(4). Centre for Global Development

Chaddna N, Das A, Gangopadhyay S et al (2017) Reassessing the impact of demonetization on agriculture and informal sector. India Development Foundation (IDF), New Delhi

Dev SM (2020) Addressing COVID-19 impacts on agriculture, food security, and livelihoods in India, IFPRI, Blog. https://www.ifpri.org/blog/addressing-covid-19-impacts-agriculture-food-security-and-livelihood-india. Accessed 8 Apr 2020

Dev SM, Sengupta R (2020) Covid-19: impact on Indian economy. Indira Gandhi Institute of Development Research, Mumbai

Dey S (2020) India's badminton stars are awarded, but Jaduberia's famous shuttlecock makers live in poverty, Get Bengal. Blog. https://www.getbengal.com/details/indias-badminton-stars-are-awarded-but-jadurberias-famous-shuttlecock-makers-live-in-poverty. Accessed 4 Sep 2020

Folbre N (1986) Cleaning house: new perspective on households and economic development. J Dev Econ 22(1):5–40

Ghosal S (2012) Shortage of duck feathers has put shuttle cock manufacturing hubs in jeopardy, The Economic Times. https://economictimes.indiatimes.com/shortage-of-duck-feathers-has-put-shuttlecock-manufacturing-hubs-in-jeopardy/articleshow/17789059.cms?from=mdr. Accessed 28 Dec 2012

Pathak N (2017) West Bengal village a hub of shuttlecock production. World Asia. https://gulfnews.com/world/asia/india/west-bengal-village-a-hub-of-shuttlecock-production-1.2137648. Accessed 8 Dec 2017

Zhikuan S, Zhang Y (2022) Strategic crisis response of shipping industry in the post Covid-19 era: a case of thetop10 shipping lines. J Marian Sci Eng 10(5):635

# Chapter 19
# India's Smart Energy Solution for Sustainable Agriculture: Solar Irrigation Systems

**Mukesh Kumar** and **Amruta Pattnaik**

**Abstract** The solution to growing energy demands in India lies in solar panels that are both efficient and affordable. Indian farmers need intelligent irrigation systems fueled by solar energy. An integrated solar-powered water pump and autonomous, microcontroller-based irrigation system with moisture sensors make up this setup. Power for the whole system comes only from the sun. The paddy field is equipped with sensors that continually monitor moisture and water levels, relaying that data to the farmer. A farmer may find out how much water is in the paddy fields without going physically. A farmer can adjust the speed of the engine according to the depth of the water. However, if the water level rises to a dangerous point, the engine will begin running without the farmer's input to maintain the correct water level at the location. The answer has been presented to address the energy issue facing Indian farmers right now. Integrated solar-powered water pumps or autonomous, microcontroller-based irrigation systems help the environment by minimizing the demand on the electrical grid and the amount of water wasted due to leaks.

**Keywords** Smart irrigation · Solar power · Solar pump · Moisture sensor · Energy crisis

## 19.1 Introduction

Sun is the primary source of the great majority of the energy that is required by the planet. Solar power presents a solution to the challenge that the world is currently confronting with its energy supply that is both sustainable over the long term and kind

M. Kumar (✉)
Department of Physics, Swami Shraddhanand College, University of Delhi, Alipur, Delhi 110036, India
e-mail: mukeshrana@ss.du.ac.in

A. Pattnaik
Assistant Professor, EE, NIST University, Berhampur, Odisha, India



S. Anand et al. (eds.), *Emerging Geosustainability Transformations in India*, Advances in Geographical and Environmental Sciences,
https://doi.org/10.1007/978-981-97-7993-2_19

to the natural world. Recently, photovoltaic production has developed as a practical method for generating usable energy from the sun (UNDP 1981). This development occurred quite recently. Solar panels, which are simply arrays of photovoltaic (PV) cells are increasingly used to provide electricity for everything from individual devices like streetlights and water heaters to entire buildings. There may be a rise in the variety of businesses that use solar panels as the prices associated with making them continue to decrease. One example of the technology's utility is its incorporation into irrigation systems for agricultural purposes.

Farmers in India, facing a lack of access to electricity, might discover that a solar-powered irrigation system is a sufficient replacement for their current method. Once the upfront costs of this eco-friendly technique of energy generation are covered, it is possible to generate a limitless amount of power at no additional expense. It would be beneficial to have a definition of solar irrigation. Permit me to reassure you that this is not a complicated issue. To oversimplify things, a solar photovoltaic array, with or without inverters, will provide electricity to a water pump, which farmers use to irrigate their agricultural area and which is a significant assist in the entire process of irrigation (and a battery bank if it is required) (Harmim et al. 2012). In this study, we suggest a solar-powered automated irrigation system in which water pumps are driven by the sun to transport water from a bore well into a storage tank, where, it may be controlled automatically by a controller, and a moisture sensor before being pumped out to the irrigation area. In other words, the sun powers water pumps to transport water from a bore well into a storage tank, where the water may be controlled automatically by a controller and a moisture sensor. In other words, solar energy is used to power water pumps, which move water from a bore well into a storage tank, where it can be automatically controlled by a controller and a moisture sensor. An automated irrigation system's primary purpose is to fulfill its primary objective, which is to supply plants with the quantity of water essential to support their continued healthy development over the entirety of their life cycles (UNDP 1981). Even when the farmers are not there, these self-sufficient irrigation systems ensure that the crops receive the appropriate amount of water. In addition to this, it monitors the level of moisture present in the soil, which enables it to make the most effective use of water. Several aspects of automating the irrigation process have been credited in several studies that have been published. As technologies for water-efficient irrigation and automation continue to progress, it is expected that farms will begin to use autonomous irrigation more frequently (Tse et al. 2002; Kumar et al. 2021). One such example is an automated irrigation water control system that is based on GSM technology.

Millions of people all around the world have inadequate access to clean drinking water. Electric water pumps that run on diesel are commonly used to draw water from the ground in many areas. Photovoltaic water pumping (PVP), often known as solar water pumping, is an alternative. It's been operationally, fiscally, and environmentally sustainable after years of research and technology advancements. In addition, the solar-powered water pumps are preventing the usage of diesel pumps for off-grid electricity, which reduces environmental consequences and greenhouse gas emissions. This is reducing the amount of energy required for irrigation and making crops more resistant to the effects of climate change. There is no need to buy fuel, there is less carbon dioxide output, the systems last for long periods, they conserve water, and so on. These factors establish solar pumping as a sustainable solution to agriculture's water problems.

## 19.2 Literature Review

The Bureau of Electrical Energy (HP) conducted research in 2011 and discovered that India creates about 500,000 new connections annually with an average capacity of five horsepower (Haley and Dukes 2007; Prakash et al. 2011; Kumar 2014; Central Statistics Office 2016). According to this information, India is home to over 18 million agricultural pump sets. The agriculture sector is responsible for 19% of annual power use, which is equivalent to 131.96 billion kWh (Kumar 2014; Central Statistics Office 2016). According to the paper, the use of solar power to power intelligent irrigation systems is both the wave of the future in agriculture and a potential solution to the problem we are currently facing with energy. Therefore, the suggested solar-powered system makes use of methodologies obtained from literature reviews, with some minor modifications made to them. The article's description of the sine PWM technique for inverter operation with minimal harmonics leads to a significant increase in the system's level of efficiency. The system rating was determined based on the specifications of the pump (Central Statistics Office 2016).

## 19.3 Irrigation and Sustainability

Sustainable agriculture is farming done in such a way that it satisfies the current demands of society for food and textiles without affecting the ability of either the current generation or future generations to satisfy their wants. It may originate from a better awareness of the roles that ecosystems play in the world. There is a wide variety of possible courses of action that can be done to make farming more friendly to the environment. Adaptable agricultural practices and business procedures are essential for making agriculture part of sustainable food systems when expanding agriculture. Agriculture has a double-edged sword effect on the environment: it both causes and is affected by phenomena such as climate change (food systems are responsible for a third of anthropogenic greenhouse gas emissions), water scarcity, water pollution, land degradation, and deforestation. The food system is responsible for one-third of the anthropogenic greenhouse gas emissions; hence, agriculture has a major impact on the environment and has significantly contributed to the advent of climate change. Sustainable agriculture entails adopting farming practices that are not only good for the environment but also for the well-being of people. By adhering to these practices, farmers can cultivate land and raise animals in a manner that is harmless to both humans and the environment (Chaitali and Pranjali 2014).

It entails preventing undesirable effects on the soil, water, biodiversity, surrounding resources, or downstream resources, as well as on the people who work or reside on the farm or in neighboring areas. Practices such as permaculture, agroforestry, mixed farming, multiple cropping, and crop rotation are all examples of methods that could be included in a sustainable agricultural system (Luciana et al. 2013).

The continued existence of the human population in the future can be ensured in part by the establishment of food distribution networks that are ecologically sound and economically viable. For instance, one of the most successful solutions for mitigating the effects of climate change is the development of sustainable food systems that are based on sustainable agriculture. The challenge of how agricultural systems may continue to meet the demands of a growing population despite changes in environmental conditions may have a potential solution in the form of the adoption of sustainable farming practices. Alterations to one's diet that are geared toward producing more sustainable meals are an interlinked strategy that, in addition to sustainable farming practices, is a method to considerably cut down on the bad effects that humans have on the environment (Awati and Patil 2012).

Some areas receive an acceptable amount of rainfall for the production of crops, whereas the vast majority of other places require irrigation. Only irrigation systems that are well-managed—that is, those that prevent salinization—and those that do not utilize more water from their source than the environment generally replenishes can be referred to as sustainable. If this occurs, the water's source will essentially turn into a resource that does not automatically refill itself. It is now possible to consistently achieve high crop yields in regions that had previously relied solely on rainfall for successful agriculture. This has led to greater levels of unpredictability in the agricultural industry. Innovations in water well drilling technology and submersible pumps, as well as the development of drip irrigation and low-pressure pivots, have all contributed to this possibility.

The practice of agriculture in India is plagued by a myriad of difficulties, some of which include: inadequate power supply to farms; inadequate irrigation, fertilization, and drainage systems; inadequate transportation networks; limited access to financial resources; inadequate farm mechanization, and many more. The practice of agriculture has become exceedingly frustrating for farmers as a result of these impediments, which have also contributed to the low level of production seen in large-scale farming. On the other hand, the purpose of this research is to develop an automated irrigation system that utilizes a source of power that is both environmentally friendly and cost-effective. Such a system is an essential component of large-scale farming in India. An Automated Irrigation System (AIS) is a technology that applies water to boost the productivity of agricultural and crop production. This can be accomplished by using the use of either intelligence or artificial means in an automated irrigation system. Because it focuses on supplying an acceptable quantity of water at the proper time for the goal of increasing the growth and development of agricultural products, this condition is essential for situations in which there is insufficient rainfall. In agricultural irrigation, the impacts of the applied quantity of water, the time of irrigation, and the utilization of water are particularly significant elements to consider. Particularly in times of water shortage, there is a pressing need for an automated water system that is equipped with scheduling functions. This will help preserve around 80% of the water, which will in turn enhance water efficiency and agricultural output in general (Kumar et al. 2020). This is of utmost significance during times of severe drought, when available water is in short supply, as the restrictions placed on irrigation systems continue to become more stringent.

## 19.4 Proposed Solution: Automated Irrigation System

The solar pumping module and the Autonomous Irrigation Module are the two main components that make up the suggested solar automated irrigation system. A solar panel that conforms to the required criteria is included in the solar pumping module. This panel is positioned close to the pumping system (Kumar et al. 2014). After that, a control circuit transfers the power to a battery. The batteries supply the submersible water pump in the well with the electricity that it needs to operate through the use of a converter circuit (Fig. 19.1). After being held in storage for a short period, the water is pumped into a big tank that is suspended over the field.

A moisture sensor is responsible for controlling the electrical operation of the water release valve on the tank when an automatic watering module is present. For the sensor to monitor the progress of the crop, it has been installed in the field. The quantity of moisture that is present in the soil is converted into a voltage reading by the soil moisture sensor so that it can perform its primary purpose of providing a voltage measurement. This distinction goes to a moisture sensor circuit that provides the grower with the ability to modify the reference voltage in order to better accommodate the needs of the specific crop that they are working with. The difference between these two voltages has a direct bearing on the quantity of water that the soil must have accessible, and it also affects how much water is needed. The stepper motor was controlled by issuing a command to the stepper motor, whose rotational angle is proportional to the voltage difference. This method was utilized to get the desired results. A stepper motor is responsible for adjusting the cross-sectional area of the

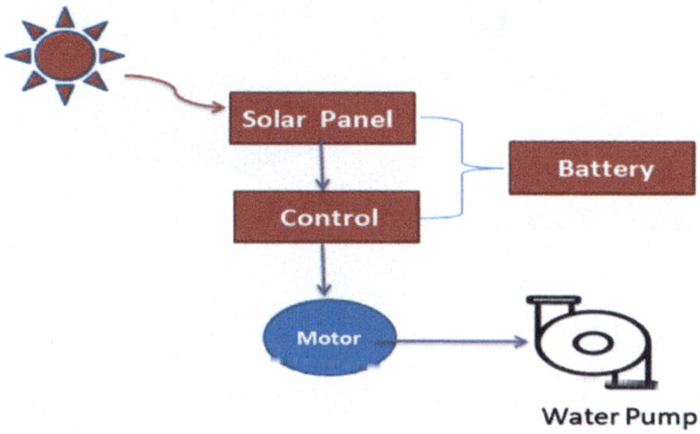


**Fig. 19.1** Proposed automated irrigation system. *Source* Authors

valve. This area controls whether or not water will flow through the valve. Therefore, there is a direct correlation between the moisture gradient and the discharge rate. The overall efficiency is also increased if we use artificial intelligence for controlling moisture content, tank storage, and pump control (Kumar et al. 2014) (Fig. 19.2).

The majority of farmers in India still use irrigation methods that are operated manually (Fig. 19.3). It requires a lot of work and a significant amount of time to use this method for monitoring the vitality and output of crops. Because it relies on human labour rather than an automated system, the cost of this method is significantly higher than that of the latter. By analyzing the level of moisture present in the ground, it is the farmer's responsibility to decide whether or not the field needs to be watered. The inaccuracy of this way of estimating water levels was one of the motivations behind the incorporation of a soil moisture sensor into our design (Gora and Dulawat 2017).

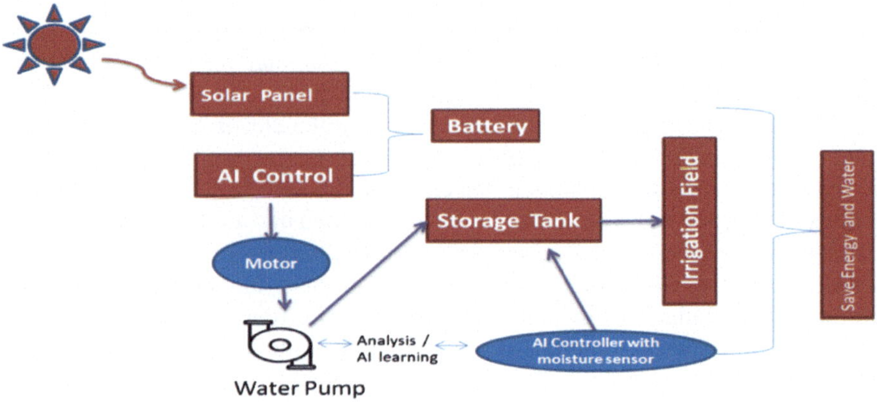


**Fig. 19.2** Sustainable layout for solar pumping unit. *Source* Authors

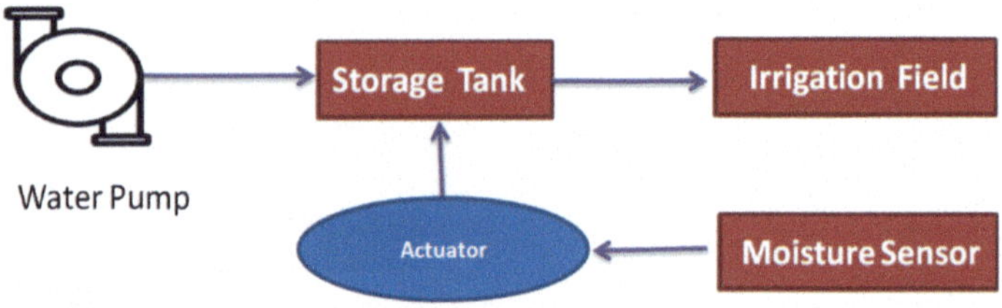


**Fig. 19.3** Tradition/existing irrigation system. *Source* Authors

## 19.5 Working Model

In the Autonomous Irrigation system, there are not just one but two different types of pumps at work: one that is powered by the sun and another that is controlled by a computer. Solar pumping equipment is responsible for converting sunlight into usable energy that powers the pump. Photovoltaic cells, which are located in close proximity to the pumping equipment, can gather sunlight and convert it into electricity. A controller circuit is in charge of monitoring and controlling the batteries. However, a smart irrigation system's electrical components enable the solenoid valve to be adjusted from afar. This valve is used to regulate water flow in response to readings from a soil-moisture sensing device. The sensors give a voltage readout that represents the amount of moisture. After receiving this voltage signal, the sensor unit compares it to a reference voltage the farmer has set up in line with the needs of the crop. The required amount of water is related to the voltage gap between the two sources. The signal is then sent to the motor, where its rotational angle is calculated based on the potential gap. Through the use of solenoid valves, the motor regulates the flow of water through them. Because of this, there is a direct correlation between the moisture difference and the volume of water that is travelling over the field.

The microcontroller is also linked to the GSMC, which enables remote, real-time monitoring and allows text messages to be transmitted to the user via the microcontroller (Fig. 19.4). If the moisture level is lower than the threshold value, the solenoid valves will open on their own automatically. The pipe's solenoid valves will remain open for a certain amount of time before they begin to close again in preparation for the next cycle. Due to the fact, that this device will be activated once every hour, it is more than sufficient for a plant to maintain the essential level of moisture. Similarly, the water level sensor in the tank will monitor the amount of water present in the tank and will activate the pump to draw water from the well if the level of water in the tank falls below a predetermined level. In either case, an SMS update is sent to the consumer, including information about the current water level, whether the engine is on or off, and the relative humidity. It is a terrific method to tackle the problem of where the energy to run everything is going to come from if every node can be powered by solar energy that comes from a single panel. The accompanying flowchart details the operation of the entire system in its entirety (Fig. 19.5). Because of this, the system is constructed in a particular way so that it can reduce the amount of money spent on sensor-to-sensor transmission.

## 19.6 Components of Smart Solar Irrigation System

The major components of a smart solar irrigation system are discussed below:

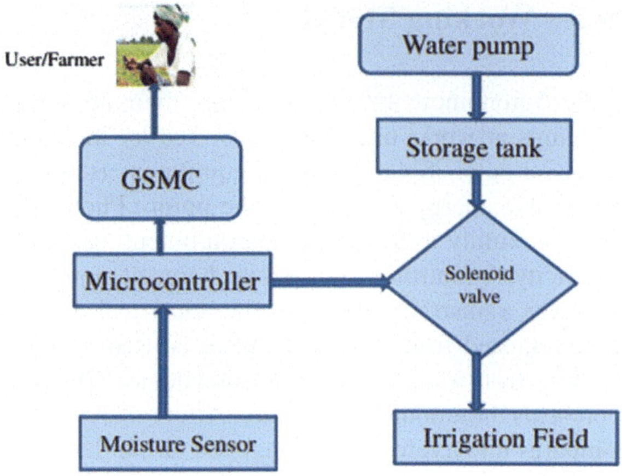


**Fig. 19.4** Layout of the smart irrigation unit. *Source* Adapted from Gora and Dulawat (2017)

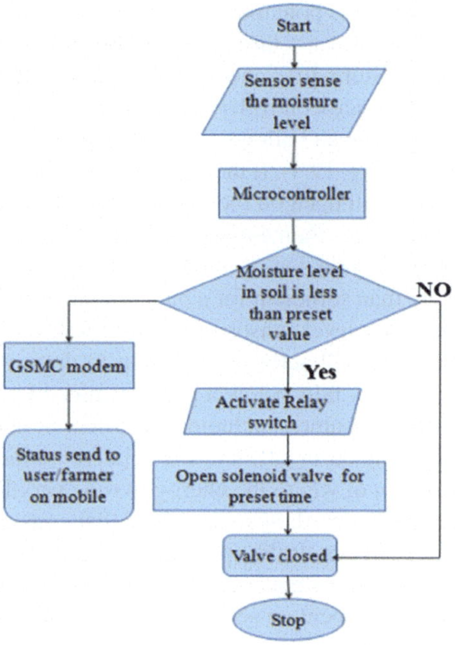


**Fig. 19.5** Flow diagram of the solar-powered smart system. *Source* Adapted from Gora and Dulawat (2017)

### *19.6.1 Solar Photovoltaic Panel*

An array of solar PV cells that have been electrically interconnected and mounted on a frame is known as a solar panel. Semiconductor materials like silicon are used to create PV cells. In essence, when light is shone on a solar cell, some of it is absorbed by the semiconductor. Direct Current (DC) refers to the electricity that can be used immediately or stored in a battery, and this is what the semiconductor produces.

### *19.6.2 Battery*

To convert chemical energy into electrical energy, batteries combine one or more electrochemical cells. In the system being proposed, solar energy is stored in batteries for later use. The energy reserve was put to use by powering the sprinklers.

### *19.6.3 GSMC (Global System for Mobile Communication)*

The GSMC is employed to inform farmers and other users of the status of their pumps via text message. The GSMC is an international benchmark for digital cellular communication. An external modem device, such as a GSM modem, is possible. Connect this modem to a free serial port and insert a SIM card from the GSMC.

### *19.6.4 Soil Moisture Sensor*

Among the components of the proposed system is a soil moisture sensor, whose job it is to monitor the moisture levels in the ground. At a predetermined depth, a cheap soil moisture sensor is buried. The soil moisture sensor merely detects the presence of moisture. The amount of current flowing through the soil can be used as an indicator of the soil's moisture content. These low-power, high-resolution sensors are ideal for a variety of applications. As a result, it's possible to create several estimates (say, hourly) over a long period of time with little battery drain.

### *19.6.5 Relay*

Electrical equipment may be managed by employing electromagnetic switches like the relay. Our suggested system employs a pair of relays for its switching needs. Relays switch both the microcontroller and the GSM signal; a different relay switches the RF signal. The relay transmits information. An electromagnet typically activates the switching mechanism in relays mechanically.

### 19.6.6 Solenoid Valve

An electromechanical device, a solenoid valve is often used in fluid or gas applications. A solenoid coil is used to control the valve using electricity. In this setup, solenoid valves are employed to regulate water flow. Solenoid valves can have anywhere from two to three ports; with a two-port valve, flow may be stopped or started, while with a three-port valve, flow can be redirected between two separate outlets. In fluidics, solenoid valves are the most commonly used control components. Their duties may include dispensing, regulating, or blending fluids. Due to their low control power requirements, small size, rapid switching times, good quality, long lifespan, medium material similarity, and high reliability, solenoids are the optimum alternative for many applications.

### 19.6.7 Microcontroller

A microcontroller is the brain of this operation. All sensors must communicate with the microcontroller. As a result, the microcontroller will need to make a call based on the circumstances. As a member of the ARM-7 family of ICs, ARM-LPC2148 sees widespread use. Phillips makes it and loads it up with a bunch of extra stuff. It's a good solution for both inexperienced and experienced programmers.

A solar photovoltaic panel is an array of interconnected solar PV cells, powered by semiconductor materials like silicon. The solar energy is stored in batteries for later use, which are used to power sprinklers. The Global System for Mobile Communication (GSMC) is used to inform farmers about their pumps' status via text message. A soil moisture sensor monitors ground moisture levels, detecting the presence of moisture and allowing for hourly estimates with minimal battery drain. A pair of relays manage electrical equipment, switching the microcontroller and GSM signal. A solenoid valve is an electromechanical device used in fluid or gas applications, to regulate water flow. A microcontroller, like the ARM-LPC2148, is the brain of the operation, requiring communication between all sensors. The ARM-7 family of ICs is widely used, making it suitable for both novice and experienced programmers.

## 19.7 Advantages of the Smart Irrigation System

Sustainability-based energy sector development includes energy generation, distribution, and use. Energy systems will greatly affect the environment in both industrialized and emerging nations. The worldwide sustainable energy system must optimize efficiency and reduce emissions. Economics underpins sustainable development. These efforts support long-term climate, clean air, and energy access goals.

However, millions lack access to water. Many towns employ diesel-powered electric water pumps to harvest groundwater. These devices release carbon dioxide and require expensive maintenance and fuel. Photovoltaic water pumping (PVP) is another option. It is economically, ecologically, and operationally sustainable. These qualities have made solar water pumping an exceptionally practical option to expand energy availability throughout underdeveloped countries and communities while generating a strong resistance to rainfall variations induced by climate change or inconsistent seasonal patterns.

Solar water pumping is ready for mainstreaming and has taken hold in some places, but communities, governments, and development organizations are unaware of its benefits. The proposed system does more than just regulate water flow; it also *ves materials by monitoring soil health in real-time and alerting administrators to any issues as they arise. Since there is no longer any need for human control, water and labour costs are reduced thanks to this automation. The flexibility of this system to work with a wide variety of crops, climates, and soil types is its greatest strength. Agricultural and horticultural areas, as well as public green spaces like parks, gardens, and golf courses, will all benefit from this set-up for their irrigation needs. Using renewable energy sources, this system is efficient, environmentally friendly, and a sustainable solution for mankind.

## 19.8 Conclusions

If the solar irrigation system that has been presented is to be implemented, it would be beneficial to a great number of individuals, including members of the government and farmers. The government is provided with a potential solution to the energy dilemma. By reducing the amount of water that is wasted and the amount of labour that is required, automatic irrigation systems help farmers save both time and money. The additional energy that is generated by solar panels may also be supplied to the grid. This not only provides the farmer with a potential source of income but also helps to improve India's energy situation. This might be accomplished by making some relatively minor adjustments to the circuitry of the system. The technique that has been recommended for watering plants does not put undue strain on the local ecosystem and is straightforward to put into practice. This technique has been utilized with great success to drill holes, which pump nonstop throughout the day. Solar pumps are another environmentally friendly choice because they do not put boreholes in danger of becoming contaminated with harmful substances. The system is largely capable of providing for its own needs and requires very little in the way of maintenance or oversight. The utilization of tracking arrays has the potential to result in increased daily pumping rates. By utilizing this technology, it is demonstrated that it is not only possible but also practical to use solar photovoltaics to power the pumps that are required for spray irrigation. Even though there will be significant costs involved in the beginning stages of creating this system, the benefits will make those costs worthwhile, and the system will be cost-effective in the long run.

## References

Awati JS, Patil VS (2012) Automatic irrigation control by using wireless sensor networks. J Exclus Manag Sci 1(6):1–7

Central Statistics Office (2016) Energy statistics India. Ministry of Statistics and Programme Implementation, Government of India, New Delhi

Chaitali RF, Pranjali KA (2014) Design and implementation of a real time irrigation system using a wireless sensor network. Int J Adv Res Comput Sci Manag Stud 2(1):401–404

Gora A, Dulawat MS (2017) Solar powered smart irrigation system: an innovative concept. Res J Agric for Sci 5(6):15–19

Haley and Dukes (2007) Evaluation of sensor-based residential irrigation water applications. In: ASABE 2007 annual international meeting, Minneapolis, Minnesota, 2007 ASABE Paper No 072251

Harmim A, Merzouk M, Boukar M et al (2012) Performance study of a box-type solar cooker employing an asymmetric compound parabolic concentrator. Energy 47(1):471–480. https://doi.org/10.1016/j.energy.2012.09.037

Kumar A, Kamal K, Arshad MO et al (2014) Smart irrigation using low-cost moisture sensors and XBee-based communication. In: IEEE global humanitarian technology conference, San Jose, CA, USA, 10–13 October, pp 333–337

Kumar M (2014) Taming waste via laws of Physics. Int J Sustain Energy Environ Res 3(3):164–170

Kumar M, Kesari JP, Rana L (2020) Flexible solar power tree: a concept. Int J Eng Appl Sci Technol 5(2):217–221

Kumar M, Rana L, Pattnaik A (2021) Solar tree: a sustainable energy approach for farmers. J Univ Shanghai Sci Technol 23(2):410–420

Luciana ML, Ramya B, Srimathi A (2013) Automatic drip irrigation unit using a pic controller. Int J Latest Trends Eng Technol 2(3):108–114

Prakash P, Sangster N, Cumberbatch E et al (2011) Investigating the feasibility of solar powered irrigation for food crop production: a Caroni case. J Assoc Prof Eng Trinidad Tobago 40(2):61–65

Tse KK, Ho MT, Chung HSH et al (2002) A novel maximum power point tracker for PV panels using switching frequency modulation. IEEE Trans Power Electron 17(6):980–989

UNDP (1981) Small-scale solar powered irrigation pumping systems. UNDP Project GLO/78/004. World Bank, Sir WilIntermediate Technology Power, London, UK. https://www.ircwash.org/sites/default/files/232.4-81SM-476.pdf. Accessed 21 Jun 2023

# Chapter 20
# Participation in Community Development and Role of State: An Exploratory Study of Mewat Area Development Project, Haryana

**Naresh Kumar**

**Abstract** The study attempts to understand the concept of participation and the role of the state in promoting sustainable development in the Mewat Area Development Project, in the Mewat region, Haryana. The study aims to problematize the concepts of 'participation' and 'community' in the development process. It tries to find out how these two concepts are contextualized and made operational in the project and what catalyst role the state plays in this process. However, these may not always align with the complex and diverse realities of the communities in situ. The romanticization of a single and (a)historical 'community' persists in our participatory approaches despite considerable evidence of the overlapping, shifting and subjective nature of dynamic and plural communities. The mainstream participatory approaches to development lay great stress on village-based solidarity within communities; processes of conflict and negotiation, in addition, exclusive inclusion and inclusive exclusion are either ignored or occasionally acknowledged but little investigated. Second, this approach ignores the contextual and socio-cultural history of the people of Mewat. The study uses a mixed methods approach and variables like embedded economic activities which include soil and water conservation, milk cooperatives, horticulture, social development including self-help groups, non-governmental organizations, and women empowerment to achieve the objectives.

**Keywords** Community · Participation in development · Role of state · Mewat

---

N. Kumar (✉)
Jawaharlal Nehru University, New Delhi, India
e-mail: jnunaresh@gmail.com



S. Anand et al. (eds.), *Emerging Geosustainability Transformations in India*, Advances in Geographical and Environmental Sciences,
https://doi.org/10.1007/978-981-97-7993-2_20

## 20.1 Problematizing Community and the Role of State

In the evolving landscape of participatory development, the withdrawal of the state from numerous social security programs has become a pivotal force shaping societal dynamics (Rahnema 1992). This deliberate retrenchment has not only deepened socioeconomic disparities but has also given rise to a palpable divide between the affluent and the economically disenfranchised.

The repercussions of this paradigmatic shift are extensive, fostering an environment where marginalized segments contend with a mounting panorama of economic inequality (World Bank 2000). Concurrently, the state's institutional apparatus, rendered functionally inept and apathetic, neglects the exigencies and aspirations of diverse social groups, compounding sentiments of alienation and disillusionment (Clarke 1992). This disconcerting tableau signifies more than just economic restructuring; it marks a profound erosion of the societal tapestry that historically binds a nation. The impact extends beyond the economic realm, permeating a broader sociocultural malaise. The conspicuous surge in crime rates serves as an emblematic manifestation of the toll exacted by policies marginalizing the welfare of the socioeconomically disadvantaged. Moreover, the vast state machinery, now a manifestation of systemic insensitivity, stands derelict in aligning with the authentic aspirations of varied social groups. As the chasm between the haves and have-nots widens, the imperative to reassess the equilibrium between fiscal prudence and social equity becomes increasingly urgent.

In the realm of development studies, participatory development assumes a pivotal role, intricately woven into the fabric of social networks, formalized structures, and grassroots initiatives. This research seeks to unravel and comprehend the nuances of participatory development by delving into the intricate social fabric of village communities and examining formalized participation in the Mewat Area Development Project (MADP) from 1995 to 2017 in the Mewat region of Haryana. Participatory development once heralded as a transformative force in critical development studies, has permeated discourse on poverty alleviation, community empowerment, indigenous knowledge, and bottom-up planning (Cook and Kothari 2001). However, the potency of terms such as "participation," "bottom-up planning," "empowerment," "civil society," and "indigenous knowledge" has collectively given rise to a new discourse in developing countries (Bastian and Bastian 1996).

The concept of participation, originally conceived as a catalyst for questioning and critical thinking, has paradoxically morphed into a catch-all phrase (Alejandro 2007), losing its ability to evoke scrutiny and focus (Sachs 1992). Additionally, Stirrat (1996) also highlights the challenges faced by non-governmental organizations (NGOs) in the participatory development process. Another significant aspect related to participation is the lack of clarity surrounding the concept of "community." The academic literature on participation in development has not properly defined the term "community", which takes on various hidden and subjective meanings and has been used uncritically in several mainstream research studies (Bastian and Bastian 1996; Cook and Kothari 2001). Ideas about local institutions are often based on problematic notions of "community" and "solidarity." The "community" in participatory development discourse is often viewed as a natural, social and cultural entity characterized by organic solidarity relations.

One of the key challenges haunting the participatory development discourse is the nebulous understanding of the term "community" (Guijt and Shah 1998). As the academic literature grapples with this elusive concept, it becomes apparent that the term carries hidden and subjective meanings, often used uncritically in mainstream research studies. The lack of a standardized definition for "community" poses a formidable obstacle, casting shadows over the very foundations of participatory development initiatives. Ideas about local institutions and solidarity are frequently built upon problematic notions of a monolithic "community," obscuring the diverse realities that exist within these supposed entities. The so-called solidarity approach upon which much development intervention is based may acknowledge social exclusion and stratification but assumes a commonality of people's interests that also helps the state to forge a 'staged' consensus (Williams 2004). At the heart of this research lies an exploration of the Mewat Area Development Project, a microcosm reflecting the broader challenges and triumphs of participatory development.

The temporal scope, spanning over two decades, provides a rich backdrop against which the evolution of participatory approaches can be scrutinized, offering insights into the shifting dynamics of state power, agency, and community participation. The new discourse in development, while initially well-intentioned, has inadvertently fostered complacency and conformity in the discourse surrounding participatory development.

The uncritical usage of terms like "participation," "bottom-up planning", and "empowerment" has obscured the heterogeneity of experiences and perspectives within communities (Kothari 2019). This research endeavors to rekindle the spirit of inquiry, peeling back the layers of rhetoric to expose the underlying realities of participatory development in the Mewat region. To comprehend the intricacies of participatory development, it is imperative to acknowledge the symbiotic relationship between the creation of development language and action research.

The language employed in the discourse shapes not only perceptions but also policies and practices on the ground (Stirrat 1996). As participatory development is woven into the very fabric of development studies, dissecting the language becomes a critical undertaking. This research aims to unravel the linguistic nuances that have contributed to the 'new discourse,' reinvigorating a space for critical reflection and reevaluation. Amidst the intricate tapestry of participatory development lies a crucial actor whose presence reverberates through the socio-economic landscape—the state. The withdrawal of the state from numerous social security programs, ostensibly justified by imperatives of fiscal discipline and structural adjustment, has played a pivotal role in shaping the dynamics of participatory development.

This calculated retrenchment has led to a discernible exacerbation of socioeconomic disparities, creating a societal landscape marked by a growing divide between the affluent and the economically disenfranchised. The consequences of the state's paradigmatic shift are manifold, contributing to an environment where marginalized segments grapple with an escalating panorama of economic inequality. Simultaneously, the institutional apparatus of the state, rendered functionally inept and apathetic, neglects to address the exigencies and aspirations of diverse social groups and communities. This neglect compounds sentiments of alienation and disillusionment, further deepening the societal divides. As the state machinery stands derelict in aligning with the authentic aspirations of varied social groups, participatory development assumes a critical role in bridging the gaps left by the retreating state The rise in crime rates serves as an emblematic manifestation of the toll exacted by policies that marginalize the welfare of the socioeconomically disadvantaged, underscoring the urgency to reassess the equilibrium between fiscal prudence and social equity (Polyani 1944).

Mewat, located in the state of Haryana, India, presents a complex political economy shaped by a confluence of historical, social, and economic factors (Maheshwari 2003). Historically marginalized and marked by high levels of poverty, Mewat grapples with issues such as inadequate infrastructure, limited access to education, and a predominantly agrarian economy. The region has been characterized by a significant Meo Muslim population, contributing to unique socio-political dynamics. Political representation and governance in Mewat have been central concerns, with efforts to address historical neglect and uplift the region (Mayaram 2004; Chowdhry 2011). As the area undergoes socio-economic transformations, there is a delicate balance between preserving cultural identity and fostering economic development (Mayaram 1997). Navigating these challenges requires nuanced policy interventions and inclusive strategies that recognize the specific needs of Mewat within the broader political and economic landscape of Haryana.

## 20.2 The Study Area

The Mewat region, Haryana formed in 2005,[1] comprises four tehsils of Nuh district and one tehsil of Palwal (Rao 2004). The region is situated between 26° and 30° north latitude and 76° and 78° East longitude. The Mewat region is surrounded by the district Gurugram on its north, Rewari from the west, and Faridabad from the east. The total population is 1.10 million spread over 1555 km$^2$ (Census of India 2011). It comprises four sub-divisions Taoru, Nuh, Ferozepur Jhirka, Punhana and Hathin. Meo is the preponderant social group (79%) followed by Scheduled Castes (SCs), Jats, Gujjars and Ahirs (Aggarwal 1971; MDA 2000a). The majority of the people live in the rural areas (89%) and 11% in the urban areas. It has a sex ratio of 906 females for every 1000 males and a literacy rate of 56% in which the male

[1] It was first carved out of Gurugram and Faridabad districts by the Om Prakash Chautala Govt. in 2004 and named as Satyamevpuram district. After that Bhupender Hooda Govt renamed it as Mewat district in 2005 and again renamed as Nuh. The reason given is that it is a cultural region spreading over Haryana, Rajasthan and Uttar Pradesh. Therefore, the Mewat region in Haryana has been renamed as Nuh to distinguish from the rest.

literacy rate is 73% and the female 37.60%. The gender gap is 35.40 which is double the state average (18.12).

The demarcation of the Mewat region, Haryana, in this study, is strictly based on the area for which the Mewat Development Board[2] (MDB) works. Official meetings of the MDB took place in Chandigarh. After the partial failure of the 'Gurgaon Experiment' (1928) and the Community Development Programme (CDP) in the 1950s, the Mewat Development Board was constituted in 1980 by the then Haryana government "with a commitment to deliver social and economic justice to the backward and underprivileged sections of society" (Amir-Ali et al. 1970; Ankit 2020). An autonomous field agency known as Mewat Development Agency (MDA) at the field level was created to support the development initiatives in the region.

Dr. S.Y. Qureshi was the first Chief Executive Officer (CEO) of MDA and played a crucial role in its formation years. Project management was the responsibility of the MDA. The project established and covered the costs of 3 field offices to facilitate coordination of field activities. Social development activities were sub-contracted to NGOs through the Mysore Resettlement and Development Agency (MYRADA[3]) and other service providers while all other project economic activities were given to the respective state government departments.

The role of the NGOs was to strengthen local community organizations so that they became capable of sustaining development activities. They were the project's main partners in awareness raising, participation and cooperation at the village level. As most members of the MDA Board had official responsibilities, the CEO of MDA was responsible for project implementation. Project implementation, however, suffered from frequently changing CEOs (10 CEOs in total from 1995 to 2002). In 2002, MDA management was restructured and a chairman was appointed to oversee project implementation in a time-bound manner (Kumar 2011, 2016). This decision improved Annual Work Plans and Budget (AWP & B), coordination processes and accelerated project implementation. There was a monitoring and evaluation (M and E) problem in MDA because of severe staff crunch.

MDA did not generate enough data so that an impact assessment could be carried out. As a result, it had only little data to compare the post-project situation with the pre-project situation since no baseline survey was conducted.[4] Funds were not provided by the Government of Haryana in 2002 (Singh 2002) which slowed down the pace of project implementation especially social development by NGOs.[5] Staff

---

[2] MDB is headed by the Cheif Minister of Haryana, the Chairperson MDA, Faridabad Commissioner is the Member Secretary and other members includes several Secretaries of the line Departments and two NGO representatives.

[3] MYRADA was chosen because there was no experience NGO in the district and execute the social development activities as envisaged in the MADP.

[4] Only one Baseline survey was sponsored by Ministry of Minority Affairs (Govt. of India) and Indian Council of Social Science Research (ICSSR) and executed by the Institute of Human Development (IHD), New Delhi in 2008.

[5] The CEO of MDA during that time Mr. P.R. Bishnoi had issued an order to stop funding to NGOs on the directions of Mewat Development Board (MDB) headed by the Chief Minister, Haryana. The international Fund for Agricultural development (IFAD) and United Nations Office for Project

salaries could not be paid and field activities were hampered a lot. The project led to institutional strengthening and service improvement. However, it did not result in changing the style of the cooperating government departments.

The most important achievement of the project in terms of institutional building was the restructuring of the Mewat Development Agency under one CEO and the management under the supervision of the Deputy Commissioner as Chairperson (MDA 2000a). These changes resulted in a better coordination approach to project implementation which was badly needed as the region was previously split into two districts at that time.

## 20.3 Theoretical Underpinning of Participatory Development and State

The concept of participation in development has been a subject of extensive exploration and analysis, particularly within the framework of neoliberal ideologies (Leal 2007). His investigation into the changing meanings and implications of "participation" during the neoliberal era sheds light on the dynamic evolution of this discourse. Taking a South Asian perspective, the study by Bastian and Bastian (1996) engages in a critical dialogue, presenting diverse viewpoints on the challenges associated with assessing participation. This collection not only offers insights into the regional dynamics of community involvement in development processes but also underscores the cultural and contextual variations influencing participation.

Bhattacharya et al. (2004) delves into the intricacies of social capital and its impact on participation. Many studies add depth to the same discourse by highlighting the complex intersections between gender, community, and participatory development by providing valuable insights into the gendered aspects of participation dynamics (Cornwall 1998; Bhattacharya et al. 2004). The study by Williams et al. (2003) on the local state in rural Bihar, Jharkhand, and West Bengal significantly contributes to understanding how the state functions at the grassroots level. By emphasizing the importance of acknowledging local dynamics and power structures in participatory development initiatives, this work challenges the notion of a universal approach and underscores the need for context-specific strategies.

Krishna (2002) introduces a theoretical framework for tracing the roots of development and democracy. It broadens the understanding of participation beyond a conceptual framework, positioning it as a dynamic force shaping development trajectories. Empirical evidence is presented in the study by Gaiha (2004) on the utilization of local knowledge in impact assessment in rural India and the Ministry of Minority Affairs', Govt. of India Baseline Survey of Minority Concentrated Districts (2008) in Mewat. These studies not only underscore the practical implications of participation but also highlight the necessity for context-specific approaches in development

---

Services (UNOPS) protested the move and subsequently the order was repealed. So, the diversion of funds was common and it hampered project activities especially community development.

interventions, reflecting the diverse needs and conditions of different communities (Gaiha 2004, Ministry of Minority Affairs and ICSSR 2008).

Nelson and Wright (1995) explore power dynamics in participatory development and research, emphasizing necessary shifts for effectiveness. Bridging theory and practice, it examines theoretical foundations and diverse case studies. The focus extends to power dynamics between researchers and communities, essential for participatory operation. Targeting academics and professionals, it serves as a valuable resource for global participatory initiatives. In addition to regional perspectives, international viewpoints on participation are provided by the World Bank's (2000) influential 'World Development Report 2000/2001: Attacking poverty', and the United Nations (2008a, b) project completion reports on the Mewat Area Development Project (2008). These documents contribute to the broader discourse by presenting overarching strategies and policies, showcasing how participation is integrated into global development agendas.

Cooke and Kothari (2001) and Stirrat's critique (1996) of participation, empowerment, and other buzzwords challenge the prevailing orthodoxy in development studies. By highlighting the potential pitfalls and unintended consequences of participatory approaches, these works advocate for a more critical stance among scholars and practitioners, encouraging a nuanced understanding of the complexities surrounding participation. Expanding the discourse, Gaiha and Kulkarni (2002) further explore the role of local governance structures in facilitating or impeding participatory development. Stirrat (1996) critically examines the new orthodoxy and old truths, scrutinizing popular concepts such as participation and empowerment.

The literature on participation in development reveals not only its dynamic and context-specific nature but also emphasizes the necessity for nuanced and critical perspectives. Understanding the intricate intersections between participation and various contextual factors is crucial for formulating effective and sustainable development interventions that genuinely empower diverse communities.

## 20.4 Data Source and Methodology

Database is an important and integral part of any research. However, it is very significant in social sciences where the probability of predicting human behaviour is relatively low (Johnston 1978). The study uses both secondary and primary data. The secondary data sources include the Primary Census Abstract (Census of India 2011), ICSSR Baseline Survey Report and various evaluation studies and associated organizations like the International Fund for Agricultural Development (IFAD) and Statistical Abstract of Haryana (Government of Haryana 2015–16). Various newspaper and magazine reports have also been consulted. There is a paucity of secondary data on different and dynamic aspects of social networks. However, the lack of secondary data has implored the researcher to undertake an extensive primary survey in the fulfillment of the objectives of the study.

Initially, key information about the villages was collected in the form of a community schedule. It included both open-ended and closed questions about participatory development and social networks in the village. Indicative and relevant questions on demographic characteristics, resource endowments, livelihood opportunities and participation in development were framed. Key informant Interviews (open-ended and closed) were held with the key stakeholders in the formulation, designing, dissemination and implementation of schemes at the state, district, block and village levels. The second step is the selection of sample villages. A Panchayat village represents some stock of social networks in the sense that it is governed by a Gram Panchayat and the Gram Panchayat in turn is connected to the block level and then to the tahsil and district level. Therefore, a Panchayat village is an ideal unit for studying social networks at the village level.

A total of five Panchayat villages[6] (revenue villages six) were selected for the detailed survey (case study and focus group discussion) based on stratified random sampling in consultation with the local people (Fig. 20.1). These five villages are Mohammadpur Ahir in Taoru, Ghasera in Nuh, Mandikhera in Nagina, Indana in Punhana and Mandkola village in Hathin Tahsil. Two villages were selected for watershed development and agriculture. One key-informant interview was done with a senior ward member as the village Sarpanch was away due to a medical emergency in the city. One beneficiary from each of the watershed works in both villages, agon and Patkhori, was interviewed. Empirical data was obtained from the Project Completion Report (2008) and the MDA website. One beneficiary from each cooperative society (a total of ten, two from each village) from the sample village was interviewed. Empirical data was obtained from the MDA website and MDA (2000b).

The key-informant interview was conducted with at least one member (total 5) of the cooperative society whomever was available. Key-informant interview with field staff at Pinangwan Nursery in Punhana as the official-in-charge was suspended over negligence pending an inquiry. Empirical data over the years was obtained from the MDA website and Project Completion Report (2008). Two SHGs (Self-Help Groups) (a total of 10) were selected from each village. Empirical data about SHGs and SHG Federation was obtained from the MDA Website. Key-informant interviews (total five) with field coordinator of NGOs Deepalya in Taoru, Development Support Team (DST) in Nuh, Mewat Development Society (MDS) in Ferozepur Jhirka including Nagina, Mewat Social and Educational Development Society (MSEDS) in Punhana and Sehgal Foundation in Hathin. Qualitative data from interviews was analyzed using MAXQDA 12.0. and the responses were analyzed. All the figures except watershed development were taken from the field. Location and sample villages map was created using ArcGIS 10.3.

---

[6] A Panchayat village is one which is governed by an elected Gram Panchayat which consists of Sarpanch, Deputy-Sarpanch and ward members Depending upon the population of the village. A panchayat village may have several revenue villages. While a revenue village may not have a Gram Panchayat but it has a defined boundary and created for the purpose of revenue collection.

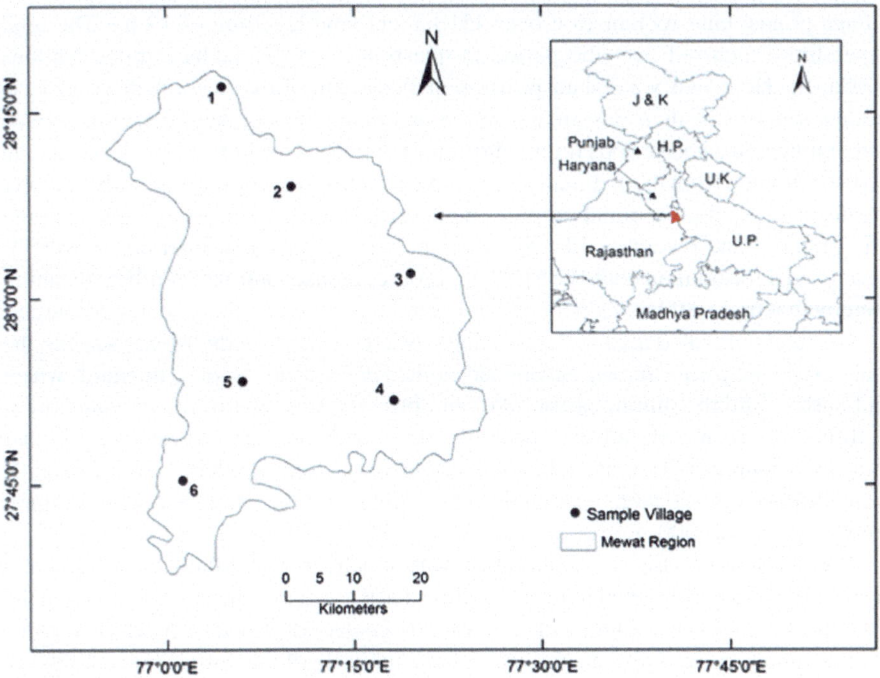


**Fig. 20.1** Location of study area and sampled villages. *Source* Authors (2021)

## 20.5 Results and Discussion

### 20.5.1 *Soil and Water Conservation*

The Mewat region is mostly semi-arid and subject to frequent droughts. Two-thirds of the study area has brackish water, restricting its use for irrigation (Government of Haryana 2015–16). As a result of excessive pumping, the water table has fallen steadily over the years. Since the Aravalli Hills are devoid of vegetative cover, the heavy run-off during the monsoon season causes acute soil erosion. Excessive grazing of animals has aggravated the drought proneness of this region beyond repair. Under these unfavorable conditions, considerable importance is attached to soil and water conservation works.

The objective of this sub-component is to increase rainfed cultivation cover with proposed better 'community' watershed management practices. In principle, the strategy is to use indigenously developed techniques with active 'community' involvement. It was proposed to form watershed management committees around the work sites.

The project or works completed include field-bunding over 964 hectares (ha), 173 check dams, 235 gully plugs, 312 diversion channels, 110 percolation ponds, 108 village ponds, land reclamation over 451 ha and land levelling of 13 ha. The total expenditure incurred over the period in question is 457.77 lakhs (United Nations 2008a, b). Here two watershed projects in Ferozepur Jhirka are taken up as case studies in terms of their design, execution and maintenance. Negligence in service and maintenance had serious implications for the life course of these projects and the benefits flowing from them. Finally, even if the benefits to the farmers are substantially greater with good monsoonal rainfall,[7] neglect of the group grazers and others living off common land resources like Shamlat and Gram Panchayat land taken over for watershed development could be a potential source of distributional conflicts meaning thereby that trade-off between the two is essential to maintain these scarce resources.

The total cultivated area in Agon Village is 2786 acres.[8] In the Kharif season, the main crops are *jowar, bajra, guwar* and in the *rabi* season, *moong,* mustard, wheat and barley. Mainly, onion, carrot, okra and brinjal are grown in the irrigated areas in different seasons. Relatively speaking, onion and carrots yield relatively higher returns. A total of 600 households live in this village out of which 480 households were farmers and 120 were near landless or landless. Of the latter, 90 are agricultural labourers (MDA 2000c).

Another sample village is Patkhori, a very much smaller village than Agon and about 95% area is cultivated here. The main Kharif crops are *bajra* and *jowar* and the *rabi* crops are wheat and mustard. The varied vegetables grown in different seasons include tomatoes, brinjals and chillies. A large majority of the sample households are dependent on farming as the main source of household income. Although tubewells are an important source of irrigation, the dependence of the beneficiaries on rainfall remained heavy. The reasons included insufficiency of well water for irrigation, decline in water level due to deficient rainfall and the time needed for recharging it. Sehgal Foundation carried out watershed works in the Patkhori village (Fig. 20.2).

### *20.5.2 Participation' in the Milk Cooperatives*

Except for the cooperative at Mandikheda village, all other cooperatives were formed by the villagers only. Those who took the initiative also had a role in their management as well as office-bearers. Detailed data on the formation and membership was available on 7 cooperatives. Out of these, two were formed in 1995–96, one in 1997–98 and two in 1999. Registration of these cooperatives was often delayed. Whether this was a result of non-compliance with procedural requirements or bureaucratic

[7] One of the major limitations of watershed development component was that these projects went to those parts where the farming was relatively better than other parts of Mewat which means there was fault with the project design, community consultation and award of the projects.

[8] During the brief tenure of L.S.M. Salins (19-11-1999 to 2000) as MDA CEO, A prominent local *jat* MLA (Member of legislative Assembly) is alleged to have diverted the funds meant for Meo area to his constituency in Hathin, although part of Mewat but in Palwal district. So, the diversion of funds was not uncommon.

**Fig. 20.2** Water storage tank built in association with Sehgal Foundation and rooftop rainwater harvesting in the village school, Patkhori, Ferozepur Jhirka. *Source* India Water Portal (2014)

inefficiency could not be found out. The functioning of the cooperatives in general was unsatisfactory and quite uneven in some respects.

Some estimates of the growth of membership were available. In one out of four, for which such estimates were available, there was a fairly rapid growth of membership from 24 in 1995–96 to 138 in 1999–2000. In the remaining, the growth was slow or moderate. Whether there was a growth of active ‘participation’, however, was not known. Some people were of the view that active membership was more or less unchanged while others were emphatic that there was a decline. The latter cannot be ruled out because of the switching of milk supplies of more than a few of the members to private buyers. Moreover, there is some evidence of a slight reduction in milk collection over the period 1997–98.

Evidently, among those, who joined the cooperatives, initially, there was an element of self-selection. These members presumably belonged to households that did not require advance cash payment and, therefore, were attracted by the higher price range of the cooperatives. On the other hand, among those who withdrew or did not join the cooperatives, a change in the economic circumstances counted for much. But there were other factors too connected with the mismanagement of the cooperatives and the lack of transparency in their functioning. Although there was a reluctance among the members to report mismanagement or lack of transparency in the decision-making, some indicators explained a similar story.

Out of the nine cooperatives, three-yearly elections of the managing committees were held in two. General Body and executive committee meetings (every third or fourth month and every month, respectively) were regularly held in just three cooperatives. While these cooperatives also maintained minutes of these meetings and were accessible to the members, in all other cases, the only records available were confined to quantities of milk purchased and the price paid. In general, there was no information on the revenues and expanse of the cooperatives. What is somewhat surprising is that nobody protested or raised questions about their unsatisfactory

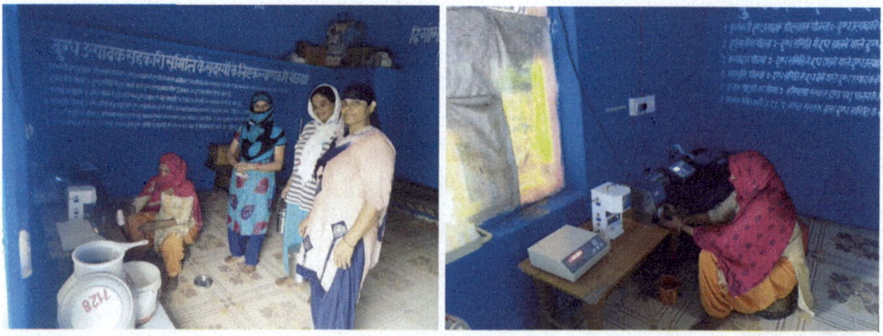

**Fig. 20.3** Cooperative members collecting milk and testing of milk fat using an electronic tester. *Source* Author (2021)

functioning. There could be many reasons like apathy or fear of retaliation from the management that could not be found out.

Nine out of the ten cooperatives used a manual milk fat testing procedure that took roughly 20 min per sample. As against this, the electronic device used in the Mandkola cooperative took half a minute (Fig. 20.3). With a large membership, the manual testing procedure involved long waiting periods. Although none of the households reported cheating by the staff at the time of testing of the samples. These responses cannot be accepted at face value. In any case, it was common knowledge that there was an adulteration of milk with water during transportation.

Initially, the Haryana Dairy Development Federation (HDDF) provided veterinary services through its network. Realizing that this duplicated the work of the Department of Animal Husbandry and Dairying (DAHD). The responsibility for these services was assigned exclusively to this department. In the absence of the provision of these services by the HDDF, the cattle owners faced a difficult situation. Government doctors were few and far between and they charged exorbitant fees for home visits. Medicines added considerably to consultation fees; government hospitals were rife with corruption and among the private practitioners, it was difficult to distinguish between a quack and a trained person. Given the acute scarcity of drinking water, fodder and unhygienic conditions in this region, the cattle were highly vulnerable to disease-especially a form of brain fever. As this could result in a substantial loss of household income, several owners were willing to pay for veterinary services in case the Federation took the initiative to organize these services.

Complaints were voiced against the negligent attitude of absenteeism among the field staff and lack of professionalism among the Secretaries of Village Cooperatives. These resulted in low collection of milk, adulteration and delays in payments. It was, for example, pointed out by some people that the field staff seldom checked milk collection and testing and were often unavailable for days to sign the bills without which milk suppliers could not be paid. It was also pointed out that even small villages had two or three cooperatives that were not viable without subsidies.

A case in point is *Mandikheda* village which had 90 households and three cooperatives. These cooperatives were set up by the field officers of the Nuh chilling center to meet ad-hoc targets linked to subsidies under the MADP. In their zeal to set up cooperatives, little, if any, attention was given to their viability. The overall impression of the functioning of the cooperatives is far from impressive despite the hesitation to express a frank opinion. One respondent, for example, was emphatic that the functioning of cooperatives in his village was messy and did not result in sustainable income and livelihoods.

### *20.5.3 Horticulture*

The soil and climatic conditions of Mewat favour horticulture. Several varieties of vegetables such as tomato, onion, cauliflower, cabbage, brinjal, chillies, radish, turnip, carrot, garlic, etc. are grown here. Among fruits, guava, mango, papaya, berries, banana, etc. are also grown in large quantities. The potential for acreage expansion, is, however, limited by the lack of irrigation and use of traditional inputs resulting in low yields. The total area under vegetables in 1993–94 was 928 ha (barely 0.62% of the total cultivable area) and under fruit trees 97.8 ha (0.06% of the total cultivable area). Yield rates were also lower than in other parts of Haryana and elsewhere about 15–20%.

Specifically, the yield for tomato, for example, was 16.1 tonnes/ha, for onion 14.8 tonnes/ha, and for cauliflower 13.4 tonnes/ha in Mewat. Two strategies are, thus, relevant here: one is to expand the acreage under horticulture by expanding irrigation and the other is to intensify horticulture in irrigated areas through the use of better-quality inputs and improved pre-harvest and post-harvest management practices. MDA has chosen to concentrate on the latter.

The horticulture development scheme was designed to expose the farmers to new seeds, fertilizers and disease-control chemicals. It was referred to as Arable Crops and Horticulture Development Scheme. Further, it aimed to expose them to pre-harvest and post-harvest management techniques through demonstrations on selected farms. Besides demonstrations, short-duration training in cultivation techniques, and the use of inputs were also imparted. Field visits to *Krishi Vikas Kendra* (KVK) of Haryana Agricultural University (HAU), Hisar were organized. A field nursery was established at Pinangwan, Punhana, to supply fruits and vegetable seeds and seedlings suitable for the agro-climatic conditions of the region (Fig. 20.4). It was claimed that the demonstrations and inputs were primarily for the benefit of small and marginal farmers. SHGs were involved in the identification of small and marginal farmers. The extension wings of MDA were supposed to be in close touch with the farming 'community' to facilitate the adoption of new inputs and practices.

As a part of the promotional strategy, the following measures were undertaken: seeds, fertilizers and pesticides for growing high yielding variety of seeds of tomatoes, onions, brinjals, etc. on an acre of land were given to 1520 farmers together with demonstrations of how to grow them; 11,330 households received mini-kits for

**Fig. 20.4** Submersible tube well at *Pinangwan* horticulture nursery farm; plantation nursery spread over and Neem, *Papdi* and baingan nursery trees. *Source* Author (2021)

growing nutritive food; 60 demonstrations for growing fruit such as guava, berries, etc., were given; 10,057 farmers received plants for growing guava, berries and citrus fruits; 2140 farmers (including women) received training in growing vegetables and fruits; expansion of the Pinangwan nursery through the installation of tube-wells, purchase of a tractor, construction of a threshing floor and a training hall, etc. (United Nations (2008a, b). A total sum of Rs. 107.26 lakh has been spent under this scheme.

While the range of promotional measures[9] undertaken has been associated with significant income gains, the intended beneficiaries have been largely bypassed in the favour of the better-off and the gap between the actual and the potential gains is huge. There was no community consultation in project design and implementation which resulted in no gains to the beneficiaries.

## *20.5.4 Self-Help Groups*[10]

As women are a particularly deprived section of the Mewat society—in terms of literacy, communal property rights and autonomy—a major thrust of the MADP is their empowerment through modest but sustained economic betterment. In pursuit of this objective, SHGs are created that encourage thrift, facilitate access to credit, promote self-employment and above all enable women to overcome sociocultural

[9] The Development Support Component covered the awareness and sensitization measures of the MADP. During the formative years of the project, the fund utilization was very slow. Funds were diverted from one component to other. In fact, it was reported that at that time, the Chef Minster rally was sponsored with the project money.

[10] SHGs here is largely refereed to women SHGs.

barriers to 'participation' in the 'community' sphere.[11] In the formation and growth of SHGs, the NGOs are assigned a key role. 492 SHGs were organized during 1998–2000, with a total membership of 6703. The savings mobilized by the SHGs were of the order of Rs. 34.30 lakhs. 331 SHGs disbursed loans totaling Rs. 42 lakhs (United Nations 2008a).

The dependence on the moneylender was significant. Three members of SHGs borrowed amounts ranging from Rs. 1000 to Rs. 5000. Another important source of credit was private milk buyers. Five members borrowed amounts ranging from Rs. 6,000 to Rs. 18,0000. Another institutional source was Kalyan Nigam—a state corporation created mainly for the welfare of Scheduled Caste and Scheduled Tribe households from which two households borrowed Rs. 1,0000 each. Although there was a wide range for the loans from SHGs as well from Rs. 1,000 to Rs. 3,0000—eight members borrowed amounts ranging from Rs. 1,000 to Rs. 1,0000 (United Nations 2008a).

What is significant is the dependence of member households on multiple sources of credit. Although institutional sources of credit have expanded in recent years in terms of disbursed amounts and their spatial spread, access to these sources, especially for the poor remains limited due to procedural and other requirements. Unavoidably, dependence on the local moneylender has still not diminished much despite exorbitant interest rates and their malpractices. Usually, no collateral is required, and loans are promptly given. Borrowing from moneylenders against contingencies is, comparatively, convenient. The rate of interest charged by the banks ranges from 12 to 14% per annum, with the lower limit applicable to SHGs and the upper to individual loans. If to the latter, transaction costs of borrowing (e.g., bribes, time wasted) from the banks are added, the cost of bank credit is markedly higher.

Local moneylenders charge considerably higher interest rates of 3–5% per month depending, not so much, on the loan amount or the urgency of the need. Loan from milk buyers and traders involved a pre-commitment to sell the entire output. The milk buyer, for example, offered a lower price to those who required advanced payment and a commitment that the entire milk output would be solely to him. Similar arrangements were reported for loans given by other traders.

A majority of the members were inducted through the initiatives of MSEDS fieldworkers. Two other members joined after hearing about successful SHGs from friends and relatives who were already members. Another joined after becoming aware of the credit facility of SHGs in a group meeting. Yet another was so impressed after attending a group meeting that she decided to start her own group. The reasons for not joining SHGs included (i) the inability to pay loans, (ii) resistance from male family members, (iii) lack of awareness of the potential benefits of joining SHGs, (iv) lack of interaction with NGO fieldworkers. Neither SHGs nor NGOs admitted social exclusion as a barrier to joining an SHG.

[11] During the fieldwork, a very interesting thing came to light. It was reported by the women SHG members that initially, MDA also supported SHGs of menfolk but they could not sustain as the members could not mobilize savings or just squandered money, therefore, after sometime, they were closed.

The process of formation of SHGs was slow and uncertain due to strong resistance from male household members. Apart from the taboos on the interaction of Meo women with strangers,[12] serious apprehensions were raised about the loss of savings. Unavoidably, therefore, NGO fieldworkers had to convince the male members first of the potential benefits of SHGs. The resistance was overcome when they were confronted with examples of successful SHGs. Usually, the formation of a group took one month or longer, depending largely on social obstacles. After some initial training, a group is formed.

As a priority, savings are encouraged. These are deposited in a bank. Bank borrowing usually begins after six months of the formation of an SHG. The group size of SHG ranged from 13 to 16 members. The majority of them were homogenous and a few were mixed (comprising both Meos and Non-Meos). The Sangam Group in Indana village, for example, consisted entirely of *Chamar* (Scheduled Caste people) *and Jats* community people. By contrast, The Mahila Sudhar Group in Mohammadpur Ahir, Ghasera was mixed, with Meos, *Chamars* and Miyan (Meo, but lower in the hierarchy) as members. All the SHGs formed showed social cohesion among themselves.

In most cases, SHG met for times a month. A member was nominated to chair the meeting on a rotational basis. The venue was usually a 'community' temple or the residence of a member. Generally, the agenda for a meeting was not fixed in advance. Various matters considered important by the members present—including personal problems—were discussed. The meetings were well attended with very few absentees. The entry and exit rules were collectively framed with some help from NGO fieldworkers.

Considerable importance was attached to saving mobilization. The monthly amounts saved by the members varied from Rs. 50 to Rs.100. Normally savings were collected between the 1st and 10th of every month. An account was opened in the name of the group in which savings were deposited. After a period (i.e., usually at least six months), given satisfactory savings mobilization, the group was able to borrow from the bank (at an annual interest rate of 12%). Disbursal of this amount among the members was decided in a group meeting following the priorities collectively agreed upon. The loans were repayable in monthly installments over a year. The monthly interest rate was 2%. The difference between the borrowing and lending rates accrued to the members. The repayment rates were very high with hardly any defaulters. Although the range for the loans was wide, most borrowers took small loans. Since there was no restriction on their use, they were used for contingencies as well as directly productive activities.

All records of loans and savings were maintained by and kept in the custody of the head (Pradhan), the group leader, or a group member designated by SHG members. NGO fieldworkers assisted in this task. All members had access to these records. Other activities included awareness building and training relating to 'community'

[12] The Meo women do like to interact with the people outside the village. While this is common among the womenfolk in every peasant village society in north India, this may be due to the practice of Islam religion among them. For more, see Yoginder Sikand (1995).

welfare such as health, hygiene and sanitation. This helped group members to broaden their sphere of activities. The Ambedkar Group in Mandikhera village, Nagina, for example, undertook cleanliness campaigns every second month. In addition, the training was imparted in income-augmenting skills e.g., tailoring. The Laxmi Group in Mandkola village, Hathin had started a *dhoop* (Incense sticks) making and tailoring center.

The drop-out rate was negligible—not more than 1% from the SHGs. Two reasons for dropping out were frequently cited: migration and employment outside the village. What is significant is that no one dropped out because of repayment difficulties and other domestic issues. Subject to minor variations, when a member left the group, her savings and interest were refunded. New members were enrolled with the consent of the existing members. The former was bound by the rules and obligations of the group. A resolution was passed including the names of the new members (instead of those who had left the group) and a copy was sent to the bank. MDA gives a matching grant of Rs. 7000 to an SHG that performs well on certain criteria, e.g., savings mobilized, loan disbursal, and repayment rate. This is credited to the bank account of the SHG.

## 20.6 Findings and Conclusion

It is clear from the discussion that there is a dire need for drastic restructuring of the allocation pattern of the MADP budget to promote a cluster of activities, as opposed to promoting specific activities in different areas. Going by the findings, a major drawback in the development of horticulture and milk producers' cooperatives is the lack of easy access to institutional credit over private moneylenders. Since institutional credit through SHGs is also a major component of MADP, it is possible that under certain conditions, greater emphasis should be placed on the formation of women SHGs in the Panchayat villages with high concentrations of horticulture and dairying would add substantially to the cost-effectiveness of MADP. The diversion of funds should stop across different components and regions. However, there are some pockets of extreme poverty in the region especially Nuh, Ferozepur Jhirka and Punhana blocks where there are few income-earning opportunities. In this belt, there is a maximum concentration of Meos. The role of SHGs in these pockets may help in lessening the impact of poverty in these villages. There are many drawbacks to the project. One of the major drawbacks is the failure to strengthen the existing village institutions in the study area. There is total or near exclusion of the statutory panchayats from project activities, justified by the project implementing agency on the ground that they were 'inactive', 'faction-ridden', and 'not trusted' by the villagers.

Since the Gram Panchayats have exclusive control over common resources, many watershed structures were built by them, and their cooperation with the implementing agencies was critical. Once the watershed structures were built, these used to be handed over to the village panchayat or a watershed development committee. The new institutions required to ensure equitable watershed development including watershed associations, watershed committees and user groups were not formed, thus, putting a question mark on the sustainability of these structures. There was an urgent need to create awareness about this because, without it, the sustainability of the structures would be in jeopardy.

There was also a full-fledged budget allocation for creating awareness regarding the MADP with the involvement of NGOs. It also points to the failure of the civil society-led progress in the study area. The civil society in the region is not fully developed to the extent that it could influence the development process including design and implementation. Therefore, to compensate for this, the whole of civil society apparatus was imported outside the region. The local NGOs were only asked to assist the outside NGOs so that some respectability and legitimacy could be derived by saying that it entailed the 'participation' of the people in letter and spirit. This study proposes a larger role of the local community at large in both project design and implementation. Only then the fruits of development can reach the poorest of the poor at the local level in Mewat.

**Acknowledgements** This paper is a chapter of my Postdoctoral research work conducted at JNU, New Delhi. For this, I thank the ICSSR, India, for providing funding assistance in the completion of research work. All the due research ethics were undertaken in the completion of primary fieldwork and data was anonymized to protect the identity of the respondents.

## References

Aggarwal PC (1971) Caste, religion, and power: an Indian case study. Shri Ram Centre for Industrial Relations, New Delhi

Alejandro LP (2007) Participation: the ascendancy of a buzzword in the neo-liberal era. Dev Pract 17:539–548. https://doi.org/10.1080/09614520701469518

Amir-Ali H, Khan R, Kumar OP (1970) The MEOS of Mewat: old neighbours of New Delhi. Oxford and IBH Publishing, Delhi

Ankit R (2020) Bureaucracy, community, and land: the resettlement of Meos in Mewat, 1949–50. J Soc Hist 54:306–329

Bastian S, Bastian N (eds) (1996) Assessing 'participation': a debate from South Asia. Konark Publishers, Delhi

Bhattacharya D, Jayal NG, Mohapatra BN et al (eds) (2004) Interrogating social capital: the Indian experience. Sage Publications, New Delhi

Census of India (2011) Primary census abstracts. Registrar General and Census Commissioner, Government of India, New Delhi

Chowdhry P (2011) The political economy of production and reproduction: caste, custom and community in north India. Oxford University Press, New Delhi

Clarke J (1992) Democratising development: NGOs and the state. Dev Pract 2(3):151–162

Cooke B, Kothari U (eds) (2001) Participation: the new tyranny? Zed books, London

Williams G, Srivastava M, Veron R, Corbridge S (2003) Making social science matter-I: how the local state works in rural Bihar, Jharkhand and West Bengal. Econ Pol Wkly 38:2377–2389

Cornwall A (1998) Gender, participation and the politics of difference. In: Guijt I, Shah M (eds) The myth of community: gender issues in participatory development. IT Publications, London, pp 46–57

Gaiha R (2004) Use of local knowledge in impact assessment: evidence from rural India. Econ Pol Wkly, 4471–4477
Gaiha R, Kulkarni V (2002) Panchayats, communities, and the rural poor in India. J Asian Afr Stud 37:38–82
Government of Haryana (2015–16) Statistical abstracts of Haryana. Department of Economic and Statistical Affairs, Chandigarh
Guijt I, Shah M (eds) (1998) The myth of community: gender issues in participatory development. IT Publications, London
India Water Portal (2014) A village becomes water secure. https://www.indiawaterportal.org/articles/village-becomes-water-secure Accessed 15 Sep 2023
Johnston RJ (1978) Multivariate statistical analysis in geography: a primer on the general linear model. Longman, New York
Kothari U (ed) (2019) A radical history of development studies: individuals, institutions and ideologies. Bloomsbury Publishing, London
Krishna A (2002) Active social capital: tracing the roots of development and democracy. Columbia University Press, New York
Kumar N (2011) Political economy of (under) development in Mewat region of Haryana. MPhil Dissertation, Jawaharlal Nehru University, New Delhi
Kumar N (2016) Social identities and cultural capital among Meos and other social groups in Mewat, Haryana, PhD Thesis, Jawaharlal Nehru University, New Delhi
Maheshwari B (2003) The Meos of Mewat: perspectives on ethnicity and nation building. Social Change 33:81–91
Mayaram S (1997) Resisting regimes: myth, memory, and the shaping of a Muslim identity. Oxford University Press, New Delhi
Mayaram S (2004) Against history, against state: counter perspectives from the margins. Orient Blackswan, Hyderabad
MDA [Mewat Area Development] (2000a) Mewat: its development, (mimeo)
MDA [Mewat Area Development] (2000b) Arable crop and horticulture development (typescript)
MDA [Mewat Area Development] (2000c) Watershed development project: village agon. (mimeo)
Ministry of Minority Affairs, Government of India and ICSSR (2008) A baseline survey of minority concentrated districts of India: Mewat. Institute for Human Development, New Delhi. https://www.icssr.org/sites/default/files/districts/Mewat[1].pdf. Accessed 15 Sep 2023
Nelson N, Wright S (1995) Power and participatory development: theory and practice. ITDG Publishing, London
Polyani K (1944) The great transformation. Rinehart, New York
Rahnema M (1992) Participation. In: Sachs W (ed) The development dictionary: a guide to knowledge as power. Zed Books, London, pp 116–131
Rao R (2004) Satyamevpuram: chautala gifts dist, wins name game. The Indian Express. https://indianexpress.com/article/news-archive/satyamevpuram-chautala-gifts-dist-wins-name-game/. Accessed on 20 January 2024
Sachs W (ed) (1992) The development dictionary: a guide to knowledge as power, Zed Books, London, pp 116–131
Sikand Y (1995) Meonis of Mewat. Econ Pol Wkly 30:490–492
Singh RS (2002) Government rescinds order on Mewat: bows to pressure from UN body. The Tribune, The Tribune, Chandigarh, India—Haryana (tribuneindia.com). Accessed on 20 January 2024
Stirrat J (1996) The new orthodoxy and old truths: participation, empowerment and other buzz words. In: Bastian and Bastian (eds) Assessing participation: a debate Irom South Asia. Konark Publishers, New Delhi
United Nations (2008a) Project completion report' Mewat area development project. International Fund for Agricultural Development, Rome
United Nations (2008b) Project completion report digest. International Fund for Agricultural Development, Rome

Williams G (2004) Evaluating participatory development: tyranny, power and (re) politicization. Third World Q 25:557–578

World Bank (2000) World development report 2000/2001: attacking poverty. The World Bank, Washington DC

# Part V
# Sustainability Through Education and Health Policy

# Chapter 21
# Higher Education Institutions and Sustainable Practices in India

**Jyoti Paul, Rajesh Kumar Abhay, and Indraj Kumar Kumawat**

**Abstract** Higher Education Institutions have huge potential to foster sustainable human development. HEIs can play a crucial role in promoting sustainable practices by incorporating the principles of teaching, research, operations, and knowledge transfer regarding sustainability. The study utilizes secondary data sources to find out the major sustainable practices in the selected higher education institutions. The study tries to explore what are the areas in which Universities work and what sustainable practices these Universities are adopting related to social, economic, environmental, and governance dimensions. An index to gauge the level of adherence to sustainable measures by different Central Universities in India is also the purpose of the study. The study has used content analysis to rate the HEIs and narrative information to help HEIs find out their strengths and focus on sustainable practices that require attention. Finally, the levels of sustainability practices adopted by select HEIs were measured using the sustainability index. The HEIs scores were divided into quartiles to get very high, high, moderate, and low scores of performing HEIs. It has been found that sustainable practices in the Indian HEIs are in the infancy stages as far as reporting and implementation are concerned in India. The accreditation and ranking requirements in India have pushed many HEIs to take up initiatives in sustainability. Further, the present study provides insights into the Indian HEIs with reference to focusing on sustainable practices. The results of the study are beneficial for the policymakers, university authorities and others concerned to make sustainable plans for the higher education institutions in India.

**Keywords** Education · Sustainability · University · Environment · Social

J. Paul
Department of Commerce, Dyal Singh College, University of Delhi, New Delhi, India
e-mail: jyotipaul@dsc.du.ac.in

R. K. Abhay (✉)
Department of Geography, Dyal Singh College, University of Delhi, New Delhi, India
e-mail: rkabhay@dsc.du.ac.in

I. K. Kumawat
Department of Chemistry, Dyal Singh College, University of Delhi, New Delhi, India
e-mail: indrajkk@dsc.du.ac.in



S. Anand et al. (eds.), *Emerging Geosustainability Transformations in India*, Advances in Geographical and Environmental Sciences,
https://doi.org/10.1007/978-981-97-7993-2_21

## 21.1 Introduction

Sustainable practices in Higher Educational Institutions (HEIs) are imperative as universities across the globe create young graduates who are going to contribute to the future. Further, the HEIs are using resources and are affecting social, environmental, economic, and governance dimensions (Alshuwaikhat and Abubakar 2008; Katiliute et al. 2014). For example, HEIs use resources like energy, water, and infrastructure, which in turn have an impact on the environment. Therefore, it becomes important to assess the status of sustainable practices on campus so that all stakeholders learn and apply these at different places to achieve the goal of sustainable communities. The role of HEIs in spreading education about sustainability practices is immense and has been emphasized in various studies in the past (Viebahn 2002, Bantanur and Mukherjee 2015). Various studies have taken the environmental, social, and economic dimensions of sustainability with emphasis on different parameters (Viebahn 2002; Wright 2002; Lozano 2006; Lukman et al. 2010; Gomes et al. 2023). Sustainability, as a complex concept, has many constructs beyond the environment (Steiner and Posch 2006). The lack of a comprehensive index and that too in the Indian context to evaluate adherence and proactive actions of HEIs in sustainability has been the major reason to choose and delve into the compliance of sustainability in HEIs in India.

The Brundtland report (WCED 1987) has brought to light the environmental emergency given the scarce resources that are used in society and leading to future uncertainties. The limited resources must be judiciously used to meet present needs and to conserve resources for future generations to come. HEIs are an integral part of society and are vital to adopting sustainable practices on campus. In 2000, the United Nations Millennium Declaration designed 8 Millennium Development Goals (MDGs) for UN Member states majorly focusing on social, economic, and environmental to be achieved by 2015. These goals are further enlarged in the form of Sustainable Development Goals (SDGs) (to be achieved by 2030) which emphasize People, Planet, Prosperity, and Partnerships between nations, both in developed and developing worlds for an inclusive society where inequalities are reduced, and equitable and sustainable community lives (Fisher and Fukuda-Parr 2019). The process of reporting, on the progress of the parameters defined by these goals is executed at the country level and globally too. HEIs are important to bring changes to achieve these goals through education, research, and sensitization of issues and further investigate progress made by them towards creating sustainable societies. With a total number of enrolments (more than 4 crores) that is humungous in universities of India (AISHE 2020; 21), there is no doubt that HEIs should inculcate sustainability concerns in youth who are the change makers for a better tomorrow. Collins (2017) has pointed out that universities have a societal responsibility to train future professionals and leaders and to stimulate public awareness of sustainability.

Studies in the past have delved into ratings of HEIs and checked different parameters related to how HEIs are working towards achieving sustainability (Parvez and

Avlokita 2018; Waheed et al 2011). There have been varied tools related to the assessment of sustainability practices used by HEIs to assist them in reducing their environmental footprint and identify areas of improvement. Each tool has adopted different indicators and has its strengths and weaknesses (Shriberg 2002; Jain et al. 2013). Despite many rating systems available predominantly for Anglo-Saxon regions those valuing sustainability in a broad sense are scarce (Orduna-Malea and Perez-Esparrells 2021; De La Poza et al. 2021).

There has been a plethora of rating instruments related to different dimensions for HEIs, that have been developed for Western countries and very few have considered non-western countries' context (Chakraborty et al. 2021). Schlickmann et al. (2021) have identified the main sustainability assessment tools in higher education and developed indicators for environmental education. The Graphical Assessment of Sustainability in Universities (GASU) tool has been used in sustainability reporting. Sustainability Tracking, Assessment and Rating System (STARS) highlighted HEI-specific assessment tool. Auditing Instrument for Sustainability in Higher Education (AISHE) has educational dimensions to it. All these famous tools have been used as parameters in a Western context and have focused on either environmental or educational dimensions.

Studies in India related to campus sustainability are still experiencing a growing phenomenon and have focused more on environmental indicators (Bantanaur et al. 2015). Few Universities in India are making a case for sustainability through curriculum enhancement (Jain et al. 2013). Sustainability alone is not a matter of mere introduction in curriculum, but it is a more holistic approach that is required to attain such practices in India. Only, a few universities have limited introduction to sustainability courses so there is a need to develop a rating system specifically for Indian universities to help them and try to find out their strengths and weaknesses (Chakraborty et al. 2021). The Indian higher education institution system is trying to inculcate sustainability in its vision and curriculum. The main purpose of this paper is to investigate how the HEIs are performing in India in terms of executing sustainable practices from the perspective of social, economic, environmental, and governance parameters. The study identifies practices adopted by the HEIs and their efforts towards achieving these at the campus level. The paper is significant for the universities, academicians, and planners, to find out their strengths and weak areas related to practices adopted to improve sustainability within the campus and their outside impact too.

## 21.2 Material and Methods

The HEIs chosen for the study are Central Universities in India. There is a total of 56 Central Universities in India under the University Grants Commission (UGC 2023a, b), out of which National Assessment and Accreditation Council (NAAC) Accreditation data is available for 30 Universities for the current study. The data has been collected from NAAC Reports, University websites, available literature, etc. It is observed that NAAC parameters have changed over the years and have become more stringent towards sustainable practices in the last five years. To maintain uniformity in comparison to universities in terms of sustainable practices, those universities that have at least completed one cycle of accreditation and have received accreditation in the year 2019 or after this year have been taken into consideration. The year 2019 has marked several changes in NAAC reporting in the Indian higher education system. Based on these two criteria, 11 Central Universities have been finally considered and assessed for adopting sustainable practices (Table 21.1).

The HEIs have been selected in such a way that they represent institutions from all parts of the country and represent regional representation. There are three universities each from the north (CUP, Punjab; JMI, Delhi and AMU, Uttar Pradesh) and south (CUK, Kerala; MANNU and UoH, both from Telangana. Four universities have been finalized from the north-east region (MzU, Mizoram; TU, Tripura; TzU

**Table 21.1** Selected universities for the study

| S. no. | HEI institution | Abbreviation | Location, State | NAAC accreditation cycle (year) | NAAC score (grade) year |
|---|---|---|---|---|---|
| 1 | University of Hyderabad | UoH | Hyderabad, Telangana | 4th cycle, 2022 | 3.26 (A+), 2022 |
| 2 | Central University of Kerala | CUK | Kasaragod, Kerala | 2nd cycle, 2022 | 3.14 (A), 2022 |
| 3 | Tezpur University | TzU | Sonitpur District, Assam | 4th cycle, 2022 | 3.41 (A), 2022 |
| 4 | Aligarh Muslim University | AMU | Aligarh, Uttar Pradesh | 2nd cycle, 2021 | 3.35 (A+), 2022 |
| 5 | Maulana Azad National Urdu University | MANUU | Hyderabad, Telangana | 3rd cycle, 2022 | 3.36 (A+), 2022 |
| 6 | Tripura University | TU | West Tripura District, Tripura | 3rd cycle, 2022 | 2.77(B++), 2022 |
| 7 | Jamia Millia Islamia University | JMI | Delhi | 2nd cycle, 2021 | 3.61(A++), 2021 |
| 8 | Vishva Bharati University | VBU | Birbhum, West Bengal | 2nd cycle, 2021 | 2.65 (B+), 2021 |
| 9 | Mizoram University | MzU | Aizawl, Mizoram | 2nd cycle, 2019 | 3.16 (A) 2019 |
| 10 | Assam University | AsU | Cachar District, Assam | 3rd cycle, 2021 | 2.39 (B) 2021 |
| 11 | Central University of Punjab | CUP | Bathinda, Punjab | 2nd cycle, 2022 | 3.3 (A+), 2022 |

*Source* Compiled by authors from respective HEI websites and NAAC reports and AsU, both in Assam) and one from West Bengal (VBU) (Fig. 21.1). HEIs in Gujarat, Madhya Pradesh and Rajasthan must look for and start working towards accreditation. HEIs from the west region majorly received accreditation in the year 2016 and have only completed one cycle of accreditation; therefore, no HEIs have been taken in the present study.

The study has reviewed available data on the assessment of sustainable practices in the past, at the national and international levels. In the present study, an innovative approach has been adopted and indicators related to economic, social, environmental, and governance are selected, particularly, applicable to HEIs in the Indian context. An index to gauge the level of adherence to sustainable measures by different Central Universities in India is also the purpose of the study. The study has used content analysis to rate the HEIs and narrative information to help HEIs find out their strengths and focus on sustainable practices that require attention in the near future. The index has adopted 33 parameters and they have been rated as score '0' if no information

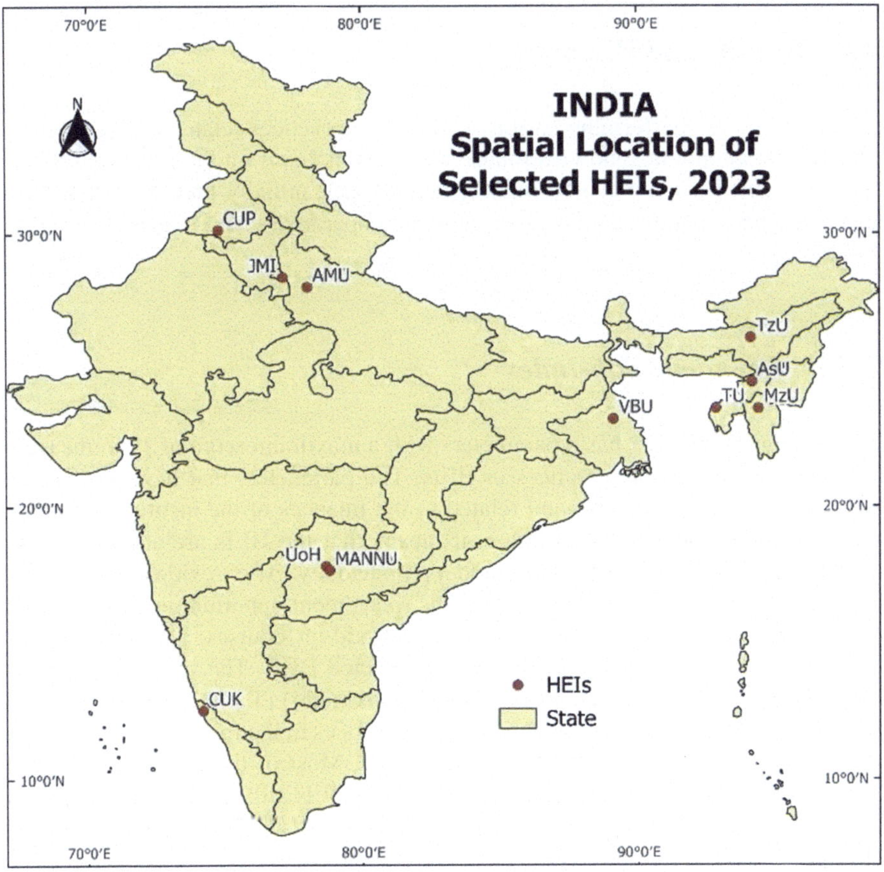


**Fig. 21.1** Selected HEIs and their spatial distribution. *Source* Authors

about the indicator exists; score '1' if partial information about the indicator is available with no substantial evidence; score '2' if full information with supporting evidence exists and available (Table 21.2). All the indicator scores range from 0 to 2, except two parameters viz. elected faculty members are part of the academic/ executive council and second HEIs has a student union. These are rated as '0' if no and '1' if there is an existence of them. The maximum score of these 33 parameters can be up to 64 points.

Finally, the levels of sustainability practices adopted by select HEIs were measured using the sustainability index. The HEIs scores were divided into quartiles to get very high, high, moderate, and low scores of performing HEIs. The basis was a score less than 32 and was coded as Low. A score of 32–37 was coded as Moderate; a score of 37–45 was coded as High and HEIs scoring more than 45 were included in Very High Category. The average scores of the HEIs are 37 and the score was divided into four categories of level of sustainability practices using quartile scores.

## 21.3 Results and Discussion

In India, HEIs are performing various sustainable practices related to the environment, social, economic, and governance. The results focus on the four sub-indices (economic, environmental, governance, and social), initially and then they have been combined to understand the overall and comparative scenario across the HEIs (Table 21.3).

### *21.3.1 Economic Sub-Index*

The economic sub-index has 6 parameters with a maximum score of 12 to measure the economic sustainable practices in HEIs. The parameters that have been taken into consideration are diverse and related to the finances of the institution and are beneficial to its stakeholders. Further, it means that the HEIs are adopting these economic practices to help the stakeholders, financially. After reviewing the extant literature, the parameters selected under this (placement opportunities for students, intra-institutional collaborations, skill-oriented add-on courses, and schemes for poor students) show wide disparities across selected HEIs. The score of the index ranges from 4 in AsU and TU to 11 in UoH (maximum) (Table 21.3). UoH shows greater contributions towards these economic practices in the form of benefitting the stakeholders' whether students, teachers, or staff. Most of the universities scored 7 or fewer scores, which reflects that most of the HEIs still need to adopt various economic practices for the benefit of their stakeholders and there is enough scope for improvement.

It has been seen that very few HEIs have taken up courses that help institutions to be financially independent; many institutions are starting the same, as it has

**Table 21.2** Selected indicators for sustainable practices in various HEIs

| S. no. | Parameters | Category |
|---|---|---|
| 1 | Schemes to support poor students from low or lower-middle-income countries/states/weaker sections, for example, offering free education or grants (0, 1, 2) | Economic |
| 2 | Collaboration internationally for SDG (0, 1, 2) | Economic |
| 3 | Number of research projects by faculty related to dimensions of sustainability (0, 1, 2) | Economic |
| 4 | Number of Certificates/Add-on courses for those not studying in the University (0, 1, 2) | Economic |
| 5 | Placements to students (less than 5, 10, and >15%) (0, 1, 2) | Economic |
| 6 | Audited financial statements are in the public domain (0, 1, 2) | Economic |
| 7 | Publications by faculty related to dimensions of sustainability in UGC CARE/Scopus/journals (0, 1, 2) | Environmental |
| 8 | Students doing research in sustainability (0, 1, 2) | Environmental |
| 9 | Policy for water recycling (0, 1, 2) | Environmental |
| 10 | Wastewater treatment on campus (0, 1, 2) | Environmental |
| 11 | Tracking of water consumption and water reuse (0, 1, 2) | Environmental |
| 12 | Rainwater harvesting (0, 1, 2) | Environmental |
| 13 | Water conservation activities off campus (0, 1, 2) | Environmental |
| 14 | Biodiversity in building plan (0, 1, 2) | Environmental |
| 15 | Process for carbon management and reducing carbon dioxide emissions (0, 1, 2) | Environmental |
| 16 | Reviews to identify areas where energy waste is highest (0, 1, 2) | Environmental |
| 17 | Recycling of waste (0, 1, 2) | Environmental |
| 18 | Policy for minimizing the use of single-use plastic (0, 1, 2) | Environmental |
| 19 | Minimize food waste by composting (0, 1, 2) | Environmental |
| 20 | Policy to reduce air, water, and sound pollution (0, 1, 2) | Environmental |
| 21 | Use of Bicycle/battery-powered vehicles on campus(0, 1, 2) | Environmental |
| 22 | Sensor-based energy conservation | Environmental |
| 23 | Elected Faculty members in AC/EC of the same or any University (0, 1) | Governance |
| 24 | Use of technology in administrative and admission, assessment/ERP 2.5.5 (0, 1, 2) | Governance |
| 25 | Elected Student Union (0, 1) | Governance |
| 26 | Strategic plan and its deployment on the website (0, 1, 2) | Governance |
| 27 | Educational outreach activities in the local community (0, 1, 2) | Social |
| 28 | Dedicated courses related to education on SDGs (0, 1, 2) | Social |
| 29 | Percentage of women who receive a degree (less than 25, 50, and >50%) (0, 1, 2) | Social |
| 30 | Childcare facilities for staff (0, 1, 2) | Social |

(continued)

**Table 21.2** (continued)

| S. no. | Parameters | Category |
|---|---|---|
| 31 | Accessible facilities for the Disabled on campus (software/signage/notices/lifts/ramps/access audit) (0, 1, 2) | Social |
| 32 | Alumni participation and linkages for Institution growth/sustainability issues (0, 1, 2) | Social |
| 33 | Policy for non-discrimination against women/transgender people (0, 1, 2) | Social |

*Source* Compiled by authors

**Table 21.3** Category-wise aggregate score of selected HEIs

| S. no. | Sub-indices | MANUU | TU | AMU | VBU | CUK | MzU | CUP | UoH | AsU | TzU | JMI |
|---|---|---|---|---|---|---|---|---|---|---|---|---|
| 1 | Economic | 8 | 4 | 7 | 8 | 5 | 6 | 5 | 11 | 4 | 9 | 8 |
| 2 | Environmental | 14 | 13 | 16 | 10 | 26 | 12 | 14 | 25 | 15 | 22 | 16 |
| 3 | Governance | 6 | 5 | 4 | 6 | 6 | 6 | 4 | 6 | 4 | 5 | 4 |
| 4 | Social | 8 | 13 | 12 | 9 | 8 | 6 | 6 | 10 | 7 | 8 | 6 |
| | **Total Score** | **36** | **35** | **39** | **33** | **45** | **30** | **29** | **52** | **30** | **44** | **34** |

*Source* Calculated by authors

become one of the requirements for national-level accreditation of HEIs. Among the international collaborations related to Sustainable Development Goals (SDGs) are also found to be less in the selected HEIs. Universities are having international collaborations on sustainability-related studies are directly or indirectly investigate SDGs. Collaborations within the country and through research projects with agencies remain high. Few HEIs are having prominent collaborations internationally too. TU has developed a collaborative model to work with UNICEF to train adolescents from different ethnic communities of Assam and to use their folk resources for community messaging against social maladies. UoH has also collaborated, many of which are concerned with sustainability issues. Further, placements, ironically, are another area in which HEIs shall work. Though being Central Universities, 6 universities showed less than 10% placements and 5 have a 15% or a little higher than 15% placement record. All the HEIs are having Post Graduate courses; still, this parameter remains low and needs attention.

## *21.3.2 Environmental Sub-Index*

The environmental sub-index has a maximum number of parameters to assess the environmental sustainability practices of the selected HEIs. It includes 16 parameters with a maximum score of 32 that an institution can score. The parameters reflect varied dimensions related to the environment and rate the institution on how well it is performing for energy conservation, water conservation, pollution, recycling, use of alternate sources of energy, and performing activities not only on campus but also outside the campus. CUK is scoring the highest (26 points) closely followed by the UoH (25 points) (Table 21.3). VBU is performing poorly on this sub-index. It has been found that some HEIs are catering to water conservation, while others are doing more recycling, banning single-use plastic, and while few are focusing more on publication in the domain of environmental sustainability.

HEIs can have direct and indirect impacts on the environment, society, and the natural environment (Findler et al. 2018). Many institutions in the recent past have worked on these dimensions and executed them in the form of best practices too. Rainwater harvesting has been widely practiced in all the HEIs in India. Some have initiated it partially, but few HEIs have been doing it for a long time. There has been a policy for wastewater recycling in many HEIs. Some have formally articulated it and disclosed it in reports and on websites. The UoH, CUK, AMU, TzU, and CUP are doing remarkable work in terms of wastewater treatment on campus. For example, UoH has installed Sewerage Treatment Plants (STPs) at various places with different capacities, and recycling water for horticulture and use in toilets. Rainwater is also collected for recharging groundwater through several open wells and for use in the botanical garden at TzU. The geographical situation of AsU provides scope for efficient use of rainwater and harvesting the same. The university is located on a hillock and a natural small-sized lake is formed at the foothill that captures all the water flowing downhill. Water harvesting using two lakes on the campus and the digging of another lake is also underway at the University (Assam University, SSR 2020). CUK and UoH are also tracking their water consumption patterns and water reuse. Biodiversity conservation in building plans was found to be there in almost all selected HEIs in India. UoH, CUK, and TzU boast of such biodiversity in their plans.

HEIs contribution towards reducing carbon dioxide emissions, sensor-based energy conservation, and review of areas where energy consumption is highest, are indicators used for energy efficiency measures. All the HEIs are using alternate sources like solar panels and windmills. For example, CUP has installed two windmills with a total capacity of 6.8 KW. UoH ensures energy conservation and is an eco-friendly campus. It has an 1140 MW solar panel system installed at the University, apart from 4 KW solar panels in the library building, separately. The University conducts regular green audits and updates energy conservation methods. CUK has a 'green' renewable sustainable energy alternative of a biogas plant generating energy from kitchen wastes from canteens, hostels, and common mess facilities. Minimizing food waste by composting was another indicator. Five HEIs, namely, CUK, UoH, TzU, CUP, and AsU have mentioned such steps adopted by them. Practices related to recycling waste are followed at CUK, UoH, and TzU. Ban on single-use plastic is being strictly followed in 5 HEIs (CUK, MzU, CUP, AsU, and JMI) that have banned single-use plastic on campus. Publications in the areas of sustainable practices and students doing research in sustainability are consistent in all HEIs and show an increasing trend.

### 21.3.3 Governance Sub-Index

Governance is highly essential and required for achieving sustainability. The governance sub-index takes care of how well the HEIs are planning and making sustainability a priority in their decision-making plans. This sub-index considers parameters like e-governance in operations, having a strategic plan and the presence of faculty members in the Academic Council, and the elected Student Union. Most of the HEIs are performing well in this index with the least being 4 and five HEIs having a maximum score of six. Most of the parameters under this index are also statutory requirements for HEIs in India. Elected faculty members form part of the Academic Council or Executive Council of the same HEIs or some other. Their presence in such councils is bound to ensure good governance. An elected student union ensures that students' concerns are well taken and addressed by HEIs. The elections are conducted each year following the Lyngdoh Committee's recommendations (Assam University, SSR 2020; Tezpur University, SSR 2022; University of Hyderabad, SSR 2021). HEIs score almost a maximum of two points in e-governance in the process of admissions, examinations, assessments, etc. as all the HEIs have dedicated ERP systems to execute the same. The presence of a strategic plan was one of the parameters in this sub-index. HEIs have such a strategic plan and its deployment is available on the websites. A strategic plan is necessary for HEIs to have a long-term action plan and to revisit, modify, and enhance in the future based on actions taken in the enforcement of the plan. Action taken reports, organogram and plan for the year ahead are major requirements for accreditation also, thus, all HEIs have such plans. Governance indicators have been included in the study as good governance is the backbone and will ensure the deployment of goals. HEIs have different goals and their organizational structure, vision and mission will make sure that goals are achieved on time.

### 21.3.4 Social Sub-Index

The objective of this index is to assess the practices adopted by HEIs towards the social welfare of their stakeholders. This sub-index measures practices like educational outreach in the local community, inclusiveness, participation of alumni in institutional growth, and so on. This sub-index covers 7 parameters, and an HEI can score a maximum of 14 points. It was found that TU received the highest score of 13 points and is closely followed by AMU with 12 points.

The social sub-index incorporates relevant and prominent parameters that are communicating and affecting various stakeholders of the institution. HEIs must ensure in their policies that all the stakeholders are taken care of concerning access to various opportunities. Indicators like educational outreach of the HEIs in the local community, percentage of women receiving degrees, facilities for the disabled, childcare facilities for staff, alumni participation, inclusion of all while making policy for transgender, etc. are very important. UoH is the first University in India to have an Online Education Policy to have open access for massive outreach (University of Hyderabad, SSR 2021). Other HEIs are reaching the local community mainly because of National Service Scheme (NSS) activities too. Dedicated courses related to Sustainable Development Goals (SDGs) are not found directly in HEIs. However, most of them are offering one course in environment studies to students under the Choice-based Credit System (CBCS), widely applied in HEIs of India. For example, in UoH students of all branches undergo a foundation course in environmental science. The University has also specialized centres for research in Earth Ocean and Atmospheric Sciences, Cognitive Sciences, High Energy Materials, Health Psychology, Nanotechnology, and Electronics Science and Technology, etc. that relate to different SDGs. Mizoram University has also three centres viz., Centre of Petroleum Exploration, Centre of Biodiversity, and Centre of Disaster Management and is conducting research related to SDGs. At Pondicherry University, the Radio Tutor series and subject-related documentaries aimed at providing Adaptive Learning Methods for the students of primary, secondary, and higher education and contributing to the local community services. Each discipline, though not having courses with names of sustainability, is looking into sustainability practices from different perspectives. Though many HEIs have started with different centres, many have to catch up with the pace to start dedicated courses, and add-on courses on sustainable development and its allied areas.

Most of the HEI campuses are disabled-friendly having tactile paths, lifts, Braille facilities, etc. However, still a lot of work can be done in this area. Most of the HEIs do not have an access audit which should be a regular feature to enhance disabled-friendly provisions. Though HEIs have at least one software for disabled friendly, many new things like a screen reader, accessibility options on the websites, and signage in Braille throughout the campus are missing. Tezpur University has an accessible website for visually challenged people too (Tezpur University, SSR 2022). Childcare facilities for staff are present in most of the HEIs except MANNU and MzU.

Alumni participation in institutional growth and linkages is another significant parameter in the social sub-index. All HEIs have registered alumni associations but few of them are actively engaged in HEIs and sustainability issues. In CUK, the alumni cell encourages students to contribute to the university through three schemes. The first is My Imprint 'by the students, for the students, the scheme encourages the passing out students to donate one of the given options through online voting. The second is the Senior Class Gift 'by the class-for-the-class' scheme. This scheme asks graduating batches to forfeit caution deposits to make an endowment in their department; the third is an Endowment Scheme which invites alumni to institute endowments, and prizes to contribute. At UoH, the alumni association has extended help to the needy and deserving students during crisis periods such as the COVID-19 pandemic, however, there are no fixed provisions related to this.

Policy for non-discrimination against women/transgender has been another rating parameter in the social sub-index. Here, three HEIs have such policies, UoH, AMU, and VBU. They have recently taken steps to recognize the transgender presence in the student community. On the parameters of the percentage of women receiving degrees, all HEIs are performing well. All the HEIs have chosen to have at least 50% or more women receiving degrees which is a progressive sign as far as social index is concerned. For example, CUK has a high gender ratio of students, consistently and significantly favourable to women. It also boasts of a 100% passing rate in examinations.

UoH has performed better in all 4 sub-indices (Fig. 21.2). The performance of CUK is also significant with reference to the environmental and governance side but moderate performance on the other indices. AsU and CUP have performed poorly, in comparison to all other selected HEIs. TzU has performed well in economics while moderate to poor on other indicators, while AMU has performed better in social practices and moderate performance in other sectors. JMI and MzU have poorly performed on social indicators while moderately on others. The overall reflection is that one or two universities are performing better in one sector while having poor performance in others. It reflects that there is a need to work in all sectors if HEIs have to contribute to all the sectors to achieve sustainability.

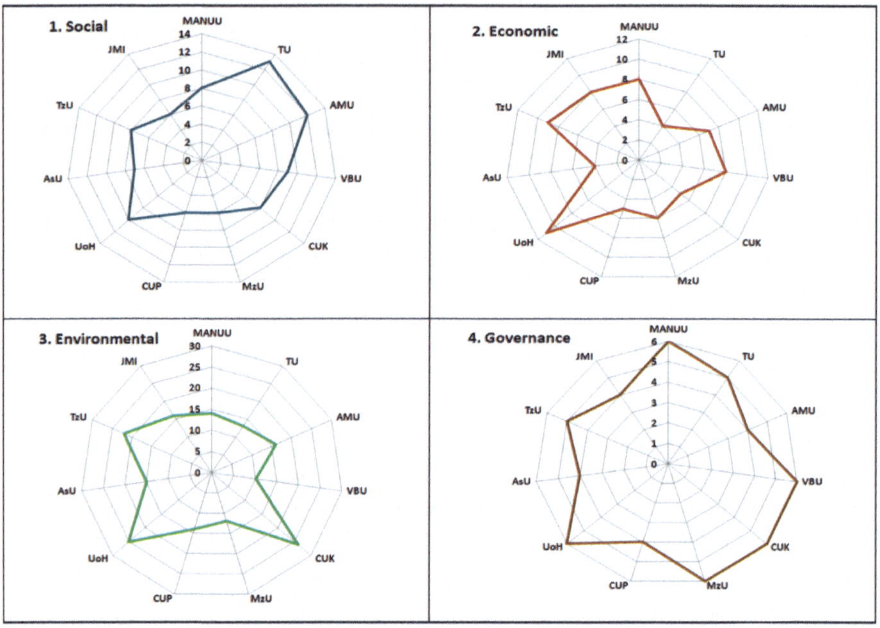


**Fig. 21.2** Performance of HEIs on various indicators of sustainability practices. *Source* Authors

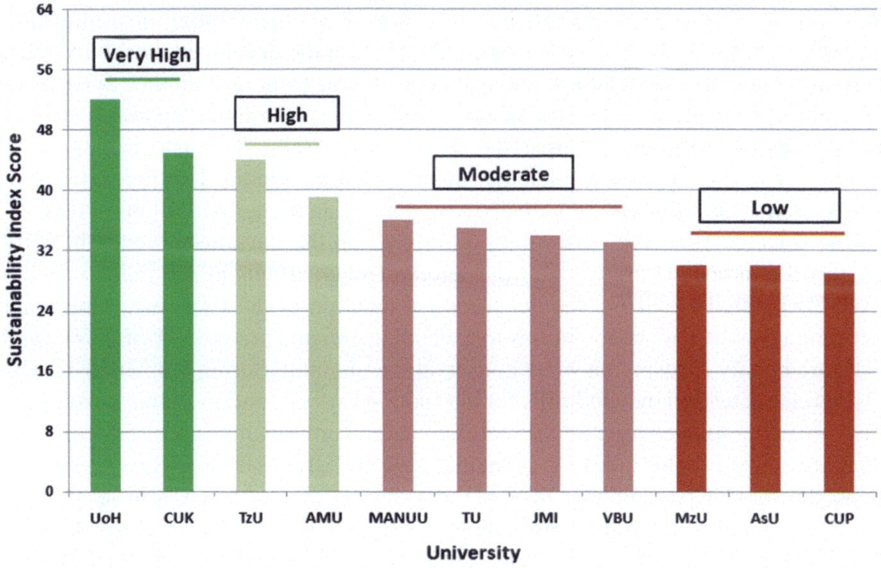


**Fig. 21.3** Level of sustainability practices across the selected HEIs in India. *Source* Authors

### *21.3.5 Levels of Sustainability in HEIs*

Based upon the selected indicators, it is found that among the selected 11 HEIs, UoH has scored the highest (52) and is a leading HEI in terms of executing sustainability practices. It is in its fourth cycle of NAAC accreditation and has tremendously worked on sustainability practices. CUK is also in the very high category according to classification and has an overall score of 45 (Fig. 21.3). Each HEI differs in the sustainability practices that are adopted by them. CUP, MzU, and AsU are scoring the lowest scores and are falling under the low level of sustainability practices adopted by them. The variation among the scores is from a Low level with a score of 29 to a very high level of score being 52. Out of a maximum score of 64, HEIs are scoring as low as less than 50% score according to the index. Improvements for all the HEIs according to the parameters taken are discussed in the last section.

## 21.4 Finding, Conclusions, and Recommendations

The present study evaluated the status of various sustainability practices adopted by different sectors. It has been seen that sustainability practices in HEIs remain in the infancy stages as far as reporting and implementation are concerned in India. The accreditation and ranking requirements in India have pushed many HEIs to take up initiatives in sustainability. The primary source of this paper is the NAAC reports and websites of these selected HEIs. There have been many rating instruments in the past, which have been used for rating the HEI, in the developed world. In India, there are 56 central universities, and each is different from one another in terms of geographical location, situations, courses offered, etc. An index approach is easy to measure and facilitates comparison too (Lozano 2006). This paper attempts to develop a comprehensive sustainability index for HEIs in India. This index has been constructed keeping in view the diversity in the Indian higher education system.

The selected HEIs are performing quite well on the governance index because most of the indicators are usually mandatory. The environment sub-index is being diversely catered by HEIs. The geographical location of the HEIs is an important factor that provides ideal conditions to execute associated practices. UoH, TzU, and CUK are mostly focusing on water conservation and outperforming other HEIs. The CUP is characterized by windmills for alternate sources of energy, while most of the HEIs are using solar energy. The climate and natural environment are also facilitating HEIs are do well on this index. A recent initiative in CUK has led to greater survival of plants through the past summer. 90% of the fruit trees planted by CUK have survived and are thriving to meet the requirements of the CUK staff and students (Central University of Kerala, SSR 2022). Recent publications related to sustainability by students in the area again differ in aspects as to which courses HEIs are offering to students. UoH has an outstanding publication record and has specialized centres for research on various environmental dimensions. The economic sub-index has clearly shown that placements, collaborations, and courses on sustainability need to be focused upon by HEIs. Being Central Universities, the placement of students is a major and healthy economically sustainable practice. The geographical location is an advantage for the environment sub-index related to sustainability practices, the same becomes a challenge for HEIs, placements, collaborations and attracting staff and foreign students to the campus. For example, the limited scope for the placement of the students in and around Mizoram is cited by Mizoram University due to the slow pace of industrialization. Limited scope for interaction, collaboration and consultancy with the industry was another challenge cited (Mizoram University, SSR 2019). CUP also states that inviting big business houses for placement is an issue on account of the smaller number of graduating students and the rural setting of the University (Central University of Punjab, SSR 2021). JMI is also trying for Industry-University linkages, and technology parks and facing challenges in connecting to alumni and global linkages (Jamia Millia Islamia, SSR 2021). The curriculum in various HEIs has courses to promote sustainability knowledge and research, but selected HEIs do not have any dedicated degree courses related to sustainability. Further, fee concessions and various scholarship schemes are being run in HEIs for weaker sections, but HEIs should look for full-fee waiver scholarships (including all costs) to attract foreign students.

The social sub-index caters to inclusion and diversity parameters. Most of the HEIs in India have good gender parity. To have inclusion on campus, UoH has recently initiated efforts to recognize transgender presence among the student community. Educational outreach in the local community is being catered to by NSS (National Service Scheme), a Central Government scheme in India. But still, many HEIs are not doing much under this criterion. Few have such programs for educational outreach, substantively. Dedicated courses on SDGs are the need of the hour to sensitize youth towards sustainability.

This study has provided insights into the Indian HEIs, regarding focusing on sustainable practices. HEIs in India have adopted many practices due to new accreditation norms too. The study highlights good sustainable practices in selected HEIs of India and evaluates weak areas to work on. The paper has ranked these select HEIs into a very high to a low number of sustainable practice categories. The setting of the HEIs in rural/urban areas has a direct impact on the adoption of sustainable practices like environmental aspects, outreach to the local community, etc. The economic and social conditions are also differing for each HEI, and all are trying to create a sustainable campus. Though the study is limited to 11 Central Universities, it has taken HEIs from all parts of India except from the West for which we could not find any reports in the years selected under study. The study provides insights into various sustainable practices that need to be initiated by other HEIs also in India.

## References


AISHE (2020–21) All India survey of higher education. https://aishe.gov.in/aishe/viewDocument.action?documentId=322. Accessed on 03 May 2023

Alshuwaikhat HM, Abubakar I (2008) An integrated approach to achieving campus sustainability: assessment of the current campus environmental management practices. J Clean Prod 16:1777–1785. https://doi.org/10.1016/j.jclepro.2007.12.002

Assam University, SSR (2020) Self Study Report. http://www.aus.ac.in/iqac/wp-content/uploads/sites/88/2022/09/ASUNGN10139.pdf. Accessed on 7 May 2023

Bantanur S, Mukherjee SMR (2015) Emerging dimensions of sustainability in institutes of higher education in India. Int J Sustain Built Environ 4(2):323–329. https://doi.org/10.1016/j.ijsbe.2015.03.004

Central University of Kerala, SSR (2022) Self Study Report. https://cukerala.ac.in/cukpdfs/IQAC/SSR.pdf. Accessed on 7 May 2023

Central University of Punjab, SSR (2021) Self Study Report. http://cup.edu.in/iqac_new/sites/default/files/users/cup/PBUNGN11423_0.pdf. Accessed on 7 May 2023

Chakraborty A, Kumar S, Shashidhara L et al (2021) Building sustainable societies through purpose-driven universities: a case study from Ashoka University (India). Sustainability 13(13):7423. https://doi.org/10.3390/su13137423

Collins TJ (2017) Review of the twenty-three-year evolution of the first university course in green chemistry: teaching future leaders how to create sustainable societies. J Clean Prod 140:93–110. https://doi.org/10.1016/j.jclepro.2015.06.136

De La Poza E, Merello P, Barberá A et al (2021) Universities' reporting on SDGs: using the impact rankings to model and measure their contribution to sustainability. Sustainability 13(4):1–30. https://doi.org/10.3390/su13042038

Findler F, Schönherr N, Lozano R et al (2018) Assessing the impacts of higher education institutions on sustainable development—an analysis of tools and indicators. Sustainability 11(1):59. https://doi.org/10.3390/su11010059

Fisher A, Fukuda-Parr S (2019) Introduction—data, knowledge, politics and localizing the SDGs J Human Dev Capb 20(4):375–385

Gomes LP, Caetano MO, Brand SM et al (2023) Maintenance of an environmental management system based on ISO 14001 in a Brazilian private university, seeking sustainable development. Int J Sust Higher Edu 24(2):361–381

Jain S, Aggarwal P, Sharma N et al (2013) Fostering sustainability through education, research and practice: a case study of TERI University. J Clean Prod 61:20–24

Jamia Millia Islamia, SSR (2021) Self Study Report. https://www.jmi.ac.in/NAAC/NAAC-SSR-C1.html. Accessed on 7 May 2023

Katiliute E, Daunoriene A, Katkute J (2014) Communicating the sustainability issues in higher education institutions World Wide Webs. Procedia Soc Behav Sci 156:106–110. https://doi.org/10.1016/j.sbspro.2014.11.129

Lozano R (2006) A tool for a graphical assessment of sustainability in universities. J Clean Prod 14:963–972

Lukman R, Krajnc D, Glavič P (2010) University ranking using research, educational and environmental indicators. J Clean Prod 18:619–628. https://doi.org/10.1016/j.jclepro.2009.09.015

Mizoram University, SSR (2019) Self Study Report. https://mzu.edu.in/iqac/. Accessed on 7 May 2023

Orduna-Malea E, Perez-Esparrells C (2021) Multidimensional taxonomy of university rankings. In: Hazelkorn E, Mihut G (eds) Research handbook on university rankings: theory, methodology, influence and impact. Edward Elgar Publishing, Gloss, pp p106-126

Parvez N, Avlokita A (2018) Review of campus sustainability rating systems for Indian campuses. In: Proceedings of the 52nd international conference of the architectural science association, Melbourne, Australia, 28 November–1 December 2018, pp 503–510

Schlickmann A, de Lima EP, Bortoluzzi SC (2021) assessment tools and performance indicators for HEI environmental and sustainable development education. In: Leal Filho W, Tortato U, Frankenberger F (eds) Integrating social responsibility and sustainable development. World Sustainability Series. Springer, Cham. https://doi.org/10.1007/978-3-030-59975-1_31

Shriberg M (2002) Institutional assessment tools for sustainability in higher education: strengths, weaknesses, and implications for practice and theory. High Edu Policy 15:153–167. https://doi.org/10.1016/S0952-8733(02)00006-5

Steiner G, Posch A (2006) Higher education for sustainability by means of trans-disciplinary case studies: an innovative approach for solving complex, real-world problems. J Cleaner Prod 14(9–11):877–890. https://doi.org/10.1016/j.jclepro.2005.11.054

Tezpur University, SSR (2022) Self Study Report. https://luit.tezu.ernet.in/naac2/ASUNGN10116.pdf. Accessed on 7 May 2023

UGC (2023a) List of 56 central universities. https://www.ugc.gov.in/oldpdf/Consolidated_CENTRAL_UNIVERSITIES_List.pdf. Accessed on 15 Jul 2023

University Grants Commission [UGC] (2023b) List of 56 central universities. https://www.ugc.gov.in/oldpdf/Consolidated_CENTRAL_UNIVERSITIES_List.pdf Accessed on 15 Jul 2023

University of Hyderabad, SSR (2021) Self Study Report. https://uohyd.ac.in/wp-content/uploads/2022/02/Final-Approved-UH-SSR.pdf. Accessed on 7 May 2023

Viebahn P (2002) An environmental management model for universities: from environmental guidelines to staff involvement. J Clean Prod 10:3–12

Waheed B, Khan FI, Veitch B (2011) Developing a quantitative tool for sustainability assessment of HEIs. Int J Sust High Edu 12(4):355–368. https://doi.org/10.1108/14676371111168278

WCED [World Commission on Environment and Development] (1987) Our common future: report of the world commission on environment and development. Oxford University Press, Oxford

Wright TSA (2002) Definitions and frameworks for environmental sustainability in higher education. Int J Sustain High Educ 3(3):203–220. https://doi.org/10.1108/14676370210434679

# Chapter 22
# Gender Disparity in Literacy Rate and Levels of Educational Attainment Among the Balmiki Caste in Jhajjar District, Haryana

**Neeraj Rani and K. V. Chamar**

**Abstract** Sustainability and education are closely related as education plays a critical role in promoting sustainable development. Education provides individuals with an understanding of the interconnectedness of social, economic, and environmental systems. In recent years, there has been a growing awareness about the importance of education in promoting sustainable development and reducing poverty, among marginalized communities particularly the Balmiki caste in rural areas of Haryana. The study uses primary data collected from selected villages of the Jhajjar district during the period of February-April 2022. Gender disparity in literacy rate and level of educational attainment up to the levels of post-graduation, among the Balmiki caste, have been identified. It has been found that about 92% of males and 89% of females got an education up to senior secondary levels among the Balmiki caste. The Jhajjar district exhibits a gender disparity of 0.11 points. The highest gender disparity is observed in the village Jahangirpur (0.44) at the level of 'senior secondary', while it is lowest in the village Silana at the level of 'post-graduation' (4.04). Further, the highest disparity is recorded at the level of 'senior secondary', while it is lowest at the level of graduation among the Balmiki caste in the Jhajjar district. The study concludes that females are significantly lagging behind males across various levels of educational attainment except up to 'graduation level' among the Balmiki caste in Jhajjar district.

**Keywords** Sustainability · Education · Literacy rate · Educational attainment · Gender disparity · Balmiki caste

N. Rani · K. V. Chamar (✉)
Department of Geography, M.D. University, Rohtak, Haryana, India
e-mail: chamar_kv@yahoo.com



S. Anand et al. (eds.), *Emerging Geosustainability Transformations in India*, Advances in Geographical and Environmental Sciences,
https://doi.org/10.1007/978-981-97-7993-2_22

## 22.1 Introduction

Sustainability and education are closely related as education plays a critical role in promoting sustainable development. Education is a key factor in achieving the goal of sustainable development because it can raise awareness about environmental issues, promote sustainable practices, and provide individuals with the skills and knowledge needed to contribute to sustainable development. Further, education can promote sustainability in various ways, because it provides individuals with an understanding of the interconnectedness of social, economic, and environmental systems. Moreover, it can help to build a culture of sustainability by promoting values such as environmental stewardship, social responsibility, and intergenerational equity. It also creates a more sustainable society by encouraging individuals and a particular community to take actions that promote sustainability in their daily lives. Education is a crucial aspect of sustainable development because it helps individuals acquire knowledge and skills to participate fully in social, economic, and political activities.

In 1987, the World Commission on Environment and Development defined sustainable development as "refers to meeting the needs of the present without compromising the ability of future generations to meet their own needs". It includes economic, social, and environmental dimensions, and it is essential for reducing poverty and promoting human well-being" (WCED 1987). "The scheduled castes in India represent the most backward and deprived section in the society. They are at the bottom of social indicators, such as health and education, and they are exploited economically, socially, and psychologically in every sphere" (Raghavendra 2020). Census of India (2011a), categorized the "educational attainment as follows: (i) not literate, (ii) literate without formal schooling, (iii) below primary, (iv) primary, (v) middle, (vi) secondary, (vii) higher secondary, (viii) diploma/certificate, (ix) graduate, and (x) postgraduate and above". It is important to note that educational attainment refers to the highest degree of education completed by an individual.

In recent years, there has been a growing awareness of the importance of education in promoting sustainable development and reducing poverty, particularly among marginalized communities and particularly among the Balmiki in rural areas of Haryana. Glick and Miller (1956) described that "educational attainment is one of the most important factors in determining the occupational and income levels to which a person can aspire". Raju (1988) highlighted that "historically, women and scheduled castes have faced significant challenges in attaining high literacy levels in India. It has been analyzed that the differentiation between non-scheduled castes and scheduled castes is more prominent than the distinction observed between the male and female members within each group". "Education is considered to be an important mechanism for development in India. It has served as a positive contribution to the process of planned change, one finds that changes have been restricted to urban areas, to particular castes, and individuals among the scheduled castes" (Wankhede 2001). Singh (2001) emphasizes that "education is one of the well-known and widely recognized as a key factor of social mobility in rural India". "Girls' education plays a pivotal role in diminishing gender disparities across various aspects of life" (Ghosh

2003). "Caste plays a significant role in social interaction, particularly when it comes to restricting educational opportunities for minority backward communities. Parental education, religion, and region were found to be important key factors that substantially influence a child's likelihood of undergoing an educational transition" (Vaid 2004). Ghosh (2007) also described that "gender disparity varies among different states and social groups, particularly evident in scheduled castes (SCs) and scheduled tribes (STs) in Jharkhand and West Bengal. It is highlighted that certain tribes, such as the Ho, Mahali, and Lodha communities, require focused attention to improve their literacy and educational levels".

Husain and Sarkar (2011) analyzed "the educational trajectories in India, focusing on changes in disparity across various levels of education. They observed an interesting reversal in gender disparity performance within certain states, namely Kerala and Punjab, where negative disparities were consistently present across all educational stages. Likewise, another state, Karnataka, demonstrated commendable performance with negative disparities evident in all educational levels except for the primary level". Chamar and Chamar (2015) examined the "levels of educational attainment among different castes in the village of Sirsa Kheri in the Jind district of Haryana state. The study was based on primary data from a total of 317 households in a village. These households included nine different castes, namely Brahmins and Jats from the general category, Bairagi Khati, Kumhar, Lohar, and Jhinwar from other backward castes, and Chura and Chamar from scheduled castes. The study revealed that 72% of persons are educated in the village. Interestingly, the Chamars emerged as the leading group in terms of educational attainment due to occupational mobility among scheduled castes". In the same year, Chamar and Rekha (2015) conducted a study on the "clan-wise levels of educational attainment of the Chura caste in Badli village, located in the Jhajjar district of Haryana state. They collected primary data from all 91 households of four clans Bagri, Lauth, Taak, and Deeka belonging to the Chura caste. The study revealed that 70% of the persons were found educated, comprised of 74% of males and 65% of females. The majority of the educated persons (44%) had attained education up to the 'middle' level. Interestingly, the study also highlighted that none of the females belonging to the Chura caste in Badli village had achieved education at the 'graduation and above' levels". Further, in a similar study conducted by Arya and Chamar (2015), "the levels of educational attainment of the Dhanak caste in rural Haryana were described. It was highlighted that nearly 56% of persons were literate, with males accounting for 63% and females 47%. The study also revealed that the highest educational attainment was observed at the primary level in the Rewari district (42%), the middle level in the Jhajjar district (38%), and the Gurgaon district boasted the highest educational attainment at the plus-two, graduation, and postgraduate levels. According to the recorded data, among the educated persons of the Dhanak caste in rural Haryana, 94% have attained education up to the plus two levels (10 + 2). However, the data reveals that only 5% of educated persons have reached up to the level of graduation".

Katiyar (2016) examined "the gender disparity in literacy rate among the states of India and UTs. He observed that gender disparity increased continuously from 1901 to 1961 and first time decreased in 1971. The study specifically highlighted

Rajasthan had the highest gender disparity in both rural and urban areas. On the other hand, Meghalaya recorded the lowest gender disparity in rural areas, while Mizoram displayed the lowest disparity in urban areas".

Similarly, Chamar et al. (2018) analyzed the "caste differentials in levels of educational attainment in the 'Jat Land' of Haryana. The study revealed that approximately 95% of educated persons had obtained education up to the 'plus two' levels, while only 5% of persons reached this level. Remarkably, the Chamar caste displayed a prominent position in terms of educational attainment followed by the Sansi among scheduled castes. On the other hand, the Banjara and the Doom castes presented low rank of educational attainment".

Likewise, Sarkar and Kasemi (2020) described "the gender gap in literacy and education among educationally disadvantaged scheduled castes in the Maynaguri block of the Jalpaiguri district of West Bengal. Among the 59 scheduled castes in Jalpaiguri district, Turi, Lohar, Bhuiya, and Ghasi identified as the most educationally backward were selected for the primary survey. The study revealed that the Bhuiyas and the Ghasis castes recorded the highest and lowest literacy rates respectively". Khan and Anwaruzzaman (2021) examined "the role of education in determining fertility levels among Muslim women in rural areas of West Bengal. They highlighted that female educational attainment and fertility have a negative relationship. Notably, the most significant decline in total fertility rate (TFR) was observed from primary, but below middle to middle, and below the secondary level of educational attainment". Choudhury et al. (2023) examined "the prevailing regional socioeconomic disparities in the accessibility of pre-primary education within India. They highlighted the significant inequalities experienced by historically disadvantaged scheduled caste and scheduled tribe children, who were found to be less likely to have access to early childhood education, especially in rural areas".

Using the same approach, several studies have been conducted to examine the literacy rate, levels of educational attainment, and gender disparity among various social groups in India. Some of the studies aimed to shed light on the educational disparities and challenges faced by marginalized communities also (see for instance Mathai 1956; Kamat 1981; Jacobs 1996; McDougall 2000; Reddy and Rao 2003; Azim 2005; Husain 2011; Chamar and Chamar 2017; Arya 2017; Chamar and Rekha 2018; Chamar and Chamar 2019; Rani and Chamar 2019; Ali 2020).

## 22.2 About the Balmiki Caste

Traditionally, the Balmiki caste involved in manual scavenging faces multiple challenges in terms of social and economic inequality. The Balmiki caste is considered one of the most marginalized and socially excluded groups in India particularly in the state of Haryana also, and their access to education and sustainable development opportunities is still limited after a long period of independence and the formation of Haryana as an independent state in 1966 and district Jhajjar in 1998. Access to education is still limited and poverty is prevalent among the Balmiki caste in rural areas of district Jhajjar. However, education is a key factor in promoting sustainable development practices, as it raises awareness about environmental issues and provides the knowledge and skills necessary to implement sustainable solutions. Singh (1994) elaborated that the Balmiki community can be found throughout the state of Haryana, where they are locally known by various names mentioned as Balmik; Bhangi; Bhangi/Chuhra; Chura; Halalkhor; Jamadar; Khakrobe; Lalbegi; Mehtar and Valmiki, etc. The main objective of the present study is to analyze the gender disparity in literacy rate and levels of educational attainment among the Balmiki caste in the Jhajjar district of Haryana.

## 22.3 Materials and Method

### *22.3.1 The Study Area*

The Jhajjar district extends between 28° 2′ 31″ to 28° 50′19″ north latitudes and 76° 17′06″ to 76° 58′ 15″ east longitudes. It is bounded by Rohtak and Sonipat districts in the north, Gurgaon and Rewari districts in the south, Bhiwani district in the west, and NCT of Delhi in the east (Fig. 22.1). The district headquarters is situated in Jhajjar town.

As per census 2011, the Jhajjar district has 260 villages including 13 uninhabited villages. The total rural scheduled caste population in Jhajjar district is 1,30,913 persons, which accounts for 18.30% of the rural population in the district (Census of India 2011b). There are three major castes namely the Chamar, the Chura/Balmiki, and the Dhanak among scheduled castes. In terms of numerical strength, the Balmikis are the second most dominant caste after the Chamars. With a total population of 24,372 persons, the Balmiki caste accounts for 18.62% of the total scheduled caste population in the rural areas of the district.

### *22.3.2 Database and Methodology*

The Balmiki caste, among the most backward scheduled castes, has been selected for the collection of primary data. The primary data was obtained through an extensive field survey conducted from February to April 2022. The selection of villages for the survey was based on three criteria. Initially, the tahsil as the administrative division of the district headquarters was selected. Subsequently, villages in the Jhajjar tahsil with a minimum of 35 households belonging to the Balmiki caste were identified by contacting telephonically with the head (Sarpanch) of each village. After that, out of the six such villages, three villages, namely Silani Pana Zalim, Silana, and Jahangirpur were purposefully selected (Fig. 22.2) for personal interviews with the respondents by adopting the purposive sampling techniques. All the

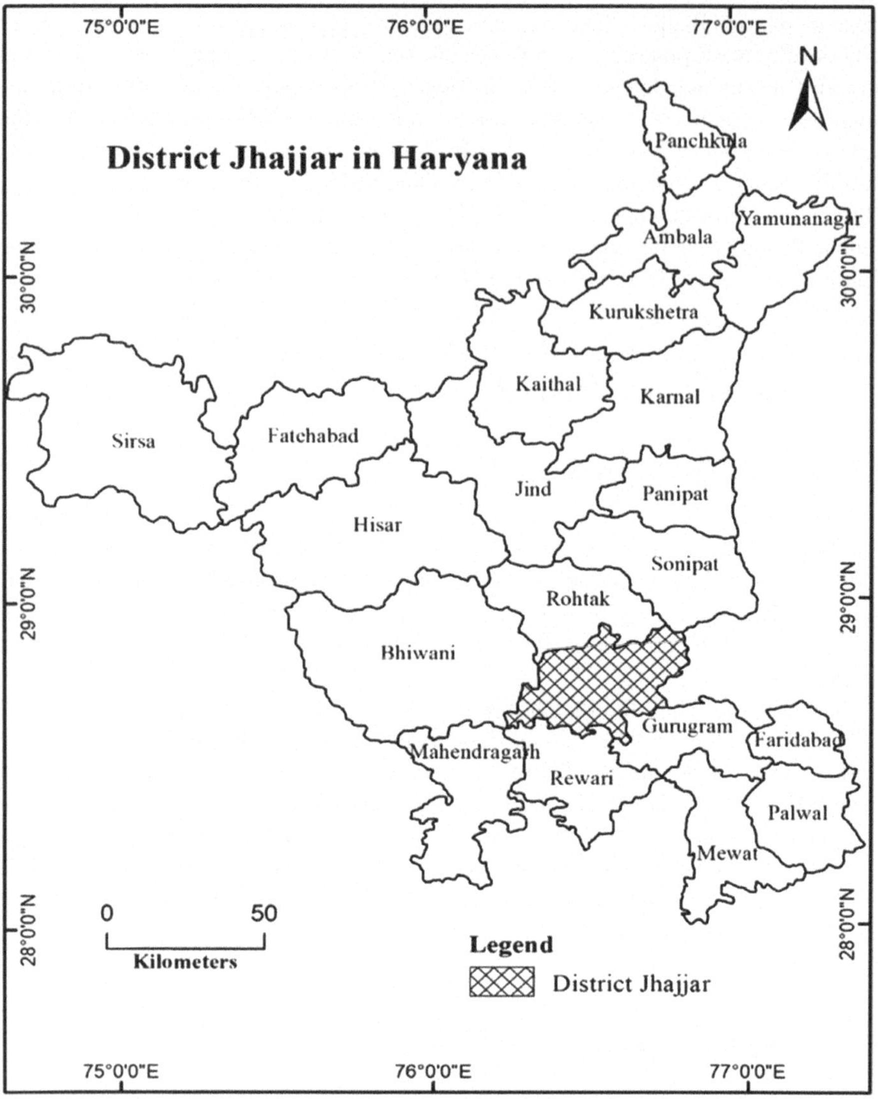


**Fig. 22.1** Location of the study area. *Source* Authors

respondents available at the time of the field survey, 141 households belonging to the Balmiki caste from the three selected villages were interviewed. Gender wise literacy rate of each household was calculated in percent. Additionally, the levels of educational attainment, including primary, upper primary, secondary, senior secondary, graduation, and post-graduation were calculated proportionally to the total educated persons of the three selected villages. To identify gender disparity in both literacy

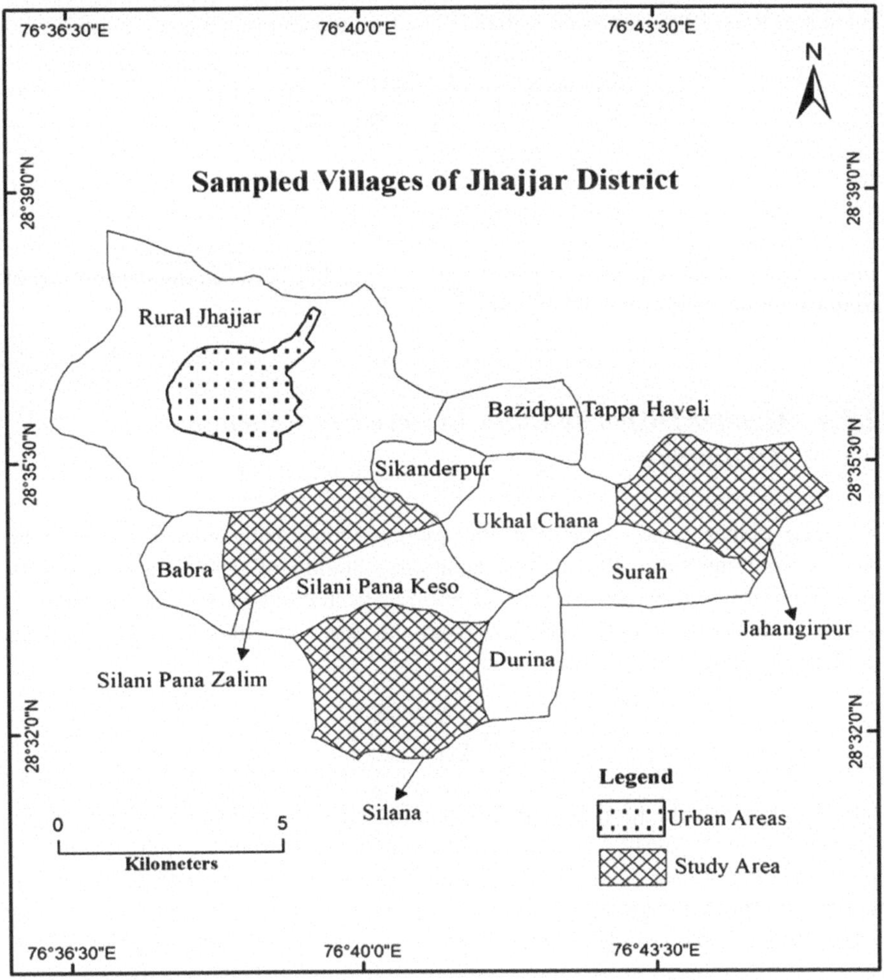


**Fig. 22.2** Selected sampled villages of Jhajjar district

rates and educational attainment levels, the Kundu and Rao (1986) formula and the Chakraborty (1999) disparity index formula were applied, respectively.

### *22.3.3 Gender Disparity Index (GDI) in Literacy Rate*

The gender disparity index in literacy rate reflects the extent by which gender is left behind the other in terms of a specific level of literacy rate. To measure the gender disparity between males and females within the three selected villages belonging to the Balmiki caste, the modified Sopher's (1974) disparity index as proposed by

Kundu and Rao (1986) has been utilized. The formula employed is as follows:

$$\mathrm{DI} = \left[\left(\frac{\log \mathrm{X}_2}{\mathrm{X}_1}\right) + \left\{\left(\frac{\log(200 - \mathrm{X}_1)}{(200 - \mathrm{X}_2)}\right)\right\}\right]$$

where DI represents the disparity index, $X_1$ denotes the literacy rate of a deprived group (females), $X_2$ signifies the literacy rate of the dominant group (males). If the value of DI equals 0, it indicates perfect equality between the genders. Consequently, a higher value of DI implies a greater level of inequality between the two groups of variables, while a lower value indicates a more balanced situation.

### *22.3.4 Gender Disparity Index in Levels of Educational Attainment*

The gender disparity index quantifies the extent to which one gender lags behind the other in achieving various levels of educational attainment. The Gender Disparity Index (GDI) devised by Chakraborty (1999) has been employed at different educational levels based on data collected from the selected villages. GDI has been calculated as under

$$\mathrm{GDI} = \left(\frac{\mathrm{Y_p} \times \mathrm{X}}{\mathrm{Y} \times \mathrm{X_p}}\right)$$

where X represents the total male population aged 7 years and above, and $X_p$ is the male population at specific levels of educational attainment. Similarly, Y denotes the total female population aged 7 years and above, and $Y_p$ is the female population at the given level of educational attainment.

The gender disparity index (GDI) ranges from 0.00 to 1.00, indicating the ratio of female-to-male education. A score less than 1.00 signifies lower female education in comparison to male education. A higher score on a scale of 0.00 to 1.00 indicates lower gender disparity between males and females at a particular educational level. On the other hand, a lower score indicates higher gender disparity.

In the present study, the gender disparity index has been calculated for primary, upper primary, secondary, senior secondary, graduate, and postgraduate levels of educational attainment among the Balmiki caste in Jhajjar district. Finally, the tables and diagrams have been prepared to interpret the results of the literacy rate and gender disparity in the study area.

## 22.4 Results and Discussion

### 22.4.1 *Population Characteristics*

The population characteristics provide insights into the gender distribution, age composition, literacy rate, work participation rate, and household size, within the Balmiki caste in the Jhajjar district of Haryana as expressed in Table 22.1. According to the sample households, the total population of the selected villages was 746 persons comprised of 390 males (52%) and 356 females (48%) among the Balmiki caste. In terms of age groups, the population aged 0–6 years accounted for 100 persons (13%), with 47 males (47%) and 53 females (53%). Based on sample households, 531 persons were literate, representing 82% of the total population. Out of the literate population, 301 were males (88%) and 230 were females (76%). In terms of sex ratio, there are 913 females for every 1000 males in the study area. The work participation rate of the Balmiki caste is recorded at about 35%. The average household size among the sample households was 5 persons per household.

**Table 22.1** Population characteristics of sample households among Balmiki caste in Jhajjar district

| Population characteristics | Balmiki caste | |
|---|---|---|
| Number of total sample households | 141 | 100.0 |
| Total population | 746 | 100.0 |
| Male population | 390 | 52.28 |
| Female population | 356 | 47.72 |
| Total population (0–6 yrs) | 100 | 13.40 |
| Male population (0–6 yrs) | 047 | 47.00 |
| Female population (0–6 yrs) | 053 | 53.00 |
| Total population (>6 yrs) | 646 | 86.60 |
| Male population (>6 yrs) | 343 | 53.10 |
| Female population (>6 yrs) | 303 | 46.90 |
| Total literate persons | 531 | 82.20 |
| Male literates | 301 | 87.76 |
| Female literates | 230 | 75.91 |
| Total workers | 259 | 34.72 |
| Male workers | 195 | 50.00 |
| Female workers | 064 | 17.98 |
| Sex ratio (female/1000 males) | 913 | |
| Child sex ratio (female/1000 males) | 1128 | |
| Household size (persons/household) | 5.3 | |

*Source* Field Survey 2022, Compiled by authors

**Table 22.2** Literacy rate and gender disparity among Balmiki caste in Jhajjar district

| Selected villages | Number of households | Literacy rate (%) | | | | Gender disparity |
|---|---|---|---|---|---|---|
| | | Total | Male | Female | Gender gap | |
| Silani Pana Zalim | 47 | 75.58 | 79.84 | 69.89 | 09.95 | 0.09 |
| Silana | 50 | 87.56 | 96.04 | 79.00 | 17.04 | 0.15 |
| Jahangirpur | 44 | 83.77 | 88.98 | 78.18 | 10.80 | 0.10 |
| Jhajjar district | 141 | 82.20 | 87.76 | 75.91 | 11.85 | 0.11 |

*Source* Compiled by authors

### 22.4.2 Literacy Rate and Gender Disparity

A comprehensive overview of the literacy rate and gender disparity among the Balmiki caste is represented in Table 22.2 in the Jhajjar district. The district has recorded an 82% literacy rate, with males accounting for 88% and females for 76%. Silana and Jahangirpur villages have significantly better literacy rates than the average of the study area. The village Silana has recorded the highest literacy rate of 87.56%, with males consist of 96.04% and females have 79%. On the other hand, village Silani Pana Zalim has the lowest literacy rate of 75.58%, with males count 79.84% and females have 69.89% of literacy rate. Further, on average, the Jhajjar district exhibits a gender disparity with 0.11 points among the Balmiki caste. The highest gender disparity of 0.15 points has been recorded in village Silana due to the low female literacy rate, which accounts for a gap of 17.04% compared to their male counterparts, while, village Silana Pana Zalim has observed the lowest gender disparity with 0.09 points. On the other hand, the village Jahangirpur has a gender disparity of 0.10 points, and the gender gap is recorded at 10.80%.

### 22.4.3 Levels of Educational Attainment

Based on sample households, it is recorded that out of total educated persons, nearly 24% have attained education up to 'primary level', 22% up to 'upper primary' level, and 27% up to the level of 'secondary/matriculation'. Nearly 9% of educated persons have reached up to the levels of 'graduation and post-graduation'. It may also be noted that about 91% of educated persons have obtained the levels of 'plus two'. An interesting correspondence between selected villages and the share of educated persons under different levels has also been indicated (Table 22.3; Fig. 22.3). The village Jahangirpur has obtained the maximum share of education up to the level of 'primary' (27.75%) and 'upper primary' (26.70%). The largest share of educated persons is observed in Silani Pana Zalim in 'secondary/matriculation and above levels'.

**Table 22.3** Educated persons in various levels of educational attainment among the Balmiki caste in Jhajjar district

| Levels of educational attainment | Silani Pana Zalim | | | Silana | | | Jahangirpur | | | Jhajjar district | | |
|---|---|---|---|---|---|---|---|---|---|---|---|---|
| | Total | Male | Female | Total | Male | Female | Total | Male | Female | Total | Male | Female |
| Primary level | 32 | 15 | 17 | 42 | 23 | 19 | 53 | 33 | 20 | 127 | 71 | 56 |
| (In %) | 19.51 | 15.15 | 26.15 | 23.86 | 23.71 | 24.05 | 27.75 | 31.43 | 23.25 | 23.91 | 23.59 | 24.35 |
| Upper primary level | 25 | 18 | 07 | 40 | 26 | 14 | 51 | 23 | 28 | 116 | 67 | 49 |
| (In %) | 15.24 | 18.18 | 10.77 | 22.73 | 26.80 | 17.73 | 26.70 | 21.90 | 32.56 | 21.85 | 22.26 | 21.30 |
| Secondary level | 49 | 29 | 20 | 47 | 28 | 19 | 48 | 26 | 22 | 144 | 83 | 61 |
| (In %) | 29.88 | 29.30 | 30.77 | 26.70 | 28.88 | 24.05 | 25.13 | 24.76 | 25.58 | 27.12 | 27.57 | 26.52 |
| Sr. secondary level | 37 | 25 | 12 | 34 | 15 | 19 | 24 | 17 | 07 | 95 | 57 | 38 |
| (In %) | 22.56 | 25.25 | 18.46 | 19.32 | 15.46 | 24.05 | 12.57 | 16.20 | 08.14 | 17.89 | 18.94 | 16.52 |
| Up to graduation level | 15 | 08 | 07 | 08 | 04 | 04 | 14 | 05 | 09 | 37 | 17 | 20 |
| (In %) | 09.15 | 08.08 | 10.77 | 04.55 | 04.12 | 05.06 | 07.33 | 04.76 | 10.47 | 06.97 | 05.65 | 08.70 |
| Up to post-graduation level | 06 | 04 | 02 | 05 | 01 | 04 | 01 | 01 | 00 | 12 | 06 | 06 |
| (In %) | 03.66 | 04.04 | 03.08 | 02.84 | 01.03 | 05.06 | 00.52 | 00.95 | 00.00 | 02.26 | 01.99 | 02.61 |
| Jhajjar district | 164 | 99 | 65 | 176 | 97 | 79 | 191 | 105 | 86 | 531 | 301 | 230 |
| (In %) | 100 | 100 | 100 | 100 | 100 | 100 | 100 | 100 | 100 | 100 | 100 | 100 |

*Source* Compiled by authors

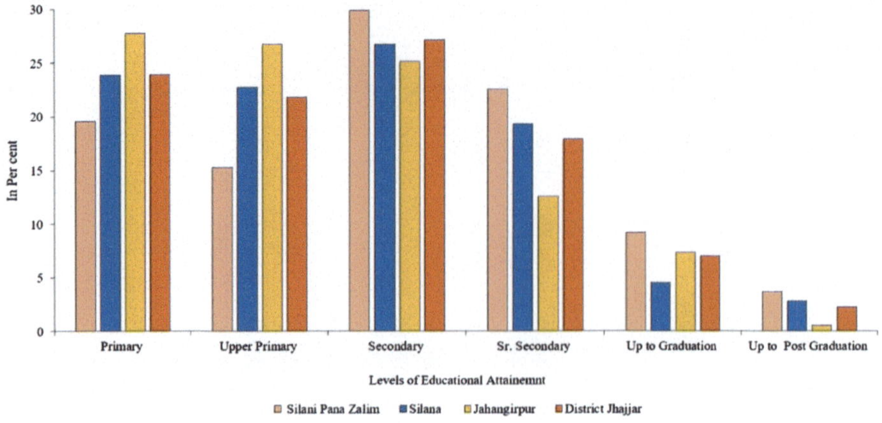


**Fig. 22.3** Levels of educational attainment among Balmiki caste in Jhajjar district, 2022

The various levels of educational attainment for males and females separately reveal another interesting picture. The study reveals that approximately 91% of the educated persons reach the senior secondary level of educational attainment, with males comprising around 92% and females 89% among the Balmiki caste in the Jhajjar district. Based on sample households, the study also reveals variation across different levels of educational attainment. The village Jahangirpur boasts the highest percentage (27.75%) at the primary level among educated persons, followed by the villages of Silana (23.86%) and Silani Pana Zalim (19.51%). Similarly, at the 'upper primary' level, Jahangirpur village records the highest percentage (26.70%), while Silani Pana Zalim witnesses the lowest position (15.24%) among the Balmiki caste.

Further, the village Silani Pana Zalim has recorded the highest percentage at the levels of 'secondary' (29.88%) and 'senior secondary' (22.56%). On the other hand, Jahangirpur has witnessed the lowest percentage (25.13% and 12.57%) at the same levels of educational attainment respectively. Notably, the Balmiki caste in the Jhajjar district gracefully maintains their positions at the secondary level of education. Likewise, Silani Pana Zalim has recorded the highest position of 9.15% and other side Silana observed the lowest position of 4.55% of the educated persons up to 'graduation level'. While, at the level up to 'post-graduation', the village Silani Pana Zalim has attained the highest position (3.66%) and lowest in Jahangirpur (0.52%) village of Jhajjar district. Moreover, approximately 9% of the literate people reach the graduation and post-graduation levels, with males accounting for about 8% and females 11% among the Balmiki caste in Jhajjar district.

It is highly noted that at the level of up to graduation and post-graduation of educational attainment results show poor condition of Balmiki caste on education among all three villages. It is noted that the candidates have obtained the degree in various courses like JBT, ITI, B.A, B. Com, B. Ed. M.A., M.Tech., and MBA belong to the Balmiki caste in the study area. However, overall, education among females which is more than 72% is encouraged only for the lower levels like primary and upper primary and secondary/matriculation in the Jhajjar district as expressed in Table 22.3.

### 22.4.4 Differentials in Levels of Educational Attainment

There is a remarkable difference has been observed in the magnitude of concentration of educated persons among the Balmiki caste in the Jhajjar district (Table 22.4). The highest difference has been recorded in the upper primary level from Jahangirpur village to Silani Pana Zalim village (11.46%)and vice-versa in postgraduate level is 3.14% (Table 22.4a). The highest difference of male educated persons has also been recorded in primary level from Jahangirpur village to Silani Pana Zalim village (16.28%) and the lowest in the level of post-graduation from Silani Pana Zalim village to Jahangirpur village is 3.09% (Table 22.4b). On the other hand, the highest difference of female educated persons has been recorded in the ' upper primary' level from village Jahangirpur . to Silani Pana Zalim (21.79%) while it is lowest in primary level from Silani Pana Zalim village to Jahangirpur village (2.9%) in the study area (Table 22.4c).

### 22.4.5 Gender Disparity in Various Levels of Educational Attainment

Finally, based on data shown in Table 22.3, gender disparity in various levels of educational attainment has also been identified (Table 22.5). It has been observed that the village Jahangirpur exhibits the highest disparity of 0.65 points at the primary level among the Balmiki caste in the Jhajjar district. On the other hand, village Silani Pana Zalim has witnessed the lowest disparity of 1.51 points between males and females at the same level. The situation is reversed in the case of the 'upper primary level'; village Silani Pana Zalim has recorded the highest disparity of 0.52 points, while Jahangirpur witnessed the lowest disparity of 1.31 points in this level of educational attainment. Likewise, at the secondary level of educational attainment; the village Silani Pana Zalim has recorded the lowest disparity of 0.92 points followed by Jahangirpur (0.91 points), whereas Silana village recorded the highest disparity of 0.69 points. Similarly, village Silana recorded the lowest disparity of 1.28 points, while Jahangirpur recorded 0.44 points of disparity at the level of senior secondary. On the other hand, Silana and Jahangirpur villages have shown reverse conditions at graduation level. Remarkably, Silana has recorded the highest disparity (1.01) and Jahangirpur recorded the lowest disparity (1.93) among the Balmiki caste in the Jhajjar district. Silana village has registered the lowest disparity of 4,04 points up to 'postgraduate level'; while Jahangirpur recorded 0.0 points of disparity, because not a single female reached up to this level of educational attainment in this village. It is

**Table 22.4** Difference in levels of educational attainment among Balmiki caste in Jhajjar district

| Educational level | Highest value | Lowest value | Difference |
|---|---|---|---|
| *(a) Difference in levels of educational attainment of total persons (%)* | | | |
| Primary level | Jahangirpur (27.75) | Silani Pana Zalim (19.51) | 08.24 |
| Upper primary level | Jahangirpur (26.70) | Silani Pana Zalim (15.24) | 11.46 |
| Secondary level | Silani Pana Zalim (29.88) | Jahangirpur (25.13) | 04.75 |
| Sr. Secondary level | Silani Pana Zalim (22.56) | Jahangirpur (12.57) | 09.99 |
| Up to graduation | Silani Pana Zalim (09.15) | Silana (04.55) | 04.60 |
| Up to post-graduation | Silani Pana Zalim (03.66) | Jahangirpur (00.52) | 03.14 |
| *(b) Difference in levels of educational attainment of male persons (%)* | | | |
| Primary level | Jahangirpur (31.43) | Silani Pana Zalim (15.15) | 16.28 |
| Upper primary level | Silana (26.80) | Silani Pana Zalim (18.18) | 08.62 |
| Secondary level | Silani Pana Zalim (29.30) | Jahangirpur (24.76) | 04.54 |
| Sr. secondary level | Silani Pana Zalim (25.25) | Silana (15.46) | 09.79 |
| Up to graduation | Silani Pana Zalim (08.08) | Silana (04.12) | 03.96 |
| Up to post-graduation | Silani Pana Zalim (04.04) | Jahangirpur (00.95) | 03.09 |
| *(c) Difference in levels of educational attainment of female persons (%)* | | | |
| Primary level | Silani Pana Zalim (26.15) | Jahangirpur (23.25) | 02.90 |
| Upper primary level | Jahangirpur (32.56) | Silani Pana Zalim (10.77) | 21.79 |
| Secondary level | Silani Pana Zalim (30.77) | Silana (24.05) | 06.72 |
| Sr. Secondary level | Silana (24.05) | Jahangirpur (08.14) | 15.91 |
| Up to graduation | Silani Pana Zalim (10.77) | Silana (05.06) | 05.71 |
| Up to post-graduation | Silana (05.06) | Jahangirpur (00.00) | 05.06 |

*Source* Compiled by authors

interesting to note that females are significantly lagging behind males across various levels of educational attainment except up to 'graduation level' among the Balmiki caste in Jhajjar district.

## 22.5 Conclusions

The study concludes that based on a sample survey, the Jhajjar district has recorded 82% literacy rate, with males accounting for 88% and females for 76% among the Balmiki caste. The Jhajjar district exhibits a gender disparity of 0.11 points. The highest gender disparity with 0.15 points has been recorded in village Silana due to the low female literacy rate, which accounts for a gap of 17.04% compared to their male counterparts. On the other hand, village Silana Pana Zalim has observed the lowest gender disparity with 0.09 points among the Balmiki caste. It is observed that 24% of the educated persons have received education up to the 'primary level', 22%

**Table 22.5** Gender disparity at various levels of educational attainment among Balmiki caste in Jhajjar district

| Levels of Educational attainment | Silani Pana Zalim | Silana | Jahangirpur | Jhajjar district |
|---|---|---|---|---|
| | Gender disparity index | | | |
| Primary level | 1.51 | 0.83 | 0.65 | 0.89 |
| Upper primary level | 0.52 | 0.54 | 1.31 | 0.83 |
| Secondary level | 0.92 | 0.69 | 0.91 | 0.83 |
| Senior secondary level | 0.64 | 1.28 | 0.44 | 0.75 |
| Up to graduation level | 1.17 | 1.01 | 1.93 | 1.33 |
| Up to post graduation | 0.67 | 4.04 | 0.00 | 1.13 |

*Source* Compiled by authors

up to the 'upper primary' level, 27% up to the 'secondary/matriculation' level, and nearly 9% have achieved 'graduation and post-graduation' levels among the Balmiki caste in the Jhajjar district. It is worth mentioning that about 91% of educated persons have reached the 'plus two' level. It is highly noted that at the level of up to 'graduation and post-graduation' of educational attainment results show poor condition of the Balmiki caste in education among all three villages. The highest difference has been recorded in the upper primary level from Jahangirpur village to Silani Pana Zalim village (11.46%) and vice-versa in postgraduate level which is 3.14%. The village Jahangirpur exhibits the highest disparity of 0.65 points at the primary level among the Balmiki caste in Jhajjar district. On the other hand, village Silani Pana Zalim has witnessed the lowest disparity of 1.51 points between males and females at the same level. Silana village has registered the lowest disparity of 4.04 points up to 'postgraduate level'; while Jahangirpur recorded 0.0 points of disparity, because not a single female reached up to this level of educational attainment in this village. It is interesting to note that females are significantly lagging behind males across various levels of educational attainment except up to 'graduation level' among the Balmiki caste in Jhajjar district.

## References


Ali A (2020) Spatial dimensions of literacy and social development in Uttar Pradesh. Punjab Geogr 16:45–62

Arya R, Chamar KV (2015) Differentials in levels of educational attainment of Dhanak caste in rural Haryana. J People Soc Haryana 5(2):14–22

Arya R (2017) Levels of educational attainment in village Sisar Khas: caste wise analysis. J People Soc Haryana 7(1):1–7

Azim S (2005) Literacy growth disparities in Karnataka. Econ Pol Wkly 40(16):1647–1649

Census of India (2011a) Special tables on scheduled castes Haryana. New Delhi

Census of India (2011b) Primary census abstract Haryana-06. New Delhi

Chakraborty G (1999) Quality of life of scheduled castes and scheduled tribes in rural India. Yojana 44(6):34–40

Chamar KV, Chamar SK, Chamar NK et al (2018) Caste differentials in levels of educational attainment in Jat land of Haryana. In: Kar BK, Bhattacharya P (eds) Environment and emerging challenges. Gauhati University (Assam, India), Department of Geography, pp 122–130
Chamar KV, Chamar SK (2015) Caste differentials in levels of educational attainment in village Sirsa Kheri. Trans Inst Ind Geogr 37(1):27–34
Chamar NK, Chamar KV (2019) Differentials in levels of educational attainment of Bazigar caste in north west Haryana: a clan wise analysis. J People Soc Haryana 9(1):13–28
Chamar SK, Chamar KV (2017) Rank differentials in levels of educational attainment among scheduled castes: a case study of Jind district. J People Soc Haryana 7(2):36–49
Chamar KV, Rekha (2015) Levels of educational attainment of Chura caste in village Badli: a clan wise analysis. J Soc Sci Res Perspect 3(1):7–14
Chamar KV, Rekha (2018) Rank differentials in levels of educational attainment of Chura caste in Jhajjar district: a clan wise analysis. J People Soc Haryana 8(2):45–58
Choudhury PK, Joshi R, Kumar A (2023) Regional and socioeconomic inequalities in access to pre-primary education in India: evidence from a recent household survey. Int J Child Care Educ Policy 17(13):1–25
Ghosh AK (2003) Female literacy and social development. Soc Chang 33(4):65–72
Ghosh AK (2007) The gender gap in literacy and education among the scheduled tribes in Jharkhand and West Bengal. Sociol Bull 56(1):109–125
Glick PC, Miller HP (1956) Educational level and potential income. Am Sociol Rev 21(3):307–312
Husain Z (2011) Gender disparities in completing school education in India: explaining geographical variations. J Popul Res 28(4):325–352
Husain Z, Sarkar S (2011) Gender disparities in educational trajectories in India: do females become more robust at higher levels? Soc Indic Res 101(1):37–56
Jacobs JA (1996) Gender inequality and higher education. Ann Rev Sociol 22:153–185
Kamat AR (1981) Education and social change amongst the scheduled castes and scheduled tribes. Econ Pol Wkly 16(31):1279–1284
Katiyar SP (2016) Gender disparity in literacy in India. Soc Chang 46(1):46–69
Khan S, Anwaruzzaman AKM (2021) Role of education in determining fertility level among Muslim women in rural areas of West Bengal. Punjab Geogr 17:109–125
Kundu A, Rao JM (1986) Inequity in educational development: issues in measurement, changing structure and its socio-economic correlated with special reference to India. In: Raja M (ed) Educational planning: a long term perspective. NIEPA and Concept Publishing Company, New Delhi, pp 435–466
Mathai S (1956) Education in India. Curr Hist 30(174):85–89
McDougall L (2000) Gender gap in literacy in Uttar Pradesh: questions for decentralised educational planning. Econ Pol Wkly 35(19):1649–1658
Raghavendra RH (2020) Literacy and health status of scheduled castes in India. Contemp Voice Dalit 12(1):97–110
Raju S (1988) Female literacy in India: the urban dimension. Econ Pol Wkly 23(44): WS57-WS59+WS61-WS64
Rani A, Chamar KV (2019) Rank differentials in levels of educational attainment of Bawaria caste in Rewari district: a clan wise analysis. Int J Anal Exp Modal Anal 11(9):4266–4278
Reddy VR, Rao R (2003) Primary education: progress and constraints. Econ Pol Wkly 38(12–13):1242–1251
Rekha CKV (2018) Differentials in levels of educational attainment of Chura caste in Jhajjar district: block wise analysis. Res Strat 8(8):9–16
Sarkar T, Kasemi N (2020) Gender inequality in literacy and education among educationally backward scheduled castes of Maynaguri block, Jalpaiguri district, West Bengal. Punjab Geogr 16:63–74
Singh JP (2001) The state of education in rural India: problems and prospects. Soc Chang 31(3):1–19
Singh KS (1994) People of India, Haryana, Anthropological survey of India. XXIII. Manohar Publishers, New Delhi, pp 33–37

Sopher DE (1974) A measure of disparity. Prof Geogr 26(4):389–392
Vaid D (2004) Gendered inequality in educational transitions. Econ Pol Wkly 39(35):3927–3938
Wankhede GG (2001) Educational inequalities among scheduled castes in Maharashtra. Econ Pol Wkly 36(18):1553–1558
WCED [World Commission on Environment and Development] (1987) Report of the world commission on environment and development: our common future. Oxford University Press, New York, p 24

# Chapter 23
# Spatial Patterns of Literacy Rate and Sex Ratio for Sustainable Education: A Case Study from Ajmer District

**Dharmender Singh Chauhan and Anamika Singh**

**Abstract** The tool of education has transformed the barbaric clan of the world into a prosperous human civilization. Sustainable education is a need of society that is directly associated with female enrollment. Pertinently, out of the total population of India, the female population is almost 50%, as such, the sex ratio plays a vital role in the sustainable educational development of the country. Literacy is a cardinal demographic element of human progress. It remains an indispensable ingredient that shapes human character, and social and economic development. Besides, the sex ratio is such a demographic parameter that reflects the real picture of society. The patriarchal dominant society has led women to face multifarious challenges in their families and society across the nation. The present study is an attempt to bring forth the correlation between the literacy rate and the sex ratio of Ajmer district for the purpose of sustainable education.

**Keywords** Sex ratio · Literacy rate · Correlation · Sustainable education

## 23.1 Introduction

Education is considered an integral ingredient for the long-term development of any region. Education transforms the life of a person by ensuring the multidimensional growth of an individual encompassing it is mental, intellectual, physical, emotional, and spiritual well-being. Sustainable Education produces an enlightened and empowered society that is capable of making the right choices for its social, economic and political welfare. Education, which is a process of acquiring knowledge and skills, discretion and values, beliefs and habits, assumes grave significance in achieving the

D. S. Chauhan
Department of Geography, University of Rajasthan, Jaipur, Rajasthan, India

A. Singh (✉)
St. Xavier's College, Jaipur, Rajasthan, India
e-mail: anamika8385@gmail.com



S. Anand et al. (eds.), *Emerging Geosustainability Transformations in India*, Advances in Geographical and Environmental Sciences,
https://doi.org/10.1007/978-981-97-7993-2_23

objective of sustainable development. The critical role of education is underlined by the fact that responses to climate change and sustainable development involve more than technological and financial policies. It requires "…broader structural, cultural, perceptual, behavioural and ideological shifts (Mochizuki and Bryan 2015).

## 23.2 Education and Sustainable Development

Education could be considered as the most powerful tool to catalyse a positive change in the society and the world. It stretches the arms of development and initiates an evolutionary process that leads to an increase in human capacity, enables him to cope with struggles, emerge with new ideas, and innovate to achieve his goals (Peet 1999). According to Reyes (2001), "development is understood as a social condition within a nation, in which the needs of its population are satisfied by the rational and sustainable use of natural resources and systems" (Reyes 2001). Todaro and Smith (2006) have presented a multifaceted definition of development, encompassing economic growth, elevation of social structures, attitudes, promotion of equality and eradicating poverty (Todaro and Smith 2006).

The very notion of education for sustainable development upholds imparting education and values that induce a fine balance between economic growth, human development, preservation of cultural tradition as well as natural resources. According to Wals and Kieft (2010), "education for sustainable development emphasizes 14 aspects of learning that enhance the transition towards sustainability including future education; citizenship education; education for a culture of peace; gender equality and respect for human rights; health education; population education; education for protecting and managing natural resources; and education for sustainable consumption (Wals and Kieft 2010).

The challenges to gain sustainability are multiple and diverse hence, only interdisciplinary and multidisciplinary processes, procedures and pedagogies empowered systems of education can successfully ensure the development of mindsets of social masses which are not only scientific but are also sensitive towards ecology and culture (Fulwari 2017).

Gopal and Anand (2006) have strived to give specific educational levels at which dedicated lessons could be imparted for environmental issues. According to their research, the higher secondary stage proves to be apt in order to brainstorm for sustainable innovation and recreate the environmental balance (Gopal and Anand 2006). A similar concept has been put forward through the studies of Bhagwat and Abhyankar (2016). Accordingly, the primary level must impart basic education and lay down a sound platform for awareness, attitudes and skills towards sustainable development. The next level of the higher education system must focus on delivering education to promote environmental and social sustainability (Bhagwant and Abhyankar 2016).

In order to achieve the aspirations laid in the 2030 agenda and Sustainable Development Goals (SDGs) the education system will play a pivotal part (UNESCO 2015).

The need of the hour is to create a transformative educational system for the world whose prime objective is to bridge the knowledge of the learners with the skills and attitudes to promote sustainable development through sustainable lifestyle (Agbedahin 2019). According to Hazra (1997), the ability to read and write is considered to be the most basic definition of Literacy (Hazra 1997). Furthermore, the trend reveals that developed nations have higher literacy rates as compared to the developing or under-developed countries of the world. Such a proposition mirrors the dominance of developed nations over the rest of the world not only in terms of economy but also in decision-making.

According to Benites et al. (2021) government plays a vital role in spearheading academic programs and vocational training for students and developing a vision for Sustainable Development. Such programs must target to address the psychosocial needs of youth and devise a mechanism to reduce absenteeism (Benites et al. 2021; UNESCO 2022).

A developed and sustainable society could only ascertain the future sustainable world. But certain sections of society with rather fewer opportunities are at greater risk. Students belonging to this marginal section face adversities of forced marriages, child marriages, meagre family income and over-pounding family responsibility (Benites et al. 2021). An increase in dropouts from schools is also evident in areas which are unable to access distant learning programs during lockdowns (Benites et al. 2021; Acevedo-Duque et al. 2023; Saiz-Álvarez et al. 2020). All such distinct concerns according to Philpot, Acevedo- Duque bring forth challenging questions before us, "What are the knowledge and skills necessary for sustainable living" and who decides which are most important? What kind of educational programs are needed to develop the knowledge and skills necessary for sustainable living and global citizenship? How will we know if these programs are working?" (Philpot 2015; Acevedo-Duque et al. 2023).

Improvement in the standards of people's lives and ultimately achieving the goals of Sustainable Development could only be commenced through quality education (Saiz-Álvarez et al. 2020). The COVID-19 pandemic and worldwide lockdowns disrupted the education system across the globe. The increase in school dropouts, and inaccessibility of distant learning programs led to the closure of educational institutions and caused a grave impact on learning, especially in the remote areas (Sáez-Gallego et al. 2019).

COVID-19 has hit the education systems hard and even checked some of the previously achieved goals of education (Mynaříková and Novotný 2021). This disruption of education due to the pandemic severely affected the learners belonging to the vulnerable and weaker section of society. But, on the other hand, the pandemic paved the way for global nexus to plan for strengthening for future of education and step towards achieving the Education 2030 Goals (Mynaříková and Novotný 2021; Kopnina 2020; Knutsson 2021).

*Sex ratio* or gender ratio is defined as the number of females per thousand males in the population. According to Broms et al. (2010), "Sex ratios are among the most basic demographic parameters and indicate both the relative survival of females and males and the future breeding potential of a population" (Broms et al. 2010).

However, the preference for male children and the elimination of girl children through infanticide and female foeticide has resulted in an imbalance in sex ratio (Joshi and Tiwari 2011). The preference for males in the male-dominated society of India and female infanticide is still prevalent. The male favourable sex ratio is also a result of improvement in health facilities whose benefits are availed for male children (Singh 2010).According to a study conducted by Chaudhuri (2012), a total of 58.28 million women went missing between 1950 and 2010. Out of these, the survey says 28% went missing due to sex-selective abortion and 72% due to post-natal mortality (Chaudhuri 2012).

The developing countries have a disturbing factor of declining sex ratio. In the case of India, there is a significant difference in the sex ratios and female literacy rates of the Southern and Northern states. Kerala has the best female literacy rate, and the same state was the only state in the country where women outnumbered men. Rajasthan, the state that had the lowest percentage of female literates had a sex ratio of 928. According to the census of 2011, the literacy rate of Tamil Nadu is 80.9%, whereas the sex ratio is 996. A similar reference can be drawn from all the southern states of India, where the female literacy rates are higher than most of the northern states and the same goes for the sex ratio too (Shetty and Shetty 2014). The census reports of past decades verify that the sex ratio across most of the states of India has improved. However, the imbalance of sex ratio and lower literacy rates in certain states such as Punjab, Haryana, and Rajasthan are matters of great concern (Barakade 2012).

## 23.3 Materials and Method

### *23.3.1 The Study Area*

Ajmer is situated at the lower slopes of the Taragarh hills of the Aravalli Range. The city is centrally located in the state of Rajasthan and lies 132 km west of Jaipur. The latitudinal and longitudinal extend of the city is 25°38'N to 26°58'N and 73°54'E to 75°22'E, respectively, covering an area of 8481 sq km. Ajmer recorded a population of 2,583,052 persons with an urban population of 1,035,410 persons (DCHB 2011). Ajmer is divided into nine tehsils namely, Ajmer, Kishangarh, Nasirabad, Beawar, Masuda, Kekri, Sarwar, Bhinay, and Peesangan (Fig. 23.1).

Ajmer is roughly in triangular shape. It is characterized by levelled plains interspersed between low hills. These low hills of the Aravalli range run from north-west to south-east lying to the western part of Ajmer. The Beawar sub-division lies to the south-west of the district and is generally hilly. The Kekri sub-division is levelled plain and forms the southeast portion of Ajmer whereas the Kishangarh sub-division forms the eastern part of the district and is sandy except for a few small isolated patches.

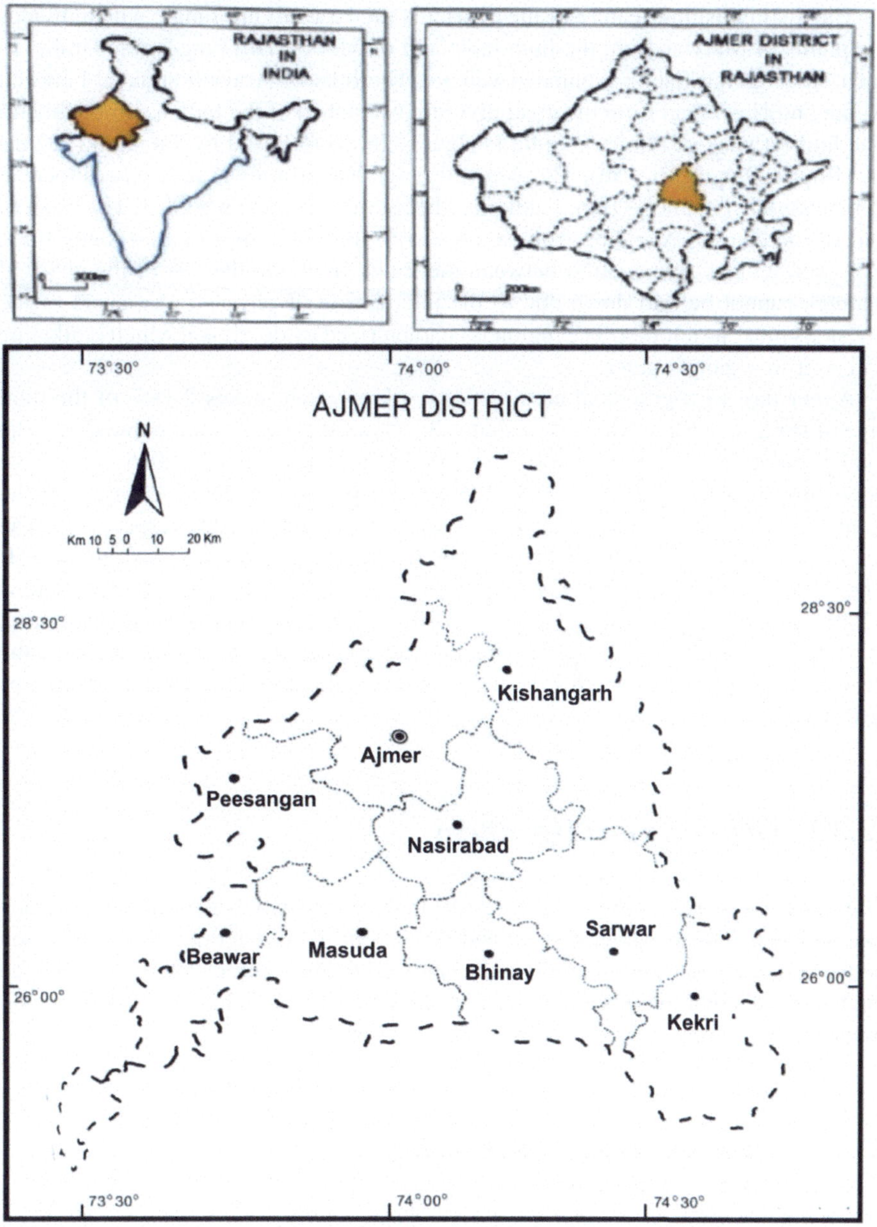


Fig. 23.1 Location of study area and Tehsils: Ajmer district. *Source* DCHB (2011)

The distinguishing feature of the district is the Aravalli hill range, which divides the plains of Marwar from the high tableland of Mewar. Hill ranges run parallel to each other giving rise to elongated valleys. The hills of Aravalli situated between Ajmer and Nasirabad form the great dividing watershed of the Indian sub-continent. The highest peak of Ajmer is Sulia Dungar (726 m) followed by Taragarh (702 m) and Bhutia Dungar (623 m). The Aravalli range has prominent four passes located in Beawar tehsil namely–Barr, Pakharia, Sheopura Ghat and Sura Ghat. The western part of the district extending from southwest to northeast appears as a sandy tract. However, a clear demarcation between the sandy tract and the rest of the physical features cannot be laid down due to the interjection of physical characters of the divisions into one another. The region is characterized by desert soil which is alkaline and poor in organic matter.

Ajmer has a geographical area of 8481 sqkm which makes 2.48% of the total area of Rajasthan State whereas, it shares 3.77% of the State's total population. The total population of Ajmer comprises 1,258,967 females and 1,324,085 males. The population density of the district is 305 persons per sq km. The percentage of the rural and urban population of the district are 59.90% and 40.10%, respectively. The sex ratio of the district is 951 whereas child (age group 0 to 6 years) sex ratio is 901. The Ajmer district has a literacy rate of 69.30% of which the male literacy rate is 82.40% and the female literacy rate is 55.70%. Ajmer recorded a decadal variation of 18.57% from 2001 to 2011 of which decadal variation in the rural areas of the district was reported to be 18.71% whereas it remained 18.38% for the urban area (DCHB 2011).

### *23.3.2 Objectives and Hypothesis*

The study has two objectives. The first is to analyze the spatial distribution of literacy rate and sex ratio in Ajmer district and the second is to determine the correlation between literacy rate and sex ratio in the study area. The research attempts to deduce the hypothesis that there is a positive correlation between literacy rate and sex ratio in the study region.

### *23.3.3 Data Source and Methodology*

The following study is primarily based on secondary sources of data. The official publications of the Census of India and District Census Handbook (DCHB) 2011, 2001 and 1991 were referred for collection of relevant data. The Spearman Rank Correlation (Lobo and Guntur 2018) method has been applied to calculate the Coefficient of Correlation between the literacy rate and sex ratio of the Ajmer district. Equation 23.1 undertakes literacy rate as an independent variable while sex ratio is considered as a dependent variable. The Spearman's rank-order correlation is the nonparametric version of the Pearson product-moment correlation. Spearman's correlation coefficient, ($\rho$, also signified by $r_s$) measures the strength and direction of association between two ranked variables.

$$\rho = 1 - \frac{6\sum d_i^2}{n\left(n^2 - 1\right)} \tag{23.1}$$

where $d_i$ = difference in paired ranks and $n$ = number of cases.

## 23.4 Results and Discussion

### *23.4.1 Spatial Distribution of Sex Ratio*

In India, the sex ratio is defined as the number of females per thousand males. It is an important demographic parameter that depends upon the rate of birth, death and migration. A lower sex ratio in a region is an indicator of many social taboos such as female infanticide, preference for the boy child, sex determination, etc. (Som and Mishra 2014). Furthermore, the change in the sex ratio is a more accurate indicator and reliable guide of socioeconomic progress than the sex ratio at a particular point in time (Jayaraj and Subramanian 2009).

The highest sex ratio in the year 2001 is recorded in Bhinay (963) followed by Sarwar (958), Beawar (955), Kekri (952), Peesangan (950), whereas the lowest sex ratio is recorded in Nasirabad (876), followed by Ajmer (913), Kisangarh (927) and Masuda (948) (Fig. 23.2). It can be inferred from that all the tehsils of the district have improved their sex ratio in the year 2011; but still, it is not up to the mark. Bhinay tehsil (971) recorded the highest sex ratio in the year 2011, followed by Kekri (970), Beawar (970), Sarwar (965), and Peesangan (964). On the contrary, Nasirabad (904) has recorded the lowest sex ratio followed by Ajmer (944), Kishangarh (944) and Masuda (953).

### *23.4.2 Spatial Patterns of Literacy*

Academic courses promote cognitive development, impart knowledge of communication and develop high-order thinking and learning processes (Flowerdew 2014; Moje 2015). According to Gee (2015), recent researchers consider literacy as a social practice putting forward an identity to the individuals (Gee 2015). Gebhard (2004) has highlighted the role of learners as active participants rather than passive attendees. Such practice has emerged in the twenty-first century and shows the changing nature of educational goals (Gebhard 2004). Academic literacy development in this sense is considered a powerful tool for knowledge generation, communication and

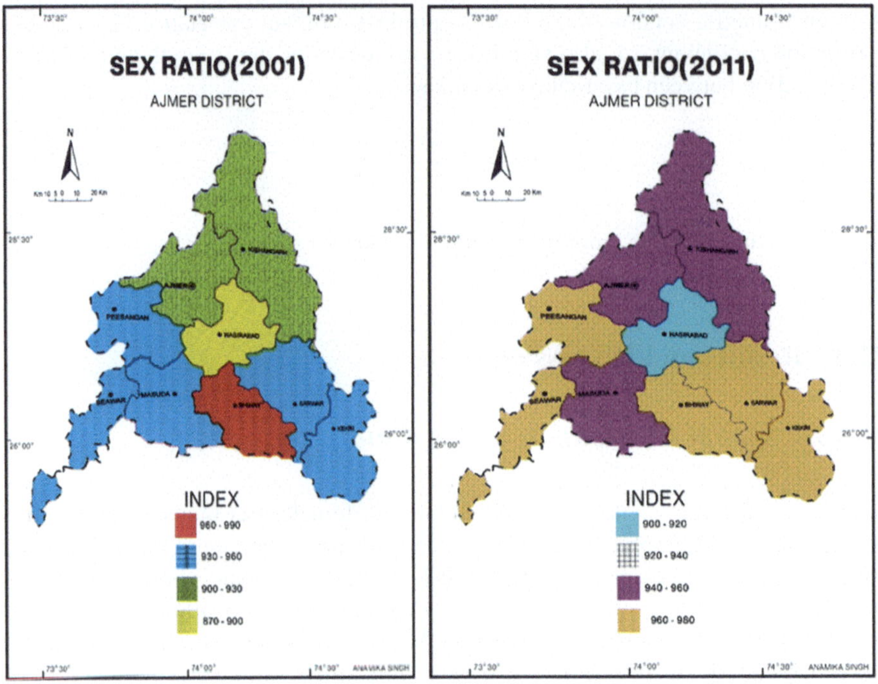


**Fig. 23.2** Sex ratio of Ajmer district 2001 and 2011. *Source* DCHB (2011)

transformation. According to the census of India, a person who can read and write in any language is considered Literate. Furthermore, "Literacy rates in India are calculated excluding the 0 to 6 age group of the population" (Shafiqullah 2011).

Kofi Annan, the former secretary general of the United Nations famously said, "Literacy is a bridge from misery to hope". And Clark has aptly marked that, "in the case of female literacy, it is both a bridge and a beacon." The past few decades have seen awareness regarding women's upliftment issues. Yet many cases of missing women, sex-selective foeticide and gender bias are alarming concerns to be worked upon. A demographic study of Indian families published facts that; firstly, there are more sons in smaller size families and secondly, in northern India, there is a higher proportion of sons, especially in the backward section of the society (Clark 2000).

### *23.4.3 Spatial Distribution of Literacy Rate*

Education not only widens the mental and intellectual capacity of a person but also cumulatively promotes the growth and progress of the entire society and nation as a whole. But for the same, it is just not enough to be literate who can only read and write but must learn through proper scholastic and academic teachings and enrich the minds for society and nation-building. The literacy rate and educational facilities of a region mirror the standard of development of that particular region.

The total literacy rate of the Ajmer district has increased over the past two decades, which proves the march of development seeping into the society of Ajmer. The total literacy rate of Ajmer in the year 1991 was 52.34% rose to 64.65% in 2001 and further rose to 69.33% in 2011 (Table 23.1). Also, the literacy rate in males and females has increased in this span of two decades. Male literacy in Ajmer was recorded at 68.75% in 1991, which increased to 79.37% in the year 2001, which again surged up to as high as 82.49% in the year 2011. In the same way, female literacy has also improved over the set of these decades. The literacy rate in females in the year 1991 was 34.50%, in 2001 it was 48.86%, whereas, in the year 2011 it reached 55.68% (Fig. 23.3). Although male literacy is much higher than female literacy in the district, figures show that the growth in the female literacy rate is more than double of growth in the male literacy rate. The growth in the male literacy rate is 3.12% whereas, the growth in the female literacy rate is 6.82%. Moreover, the growth of the total literacy rate in the last decade is 4.68%. The gap between male and female literacy rates throws light on the bleak side of the traditional mindset of society. Various schemes are being run by the government such as *Sarva Shiksha Abhiyan, Beti Bachao Beti Padhao,* etc. which have promoted to enhancement of girls' education but still much has to be done.

The given table of literacy rate shows that the literacy rate of Ajmer has increased from 52.34% (1991) to 64.65% (2001), which further increased to 69.33% (2011). The increase in literacy rate in the last decade was reported to be 4.68%. Moreover, the literacy rate of the district is higher than that of the total literacy rate of the Rajasthan state. The total literacy rate of Rajasthan state is 66.11%, which is 3.22% lower than that of Ajmer district.

Tehsil-wise data reveals that maximum literacy is found in Ajmer tehsil. In the year 1991, the literacy rate of Ajmer tehsil was 65.12%, which increased to 76.64%

**Table 23.1** Tehsil-wise literacy rate of Ajmer district, 1991–2011

| S. No | Tehsil | 1991 | 2001 | 2011 |
|---|---|---|---|---|
| 1 | Ajmer | 65.12 | 76.64 | 80.38 |
| 2 | Nasirabad | 51.72 | 61.56 | 67.85 |
| 3 | Kishangarh | 40.83 | 59.11 | 64.68 |
| 4 | Beawar | 50.64 | 78.27 | 75.78 |
| 5 | Masuda | – | 58.75 | 61.29 |
| 6 | Kekri | 41.00 | 52.52 | 60.44 |
| 7 | Sarwar | 32.15 | 45.04 | 52.60 |
| 8 | Bhinay | – | 44.73 | 54.84 |
| 9 | Peesangan | – | 51.83S | 61.03 |
|  | Ajmer | 52.34 | 64.65 | 69.33 |

*Source* DCHB 1991, 2001 and 2011

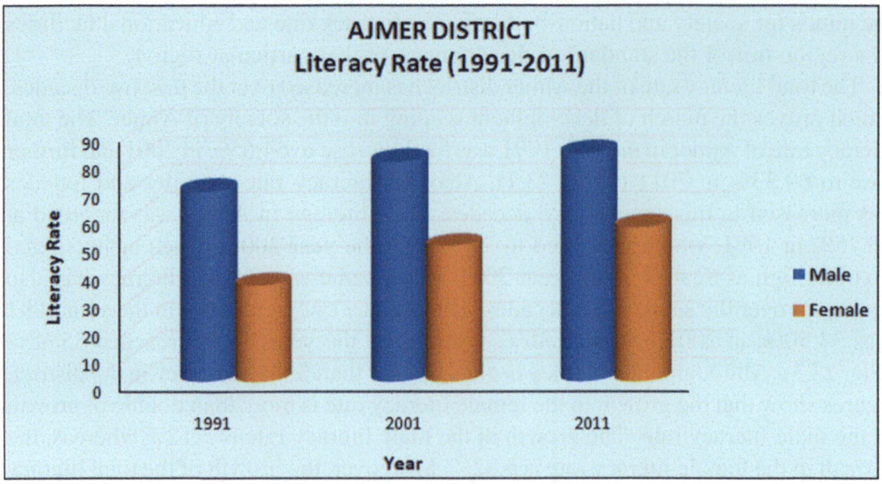


**Fig. 23.3** Literacy rate of Ajmer district, 1991–2011. *Source* DCHB 1991, 2001 and 2011

(2001) and further rose to 82.38% (2011). Over two decades, the literacy rate of Ajmer tehsil has gone up by 15.26%. Beawar tehsil remains in second position in terms of literacy rate followed by Nasirabad and Kishangarh. The literacy rate of Beawar in 1991 was 50.64%, which increased to 78.24% in 2001 whereas, a slight increase of about 2% was reported in the year 2001 with a literacy rate of 75.78%. The literacy rate in Nasirabad was 51.72% (1991) increased to 61.56% (2001) and further rose to 67.85% (2011). A similar trend of increase in literacy rate is seen in Kishangarh tehsil. The literacy rate in Kishangarh is observed as 48.83%, 59.11% and 64.68% for the years 1991, 2001 and 2011, respectively. Further the tehsils of Masuda, Peesangan and Kekri have an average literacy rate. The literacy rate of Masuda tehsil in the year 2001 was 58.57%, which rose to 61.29% in 2011. Similarly, in 2001, the literacy rate of Peesangan tehsil was 51.83% which increased to 61.03% in 2011. Whereas literacy rates of Kekri tehsil for these three decades were reported to be 41%, 52.52% and 60.44%, respectively.

The tehsils of the Sarwar and Bhinay have recorded the lowest literacy rates amongst all the tehsils. The literacy rate of Bhinay in the year 2001 was 44.73% which increased to 54.84% in 2011. Whereas the literacy rate of Sarwar was observed as 32.15%, 45.04% and 52.60% for the years 1991, 2001 and 2011, respectively. The tehsils of Server, Bhinay and Peesangan are dominantly rural and thus, the diffusion of education could not propel due to superstition, rudimentary thinking and economic backwardness.

According to the census findings Beawar recorded the maximum literacy rate (78.27%) for the year 2001, whereas for the year 2011, Ajmer recorded a maximum literacy rate of 80.38% (DCHB 2011). Such a scenario indicates a larger number of educational institutions and better means of communication in these particular regions. High literacy rates were recorded in the tehsils of Nasirabad and

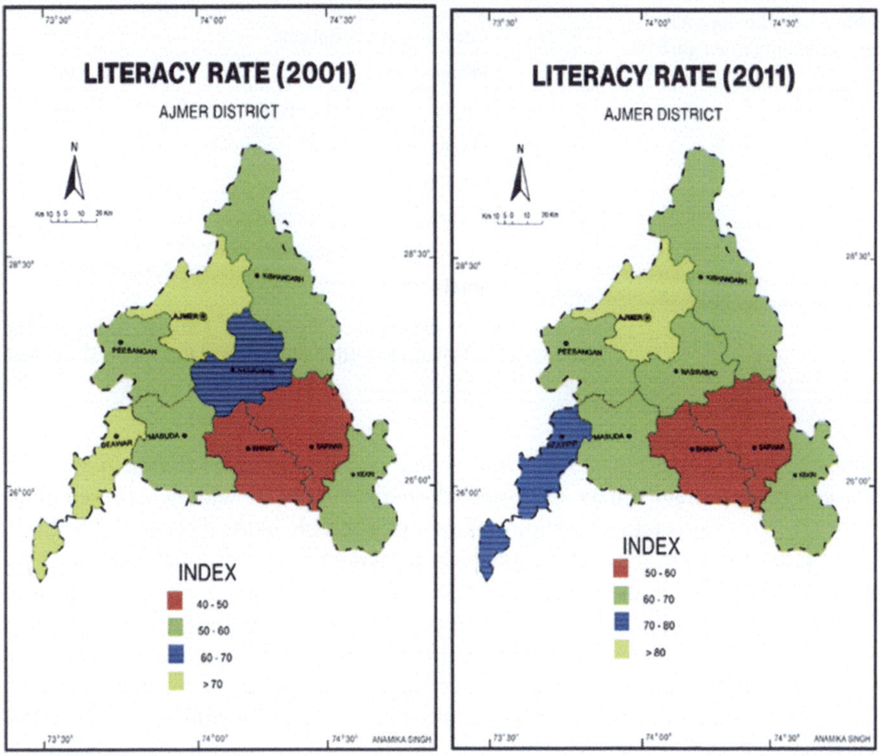


**Fig. 23.4** Spatial patterns of literacy rate in Ajmer district, 1991–2011. *Source* DCHB, 1991, 2001 and 2011

Kishangarh. The tehsils of Masuda, Kekri and Peesangan were recorded with moderate literacy rates whereas low literacy rates were recorded in Bhinay and Sarwar tehsils (Fig. 23.4).

The increase in literacy rate in Ajmer district is due to the development of educational institutions, and an increase in the number of primary and secondary Schools. Also, an improved transportation system has increased the reach of masses, even from the remotest areas for access to education thereby improving the literacy rates of the tehsils of Ajmer district.

### *23.4.4 Growth of Educational Institutions*

Advancement in standards of education becomes the prime concern of novice innovations and technological progress, which helps in gaining new observation through experiments further strengthening the economic standings of a region. The expansion of education makes a society socially, politically, economically and scientifically

**Table 23.2** Educational institutions in Ajmer district (2015)

| S. No | Educational institutions | Number |
|---|---|---|
| 1 | Primary schools | 896 |
| 2 | Upper primary schools | 1563 |
| 3 | Secondary and higher secondary schools | 844 |
| 4 | Colleges | 40 |
| 5 | Technical and specialized colleges | – |
| 6 | Universities | 3 |
| 7 | Total | 3346 |

*Source* Office of District Education Officer (Primary and Secondary), Ajmer and Office of College Directorate Education, Ajmer

sound and strong. To evaluate the current scenario of academic development of Ajmer district it is clear that Primary and Upper Primary schools are widely scattered in the district, whereas specialized institutions are not available in the district (Table 23.2).

In the past two decades, there has been a marked increase in educational institutions. Increasing population, higher educational standards, increase in demands, progress in urban and rural areas, availability of learned faculties and aid given by the government have caused the opening up of many government and private educational institutions. All these have made the roots of education go deep, making society aware, advanced and futuristic. The trends of educational institutions in the Ajmer district (Fig. 23.5).

The total number of educational institutions was 1939 in the year 1991, which hiked to 3343 in the year 2011. The main reason behind it could be the inclination of people towards education, change of traditional and rudimentary behaviour, acknowledging the importance of education, reservation and other policies run by the government such as Sarva Shiksha Abhiyan, Mid-Day Meal, etc. The number of primary schools in 1995—96 was 1254, which increased to 1561 in 2005–06, but later on reduced to 1270 in 2013–14 and further to 896 in 2014–15. The prime reason for this decrease in the number of primary schools is the opening up of more Secondary and Higher Secondary Schools and driving more attention to these institutions.

The number of Upper Primary Schools has also increased every year in the past two decades. In 1995–96 this number was 433 which grew up to 859 in 2005–06 and later surged up to 1563 in the year 2014–15. Following the same trend Secondary and Higher Secondary Schools have also increased over the past two decades. Their number in 1995–96 was 235, later aggraded to 407 and 844 in 2004–05 and 2015–16 respectively.

These figures reflect the dynamic development of academics in the district. On average there has been a 4 to 6% increase in the number of educational institutions. Initially, the colleges for higher education were only 17 in 1995–96 and now in 2014–15, this number has gone up to 40. Along with these, various institutes of varied specialized streams such as Management, Engineering Polytechnic, and ITI are also functional in the district. All these facts cumulate to the conclusion that there

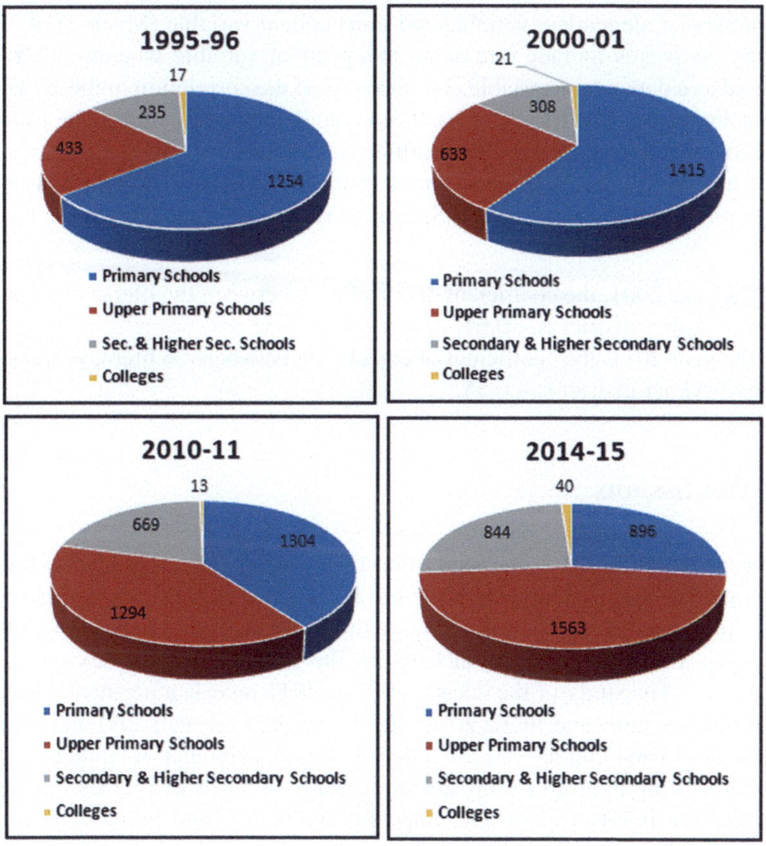


**Fig. 23.5** Qualitative and quantitative trend of educational institutes of Ajmer district (1995–2015). *Source* Office of District Education Officer (Primary and Secondary), Ajmer and Office of College Directorate Education, Ajmer

has been a positive change in the educational institutions. These institutes would help to improve the intellect of the people and thereby produce assets for the nation. This analysis projects the impact of educational institutions on the urbanization and economic development of the Ajmer district.

### *23.4.5 Analysing Correlation Between Literacy Rate and Sex Ratio*

The correlation is a statistical tool that studies the relationship between two variables and involves various techniques for studying and measuring the extent of the relationship between two variables. Correlation simply expresses the relation between two variables i.e. dependent variable and independent variable (Myers et al. 2013). The study envisages literacy rate as an independent variable whereas sex ratio is considered as a dependent variable. The analysis of the correlation of these two variables for the decade 2001 to 2011 has been calculated in this study. The following study is conducted by applying the Spearman Rank Correlation method to deduce the coefficient of correlation between literacy rate and sex ratio. The following results were derived after the calculation of Spearman rank coefficient correlation for 2001 and 2011:

1. For the year 2001, the coefficient of correlation between the literacy rate and sex ratio of Ajmer district is −0.61.
2. For the year 2011, the coefficient of correlation between the literacy rate and sex ratio of Ajmer district is −0.55.

## 23.5 Discussions

It has been hypothesized in the study that there is a positive correlation between literacy rate and sex ratio, and the same has been tested using the Spearman coefficient of rank correlation. The following method proves the hypothesis to be wrong and presents a negative correlation between the literacy rate and sex ratio in the Ajmer district. The study of the decade 2001 to 2011 reveals a negative correlation between the sex ratio and literacy rate (0.61 and 0.55 respectively) and indicates that the sex ratio and literacy rate are independent of each other. It could be inferred from the following trend of results that an increase in levels of literacy has flunked in changing the mindset of people. Superstitious beliefs and preferences of male children are prime causes for lower rates of sex ratio. Such societal notions cause gender inequality, illegal sex determination, female foeticide and infanticide. The economically advanced regions are misusing medical aid and modern technology for sex determination and foeticide. This has led to a decline in the sex ratio in areas of high literacy. Such characteristics of demographic imbalance lead to dire social consequences.

Imbalance of the sex ratio leads to societal damage. Youths of marriageable age, due to the lack of female population often engage themselves in anti-social activities. Moreover, unemployed, or illiterate men who do not have such family attachments and moral backing, end up committing gruesome acts. A report based on data analysed from the Interpol database of 70 countries reveals that higher crime rates of rapes, assaults, homicide, etc. were found in societies with an imbalance of sex ratio (Barber 2000). Messner (1991) also pointed out that young and unmarried men belonging to weaker socio-economic status commit more violent crimes (Messner and Sampson, 1991). Lim et al. (2005) have published that "the homicide rates in 56 countries were examined and it was discovered that the homicide rate could be predicted by a linear equation combining measures of economic inequality, negative per capita income growth and negative sex ratio of men to women" (Lim et al. 2005). Various measures to improve the sex ratio have been designed and implemented with varying degrees of success. Numerous measures have been implemented to improve the sex ratio. Some of such measures include monetary help on the birth of a girl child, subsidised education for females and strict prohibition and punishment for antenatal sex determination. A report from the 1990s reveals a highly grim picture of India, wherein about 1,00,000 sex-selective abortions were done, particularly reported from the states of Punjab, Haryana, and Gujarat. Enforcement of strict laws has helped curb such situations to some extent, but a far-reaching effect can be seen through societal awareness and mind shifts (Arnold et al. 2002).

Furthermore, such a grim situation destroys the very idea of sustainable education that is key to the sustainable development of a nation. To minimize this imbalance of sex ratio plans and programs must be formulated and strictly implemented to ensure societal well-being. Furthermore, mere bookish knowledge will not change the mindset of the masses, morality with education must be taught and practiced to eliminate such social issues.

## 23.6 Conclusions

The connotation of sustainable education emphasizes improving the quality of life through acquiring knowledge, boosting the economy, protecting the environment, and creating balance and harmony in society. A sustainable approach to education will ensure quality in education, and gender equality and reduce inequalities through a change in the mindset of the societies. An unbalanced sex ratio and lower literacy rate among the females projects a retarded society. Such a society will fail to progress and will hatch social evils and crimes that further impede its growth. Industrious efforts must pour in through government policies, legal framework and community awareness and participation to ensure a favourable sex ratio and hundred percent literacy. Such a practice would pave the way for sustainable education which would serve as a stepping stone for nation building.

## References


Acevedo-Duque Á, Jiménez-Bucarey C, Prado-Sabido T et al (2023) Education for sustainable development: challenges for postgraduate programmes. Int J Environ Res Public Health 20(3):1759

Agbedahin AV (2019) Sustainable development, education for sustainable development, and the 2030 agenda for sustainable development: emergence, efficacy, eminence, and future. Sustain Dev 27(4):669–680. https://doi.org/10.1002/sd.1931

Arnold F, Kishor S, Roy TK (2002) Sex-selective abortions in India. Popul Dev Rev 28(4):759–785

Barakade AJ (2012) Declining sex ratio: an analysis with special reference to Maharashtra state. Geosci Res 3(1):92–95

Barber N (2000) The sex ratio as a predictor of cross-national variation in violent crime. Cross-Cult Res 34(3):264–282

Benites M, González-Díaz RR, Acevedo-Duque Á et al (2021) Latin American micro-entrepreneurs: trajectories and meanings about informal work. Sustain 13(10):5449. https://doi.org/10.3390/su13105449
Bhagwat N, Abhyankar P (2016) Sustainable development and Indian higher education: a constructive alliance. Sch Res J Interdiscip Stud 3(22):1503–1513
Broms K, Skalski JR, Millspaugh JJ et al (2010) Using statistical population reconstruction to estimate demographic trends in small game populations. J Wildl Manag 74(2):310–3170
Chaudhuri S (2012) Female infant mortality disadvantage in India: a regional analysis. Rev Radic Polit Econ 44(3):321–326
Clark S (2000) Son preference and sex composition of children: evidence from India. Demography 37(1):95–108
DCHB [District Census Handbook] (2011) Census of India-Rajasthan-series 09-part XII A, Ajmer. Directorate of Census Operations, Rajasthan
Flowerdew J (2014) Introduction: approaches to the analysis of academic discourse in English. In: Flowerdew J (ed) Academic discourse. Routledge, London, pp 11–28
Fulwari A (2017) Role of education for sustainable development. Int Educ Res J 3(5):546–549
Gebhard M (2004) Fast capitalism, school reform, and second language literacy practices. Mod Lang J 88(2):245–265
Gee JP (2015) Literacy and education. Routledge
Gopal GV, Anand VV (2006) Environmental education in school curriculum an overall perspective. In: Proceedings of symposium on environment education and ecosystem conservation. Indian Institute of Science, pp 28–30
Hazra J (1997) Women and literacy. Geogr Rev India 59(1):62–74
Jayaraj D, Subramanian S (2009) The wellbeing implications of a change in the sex ratio of a population. Soc Choice Welfare 33(1):129–150
Joshi A, Tiwari N (2011) Sex ratio in India: embarrassing to be honest. Curr Sci 101(8):1006–1008
Knutsson B (2021) Segmented prizing: biopolitical differentiation in education for sustainable development. Comp: J Comp Int Educ 51(3): 431–447. https://doi.org/10.1080/03057925.2019.1629276
Kopnina H (2020) Education for the future?: critical evaluation of education for sustainable development goals. J Environ Educ 51(4):280–291
Lim F, Bond MH, Bond MK (2005) Linking societal and psychological factors to homicide rates across nations. J Cross Cult Psychol 36(5):515–536
Lobo M, Guntur RD (2018) Spearman's rank correlation analysis on public perception toward health partnership projects between Indonesia and Australia in East Nusa Tenggara Province. J Phys: Conf Ser 1116(2):022020. https://doi.org/10.1088/1742-6596/1116/2/022020
Messner SF, Sampson RJ (1991) The sex ratio, family disruption, and rates of violent crime: the paradox of demographic structure. Soc Forces 69(3):693–713. https://doi.org/10.1093/sf/69.3.693
Mochizuki Y, Bryan A (2015) Climate change education in the context of education for sustainable development: rationale and principles. J Educ Sustain Dev 9(1):4–26
Moje EB (2015) Doing and teaching disciplinary literacy with adolescent learners: a social and cultural enterprise. Harv Educ Rev 85(2):254–278
Myers JL, Well AD, Lorch RF Jr (2013) Research design and statistical analysis. Routledge, New York
Mynaříková L, Novotný L (2021) The current challenges of further education in ICT with the example of the Czech Republic. Sustain 13(8):4106
Peet R (1999) Theories of development. Guilford Press, New York
Philpot R (2015) Critical pedagogies in PETE: an antipodean perspective. J Teach Phys Educ 34(2):316–332
Reyes GE (2001) Four main theories of development: modernization, dependency, word-system, and globalization. Nómadas. Revistacrítica De Ciencias Sociales y Jurídicas 4(2):109–124

Sáez-Gallego NM, Abellán J, Hernández-Martínez A (2019) Attitudes toward inclusion of pre-service physical education teachers. ESHPA-Educ, Sport, Health Phys Act 3(1):73–85. http://hdl.handle.net/10481/53222

Saiz-Álvarez JM, Vega-Muñoz A, Acevedo-Duque Á et al (2020) B corps: a socioeconomic approach for the COVID-19 post-crisis. Front Psychol 11:1867. https://doi.org/10.3389/fpsyg.2020.01867

Shafiqullah S (2011) Regional analysis of urban-rural differentials in literacy in Uttar Pradesh, India. J Geogr Reg Plan 4(5):287

Shetty A, Shetty S (2014) The influence of female literacy on the sex ratio in Indian states. Int J Med Sci Public Health 3(10):1219–1224

Singh JP (2010) Socio-cultural aspects of the high masculinity ratio in India. J Asian Afr Stud 45(6):628–644

Som KS, Mishra RP (2014) Gender imbalance: trends, pattern and its impact on West Bengal. Int J Sci Res Publ 4(7):1–10

Todaro MP, Smith SC (2006) Economic development, 8th edn. Pearson South Asia Pvt. Ltd., Manila, Philippines

UNESCO (2015) Thematic indicators to monitor the education 2030 agenda. Technical Advisory Group Proposal. https://apa.sdg4education2030.org/sites/apa.sdg4education2030.org/files/2020-12/06_43-indicators-to-monitor-education2030.pdf. Accessed 24 June 2016

UNESCO (2022) Sustainable development report 2022. https://www.sustainabledevelopment.report/reports/sustainable-development-report-2022/. Accessed 30 Nov 2022

Wals AE, Kieft G (2010) Education for sustainable development: research overview. Discussion Paper. https://research.wur.nl/en/publications/education-for-sustainable-development-research-overview. Accessed 30 Nov 2022

# Chapter 24
# Education for Sustainability Transformation: Evaluating Educational Attainment Among Females in Rural India

**Ashu Rani, K. V. Chamar, and Vinod Kumar Mayala**

**Abstract** The present research work examines the levels of educational attainment among females in rural India. For this study, district-wise census data of 2011 has been used. Initially, the literacy rate of rural female population has been worked out in percentage. Based on census data, various levels of educational attainment in proportion to total educated females have been worked out in percentage *for a uniform pattern, i.e. 100%*. Thereafter, weighted scores have been derived by assigning one to 9 to the respective categories. The composite weighted scores were worked out and divided into 5 broad categories to assess the levels of educational attainment among females in rural India. The study reveals that 58% of females have formally acquired education in rural India. It has been recorded that out of the total educated females, more than a fourth-fifth of females obtained education up to 'secondary/matriculation levels'. About 7% reached up to the level of 'senior secondary'. On the other hand, nearly 5% of females shared in the categories of literate without educational level and unclassified, while only 3% received education up to graduation and above levels. Finally, it has been recorded that the East Delhi district of NCT of Delhi has obtained the highest composite weighted scores (651), while the Shajapur district of Madhya Pradesh has obtained the lowest composite weighted scores (199) in rural India. Further, it has been found that extremely northern and southern parts of India show higher levels of educational attainment, while low levels are represented by western and central parts of the study area.

---

A. Rani
Department of Geography, Government College Baund Kalan, District Charkhi Dadri, Haryana, India
e-mail: ashigeo90@gmail.com

K. V. Chamar (✉)
Department of Geography, M. D. University, Rohtak, Haryana, India
e-mail: chamar_kv@yahoo.com

V. K. Mayala (✉)
Department of Economics, Hansraj College, University of Delhi, Delhi, India
e-mail: vinodmayala@gmail.com

 
S. Anand et al. (eds.), *Emerging Geosustainability Transformations in India*, Advances in Geographical and Environmental Sciences,
https://doi.org/10.1007/978-981-97-7993-2_24

**Keywords** Educated females · Literacy rate · Educational attainment · Composite weighted scores

## 24.1 Introduction

Education and sustainable development have an unbreakable connection with each other. Education is the economic spinal column, and sustainable development is the foundation for healthy economic development. Glick and Miller (1956) highlighted that "educational attainment is one of the most important factors in determining the occupational and income levels to which a person can aspire". Raju (1988) stated that "historically, women and scheduled castes have faced significant challenges in attaining high literacy levels in India". Sen (1998) described that "a person who cannot read or write or count, and cannot follow written instructions faces several problems and barriers in everyday life". Ghosh (2003) described that "girls' education plays a pivotal role in diminishing gender disparities across various aspects of life". Tiwari (2013) also highlighted that "education is the single most important instrument for the social and economic transformation of the country". Vaid (2004) described that "education is believed to be an important direction for bringing about social change and attaining captivate into prestigious occupations in India". Roy (2008) explained that "educational level, thus, is an important demographic indicator as well as an essential tool for measuring human progress".

Educational attainment is a word that refers to an individual's highest level of education. Data about educational attainment for a certain area's population is regarding various levels of education. Indian Census defines the different levels of educational attainment. It includes 10 levels, namely, (i) literate without educational level, (ii) below the primary stage, (iii) primary stage, (iv) Middle/upper primary stage, (v) matric/secondary stage, (vi) higher secondary/ intermediate pre-university/ senior secondary, (vii) non-technical diploma or certificate not equal to a degree, (viii) technical diploma or certificate equal to degree, (ix) graduate & above, and (x) unclassified (Census of India 2011). It should be noted that if a person completes successfully the final year of a particular level of education, the person is regarded to have attained that level of education.

The study of education attainment is not a new field but each scholar has a different way of study. It has been studied by various scholars like educationists, economists, demographers and geographers etc. Geographers and other scholars have also carried out many studies about education attainment from time to time, like Mathai (1956), Jacobs (1996), Ghosh (2007), Husain and Sarkar (2011), Chamar and Chamar (2015), Chamar and Rekha (2015), Arya and Chamar (2015), Chamar and Chamar (2017), Arya (2017), Chamar et al. (2018), Chamar and Rekha (2018), Rekha and Chamar (2018), Chamar and Chamar (2019), Rani and Chamar (2019), Choudhury et al. (2023), etc. Very few studies have been carried out to examine the levels of educational attainment among females at regional or state levels in rural India. Therefore, to fill this gap, the present study has been made to examine the district-wise levels of educational attainment among rural females in the country. Thus, the objective of the present study is to assess the levels of educational attainment among females in rural India.

## 24.2 Material and Methods

### *24.2.1 The Study Area*

India extends from 8° 4' north to 37° 6' north latitudes and 68° 7' east to 97° 25' east longitudes, covering 32,87,263 $km^2$ of area. India is a federal union comprising 28 states and 7 union territories. As per the Census, 2011, there are 640 districts in India, out of which 9 districts are without rural population. The total rural female population is 405.96 million people, out of which 0 to 6 years is 58.23 million (14.35%), while more than 6 years population is 347.73 million people (85.65%). The average literacy rate among females is nearly 58% in rural India.

### *24.2.2 Data Source and Research Methodology*

The district-wise census data of 2011 has been used for the present research work. Initially, the literacy rate of the rural female population has been worked out in percentage. After that, to assess the levels of educational attainment, educated females have been divided into 9 categories that are (i) literate without educational level and unclassified, (ii) below the primary stage, (iii) primary stage, (iv) middle/upper primary stage, (v) secondary stage, (vi) higher secondary/ intermediate pre-university/senior secondary, (vii) non-technical diploma or certificate not equal to a degree, (viii) technical diploma or certificate equal to degree and (ix) graduate & above and calculated into percentage in proportion to total educated females at districts level in rural India *for uniform pattern, i.e. 100%.* Thereafter, a weighted score has been derived by assigning one to 9 to the respective above said categories *as adopted by* Chamar and Chamar (2015). *Further, to determine the level of different districts, a weighted score has been applied to the proportion of educated females in various levels of educational attainment.* The composite weighted scores were worked out and divided into 5 broad categories to assess the levels of educational attainment among females in rural India. Finally, tables and maps have been prepared with the help of suitable statistical methods and using the Arc GIS 10 software (Arc MAP Version 10.3).

## 24.3 Results and Discussion

### *24.3.1 Levels of Educational Attainment*

It is found that out of the total educated females, more than a fourth-fifth of females obtained education up to the levels of 'secondary/matriculation'. About 7% of females reached the level of 'senior secondary'. On the other hand, nearly 5% of females got an education in the categories of literate without educational level and unclassified, while only three % got education up to graduation and above levels in 2011 as mentioned in Table 24.1.

#### 24.3.1.1 State-Wise Rural Female Literacy Rate

As per the Census of India (2011), the highest % of rural female literacy rate has been recorded by Kerala state followed by the states of Goa, Mizoram, and Tripura, where female literacy rates have been more than 75%. Apart from these, there are states like Himachal Pradesh, Sikkim, Nagaland, Maharashtra, Meghalaya, Manipur, Uttarakhand, Punjab, West Bengal, Tamil Nadu, Assam, Gujarat, Odisha, Haryana and Karnataka, where female literacy rates have been more than the national average of 57.93%. On the other hand, the lowest female literacy rate has been observed in the state of Rajasthan (45.80%) followed by Jharkhand (48.91%) and Bihar (49%). Among the Union Territories, the percentage of rural female literacy rate has been highest in Lakshadweep (88.50%) and lowest in Dadra & Nagar Haveli (49.58%; Table 24.2).

**Table 24.1** Levels of educational attainment among females in rural India, 2011

| Levels of educational attainment | Educational attainment (%) |
|---|---|
| Literate without educational level and unclassified | 05.43 |
| Below primary stage | 25.09 |
| Primary stage | 29.02 |
| Middle/upper primary stage | 18.11 |
| Secondary/matriculation stage | 11.24 |
| Higher secondary/intermediate pre-university/senior secondary | 07.32 |
| Non-technical diploma or certificate not equal to degree | 00.07 |
| Technical diploma or certificate not equal to degree | 00.39 |
| Graduate & above | 03.33 |
| Rural India | 100 |

*Source* Compiled by authors

**Table 24.2** State/Union territory-wise female literacy rate in rural India, 2011

| States | Literacy rate (%) | States | Literacy rate (%) |
|---|---|---|---|
| Kerala | 90.81 | Chhattisgarh | 55.06 |
| Goa | 81.63 | Uttar Pradesh | 53.65 |
| Mizoram | 79.81 | Madhya Pradesh | 52.43 |
| Tripura | 79.49 | Arunachal Pradesh | 52.04 |
| Himachal Pradesh | 74.62 | Jammu & Kashmir | 51.64 |
| Sikkim | 72.45 | Andhra Pradesh | 51.54 |
| Nagaland | 71.51 | Bihar | 49.00 |
| Maharashtra | 68.54 | Jharkhand | 48.91 |
| Meghalaya | 68.37 | Rajasthan | 45.80 |
| Manipur | 66.34 | Union territories | |
| Uttarakhand | 66.18 | Lakshadweep | 88.50 |
| Punjab | 65.74 | Andaman & Nicobar Islands | 79.85 |
| West Bengal | 65.51 | Chandigarh | 73.17 |
| Tamil Nadu | 65.05 | NCT of Delhi | 73.10 |
| Assam | 63.03 | Puducherry | 73.02 |
| Gujarat | 61.36 | Daman & Diu | 71.93 |
| Odisha | 60.74 | Dadra & Nagar Haveli | 49.58 |
| Haryana | 60.02 | | |
| Karnataka | 59.71 | National average | 57.93 |

*Source* Compiled by authors

#### 24.3.1.2 Positions in Levels of Educational Attainment

The highest and lowest positions in levels of educational attainment in districts of different states among rural females in 2011 are presented in Table 24.3. It has been found that the Kishanganj district of Bihar witnessed the highest percentage in below primary stage (49.09%) followed by the East Delhi district of NCT of Delhi in graduation and above levels of educational attainment (47.40%) among females in rural India. The study further reveals that district Shajapur of Madhya Pradesh has recorded the highest position in literate without educational level and unclassified categories (45.37%). Similarly, the Jamnagar district of Gujarat has secured the highest position in the primary stage with 40.78%. There are other districts like Thoubal of Manipur (29.58%); Hoshiarpur of Punjab (23.66); Pathanamthitta (20.62%), and Pathanamthitta (4.72%) of Kerala which have respectively recorded the highest position in the middle stage, secondary/ matriculation stage, and higher secondary/ intermediate/ pre-university/ senior secondary stage and technical diploma or certificate equal to degree respectively. The highest position has been attained by the district Muzaffarnagar of Uttar Pradesh in the category of a non-technical diploma or certificate not equal to a degree.

**Table 24.3** Districts recording highest and lowest positions in levels of educational attainment among females in rural India, 2011

| Levels of education attainment | Districts and levels of education attainment | | | |
|---|---|---|---|---|
| | District | Highest | District | Lowest |
| Literate without educational level females and unclassified | Shajapur | 45.37 | Serchhip | 01.39 |
| Below primary stage | Kishanganj | 49.09 | East Delhi | 06.20 |
| Primary stage | Jamnagar | 40.78 | East Delhi | 10.41 |
| Middle/upper primary stage | Thoubal | 29.58 | East Delhi | 06.20 |
| Secondary/matriculation stage | Hoshiarpur | 23.66 | Shajapur | 01.88 |
| Higher secondary/intermediate/pre-university/ senior secondary stage | Pathanamthitta | 20.62 | Shajapur | 00.62 |
| Non-technical diploma or certificate not equal to degree | Muzaffarnagar | 01.25 | Anantnag | 00.01 |
| Technical diploma or certificate equal to degree | Pathanamthitta | 04.72 | Saharsa | 00.01 |
| Graduate & above | East Delhi | 47.40 | Shajapur | 00.34 |

*Source* Compiled by authors

On the other hand, the East Delhi district of NCT of Delhi has recorded the lowest percentage in the levels of the primary stage (10.41%), below the primary stage (6.20%) and middle stage (6.20%). Similarly, Shajapur district of Madhya Pradesh has registered the lowest position in levels of secondary/ matriculation stage (1.88%), higher secondary/intermediate/pre-university/senior secondary stage (0.62%) and graduate & above (0.34%). Likewise, Serchhip district of Mizoram; Anantnag district of Jammu & Kashmir; and Saharsa district of Bihar have recorded the lowest positions in the levels of literate without educational level and unclassified (1.39%); non-technical diploma or certificate not equal to degree (0.01%) and technical diploma or certificate equal to degree (0.01%), respectively.

Thus, the study highlights that among the female, Kishanganj district of Bihar holds the top position in below primary stage (49.09%), while Muzaffarnagar district of Uttar Pradesh witnessed the lowest position in the category of non-technical diploma or certificate not equal to degree (Table 24.3).

#### 24.3.1.3 Pattern of Rural Female Literacy Rate

As per census 2011, nearly 58% of females were educated in rural India. The highest literacy rate has been found in Serchhip district of Mizoram (96.89%). The very high and high literacy rate (more than 65%) is represented by 201 districts comprising about 32% of the total districts (Table 24.4). More than half of the total districts

**Table 24.4** Literacy rate of rural female population in India, 2011

| Level of educational attainment | Literacy rate (%) | Number of districts | District (%) |
|---|---|---|---|
| Very high | More than 75.00 | 070 | 11.09 |
| High | 65.01–75.00 | 131 | 20.76 |
| Moderate | 55.01–65.00 | 196 | 31.06 |
| Low | 45.01–55.00 | 146 | 23.14 |
| Very low | Less than 45.01 | 088 | 13.95 |

*Source* Compiled by authors

falling in these categories belong to only 8 states like Himachal Pradesh, Assam, West Bengal, Odisha, Gujarat, Maharashtra, Kerala, and Tamil Nadu and are located in the form of scattered patches in northwestern, northeastern, and eastern parts, forming a linear belt along the western coastal areas of India (Fig. 24.1). The districts like Kottayam, Pathanamthitta, Alappuzha, Ernakulam, Kollam, Kozhikode, Thrissur, Kannur and Malappuram of Kerala state; Sechhip, Aizwal and Champhai of Mizoram state; Mokokchung of Nagaland state; and East district of NCT of Delhi have recorded literacy rate more than 90%. It may also be noted that all the districts of Kerala state and except the Chamba district of Himachal Pradesh have a female literacy rate of more than 75%. Among the union territories, it has been observed that all the rural districts of NCT of Delhi, Chandigarh, Daman & Diu, Lakshadweep, Puducherry and Andaman & Nicobar Islands have a female literacy rate of more than 65%. None of the districts from the states of Rajasthan, Bihar and Jharkhand come under the categories of very high and high literacy rate.

The moderate literacy rate (55.01 to 65%) is found in 188 districts. More than three-fifths of the total districts falling in this category belong to 7 states such as Haryana, Uttar Pradesh, Madhya Pradesh, Gujarat, Assam, Karnataka and Tamil Nadu and are located in the form of scattered patches in India. The very low and low literacy rate is recorded in 234 districts belonging to 20 states, and mainly confined in north-western and south-eastern parts of the study area. About 86% of districts of these two categories belong to only 9 states.

It has been observed that one district Mon of Nagaland; Lawngtlai district of Mizoram; Dhubri district of Assam; Nandurbar district of Maharashtra; the union territory of Dadra & Nagar Haveli fall in this category. The lowest literacy rate has been recorded by Dakshin Bastar Dantewada district of Chhattisgarh (24.58%), followed by Alirajpur district of Madhya Pradesh (26.45%); Bijapur districts of Chhattisgarh (26.71%); Jhabua district of Madhya Pradesh (29.33%), etc.

#### 24.3.1.4 Spatial Patterns in Levels of Educational Attainment

Spatial patterns of various levels of educational attainment at the district level in rural India have been discussed in the forthcoming text.

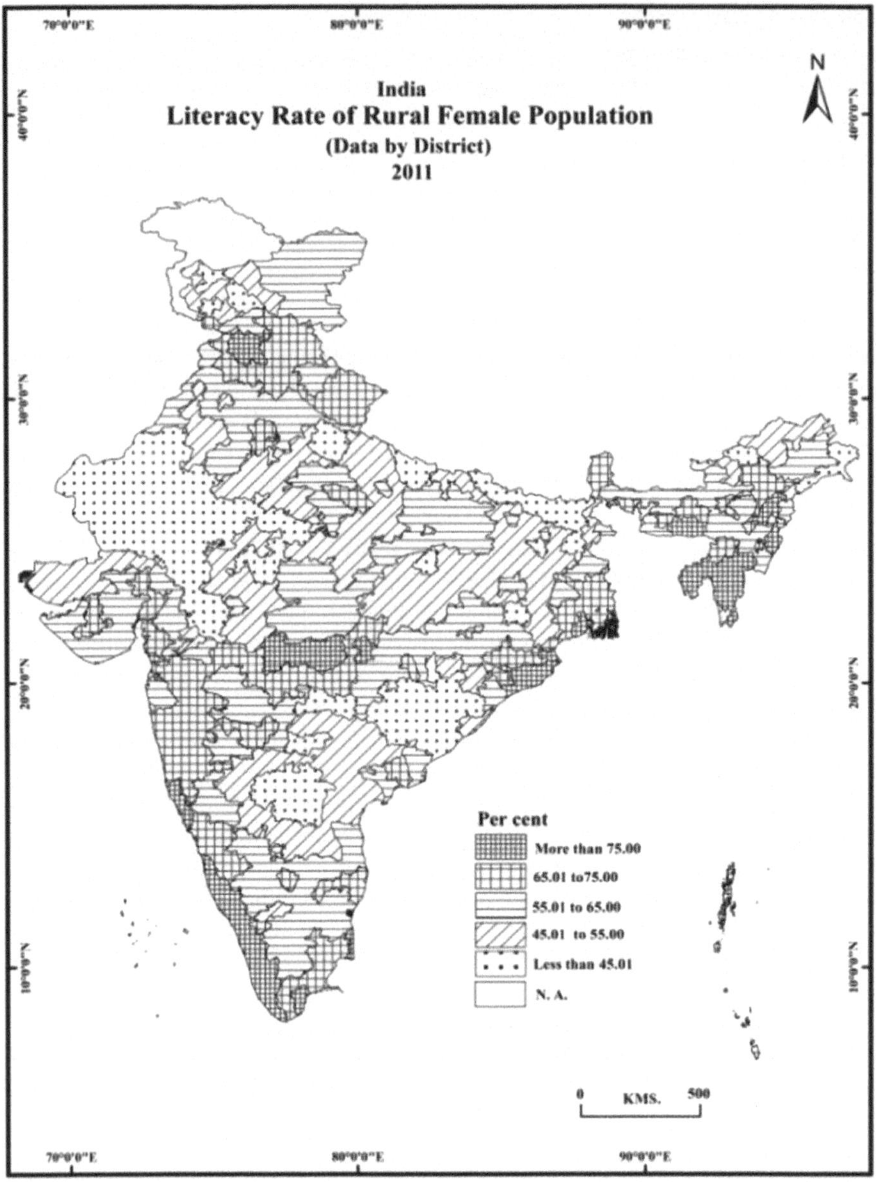


**Fig. 24.1** Literary rate of rural female population. *Source* Census of India (2011) Prepared by Authors

### (i). Levels of educational attainment: (Literate without educational level and unclassified)

Out of the total educated females, about 5% of females shared in the categories of literate without educational level and unclassified, and represented by 31 districts. Among the districts, the highest percentage has been recorded by Shajapur district of Madhya Pradesh, followed by Hingoli district of Maharashtra (17.04%). The study reveals that more than three-fifths of the total districts falling in this category belong to only three states Maharashtra, Bihar and Gujarat located in scattered patches in western and eastern parts of India (Fig. 24.2a). The moderate level ranges between 5 and 10% has been observed in 229 districts comprising 36% of the total districts of the country. More than half of the total districts falling in this category belong to 6 states such as Rajasthan, Uttar Pradesh, Bihar, Maharashtra, Andhra Pradesh and Karnataka and are located in the northern, western, southern parts and some small patches in north-eastern areas of India (Fig. 24.2a). It may be noted that none of the district from the states of Haryana, Mizoram, and Goa; and union territories of Chandigarh, Puducherry and Andaman & Nicobar Islands come under this category.

The districts registering less than 5.01% (low level) has been found in 371 districts accounting for about 59% of the total districts in rural India (Table 24.5a) and mainly confined to north-western, northern, eastern and north-eastern parts of the study area. The remaining districts of this category are found in the form of scattered small patches in various parts of India (Fig. 24.2a). Out of 371 districts, more than half of the districts are located in the states of Jammu & Kashmir, Himachal Pradesh, Punjab, Uttarakhand, Haryana, Uttar Pradesh, Madhya Pradesh, West Bengal, Odisha, and Chhattisgarh. The lowest percentage is recorded in Serchhip district of Mizoram (1.39%) and followed by Bilaspur of Himachal Pradesh (1.41%), North & Middle Andaman (1.49%), Chandigarh (1.50%), etc.

### (ii). Levels of educational attainment: (Below primary stage)

It has been recorded that about one-fourth of the educated females have attained education at below primary stage in rural India. Kishanganj district of Bihar (49.09%) has recorded the highest percentage, followed by Mewat district of Haryana (47.71%) in the category of below primary stage. The high level of educational attainment (more than 35%) in below primary stage has been found in 72 districts and about 62% of the total districts falling in this category belong to only four states such as Rajasthan, Bihar, Madhya Pradesh and Jharkhand and mostly located in the form of small patches in the northern and eastern parts of the study area (Fig. 24.2b). The moderate level of educational attainment (15.01 to 35%) in below primary stage has been recorded in more than three-fourths of the total districts (Table 24.5b) and belonging to all states and union territories except the union territory of Puducherry. On the other hand, the low level (less than 15.01%) in the category of below primary stage has been observed in 73 districts and confined to north-western and southern parts of India. It may be noticed that more than half of the total districts belong to only two states namely Punjab and Tamil Nadu. The lowest percentage in the category of below primary stage has been recorded by the East district of NCT of Delhi (6.20%) and

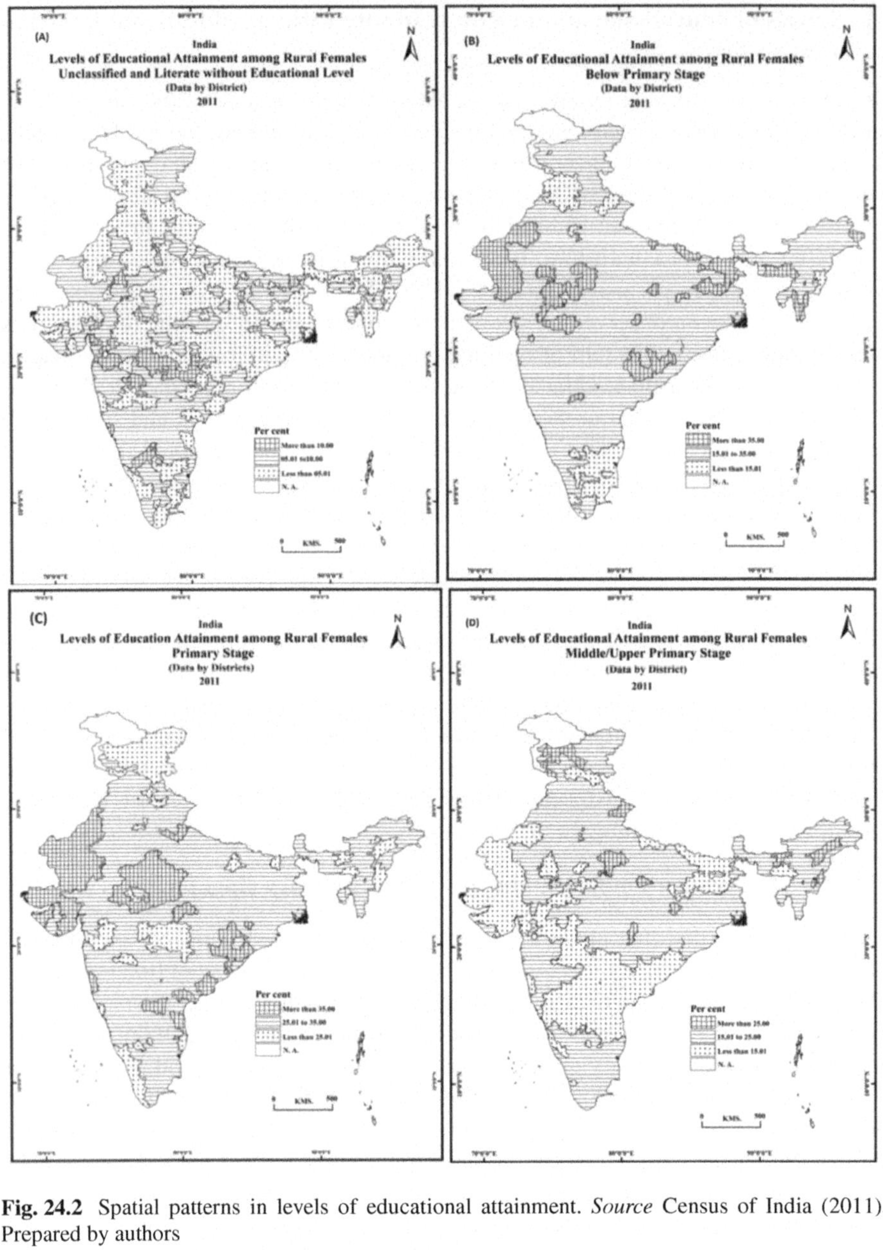


**Fig. 24.2** Spatial patterns in levels of educational attainment. *Source* Census of India (2011), Prepared by authors

**Table 24.5** Educational attainments among females at various levels in rural India, 2011

| A. Rural India: educated females, without educational and unclassified level | | | |
|---|---|---|---|
| Level of educational attainment | Educated females (%) | Number of districts | District (%) |
| High | More than 10.00 | 031 | 04.91 |
| Moderate | 5.01 to 10.00 | 229 | 36.29 |
| Low | Less than 05.01 | 371 | 58.80 |
| B. Rural India: educated females, below primary stage, 2011 | | | |
| Level of educational attainment | Educated females (%) | Number of districts | District (%) |
| High | More than 35.00 | 072 | 11.41 |
| Moderate | 15.01.01 to 35.00 | 486 | 77.02 |
| Low | Less than 15.01 | 073 | 11.57 |
| C. Rural India: educated females, primary stage, 2011 | | | |
| Level of educational attainment | Educated females (%) | Number of districts | District (%) |
| High | More than 35.00 | 065 | 10.30 |
| Moderate | 25.01.01 to 35.00 | 463 | 73.38 |
| Low | Less than 25.01 | 103 | 16.32 |
| D. Rural India: educated females, middle/upper primary stage, 2011 | | | |
| Level of educational attainment | Educated females (%) | Number of districts | District (%) |
| High | More than 25.00 | 042 | 06.66 |
| Moderate | 15.01.01 to 25.00 | 445 | 70.52 |
| Low | Less than 15.01 | 144 | 22.82 |
| E. Rural India: educated females, secondary stage, 2011 | | | |
| Level of educational attainment | Educated females (%) | Number of districts | District (%) |
| High | More than 20.00 | 025 | 03.96 |
| Moderate | 10.01 to 20.00 | 328 | 51.98 |
| Low | Less than 10.01 | 278 | 44.06 |

*Source* Compiled by authors

followed by Hamirpur (9.15%) and Kangra (9.53%) districts of Himachal Pradesh. It may be noted that none of the districts from the states of Sikkim, Nagaland, Tripura, Assam, Maharashtra Andhra Pradesh and Goa and union territories of Chandigarh, Daman & Diu, Dadra & Nagar Haveli and Lakshadweep come under the categories of high and low level of educational attainment of below primary stage.

(iii). **Levels of educational attainment: (Primary stage)**

About 29% of educated females have obtained education up to the primary stage. Among the districts, the highest percentage has been recorded by Jamnagar (40.78%), followed by Surendranagar (40.21%) and Bhavnagar (39.76%) districts of Gujarat. The high level of educational attainment (more than 35%) in the primary stage has been recorded in 65 districts and nearly 68% of the total districts falling in this category belong to only three states such as Rajasthan, Madhya Pradesh and Odisha and are located in the form of large patches in the northern, eastern and small scattered patches along the eastern coast of the study area (Fig. 24.2c). The moderate level of educational attainment among rural females (25.01 to 35%) in primary stage has been recorded in nearly three-fourths of the total districts (Table 24.5c). It may also be noted that all the districts of the states of Punjab, Sikkim, Mizoram, West Bengal, Jharkhand and Chhattisgarh; and union territories of Daman & Diu, Dadra & Nagar Haveli and Lakshadweep have fallen in this category.

On the other hand, the low level of educational attainment (less than 25.01%) in the category of primary stage has been found in 103 districts, and mostly confined to north-western and southern parts of India in the form of small scattered patches. The lowest percentage in the category of primary stage has been recorded by the East Delhi district of NCT of Delhi (10.41%) and followed by Pathanamthitta (14.66%), Kottayam (15.95%) and Ernakulam (9.53%) districts of Kerala. It may be noted that more than half of the total districts belong to only four states namely Jammu and Kashmir, Assam, Maharashtra and Kerala.

(iv). **Levels of educational attainment: (Middle/Upper primary stage)**

Nearly 18% of educated females got an education up to the upper primary stage. It has been found that Thoubal district of Manipur (29.58%) and East district (6.20%) of NCT of Delhi recorded the highest and lowest percentage among rural females in the category of upper primary stage. Among the districts, the high level of educational attainment (more than 25%) of rural females in the category of upper primary stage has been recorded in 42 districts belonging to 8 states, comprising about 7% of the total districts of rural India. Most of the districts are located in the form of very small patches in the study area (Fig. 24.2d).

The moderate level of educational attainment among rural females in the category of upper primary stage (15.01 to 25%) has been observed in 445 districts comprising about 71% of the total districts (Table 24.5d). These districts belong to all states and union territories except the state of Andhra Pradesh. The study reveals that more than two-thirds of the total districts falling in this category belong to only 12 states like Punjab, Haryana, Rajasthan, Uttar Pradesh, Madhya Pradesh, West Bengal, Odisha, Maharashtra, Kerala, Tamil Nadu, Arunachal Pradesh and Assam and most of the districts confined to north-western, northern, eastern and southern parts of the study area (Fig. 24.2d).

The districts registering less than 15.01% (low level) have been found in 144 districts accounting for about 23% of the total districts in rural India and mainly confined to eastern, western and southern parts of the study area. Out of 144 districts, more than four-fifths of the districts are located in the states of Bihar, Jharkhand, Madhya Pradesh, Gujarat, Maharashtra, Andhra Pradesh and Karnataka.

(v). **Levels of educational attainment: (Secondary stage)**

It has been recorded that about 11% of educated females have obtained education up to secondary stage. Among the districts, the highest percentage of rural females has been recorded by Hoshiarpur district (23.66%), followed by districts of Gurdaspur (23.51%) and Jalandhar (21.94%) of Punjab. The high level of educational attainment (more than 20%) has been recorded in 25 districts and all districts falling in this category belong to only four states such as Punjab (11 districts), Himachal Pradesh (6 districts), Jammu & Kashmir (4 districts), Jhajjar and Sonipat districts of Haryana; and West and South West districts of union territory of NCT of Delhi, mostly confined to north-western parts in the form of two large patches in the study area (Fig. 24.3a).

The moderate level of education among rural females in the secondary stage (10.01 to 20%) has been observed in 328 districts, (Table 24.5e) comprising more than half of the total districts of India. It has also been observed that nearly two-thirds of the total districts falling in this category belong to only 9 states Jammu & Kashmir, Haryana, Uttar Pradesh, Bihar, Maharashtra, Andhra Pradesh, Karnataka, Kerala and Tamil Nadu. None of the districts fall in this particular category from the states of Mizoram, Tripura, Meghalaya, West Bengal and the union territory of Dadra & Nagar Haveli. On the other hand, the low level of educational attainment (less than 10.01%) in the category of secondary stage has been found in 278 districts and confined to a large continuous patch from west to east of the study area. The lowest percentage has been recorded in the Shajapur district of Madhya Pradesh (1.88%) and followed by the Jaisalmer district of Rajasthan (2.57%), Malkangiri district of Odisha (3.07%) and Narayanpur district of Chhattisgarh (3.43%). Further, it may also be noted that nearly 72% of the total districts falling in this particular category belong to only 8 states such as Rajasthan, Uttar Pradesh, Madhya Pradesh, West Bengal, Jharkhand, Chhattisgarh, and Assam.

(vi). **Levels of educational attainment: (Higher secondary/intermediate pre-university/senior secondary stage)**

About 7% of educated females obtained education up to the stage of higher secondary/ intermediate pre-university/senior secondary in rural India in 2011. The highest percentage of rural females in this category has been recorded in Pathanamthitta (20.62%), followed by Kannur (20.28%), Kottayam (20.20%), Kozhikode (19.68%) and Kollam (19.35%) districts of Kerala state. Among the districts, the high level of educational attainment (more than 10. 00%) of rural females in the stage of higher secondary/intermediate pre-university/senior secondary has been recorded in 137 districts belonging to 21 states, comprising about 22% of the total districts of rural India. Most of the districts are located in the form of large patches in the north-western and southern parts and a few small scattered patches in the north-eastern parts of the study area (Fig. 24.3b). It has been noted that about 64% of the total districts falling in this category belong to only 6 states such as Himachal Pradesh, Punjab, Karnataka, Kerala, Tamil Nadu and Assam.

A moderate level of educational attainment among rural females (5.01 to 10%) has been observed in 288 districts, comprising about 46% of the total districts (Table 24.6a). It has also been observed that about three-fifths of the total districts falling in this category belong to only 8 states Jammu & Kashmir, Haryana, Uttar Pradesh, Bihar, West Bengal, Assam, Maharashtra and Andhra Pradesh. None of the districts fall in this particular category from the states of Mizoram, Goa and Kerala and the union territories of Chandigarh, Dadra & Nagar Haveli and Puducherry.

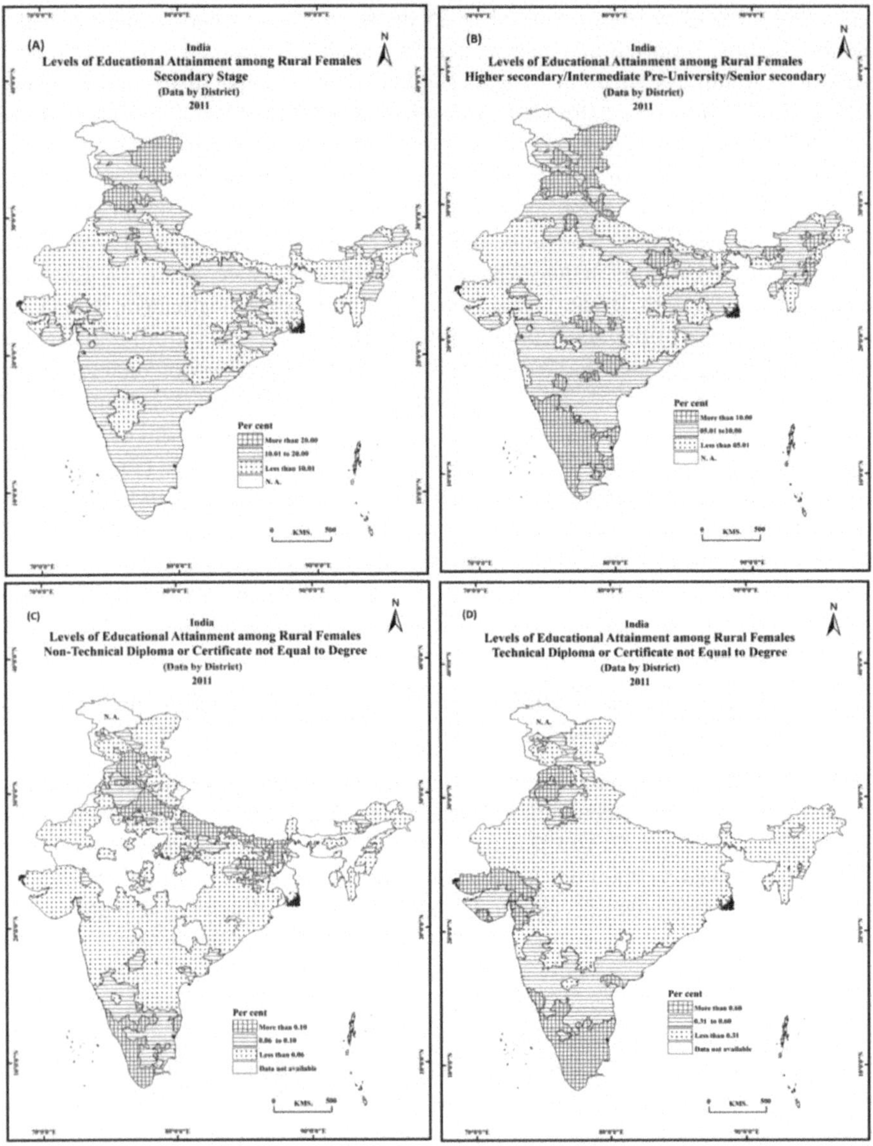


**Fig. 24.3** Spatial patterns of educational attainment among rural females. *Source* Census of India (2011), Prepared by authors

**Table 24.6** Educated females educational attainment at various levels in rural India, 2011

| A. Rural India: educated females: higher secondary/pre-university/senior secondary stage | | | |
|---|---|---|---|
| Level of educational attainment | Literate (%) | Number of districts | District (%) |
| High | More than 10.00 | 137 | 21.71 |
| Moderate | 05.01.01 to 10.00 | 288 | 45.64 |
| Low | Less than 05.01 | 206 | 32.65 |
| B. Rural India: educated females: non-technical diploma or certificate not equal to degree | | | |
| Level of educational attainment | Educated females (%) | Number of districts | District (%) |
| High | More than 00.10 | 113 | 17.91 |
| Moderate | 00.06 to 00.10 | 106 | 16.80 |
| Low | Less than 00.06 | 284 | 45.00 |
| Nil | Nil | 128 | 20.29 |
| C. Rural India: educated females, technical diploma or certificate not equal to degree | | | |
| Level of educational attainment | Educated females (%) | Number of districts | District (%) |
| High | More than 00.60 | 117 | 18.54 |
| Moderate | 00.31 to 00.60 | 091 | 14.42 |
| Low | Less than 00.31 | 420 | 66.56 |
| Nil | Nil | 003 | 00.48 |
| D. Rural India: educated females, graduate and above level | | | |
| Level of educational attainment | Literate (%) | Number of districts | District (%) |
| High | More than 06.00 | 059 | 09.35 |
| Moderate | 03.01 to 06.00 | 237 | 37.56 |
| Low | Less than 03.01 | 335 | 53.09 |

*Source* Compiled by authors

On the other hand, a low level of educational attainment (less than 5.01%) has been found in 206 districts comprising about one-third of the total districts of India and mostly confined to northern, eastern and western parts of the study area. The lowest percentage has been recorded in Shajapur district of Madhya Pradesh (0.62%), followed by Jaisalmer (1.28%) and Jalor (1.33%) districts of Rajasthan etc. Further, it may also be noted that nearly 86% of the total districts falling in this particular category belong to the states of Rajasthan, Uttar Pradesh, Madhya Pradesh, Bihar, Jharkhand, Chhattisgarh, Odisha and Gujarat.

(vii). **Levels of educational attainment: (Non-technical diploma or certificate not equal to degree**

It has been found that nearly 1.47 lakh educated females have attained education up to the level of 'non-technical diploma or certificate not equal to degree' which constitutes only 0.07% of the total educated females in rural India. Muzaffarnagar district of Uttar Pradesh (1.25%) has recorded the highest percentage, followed by Kishanganj district of Bihar (1.18%) in this category. Comparatively high levels of educational attainment (more than 0.10%) in this particular category has been recorded in 113 districts and about 90% of the districts falling in this category belong to only 8 states such as Himachal Pradesh, Haryana, Uttar Pradesh, Bihar, Jharkhand, Karnataka, Kerala and Tamil Nadu and located in the form of large patches in the north-western, northern, eastern and southern parts of the study area (Fig. 24.3c). The moderate level of educational attainment (0.06 to 01.0%) has been recorded in 106 districts, comprising about one-fourth of the total districts. Nearly two-thirds of the total districts falling in this category belong to only 5 states Punjab, Haryana, Uttar Pradesh, Karnataka, and Tamil Nadu.

On the other hand, a low level of educational attainment (less than 0.06%) has been found in 45% of the total districts and most of the districts are confined to north-western, eastern and western parts of the study area. None of the districts fall in this particular category from the states of Goa, Kerala and the union territories of Chandigarh, Lakshadweep and Puducherry. As per Census 2011, the lowest percentage that is 0.01% has been recorded in 100 districts in rural India. Further, it may also be noted that not a single female has obtained education in the category of a non-technical diploma or certificate not equal to a degree in 128 districts (Table 24.6b), comprising about one-fifth of the total districts in rural India, and nearly three-fourths of the total districts falling in this category belong to only four states such as Rajasthan, Assam, West Bengal and Madhya Pradesh.

(viii). **Levels of educational attainment: (Technical diploma or certificate not equal to degree)**

Likewise, nearly 7.84 lakh educated females have attained education up to the level of 'technical diploma or certificate not equal to a degree' which constitutes only 0.40% of the total educated females in rural India. Among the districts, Pathanamthitta district (4.72%) has recorded the highest percentage in this category, followed by Ernakulum (4.52%), Kottayam (4.41%), Alappuzha (3.70%), Kollam (3.23%), Idukki (3.14%) and Thiruvananthapuram (2.82%) districts of Kerala. Comparatively high levels of educational attainment (more than 0.60%) in this particular category have been recorded in about one-fifth of the total districts and more than three-fourths of the total districts falling in this category belong to only 5 states such as Punjab, Gujarat, Karnataka, Kerala and Tamil Nadu and confined to western and southern parts of the study area (Fig. 24.3d). The moderate level of educational attainment among rural females (0.31 to 0.60%) has been observed in 14% of the total districts and about three-fourth of the total districts falling in this category belong to only 5 states like Haryana, Gujarat, Maharashtra, Andhra Pradesh and Karnataka. On the other hand, the low level of educational attainment (less than 0.31%) in the category of technical diploma or certificate not equal to a degree has been found in a very large number of districts (Table 24.6c), comprising about two-thirds of the total districts. It has also been observed that all districts of Uttar Pradesh, Bihar, Nagaland, Mizoram, Tripura, Meghalaya, Assam, West Bengal, Jharkhand and Odisha states have fallen in this category.

While none of the districts fall in this particular category from the states of Punjab, Gujarat, Goa, Kerala, Tamil Nadu and union territories of Chandigarh, Daman and

Diu, Dadra & Nagar Haveli, Lakshadweep, Puducherry and Andaman & Nicobar Islands. The lowest percentage has been recorded in the Saharsa district of Bihar (0.01%) and followed by the Shrawasti district of Uttar Pradesh (0.01%). Further, it may also be noted that not a single female has obtained education in the category of a non-technical diploma or certificate not equal to a degree in the three districts namely Bijapur and Narayanpur districts of Chhattisgarh and Shajapur districts of Madhya Pradesh.

(ix). **Levels of educational attainment: (Graduate & Above)**

As per Census 2011, about 66.81 lakh educated females have attained education up to the level of 'graduation and above constituting only 3.33% of the total educated females in rural India. Among the districts, highest percentage has been recorded by East Delhi district of NCT of Delhi, followed by South Delhi district of NCT of Delhi (14.87%), Dehradun district of Uttarakhand (14.83%), Imphal West district of Manipur (11.06%), Nainital district of Uttarakhand (10.71%), Ernakulam and Kottayam districts of Kerala. The high level of educational attainment (more than 6%) has been represented by 59 districts. The study reveals that nearly 53% of the total districts falling in this category belong to only 4 states Himachal Pradesh, Uttarakhand, Manipur and Kerala, mostly located in scattered patches in north-western and southern parts of India (Fig. 24.4).

A moderate level of educational attainment ranging between 3.01 to 6% has been observed in 237 districts comprising 37% of the total districts of the country. About three-fifths of the total districts falling in this category belong to 6 states such as Jammu & Kashmir, Punjab, Haryana, Uttar Pradesh, Andhra Pradesh and Tamil Nadu, and mostly located in the north-western, and southern parts and some small patches in north-eastern areas of India (Fig. 24.4). It may be noted that none of the district from the states of Mizoram, Tripura, Chhattisgarh and Goa; and union territories of Chandigarh, Dadra & Nagar Haveli, Lakshadweep and Puducherry comes under this category. The districts registering less than 3.01% (low level) have been found in 335 districts accounting for about 53% of the total districts in rural India (Table 24.6d) and mainly confined to northern, eastern, north-eastern and western parts of the study area. It has also been noticed that out of 335 districts, more than 87% of the districts are located in the states of Rajasthan, Uttar Pradesh, Bihar, Assam, West Bengal, Jharkhand, Odisha, Chhattisgarh, Madhya Pradesh, Gujarat, Maharashtra and Karnataka. The lowest percentage is recorded in the Shajapur district of Madhya Pradesh (0.34%) and followed by Kishanganj of Bihar (0.35%), South Garo Hills (0.39%) and + East Garo Hills (0.48%) districts of Meghalaya etc.

Finally, based on the total weighted scores, the levels of educational attainment among rural females in India have been identified.

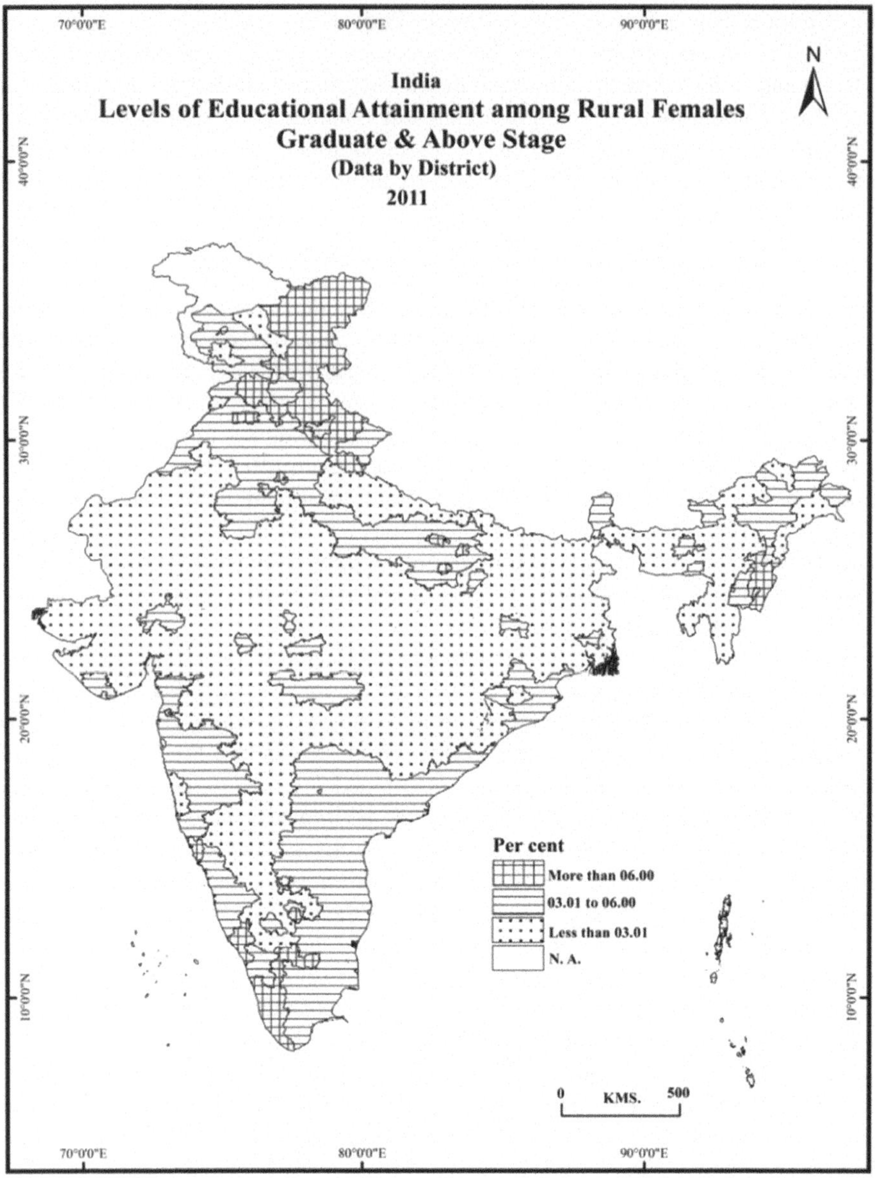


**Fig. 24.4** Levels of educational attainment among rural females (Graduate & Above). *Source* Census of India (2011), Prepared by Authors

### *24.3.2 Levels of Educational Attainment Among Rural Females in India*

The districts recorded weighted scores of more than 361 have been included in the categories of areas of high to very high levels of educational attainment among females in rural India. These categories have been represented by 237 districts, comprising about 38% of the total rural districts, and confined to north-western, southern and some small patches in western and north-eastern parts of the study area (Fig. 24.5). The highest weighted scores have been obtained by East Delhi district of NCT of Delhi (651) and followed by Pattanamthitta (466), Kottayam (460), Ernakulam (456) districts of Kerala. It may be noted that all the districts of Himachal Pradesh, Punjab, Goa, Kerala, Tamil Nadu and except Hardwar district of Uttarakhand; Churachandpur district of Manipur; Palwal and Mewat districts of Haryana; Udhampur, Reasi and Ramban districts of Jammu & Kashmir; and union territories of Chandigarh, NCT of Delhi, Daman & Diu, Puducherry and Andaman & Nicobar Islands fall under the categories of high to very high levels of educational attainment among rural females in India. A moderate level of educational attainment (weighted scores of 331 to 360) has been found in 150 districts, comprising about one-fourth of the total districts of the country. More than three-fourths of the total districts falling in this category belong to 7 states such as Uttar Pradesh, Bihar, Assam, Odisha, Maharashtra, Andhra Pradesh, Karnataka and are located in the form of scattered patches in India. On the other hand, low to very low levels of educational attainment among rural females (weighted scores less than 331) have been recorded in 244 districts, comprising about two-fifths of the total rural districts (Table 24.7) and belonging to 20 states of India. Most of the districts confined to the northern, eastern and northeastern parts of the study area. The lowest weighted scores has been recorded in the Shajapur district of Madhya Pradesh (199) and followed by the Kishanganj district of Bihar (261), Jaisalmer district of Rajasthan (267), Jamtara district of Jharkhand (271), etc.

## 24.4 Conclusions

The study reveals that 58% of females have formally acquired education in rural India. "It is recorded that out of the total educated females, more than a fourth-fifths of females obtained education up to 'secondary/matriculation levels', while 7% reached up to the level of 'senior secondary'. On the other hand, only three % got education up to graduation and above levels. It has been found that the East Delhi district of NCT of Delhi has obtained the highest weighted scores, while the Shajapur district of Madhya Pradesh has obtained the lowest weighted scores in rural India. Further, areas falling under various levels of educational attainment among rural females are marked by striking regional variations in India. The high to very high levels of educational attainment among females have been recorded in about

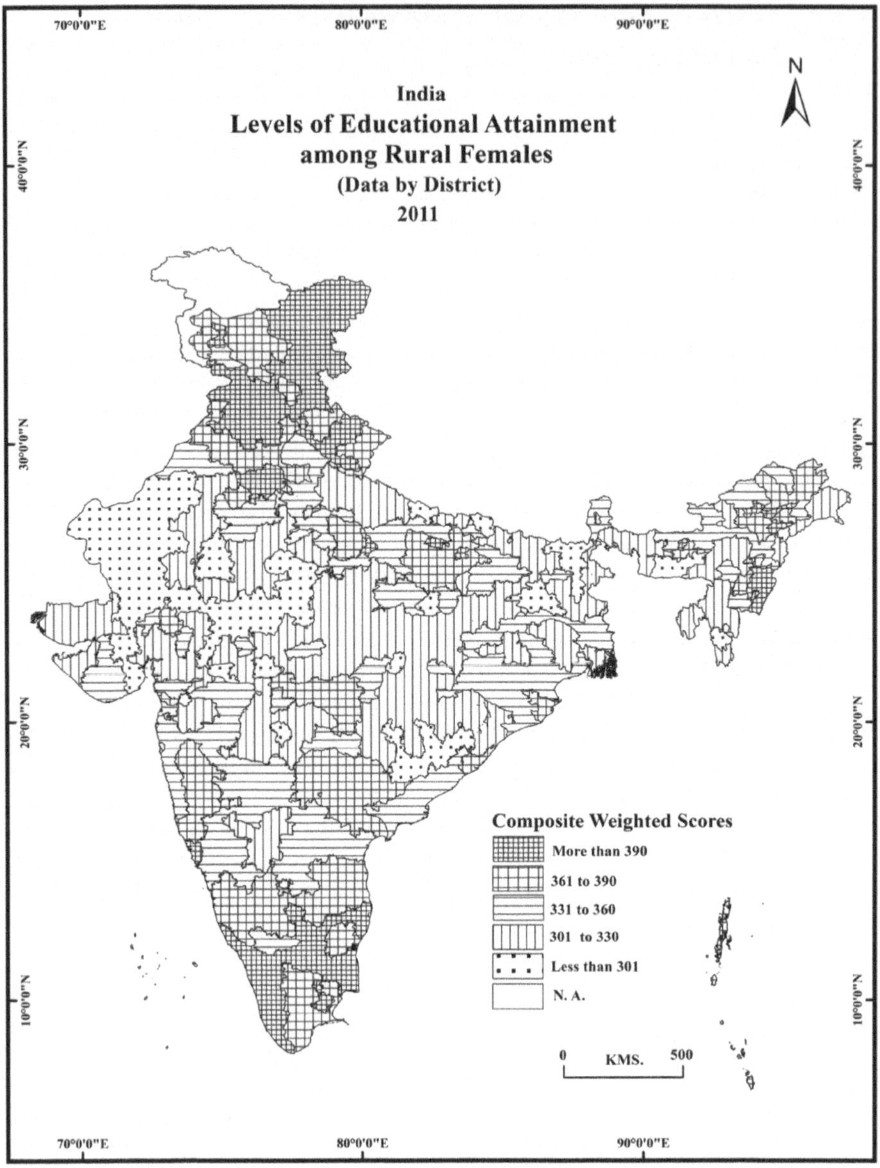


**Fig. 24.5** Educational attainment among rural females of India. *Source* Census of India (2011) Prepared by authors

**Table 24.7** Levels of educational attainment among females in India, 2011

| Level of educational attainment | Weighted scores | Number of districts | Districts (%) |
|---|---|---|---|
| Very high | More than 390 | 110 | 17.43 |
| High | 361 to 390 | 127 | 20.13 |
| Moderate | 331 to 360 | 150 | 23.77 |
| Low | 301 to 330 | 180 | 28.53 |
| Very low | Less than 301 | 064 | 10.14 |

*Source* Compiled by authors

two-fifths of the total districts, mostly confined to north-western, southern, and some small patches in western and north-eastern parts of the study area. A moderate level of educational attainment has been found in one-fourth of total districts in rural India and more than three-fourths of the total districts falling in this category and located in the form of scattered patches in India. On the other hand, the areas of low to very low levels of educational attainment among rural females have been represented by about two-fifths of the total rural districts, mostly confined to northern, eastern, and northeastern parts of the study area. Finally, as per census 2011, the uneven levels of educational attainment across the country have been recorded.

## References

Arya R (2017) Levels of educational attainment in village Sisar Khas: caste wise analysis. J People & Soc Haryana 7(1):1–7

Arya R, Chamar KV (2015) Differentials in levels of educational attainment of Dhanak caste in rural Haryana. J People & Soc Haryana 5(2):14–22

Census of India (2011) Registrar General of India. New Delhi

Chamar KV, Chamar SK (2015) Caste differentials in levels of educational attainment in village Sirsa Kheri. Trans, Inst Indian Geogr 37(1):27–34

Chamar SK, Chamar KV (2017) Rank differentials in levels of educational attainment among scheduled castes: a case study of Jind district. J People & Soc Haryana 7(2):36–49

Chamar NK, Chamar KV (2019) Differentials in levels of educational attainment of Bazigar caste in north west Haryana: a clan wise analysis. J People & Soc Haryana 9(1):13–28

Chamar KV, Chamar SK, Chamar NK, Chamar R (2018) Caste differentials in levels of educational attainment in Jat land of Haryana. In: Kumar B (ed) Environment and emerging challenges. Department of Geography, Gauhati University (Assam, India), Kar and Prasanta Bhattacharya, pp 122–130

Chamar KV, Rekha (2015) Levels of educational attainment of Chura caste in village Badli: a clan wise analysis. J Soc Sci Res Perspect 3(1):7–14

Chamar KV, Rekha (2018) Rank differentials in levels of educational attainment of Chura caste in Jhajjar district: a clan wise analysis. J People & Soc Haryana 8(2):45–58

Choudhury PK, Joshi R, Kumar A (2023) Regional and socio-economic inequalities in access to pre-primary education in India: evidence from a recent household survey. Int J Child Care Educ Policy 17(13):1–25

Ghosh AK (2003) Female literacy and social development. Soc Chang 33(4):65–72

Ghosh AK (2007) The gender gap in literacy and education among the scheduled tribes in Jharkhand and West Bengal. Sociol Bull 56(1):109–125
Glick PC, Miller HP (1956) Educational level and potential income. Am Sociol Rev 21(3):307–312
Husain Z, Sarkar S (2011) Gender disparities in educational trajectories in India: do females become more robust at higher levels? Soc Indic Res 101(1):37–56
Jacobs JA (1996) Gender inequality and higher education. Ann Rev Sociol 22:153–185
Mathai S (1956) Education in India. Curr Hist 30(174):85–89
Raju S (1988) Female literacy in India: the urban dimension. Econ Polit Wkly 23(44):WS57-WS59+WS61-WS64
Rani A, Chamar KV (2019) Rank differentials in levels of educational attainment of Bawaria caste in Rewari district: a clan wise analysis. Int J Anal Exp Modal Anal 11(9):4266–4278
Rekha, Chamar KV (2018) Differentials in levels of educational attainment of Chura caste in Jhajjar district: block wise analysis. Res Strat 8(8):9–16
Roy K (2008) Socio-economic characteristics of slum population of Kolkata. Geogr Rev India 70(1):96–104
Sen A (1998) Radical needs and moderate reforms. In: Dreze J, Sen A (ed) Indian development: selected regional perspectives. Oxford University Press, New Delhi
Tiwari KN (2013) Rural education-the way forward. Kurukshetra 61(11):3
Vaid D (2004) Gendered inequality in educational transitions. Econ Pol Wkly 39(35):3927–3938

# Chapter 25
# Advances in Public Health Policy for Sustainable Development in India

**Rosy Bansal, Gorakh Singh Teja, and Monika Hans**

**Abstract** The perceived value of the programme and the support that people and organizations are willing to give for ongoing operations are key factors in determining the sustainability of health promotion and disease prevention programmes. Sustainability can be attained by maintaining a dedication to the program's objectives and mission, building local system capacity, modifying knowledge and attitudes, collaborating continuously, enhancing service models, and enacting new laws and policies that support programme impact. COVID-19 has demonstrated how health institutions frequently fail to handle surges and quickly collapse worldwide. Infectious diseases, non-communicable diseases, emerging and re-emerging diseases and disasters place a triple burden on low- and middle-income countries, widening inequality and driving up healthcare expenses. Many non-health partners, including employment agencies, energy and security, the Department of Transportation, Housing, Social Protection, Education, Law and Justice, and Environmental Departments, are directly and indirectly responsible for the outcomes of social health. The majority of the world's line ministries are not coordinated, and they must work together to achieve the shared aim of civilization and humankind. The World Health Organization's Adelaide Statement defined health in all policies as assisting leaders to integrate health policies and services. The statement has evolved to become historical and representative without any actual application in creating better futures.

**Keywords** Public health · Sustainability · Infectious diseases · Health promotion · Environment

R. Bansal (✉) · G. S. Teja
GSSDGS Khalsa College, Patiala, Punjab 147001, India
e-mail: drrosysingla@gmail.com

G. S. Teja
e-mail: teja.gorakhsingh27@gmail.com

M. Hans
Government PG College for Women Gandhi Nagar, Jammu, Jammu and Kashmir 180003, India



S. Anand et al. (eds.), *Emerging Geosustainability Transformations in India*, Advances in Geographical and Environmental Sciences,
https://doi.org/10.1007/978-981-97-7993-2_25

## 25.1 Introduction

Understanding the durability of successful public health initiatives is a topic that researchers, evaluators, funders, and community partners are working on more and more. Despite these efforts, understanding sustainability is hampered by several issues like disparities in how sustainability is conceptualized, a lack of commonly accepted measurement frameworks, challenging reporting approaches and deadlines, and methodological issues connected to sustainability. The objective of this study is to examine these issues and suggest solutions to make public health programmes more sustainable. Failure to do so may lead to the loss of money and missed chances to enhance the health and well-being of the populace.

The economy, livelihoods, environmental conditions, our cities and infrastructure, and social relations are frequently treated as ends in themselves in discussions of sustainability as if they are the reason, we seek sustainability. However, their main worth is in serving as the cornerstones on which our survival and longer-term health depend. More than merely maintaining the economic system's reliance on the natural world's resource flows while preserving species diversity and ecosystem health are involved in ecological sustainability. The huge systems that support health and life must also be maintained, which is another aspect of sustainability. We should first define public health before looking at how non-sustainable development affects it.

### *25.1.1 Conceptualising Public Health*

Public health has been defined in many ways. As per the American Public Health Association, "Public health is the practice of preventing disease and promoting good health within groups of people, from local towns to entire countries," (APHA 2023). Public health refers to everyone's well-being as a fundamental human right, a shared interest, and a worthwhile goal. Further, public health is concerned with a broad range of multi-sectoral health factors and government policies at all levels. It provides crucial public goods, safeguards community health, tackles risk factors, and helps lessen health result inequities. According to the World Health Organization (WHO) constitution, achieving the best standard of health is one of every person's fundamental rights (WHO 2009). WHO also states that public health refers to all organized actions (whether public or private) to promote health, prevent disease, and extend life among the population as a whole. Its operations concentrate on entire communities rather than specific individuals or illnesses to foster a healthy living environment for everyone.

The Centres for Disease Control and Prevention (CDC) states that "all governmental, commercial, and voluntary institutions that contribute to the delivery of vital public health services within a jurisdiction are usually referred to as public health systems" (CDC 2017). Through the prevention and management of disease and injury, the Illinois Department of Public Health (IDPH) is responsible for safeguarding the state's 12.4 million citizens as well as countless tourists. Further, it is described as "the science and art of improving health, preventing sickness, and prolonging life via coordinated efforts of society", by the New Zealand Ministry of Health (Porritt 2005).

Public health, according to the National Resources Defence Council, an NGO committed to environmental protection, is "the health or physical well-being of a complete community" (Porritt 2005). Public health is the science and art of avoiding illness, extending life, and improving physical and mental health, as well as sanitation, personal cleanliness, the management of infectious illnesses, and the planning of health services. An awareness of the significance of community activity in the promotion of health and the prevention and treatment of disease has emerged from the regular human contacts involved in addressing the myriad challenges of social life, and this is embodied in the idea of public health.

Public health is a discipline that significantly incorporates medical science and philosophy and focuses on regulating and controlling the environment for the general good. As a result, it is concerned with housing, water, and food. Toxic substances can be introduced into these by farming, fertilizers, improper drainage and sewage disposal, building, defective HVAC systems, machinery, and hazardous chemicals. Public health medicine is part of a bigger endeavour to safeguard and improve public health. Community physicians collaborate with a variety of organizations, including toxicologists, sanitary and heating and ventilating engineers, factory and food inspectors, psychologists and sociologists, builders, architects, sanitary and heating and ventilation engineers, chemists, and physicists. The health, safety, and well-being of people at work are issues that occupational medicine addresses. It might be considered a specialist branch of public health medicine because its goal is to lower the dangers in the workplace (Adshead et al. 2006).

Epidemiology is a specialized way of information collection that is necessary for the project of protecting, maintaining, and actively promoting public health. It also calls for corporate arrangements to act on major discoveries and put them into effect. By linking variables like nutrition, environment, radiation exposure, or cigarette smoking with the incidence and prevalence of disease, epidemiologists aim to describe and explain the occurrence of disease in a population using statistics. The government establishes authorities to control and formally examine and monitor water sources, food processing, sewage treatment, drains, and pollutants through laws and regulations. The management of epidemic and pandemic diseases, the development of protocols for proper medical responses and isolation procedures, and the issuance of travel advisories to avoid the spread of disease from impacted areas are all issues that governments are concerned with (Corvalan et al. 1999).

On a national and international basis, numerous public health organizations have been founded to aid in the control and monitoring of disease within communities. For instance, the Public Health Act of 1848 in the United Kingdom created a unique public health ministry for England and Wales. The Centres for Disease Control and Prevention in the US conducts national-level research and public health coordination. WHO performs a similar function on a global scale (WHO 2009). WHO plays a crucial role in helping less developed nations around the world to build organizational and administrative procedures to deal with issues related to health and disease. Designing health services systems in these nations must take into account a variety of elements, including health issues, resource constraints, the education of healthcare workers, and others.

The development of vaccines and antibiotics, among other advances in science and medicine in affluent nations, has been essential in bringing lifesaving relief to nations with a heavy burden of disease. The incidence of avoidable disease and neglected tropical disease, however, continues to be very high globally despite the growth of resources and advances in the mobilization of these resources to the most severely affected areas. One of the main objectives of global public health is to lessen the severity and prevalence of chronic diseases. However, the continued existence of such diseases around the globe is a significant indicator of the challenges that society and health agencies continue to face.

A comprehensive system of primary, secondary, and tertiary prevention through the healthcare system that is available to everyone without discrimination and is financially accessible to all is what public health ensures. These include clean water, sanitation, clean air, healthy food, nutrition, and housing. This calls for the involvement of numerous non-health sector actors, including those involved (among others) in the education, urban planning, agricultural, transportation welfare sectors, and so on.

### *25.1.2 Public Health: Historical Background*

Reviewing the evolution of different public health concepts in light of the field's long history, which dates back to antiquity, is important. Leprosy patients were quarantined during the Middle Ages, and during the plague epidemics of the fourteenth century, attempts were made to enhance hygiene. As Europe's population increased, so did hospital construction and knowledge of infant mortality. Modern public health agencies and organizations were created as a result of these advancements to manage disease within communities and to supervise the availability and distribution of medications (Zodpey et al. 2011).

Most ancient peoples of the world were clean and well-groomed, frequently for religious purposes, including, it seems, a desire to be pure in the eyes of their God. Law, custom, and religion were all intertwined. For thousands of years, societies viewed epidemics as heavenly punishments for humankind's transgressions. However, the notion that plague is brought on by natural factors like the climate and physical surroundings eventually emerged. The first effort at a logical, scientific theory of disease causation was made during the fifth and fourth century BCE in Greece when philosophy had made significant strides. For instance, even if the causes for the link were unknown, malaria and marshes were associated fairly, early (503–403 BCE) (WHO 2009). The first systematic attempt to establish a causative relationship between human diseases and the environment was made in the book

Airs, Waters, and Places, which is thought to have been authored by Greek physician Hippocrates in the fifth or fourth century BCE (WHO 2009). This book offered a theoretical framework for understanding endemic sickness (that persisted in a specific location) and epidemic disease before the new sciences of bacteriology and immunology, appeared far into the nineteenth century (that affected several people within a relatively short period).

The development of hospitals, the provision of medical treatment and social support, attempts to deal with the filthy conditions of the cities and, through quarantine, to control the spread of disease, and a variety of other first steps in public health were made during the Middle Ages. In the sixteenth and seventeenth centuries, years of technological advancement culminated in several scientific discoveries. Leaders who had received an education at that time understood that maintaining the population's health was essential to the state's political and economic power. But, neither in England nor on the Continent were national health programmes created because the government lacked the administrative capacity and expertise to implement them (NHP 2017) As a result, just as in the Middle Ages, local communities continued to tackle public health issues.

The foundations of anatomy and physiology were laid by scientific discoveries in the sixteenth and seventeenth centuries. The more accurate identification of diseases was made feasible by observation and classification. The hypothesis that microscopic creatures may spread contagious diseases was starting to take shape. English statistician John Graunt was one of the early innovators in public health medicine. In 1662, he released a book of statistics that included numbers for deaths and occasional causes of death that had been collated by parish and municipal governments. Despite the inevitable inaccuracy of the data, epidemiology got off to a good start (Pilot et al. 2017).

The Poor Law Commission, established in 1834, investigated issues related to community health and offered solutions. "The expenses necessary to the adoption and maintenance of measures of prevention will ultimately amount to less than the cost of the disease presently constantly engendered," the report's authors argued in 1838 (PHE 2012). According to sanitary surveys, there is a connection between environmental pollution and communicable diseases, and protecting the public's health lies more within the purview of engineers than physicians.

A General Board of Health was founded by the Public Health Act of 1848 to assist local authorities with sanitary concerns and provide guidance, as their earlier efforts had been hampered by the absence of a centralized authority (DH PHE Transition Team 2012). The board had the power to create local boards of health and to look into the hygienic conditions in specific districts. Since then, several public health acts have been passed to govern sewage and refuse disposal, animal housing, water supply, disease prevention and control, registration and inspection of private nursing homes and hospitals, birth notification, and the delivery of maternity and child welfare services.

The development of public health in England had a significant impact on the United States, where appropriate administrative systems for the supervision and regulation of community health were one of the fundamental issues. Yellow fever, cholera, smallpox, typhoid, and typhus outbreaks that recurred in America made the need for efficient public health administration urgent (WHO 2017). The Sanitary Commission of Massachusetts (SCM 1949a; b) so-called Shattuck report, which was released in 1850, examined Boston's severe health issues and egregiously subpar living conditions. Its recommendations included an outline for a sound public health organization based on a state health department and local boards of health in each town. In New York City (in 1866) such an organization was created for the first time in the United States (SCM 1949a; b).

Developments in Germany and France during the nineteenth century paved the basis for later public health initiatives. In terms of political and social thought, France dominated. As a result, a spirit of public reform had a significant impact on the public health movement in France. The French considerably aided the application of scientific techniques for the diagnosis, treatment, and control of communicable diseases.

The United States Department of Health and Human Services established the Centres for Disease Control and Prevention (CDC; formerly the Communicable Disease Centre) in 1946 with the mandate of preventing, regulating, and promoting public health. The CDC plays a crucial role in gathering and providing the public with information on disease and disease prevention. It is become a renowned epidemiological hub.

While environmental sanitation advancements throughout the first decade of the twentieth century were helpful in addressing some issues, they were only marginally helpful in addressing the numerous health issues that were prevalent among the poor. Malnutrition, genital illness, drunkenness, and other illnesses were pervasive in the slums of England and the United States. Economic liberalism in the nineteenth century believed that greater production of goods would eventually put an end to poverty, suffering, and shortages. By the turn of the century, it became inevitable that reform-minded organizations, such as the state, would also need to intervene, actively. This is why social action was encouraged by a large number of medical professionals, clerics, social workers, civic-minded individuals, and government officials. The first half of the twentieth century was characterized by the creation of volunteer health agencies, health education campaigns, and occupational health programmes in addition to advancements in school children's health and mother and child care. The public health nurse also made a comeback during this time (Haines et al. 2006).

There were two notable attempts to offer healthcare for large populations in the second half of the nineteenth century. First implemented by Russia and was in the form of a system of medical services in rural areas. Following the communist revolution, this was expanded to include full government-supported medical and public health care for everyone. Since then, numerous nations in Europe and Asia have established systems similar to this. The other attempt involved prepayment for medical treatment, a type of social insurance that was first implemented in Germany toward the end of the nineteenth century where prepayment for medical care was already widespread. Many other European nations adopted comparable insurance policies.

A proposal for a single state medical service was executed in the United Kingdom in 1909 because a Royal Commission assessed the Poor Law. This programme served as a precursor to the National Health Service Act of 1946, which marked an effort by a contemporary industrialized nation to make services available to everyone.

## 25.2 Public Health Policy

Public health policy is described as the laws, rules, deeds, and choices made by society to advance wellness and guarantee the accomplishment of particular health objectives. It is possible to implement formal laws and community engagement programmes as public health policy. There are many industries where public health policy is important, including health care, insurance, education, agriculture, and business.

Public health policies are influenced by organizations like WHO, CDC and Prevention, Food and Drug Administration, and other governmental and non-governmental organizations. These organizations do research and administer health and education programmes for a population, resulting in rules and regulations that guarantee that society has access to health care, clean water to drink, wholesome food to eat, and immunizations for the sick. Because it applies public health theory and research to real-world situations, public health policy is extremely important. Research gets translated into action through public health policies, which also seek out widespread fixes for previously identified issues. The fact that funding for public health is frequently perceived as being insufficient, both in absolute terms and in comparison, to the money allotted to health care, presents an issue of particular relevance. This mismatch is partly, caused by strong entrenched interests that frequently prevent more money from being invested in public health initiatives, as well as a governmental, administrative, and public obsession with treatment-focused medical care services. This is true even though improved health promotion and disease prevention initiatives are increasingly, acknowledged in national policy statements and documents due to the epidemic of non-communicable diseases and, in some countries, the dual burden posed by the persistence of communicable diseases.

### *25.2.1 Establishing Socially Protective and Risk-Aware Public Policies*

Natural disasters, climate change, and ramifications for public health brought on by endemics, pandemics, and existing disease threats provide cascading hazards to our planet. A region has a watershed moment for development when cascading risk strikes. All nations must adopt risk-informed policies that consider social protection and encompass health-related aspects in every department, with clearly defined roles for all stakeholders to play. This is essential as we design the future of global public health. Community organizations and civil society can make a substantial contribution to public health and health systems with regard to policy development, service delivery and governance for health and well-being. In this situation, the community may include both societies, with its institutions of civil society, and the specific segments of the population who are most negatively impacted by such policies and activities. The provision of evidence, advocacy, mobilization, consensus building, the delivery of public health services and services related to the socio-environmental determinants of health, standard-setting, self-regulation, and the promotion of social partnerships are all included in this contribution (McMichael 2006).

### *25.2.2 Establishing Public Health Ministries at the National and Sub-National Levels*

Ministries of Health sometimes operate with zero funds, no clear instructions or preventive health plans, and no community-level preventive health objectives while working under fixed mandates focused on curative health outcomes. A special ministry of public health will be required to promote social health policies that benefit people to reverse this trend and mainstream health as an agenda for all policies. A risk-informed planning strategy that mainstreams social policy and places collaboration between line ministries and civil society integration at the core of social progress is required for public health penetration at the national and sub-national levels (Dalal-Clayton and Bass 2002).

A strong public health policy serves as a catalyst, integrating public health goals into every aspect of the governing structure. It is based on the recognition of health as a human right and the understanding that governments have a responsibility to safeguard and guarantee the conditions necessary for citizens' health and well-being. The multisectoral, multiagency, multi-professional, and multi-stakeholder collaborations required to deliver effective public health depend on strengthened governance for health and well-being. An improved structure of governance for health and well-being ensures that the activities conducted across the many sectors are appropriate for the purpose, which is a significant outcome. The overarching goal of good governance for health and well-being is to guarantee that processes have produced results that are consistent with expectations and that the feedback loop between policy development and implementation is effective. The best way to guarantee that the needs of every member of society are sufficiently met is to create broad coalitions for public health at the national and local levels (involving a variety of stakeholders) and establish mechanisms for shared responsibility for policy-making and practice implementation while enabling a system of continuous learning from participatory monitoring and evaluation. These coalitions are primarily driven by co-creation and participatory procedures, thus, public health systems must have the capacity to oversee, manage, and reap the rewards of these processes.

To boost the healthcare industry, the Indian government adopted the National Health Policy in March 2017, to attain the highest possible level of health and well-being for all at all ages through a preventive and primitive healthcare orientation in all developmental policies, and universal access to good quality health care services without anyone having to face financial hardship as a consequence as stated in India's recently adopted National Health Policy-2017, the UN is helping the country (NHP 2017). The public health field has historically been India's public health system and is largely guided by the Ministry of Health and Family Welfare (MOHFW) in India.

With the purpose of 'leaving no one behind', the UN General Assembly adopted the universal, integrated and transformative 2030 Agenda for Sustainable Development combined with a set of 17 Sustainable Development Goals (SDGs) in 2015. India's Prime Minister Narendra Modi stated that the globe and our beautiful planet will greatly benefit from the "Sustainable development of one-sixth of humanity" at the 2015 United Nations conference (NHP 2017). The SDGs prioritize achieving "healthy lifestyles and promoting wellbeing for all at all ages," with goal 3 being one of their core pillars. 10 of the remaining 16 goals include health-related indicators, which are those that directly relate to health services, health outcomes, and environmental, occupational, behavioural, or metabolic risks having known links to health (Eurostat 2023).

The main goal of the NHP-2017 is to better inform, clarify, prioritize, and strengthen the government's role in forming health systems in all their aspects. It includes investments in healthcare, the structuring and financing of healthcare services, the prevention of diseases and promotion of good health through cross-sectoral action, access to technologies, human resource development, medical pluralism, and the development of the knowledge base necessary for better health. There is a hope that India will achieve sustainable progress toward achieving "Health for All" and the SDG targets well within time and NHP-2017 and related regulatory frameworks will be well executed.

### *25.2.3 Recommendations for Public Health Policies*

Policymakers and authorities will conduct comprehensive research before implementing public health policies to identify the problems that need to be solved and provide the best follow-up remedies. Here, a few typical public health measures are now widely used in our society:

(a) **Food Safety Policy**: For the general well-being of individuals and members of society, food safety policies are crucial. Population health officials must develop regulations to make sure that only food that is safe for eating is made available to the general public because food-related illnesses are a serious concern.

(b) **Tobacco Use**: Numerous public health officials have implemented efforts to lower the dangers of illnesses and fatalities brought on by smoking and using tobacco. To protect non-smokers from the negative consequences of second-hand smoke, some of these policies include raising tobacco pricing and establishing tobacco-free zones in public places.

(c) **Human Immunodeficiency Viruses (HIV)**: Public health policy is crucial in teaching the public about HIV prevention, how to live with it, available treatments, and ways to avoid stigma. To lessen the effect and transmission of the virus in diverse areas, public health initiatives have been put in place, such as expanded access to testing, birth control, and medication.

(d) **Alcohol**: Abuse of alcohol is a major contributor to violent crime, sexual assault, auto accidents, and other problems. In the struggle to combat alcoholism and lessen its harmful consequences on users, public health strategies are essential. The most prevalent example of alcohol policies, which vary based on the community, is age limitations on both alcohol consumption and purchases.

## 25.3 Conceptualizing Health Attaches for Global Cooperation and Diplomacy

Every nation's Ministry of External Affairs is required to assign health attaches to its diplomatic missions abroad and set aside Hon Consul ranks to develop bilateral mandates on overarching global health objectives taking into account national security concerns and regional cooperation. Diplomacy today is a craft that requires humanitarian elements, must adapt to changing times that put people at the centre of shared progress, and must also make global health a concept to implement in everyday diplomacy. Diplomacy cannot be the traditional type of guarded diplomacy that was a norm in the past.

To conduct discussions that shape and manage the global (health and non-health) policy environment, integration—the combining of the disciplines of public health, international affairs, management, law, and economics—is essential to the effectiveness of GHD (Global Health Diplomacy). The understanding that concerns of international trade have an impact on both health and diplomacy, frequently in major ways, is an example of the cutting edge of GHD in the link between trade, foreign policy, and health. International patent agreements, the critical requirement to control vaccination access, and the list of essential medications are examples of related topics (Chattu et al. 2019).

## 25.4 Sustainable Development and Public Health

The goal of sustainable development is to raise people's physical, social, and personal standards of living without harming present or future generations. One's health will inevitably suffer in environments that are dominated by air pollution, exposed to harmful substances, or mitigated by several other low standards of living. Therefore, in civilizations characterized by enduring socioeconomic inequality, extensive environmental degradation, or pervasive sickness, sustainable development cannot exist. Development and health are closely related. Serious environmental health issues can arise in both developing and wealthy countries as a result of both inadequate development leading to poverty and incorrect development leading to overconsumption, combined with a growing global population (Patz et al. 2005).In this sense, investment in environmental and health improvements has been acknowledged at the highest decision-making level as a requirement for sustainable development.

## 25.5 Summary and Conclusions

Important aspects within the health sector are public health regulation, human resource development, capacity building, and health system strengthening. Social determinants of health, such as housing, food, safe drinking water, sanitation, early childhood development, education, and social security programmes, also have an impact on a population's health. Other crucial areas for action include population stabilization, gender mainstreaming and empowerment, lessening the negative health effects of climate change and disasters, enhancing community involvement, and governance-related challenges. It is politically difficult to make public health a shared value across the many sectors, yet such coordinated action is essential. In India, public health has been a vibrant field that has faced numerous obstacles in its quest to improve the lives of its citizens. Major public health issues like as leprosy, malaria, TB, high rates of mother and child mortality, and, more recently, HIV have been tackled by the government through coordinated efforts since the country's independence. Science, health care, and social development have all contributed to a decline in both birth and death rates. The goal of public health is to prevent and control disease at the population level by coordinated efforts and well-informed decisions made by people, groups, the public and private sectors, and society at large. However, to overcome these obstacles and achieve health fairness, the government's intervention is essential.

## References


Adshead F, Thorpe A, Rutter J (2006) Sustainable development and public health: a national perspective. J R Inst Public Health 120:1102–1105

APHA (2023) What is public health? https://www.apha.org/what-is-public-health. Accessed on 21 Jan 2024

CDC [Centres for Disease Control and Prevention] (2017) Enhancing real-time sentinel infectious disease surveillance in India. Background information for preparation of state PIP

Chattu VK, Knight AW, Kevany S, Sehovic AB (2019) Global health diplomacy, health and human security: the ascendancy of enlightened self-interest. J Educ Health Promot 8:107. https://doi.org/10.4103/jehp.jehp_391_18

Corvalan CF, Kjellström T, Smith KR (1999) Health, environment and sustainable development. Identifying links and indicators to promote action. Epidemiology 10:656–660. https://doi.org/10.1097/00001648-199909000-00036

Dalal-Clayton B, Bass S (2002) Sustainable development strategies: a resource book. Earthscan Publications Ltd., London

DH PHE Transition Team (2012) Public health surveillance: towards a public health surveillance strategy for England. Department of Health. https://assets.publishing.service.gov.uk/media/5a7ce62eed915d7c849ade9d/Towards-a-Public-Health-Surveillance-Strategy.pdf. Accessed on 15 Jan 2024

Eurostat (2023) Sustainable development in the European Union—monitoring report on progress towards the SDGs in an EU context—2023 edition. https://ec.europa.eu/eurostat/en/web/products-flagship-publications/w/ks-04-23-184. Accessed on 15 Jan 2024

Haines A, Kovats RS, Campbell-Lendrum D et al (2006) Climate change and human health: impacts, vulnerability and public health. J R Inst Public Health 120:585–596

McMichael AJ (2006) Population health as the 'bottom-line' of sustainability: a contemporary challenge for public health researchers. Eur Jr Public Health 16(6):579–582

NHP (National Health Policy) (2017) Ministry of health and family welfare. Government of India. https://main.mohfw.gov.in/sites/default/files/9147562941489753121.pdf. Accessed on 15 Jan 2024

Patz JA, Campbell-Lendrum D, Holloway T, Foley JA (2005) Impact of regional climate change on human health. Nature 438:310–317

Pilot E, Roa R, Jena B et al (2017) Towards sustainable public health surveillance in India: using routinely collected electronic emergency medical service data for early warning of infectious diseases. Sustainability 9(4):604. https://doi.org/10.3390/su9040604

Porritt J (2005) Healthy environment-healthy people: the links between sustainable development and health. J R Inst Public Health 119:952–953

SCM (1949a) Report of the sanitary commission of Massachusetts 1850. JAMA 140(6):576. https://doi.org/10.1001/jama.1949.02900410072033

SCM (1949b) Report of the sanitary commission of Massachusetts 1850. JAMA 140(6):576. https://doi.org/10.1001/jama.1949.02900410072033

WHO [World Health Organization] (2009) Climate change and human health. Fact sheet, July 2005

WHO (2017) Global task force on cholera control overview of ending cholera—a global roadmap to 2030

Zodpey S, Sharma K, Negandhi H (2011) Master of public health programs in India. Public Health Foundation of India, New Delhi

# Chapter 26
# Public Health Policy and Sustainability

**Veena M. Nare, Subhash N. Nikam, and Deepak N. Thakare**

**Abstract** Sustainability is a concept that refers to the ability to maintain ecological balance over time. It involves meeting the needs of the present without compromising the ability of future generations to meet their own needs. Sustainability is a critical issue in public health policy, as it is closely linked to the health and well-being of individuals, communities, and the planet as a whole. In recent years, there has been growing recognition of the importance of the intersection between public health policy and sustainability. This intersection involves understanding how public health policies can affect sustainability and vice-versa. By promoting sustainable practices, public health policies can contribute to the long-term health and well-being of both individuals and the planet. This chapter aims to explore the intersection between public health policy and sustainability, and how they can be integrated to promote health and well-being while preserving the environment. The chapter begins by defining key terms and concepts related to public health and sustainability, followed by a discussion on the importance of integrating these two fields. It then examines case studies and examples of successful integration of public health policy and sustainable practices in various settings, including urban planning, transportation, food systems, and healthcare. Finally, the chapter concludes with recommendations for future research and policy initiatives to promote sustainable public health. Overall, this chapter highlights the potential for a more holistic and integrated approach to public health policy that takes into account the long-term sustainability of our communities and planet.

V. M. Nare
Department of English, Mahilaratna Pushapatai Hiray Arts, Science and Commerce Mahila Mahavidyalaya, Malegaon District, Nashik, India
e-mail: vmnare@gmail.com

S. N. Nikam
Arts, Science and Commerce College, Manmad, Nashik, India

Deepak N. Thakare (✉)
Department of Geography, Mahilaratna Pushapatai Hiray Arts, Science and Commerce Mahila Mahavidyalaya, Malegaon District, Nashik, India
e-mail: dr.dnthakre@gmail.com

 
S. Anand et al. (eds.), *Emerging Geosustainability Transformations in India*, Advances in Geographical and Environmental Sciences,
https://doi.org/10.1007/978-981-97-7993-2_26

**Keywords** Public health · Sustainability · Policy · Integration · Urban planning · Transportation

## 26.1 Introduction

General public health policy and sustainability are two basic ideas that are progressively perceived as related. General health strategy refers to the actions and choices made by the government and different associations to improve and protect the strength of the population. Sustainability, then again, alludes to the capacity to address the issues of the present without compromising the capacity of people in the future to address their issues. In recent years, it has become evident that the options we make about how we live, work, and consume altogether affect the climate, which influences general public health. Natural factors, for example, air and water quality, climate change, and pollution can antagonistically affect human health, including respiratory illnesses, coronary illness, and disease, among people (Schell et al. 2013).

Public health policies that mean working on the strength of the population should consider the manageability of the climate. It implies that policymakers should consider their choices' drawn-out impacts on both public health and the climate. Thus, they can guarantee that their policies not only locate the prompt public health needs of the population but also additionally, add to a more feasible future. This chapter will investigate the convergence between public health policy and sustainability. We will examine the significance of these two ideas together, the central questions and challenges, and examples of fruitful general public health policy approaches that advance sustainability. Moreover, we will inspect the connection between the general health strategy and the Sustainable Development Goals (SDGs), and how policymakers can add to accomplishing these objectives through feasible public health approaches (Chalmers et al. 2014). Toward the finish of this section, we desire to exhibit the significance of incorporating public health policy and sustainability and give policymakers a system for settling on choices that advance both public health policy and ecological manageability.

## 26.2 Intersection Between Public Health Policy and Sustainability

The convergence between public health policy and sustainability lies in their common spotlight on further developing the prosperity of people and communities. Public health policy is worried about the soundness of the population, while sustainability plans to address the inter-generational and intra-generational issues. In recent years, it has become progressively certain that ecological elements, for example, air and water quality, climate change, and pollution, can altogether affect public health

(Shediac-Rizkallah and Bone 1998). For example, poor air quality can prompt respiratory illnesses like asthma, while exposure to ecological poisons can cause disease and other medical issues. In this way, public health approaches should consider the effect of human exercises on the climate to advance the prosperity of people and communities in the present and future too (Pluye et al. 2004).

Sustainability and public health policy converge in more than one way. To begin with, manageability contemplations can direct the improvement of public health policies. For example, policies that advance clean energy and sustainable assets can further develop air quality and reduce greenhouse gas emissions, which can assist with forestalling climate change and its negative health influences. Moreover, policies that advance manageable transportation choices like strolling, cycling, and public travel can reduce air pollution and advance active work, which can work on public health (Scheirer 2005).

Second, general public health policies can add to ecological manageability. For example, approaches that support sound and feasible food frameworks, like neighborhood and natural agribusiness, can lessen the carbon impression of the food business while further developing admittance to solid and nutritious food. In addition, approaches that address climate change, public health and sustainability can add to worldwide endeavours to moderate the impacts of climate change and advance natural manageability (La Pelle 2006).

## 26.3 Importance of Public Health Policy and Sustainability

Public health policy and sustainability are two basic issues that are critical in our modern world (Scheirer and Dearing 2011). Here are a few motivations behind why these issues are so fundamental.

### *26.3.1 Promoting the Health and Health of Individuals and Communities*

Public health policies are fundamental for advancing the health and prosperity of people and communities. These policies mean to forestall and control infections, advance solid ways of life, and guarantee admittance to quality health administrations. Then again, manageability plans to address the issues of the present without compromising the capacity of people in the future to address their issues. By advancing supportable practices, we can guarantee that people in the future approach the assets and climate they need to remain solid and flourish.

### *26.3.2 Addressing Global Challenges*

Public health policy and sustainability are addressing the absolute biggest worldwide challenges in recent times, including climate change, food security, and non-communicable diseases. These challenges are firmly interconnected. For example, addressing climate change requires decreasing emanations of greenhouse gas substances, which can be accomplished by advancing sustainable practices like sustainable power and feasible transportation. In addition, advancing maintainable food frameworks can assist with addressing food weaknesses and reduce the predominance of non-transmittable infections.

### *26.3.3 Ensuring Social Equity and Justice*

Public health policy and sustainability are basic in ensuring social value and equity. Disparities in public health and admittance to assets are frequently firmly connected to ecological debasement and impractical practices. By advancing feasible public health policy arrangements, it will be ensured that everybody approaches the assets they need to remain healthy and thrive.

### *26.3.4 Protecting the Environment and Natural Resources*

Public health policy and sustainability are fundamental for protecting the climate and regular assets. The effect of human exercises on the climate and normal assets can have huge unfortunate results on public health policy, like air and water pollution, harmful exposure, and climate change. By advancing supportable practices, we can decrease the adverse consequence of human exercises on the climate and normal assets, accordingly protecting public health. By advancing maintainable public health policies, we can address worldwide challenges, advance the public health and prosperity of people and communities, guarantee social value and equity, and protect the climate and normal assets.

## 26.4 Impact of Environmental Sustainability on Public Health

Environmental sustainability essentially affects public health policy (Moore et al. 2017). Here are some ways in which environmental sustainability affects public health:

### 26.4.1 Air and Water Quality

Air pollution is a huge environmental issue that can adversely affect public health. Exposure to air pollution has been connected to respiratory illnesses, coronary illness, stroke, and disease. By advancing feasible practices like the utilization of clean energy, maintainable transportation, and lessening emanations from ventures, we can further improve air quality and reduce the adverse consequences on public health.

Access to clean water is essential for public health. However, unsustainable practices such as industrial pollution, overuse of water resources, and inadequate waste water treatment can compromise water quality and pose health risks to communities. By promoting sustainable practices such as water conservation, proper disposal of hazardous waste, and investment in water infrastructure, we can ensure that communities have access to clean water and reduce the risk of waterborne diseases.

### 26.4.2 Food Systems

Sustainable food systems are fundamental for advancing public health. Unsustainable practices like the utilization of compound pesticides and composts, escalated animal cultivating, and the utilization of Genetically Modified Organisms (GMOs) can present public health dangers to purchasers and adversely affect the climate. By advancing supportable food frameworks like natural cultivating, agroforestry, and enhanced crop creation, we can guarantee that people approach solid and nutritious food and will become more resilient toward health challenges.

### 26.4.3 Climate Change

Climate change is a critical ecological test that can seriously affect public health. Extreme weather events, for example, heat waves, droughts, and floods can cause injuries and death, and the spread of vector-borne diseases, such as dengue fever, malaria, and Zika virus can increase due to changes in temperature and rainfall patterns. By promoting sustainable practices that reduce greenhouse gas emissions, such as the use of renewable energy, sustainable transportation, and reducing waste, we can mitigate the impacts of climate change and also on public health.

## 26.5 Role of Public Health Policy in Ensuring Environmental Sustainability

Public health policy plays a critical role in ensuring environmental sustainability (Lennox et al. 2018). Here are some ways in which public health policy promotes environmental sustainability.

### *26.5.1 Regulation and Enforcement*

Public health policy can lay out guidelines and principles to guarantee that enterprises and organizations work in a naturally sustainable way. By upholding these guidelines, public health policy can assist with decreasing pollution and guarantee that normal assets are utilized dependably and practically.

### *26.5.2 Education and Awareness*

Public health policy can promote education and awareness about the importance of environmental sustainability for public health. By educating the public about the impacts of environmental degradation on public health, policies can help in promoting sustainable practices and encourage individuals to adopt sustainable behaviors.

### *26.5.3 Research and Innovation*

Public health policy can promote research and innovation in sustainable technologies and practices. By putting resources into innovative work of sustainable practices, public health policy can advance the reception of supportable practices by enterprises, organizations, and people.

### *26.5.4 Partnership and Collaboration*

Public health policy can promote partnership and collaboration among stakeholders to promote sustainable practices. By working with industry, common society, and government, public health policy can advance economic practices and empower the reception of sustainable practices.

Public health policy plays a critical role in ensuring environmental sustainability. By establishing regulations, promoting education and awareness, investing in research and innovation, promoting partnership and collaboration, and advocating for sustainable practices, public health policy can help in promoting environmental sustainability and ensure that individuals and communities have access to the resources, they need to stay healthy and thrive.

## 26.6 Importance of Considering the Long-Term Effects of Policies on Public Health and Sustainability

It is essential to consider the long-term effects of policies on both public health and sustainability (Chambers et al. 2013). Here are some reasons why this is important.

### *26.6.1 Interconnectedness*

General public health policy and sustainability are interconnected and arrangement that advances one, can benefit the other. For example, approaches that advance sustainable transportation can likewise further improve air quality, which can decidedly affect public health. By taking into account the drawn-out impacts of policies on both public health and sustainability, we can advance arrangements that have different advantages.

### *26.6.2 Prevention*

By taking into account the protracted impacts of policies, we can focus on preventing natural issues and medical problems as opposed to treating them after they have happened. For example, policies that promote sustainable agriculture can prevent soil degradation, which can have positive impacts on public health by ensuring access to healthy and nutritious food.

### *26.6.3 Financial Advantages*

Arrangements that advance sustainability can likewise have monetary advantages in the long term. For example, investing in renewable energy can create jobs and reduce reliance on non-renewable energy sources. By taking into account the drawn-out impacts of policies, we can advance approaches that are both naturally supportable and financially feasible.

### 26.6.4 *Responsibility*

Taking into account the impacts of policies is likewise a dependable way to deal with administration. Policies that only consider short-term gains can have negative impacts on public health and sustainability in the long run. By taking into account the drawn-out impacts of policies, we can advance capable administration and guarantee that we leave a sound and supportable planet for people in the future.

Thus, it is essential to consider the long-term effects of policies on both public health and sustainability. By promoting policies that have multiple benefits, focusing on prevention, considering economic benefits, and promoting responsible governance, we can ensure a healthy and sustainable planet for future generations (Michie et al. 2018).

## 26.7 Key Issues and Challenges

There are a few main points of key issues and challenges connected with public health policy and sustainability. There are several key issues and challenges related to public health policy and sustainability. Limited funding, lack of political will, resistance to change, complex and interconnected issues, and disparities are all factors that need to be addressed in promoting policies and initiatives that promote both public health and sustainability. Some of them are discussed below.

### 26.7.1 *Limited Funding and Political Will*

Public health policy and sustainability drives frequently require huge financing, and state-run administrations and associations might have restricted assets to dispense to these endeavors. This can be a test in advancing policies and drives that are both, naturally maintainable and advance public health. Public health policy and sustainability drives may not be fundamentally important for every single political pioneer or partner. The absence of a political will can prevent the reception of Policies and drives public health and sustainability.

### 26.7.2 *Complex and Interconnected Issues*

Public health policy and sustainability issues are in many cases perplexing and interconnected, making it challenging to address them in segregation. Powerful strategy improvement and execution requires a multi-disciplinary methodology that thinks about the crossing points between public health policy and sustainability.

### 26.7.3 Disparities and Access to Natural Resources

Disparities in access to resources and infrastructure can result in unequal impacts of environmental degradation and health issues. Addressing these differences requires an emphasis on value and civil rights in strategy improvement and execution. Access to natural resources, like clean air and water, is a basic issue connected with both public health policy and sustainability. Resolving this issue requires an emphasis on advancing public health, protecting the environment, addressing disparities, addressing worldwide challenges, and perceiving the monetary advantages of clean air and water (Thompson et al. 2000).

### 26.7.4 Pollution and Waste Management

Pollution and waste management are critical issues related to both public health policy and sustainability (Health Tapestry Program 2017). The food we eat fundamentally affects our health, and practical farming can assist with advancing sound eating regimens and reducing the predominance of diet-related illnesses. Food frameworks and manageable agribusiness are fundamental for advancing public health and forestalling diet-related diseases.

### 26.7.5 Pollution and Waste Management

Pollution and waste management are critical issues related to both public health policy and sustainability (Health Tapestry Program 2017). The food we eat fundamentally affects our health, and practical farming can assist with advancing sound eating regimens and reducing the predominance of diet-related illnesses. Food frameworks and manageable agribusiness are fundamental for advancing public health and forestalling diet-related diseases.

The above-discussed issues need to be tackled in a time-bound manner as they can have multiplier impacts. For example, polluted air and water can fundamentally affect public health, including respiratory illnesses, cardiovascular infection, and disease. Pollution from sources, for example, industry and farming can have huge ecological effects, including biodiversity misfortune and climate change. Guaranteeing admittance to clean air and water is fundamental for advancing ecological manageability and protecting normal environments. Ensuring access to clean air and water is a worldwide test that requires global participation and joint effort. Climate change and natural debasement have influenced that cross public limit, and addressing these challenges requires an organized and helpful methodology. Promoting access to clean air and water can likewise have monetary advantages, including advancing the travel industry and decreasing medical services costs related to pollution-related illnesses. Food frameworks and maintainable farming are basic issues connected with public health policy and sustainability. Resolving these issues requires an emphasis on advancing public health, protecting the climate, addressing variations, addressing worldwide challenges, and perceiving the monetary advantages of Sustainable farming and quality food frameworks (Luke et al. 2014).

## 26.8 Public Health Policies that Promote Sustainability

Many public health policies contribute to promoting sustainability. These policies are linked to energy resources, transportation, food, climate change, etc. Some of them are discussed with examples.

### *26.8.1 Policies Promoting Clean Energy and Renewable Resources*

Policies promoting clean energy and inexhaustible assets are basic to advancing sustainability and protecting public health (WHO 1991; Strauss and Thomas 1998). Here are a few examples of policies that have been effective in advancing clean energy and sustainable assets. For example *Renewable Portfolio Standard* (RPS) related policies. RPS are policies that expect utilities to create a specific level of their energy from sustainable assets, like breeze, sun based, or geothermal. RPS has been effective in driving the improvement of an environmentally friendly power and reducing greenhouse gas emissions (Sen 1999). Another example is the *net metering approach*. The net metering approach permits people and organizations to sell an abundance of energy produced by sunlight-based chargers back to the matrix that can advance the reception of sustainable power and reduce dependence on petroleum derivatives. This arrangement has been effective in advancing the utilization of roof solar panels and different types of dispersed energy (Bloom et al. 2004).

*Energy efficiency standard*-related policies require apparatuses, structures, and vehicles to satisfy energy effectiveness guidelines that lessen energy utilization and advance the reception of environmentally friendly power assets. *Clean energy tax incentives* based assessment are impetuses for organizations and people who put resources into clean energy innovations can further energize the turn of events and reception of environmental-friendly power assets. *Feed-in tariffs* are policies that ensure a set cost for power produced from sustainable sources. This approach has been fruitful in advancing the improvement of huge-scope sustainable power projects. *Tax incentives*, such as tax credits for solar panel installations or electric vehicle purchases, can help to promote the adoption of clean energy technologies. *Green energy bonds* are policies that permit national-run administrations and organizations to raise assets to fund clean energy projects. *Community solar programs* permit people and communities to put resources into shared sun-powered ventures that can advance the reception of environmentally friendly power assets and decrease dependence on non-renewable energy sources. *Renewable Energy Credits (RECs)* are a market-based instrument that permits utilities to buy sustainable power credits to meet their RPS commitments. This strategy has been fruitful in advancing the improvement of sustainable power projects and making a business opportunity for environmentally friendly power (WHO 2006).

These policies have been successful in promoting the use of clean energy and renewable resources, reducing greenhouse gas emissions, and improving public health outcomes.

### 26.8.2 Transportation Policies that Reduce Pollution and Promote Active Living

Transportation policies play a significant part in advancing sustainability and public health by decreasing air pollution, and greenhouse gas emissions, and advancing actual work (WHO 1992). Here are a few examples of transportation policies that can accomplish these objectives. For example, *active transportation infrastructure*-based policies focus on the advancement of strolling and trekking framework, like bicycle paths, walker intersections, and multi-use ways that can advance active work and lessen the utilization of vehicles for little excursions. Further, *public transportation-based approaches* help and extend public transportation choices, like transport, light rail, and metros. These can decrease air pollution and greenhouse gas substances and advance dynamic living by empowering individuals to walk or bicycle to travel stops. *Congestion pricing* means policies that charge drivers an expense to enter clogged metropolitan regions that can lessen traffic, air pollution, and emanations of greenhouse gas substances while uplifting the utilization of public transportation and dynamic transportation modes. *Low-Emission Vehicle Standards-related policies* expect vehicles to satisfy low-outflow guidelines like electric or mixture vehicles can lessen air pollution and greenhouse gas substance discharges. *Car-free zones* are arrangements that assign specific regions as vehicle-free zones, like passers-by just roads or parks that can advance actual work and reduce air pollution and clamor.

### 26.8.3 Policies that Advance Sound and Sustainable Food Frameworks

Food frameworks are a basic part of public health policy and sustainability (EU 1992). Here are a few examples of policies that can advance solid and manageable food frameworks. First is the important one i.e. *sustainable agriculture practices.* This policy approach advances in sustainable cultivating rehearses, like harvest turn, protection culturing, and the utilization of cover crops. It lessens soil disintegration, water use, and the utilization of pesticides and manures. Another one is *local food systems.* These arrangements help in food frameworks, for example, farmers' markets, community-supported agriculture projects, and farm-to-school programs that can elevate admittance to new and quality food varieties while reducing transportation emanations. *Food labeling and transparency*-based policies, require clear and precise food naming, like sustenance data and nation of beginning, which can assist purchasers with pursuing informed decisions about the ecological and public health effects of their food choices. *Food waste reduction* policies improve food-squandering, can decrease and treat the soil and reduce greenhouse gas emissions and limit how much food squanders are shipped off landfills. *Plant-based diet* arrangements advance plant-based slim down, like through dietary rules, that can advance good dieting while at the same time decreasing the ecological effect of creature horticulture. By promoting healthy and sustainable food systems, these policies can improve public health outcomes while also protecting the environment and supporting local communities.

### 26.8.4 Policies Addressing Climate Change and Its Impact on Public Health

Climate change is a critical public health danger that can prompt an expansion in outrageous climate occasions, air pollution, and the spread of disease (WDR 1993). Here are a few examples of policies that can address climate change and its effect on public health. *Renewable energy standards*-based policies set sustainable power targets like the utilization of wind, sun-based, and geothermal energy and can contribute to reducing the outflow of greenhouse gas substances and further improve air quality.

*Energy efficiency standards a*rrangements are those that advance energy productivity in structures, apparatuses, and transportation and reduce energy use and greenhouse gas emissions discharges. *Climate adaptation planning*-related policies help in environmental variation arranging, for example, crisis readiness plans and heat wave reaction plans. It can assist communities with better planning for the health effects of outrageous weather events. *Active transportation infrastructure,* as mentioned earlier, are arrangements that focus on and reserve the improvement of dynamic transportation foundations, can advance active work, and reduce the utilization of vehicles for little excursions, which thus can decrease ozone-depleting substance discharges. Approaches that require *Health Impact Assessments* (HIAs) to be led for significant framework ventures and policies that can assist with distinguishing potential health influences, including those connected with climate change, and illuminate direction. By addressing climate change and its effect on public health, these arrangements can assist with protecting weak populations and further develop largely public health results.

## 26.9 Successful Public Health Policies that Promote Sustainability

Here are a few examples of fruitful public health policies that advance sustainability. The Clean Air Act is a US government regulation that directs air emissions from fixed and portable sources. Since its order in 1970, the *Clean Air Act* has assisted with lessening destructive air toxins like particulate matter, ozone, and lead, bringing about critical enhancements in air quality and decreases in public health risks. A few nations and urban cities have carried out restrictions on *single-utilized plastic bags*, which can assist with decreasing plastic waste and litter. This approach has been fruitful in reducing the use of plastic packs and advancing the utilization of reusable sacks.

*Bike-friendly cities* is another policy in which urban areas focus on and put resources into a dynamic transportation framework, for example, bicycle paths and bicycle sharing projects, etc. It can empower cycling and lessen the utilization of vehicles for brief excursions. Numerous urban communities have carried out policies and foundations to advance cycling for transportation. This not only decreases air pollution and greenhouse gas substance outflows yet in addition advances actual work and can further develop public health policy and sustainability results.

*Farm-to-school programs* mean associating schools with neighbourhood farmers to give new and good food choices for understudies. This approach has been effective in advancing healthy eating propensities and supporting local agriculture. *Carbon pricing* is a policy tool that puts a cost on greenhouse gas emissions, giving a monetary motivator to organizations and people to reduce their discharges. This arrangement has been fruitful in elevating the progress to cleaner and more manageable energy sources.

Approaches that advance *maintainable horticulture rehearses*, like natural cultivating and regenerative farming, can assist with protecting the climate, advance biodiversity, and further develop public health policy and sustainability results by reducing exposure to unsafe synthetic compounds and advancing solid weight control plans. *Access to green spaces* approach elevates admittance to green spaces, for example, parks and local area gardens, can advance active work, lessen pressure, and further develop psychological public health policy and sustainability results. Policies that advance waste decrease and reusing can lessen how much waste is shipped off landfills, reduce greenhouse gas emissions, and advance a round economy.

These are only a couple of examples of effective public health policies that advance manageability. These policies do not just advance public health and sustainability of the climate yet, in addition, can also have financial advantages and advance social value. These successful policies demonstrate the potential for public health policies to promote sustainability and improve both public health and environmental outcomes.

## 26.10 Conclusions

In conclusion, the intersection between public health policy and sustainability is a critical area of focus for policymakers and researchers alike. Environmental degradation and unsustainable practices can have significant impacts on public health, and addressing these issues requires a comprehensive and integrated approach. Policies that promote clean energy and renewable resources, sustainable agriculture, and active transportation can contribute to both environmental sustainability and improved public health outcomes. Immunization programs, universal health coverage, and policies that promote access to clean water and sanitation facilities are examples of public health policies that can contribute to achieving the SDGs. However, addressing the key issues and challenges related to public health policy and sustainability will require a concerted effort from all sectors of society. Collaborative and interdisciplinary approaches that involve stakeholders from government, academia, industry, and civil society will be necessary to achieve the changes needed to promote sustainability and improve public health outcomes. By working together and implementing evidence-based policies and practices, we can create a more equitable and sustainable future for all.

## References

Bloom DE, Canning D, Sevilla J (2004) The effect of health on economic growth: a production function approach. World Dev 32(1):1–13

Chalmers I, Bracken MB, Djulbegovic B et al (2014) How to increase value and reduce waste when research priorities are set. Lancet 11 383(9912):156–165. https://doi.org/10.1016/S0140-6736(13)62229-1

Chambers DA, Glasgow RE, Stange KC (2013) The dynamic sustainability framework: addressing the paradox of sustainment amid ongoing change. Implement Sci 8:1

EU (1992) Treaty on European Union. Luxembourg, Office for Official Publications of the European Communities. https://eur-lex.europa.eu/legal-content/EN/TXT/PDF/?uri=CELEX:11992M/TXT. Accessed 21 Dec 2018

Health Tapestry Program (2017) Sustainability frameworks and tools. www.implementationguide.healthtapestry.ca/other-aspects-of-a-successful-implementation/sustainability-scaleability/. Accessed 21 Dec 2018

La Pelle NR, Zapka J, Ockene JK (2006) Sustainability of public health programs: the example of tobacco treatment services in Massachusetts. Am J Public Health 96(8):1363–1369

Lennox L, Maher L, Reed J (2018) Navigating the sustainability landscape: a systematic review of sustainability approaches in healthcare. Implement Sci 13:27

Luke DA, Calhoun A, Robichaux CB et al (2014) The program sustainability assessment tool: a new instrument for public health programs. Prev Chronic Dis 11:E12

Michie S, West R, Sheals K et al (2018) Evaluating the effectiveness of behaviour change techniques in health-related behaviour: a scoping review of methods used. Transl Behav Med 8(2):212–224

Moore JE, Mascarenhas A, Bain J et al (2017) Developing a comprehensive definition of sustainability. Implement Sci 12:110

Pluye P, Potvin L, Denis JL (2004) Making public health programs last: conceptualizing sustainability. Eval Prog Plann 27(2):121–133

Scheirer MA (2005) Is sustainability possible? A review and commentary on empirical studies of program sustainability. Am J Eval 26(3):320–347

Scheirer MA, Dearing JW (2011) An agenda for research on the sustainability of public health programs. Am J Public Health 101(11):2059–2067

Schell SF, Luke DA, Schooley MW et al (2013) Public health program capacity for sustainability: a new framework. Implement Sci 8:15. https://doi.org/10.1186/1748-5908-8-15

Sen A (1999) Health in development. Bull World Health Organ 77(8):619–623

Shediac-Rizkallah MC, Bone LR (1998) Planning for the sustainability of community-based health programs: conceptual frameworks and future directions for research, practice and policy. Health Edu Res 13(1):87–108

Strauss J, Thomas D (1998) Health, nutrition and economic development. J Econ Lit 36(2):766–817

Thompson B, Lichtenstein E, Corbett K et al (2000) Durability of tobacco control efforts in the 22 Community Intervention Trial for Smoking Cessation (COMMIT) communities 2 years after the end of the intervention. Health Edu Res 15(3):353–366

WDR [World Development Report] (1993) Investing in health. Oxford University Press, New York

WHO [World Health Organization] (1991) Health for all targets: the health policy for Europe. European Health for All Series, No. 4, WHO Regional Office for Europe, Copenhagen

WHO [World Health Organization] (1992) Our planet, our health. Report of the WHO Commission on Health and Environment, Geneva

WHO [World Health Organization] (2006) Constitution of the world health organization: basic documents, 45th ed, Supplement October 2006

# Chapter 27
# Geosustainability Transformations: Emerging Research Frontiers

**Rajesh Kumar Abhay, Subhash Anand, Virendra Nagarale, and Shivam Chaurasiya**

**Abstract** In today's rapidly changing world where pressing challenges such as climate change, social inequality, and environmental degradation are at their peak, the need for sustainable transitions has never been more paramount than today. This chapter serves as a timely and insightful contribution to the discussions on sustainable transitions as it offers a comprehensive exploration of sustainable transitions across various facets including social, economic, environmental, and technological spheres. It not only explores the interconnectedness of these dimensions but also emphasizes the importance of integrating geographical perspectives. This chapter highlights the emerging field of sustainable transitions, particularly through a geographical lens. It tries to address the gaps in existing literature by offering novel insights into the spatial dimensions of sustainability. By bringing various research from diverse disciplines, the chapter tries to provide an overarching understanding of all the challenges and opportunities associated with sustainable transformations. By encouraging interdisciplinary dialogue and suggesting practical insights, timeserver is a valuable resource for stakeholders engaged in sustainability initiatives. The chapter endeavors to drive meaningful change toward a more sustainable and resilient future for generations to come through collaboration and coordinated efforts.

**Keywords** Sustainable transitions · Social sustainability · Economic sustainability · Environmental sustainability · Digital transformation · Interdisciplinary

---

R. K. Abhay (✉)
Department of Geography, Dyal Singh College, University of Delhi, New Delhi, India
e-mail: rkabhay@dsc.du.ac.in

S. Anand · S. Chaurasiya
Department of Geography, Delhi School of Economics, University of Delhi, New Delhi, India

V. Nagarale
Department of Geography, SNDT Women's University, Pune Campus, Pune, Maharashtra, India

 
S. Anand et al. (eds.), *Emerging Geosustainability Transformations in India*, Advances in Geographical and Environmental Sciences,
https://doi.org/10.1007/978-981-97-7993-2_27

## 27.1 Introduction

The study of sustainable transitions has become a burgeoning topic of interest in the last decade. This concept represents a paradigm shift toward a holistic approach to address the interconnected challenges of environmental degradation, social disparities, economic instability, political vulnerability, and technical susceptibility. The study of sustainable transitions concerning spatial dimensions has not been explored thoroughly. It is important to understand the interconnectedness of sustainable transitions and what is a rather a new concept in the geographical sphere-"space."

The research on the geography of sustainable transformations has garnered momentum and spread to various fields. The study entails using different approaches to study the transition in sustainability and interpreting them through a geographical perspective. The advantageous factor in the increasing attention to this field has led to insights interlinking various fields of study. The cross-connection in thoughts and ideas will eventually lead to the required changes in sustainable goals and their implementation process. Facing troubles in various sectors be it in energy supply, industrialization, depletion in resources, and various others have led to the change in conversation and the need to reimagine the already laid down framework for sustainable development. There is indeed very little literature available on the idea of transition of sustainability especially through the lens of geography. The three latest conferences on sustainable transitions in Lund, Copenhagen, and Zurich discussed the topic of urban sustainable transitions through paper presentations and discussions on the topic. A lot of novel theory development within the transitions literature originated at the intersection of evolutionary economics and more constructivist approaches in Studies of Technology and Science (STS), resulting in the key conceptual frameworks of Technological Innovation Systems (TIS), the multi-level perspective, and Strategic Niche Management (SNM) (Hansen et al. 2015).

## 27.2 Social Sustainability Transformation

The basis for having sustainability is to create a good life. The perspective of a good life was driven by the concept of sustainability for enhancing community values and society. The post-capitalist society moved toward finding ways to fight against climate change through the sharing of resources be it technological innovation, freely available information/data, or provision of free goods and services. Post-capitalism is related to a vision of technological innovation and progress that makes a sustainable contribution to the fight against climate change by utilizing inexhaustible digitally shared information. It is thus in favor of sustainable growth and sympathizes with the idea of ecological modernization (Adloff and Neckel 2019). The need to discuss social sustainability transformation arises from the most pressing issues in the social arena like growing inequalities, social injustices, and discrimination against marginalized communities. The idea of a post-capitalist society where there are no market constraints is difficult to imagine let alone implement.

Social transformations are aimed at an incremental discontinuing of accustomed behavior, overcoming consumerist mentalities, and experimenting with new ways of life. A sociological analysis of these phenomena requires an action-theoretical perspective that also accounts for the affective dimension of transformations (Adloff and Neckel 2019). Today, social counter movements claim the concept of "transformation" to convey their goal of a fundamental social change that promises to prevent the total exploitation of people and the environment by unbridled capitalism (Jonas 2017). To address the dire issues we must foster inclusive communities, promote social equity, and ensure access to basic services and opportunities for all. The idea of a society free from impulsions of dynamic growth and a critique of the idea of capitalistic growth form the concept of a post-growth society. The concept of post-growth society is closely related to degrowth, where a future society is imagined creating and implementing institutions, technologies, and innovations that sustain the whole idea of sustainability. The central concern of the degrowth movement is to strengthen networks between individual projects and cooperatives and also to build up alternative, disturbing, and antagonistic political experiments (Asara et al. 2015).

The transformation of social sustainability is an integral part of geo-sustainability transformation. For sustainable transformation in the sphere of geographical or spatial dimensions, we must start with social transformation. The transformation of society is the first step in the multi-level perspective of transformation that will eventually lead to the transformation to the concept of "Geosustainability".

## 27.3 Economic Sustainability Transformation

The agricultural revolution is considered the first step by human beings to transform their surroundings through human environmental actions. However, the Industrial Revolution fastened this process of interaction and made nature reach the receiving end. The process of sustainable transformation starting from an urban space is extremely helpful in expanding the transition phase. Many scholars have already studied the importance of space specificity when it comes to transformation (Asara et al. 2015; Vial 2019; Adloff and Neckel 2019; Feroz et al. 2021). They state that the urban space is ideal for sustainable transformation as the changes it will bring forth be it in policies or laws will later seep down into the semi-urban spaces and finally to rural areas. The transferability of technology, policies, and innovation from tier 1 cities to tier 3 regions will bring forth rapid development in the area. The core idea behind these approaches is that similarly to technologies and business models, policies and other institutional settings can also be adopted and reconfigured in other places. Other scholars emphasize the need for further consideration of endogenous processes at the city level and how exogenous factors have an impact on local processes, especially as urban sustainability transitions receive further attention (Fastenrath and Braun 2018).

The formulation of new policies aiming to fulfill sustainability goals must be laid down concretely keeping in mind their implementation and the initial resistance to it. The concept of economic sustainability involves adopting practices that balance economic growth with social equity and environmental stewardship. This includes embracing circular economic principles, investing in renewable energy, and promoting fair trade practices. Circular economy models aim to minimize waste and maximize resource efficiency by designing products for longevity, reuse, and recycling. Renewable energy investments reduce reliance on fossil fuels, mitigate climate change impacts, and create green jobs. Fair trade practices ensure equitable compensation for workers and producers, promoting social justice and sustainable development. Economic sustainability is crucial for geo-sustainability transformation because it underpins the financial resources and incentives needed to implement sustainable practices. By fostering economic resilience and equitable growth, economic sustainability supports investments in renewable energy, green infrastructure, and sustainable agriculture, which are essential for mitigating environmental impacts and promoting well-being within a geo-sustainable framework.

## 27.4 Environmental Geosustainability Transformation

Environmental sustainable transformation originated from the growing concerns over pollution, habitat destruction, and climate change. The Industrial Revolution and subsequent waves of industrialization led to widespread environmental degradation, threatening ecosystems and human health. In this digital era where human and the environment are interconnected at the global level, for environmental sustainability and to work on it would mean policy formulation that would affect our easy lifestyle. Due to rising concerns about global warming and climate change, there is huge pressure on governments and institutions to adopt sustainable methods to mitigate the impacts. Environmental sustainability requires adopting practices that mitigate climate change. These include conservation measures, restoring degraded ecosystems through reforestation, habitat restoration, and marine conservation initiatives. Transitioning to renewable energy resources such as solar, wind, and hydroelectric power reduces greenhouse gas emissions, improving air quality and promoting environmental sustainability.

Environmental sustainability is essential for geo-sustainability transformation because it ensures the health and resilience of ecosystems, which form the foundation of human well-being and economic prosperity. By protecting biodiversity, conserving natural resources, and developing both adaptation and mitigation strategies for climate change impacts, environmental sustainability supports the long-term viability of human societies within a geo-sustainable framework. Suggestions for an environmental state as the next logical step for the progress of humankind have been put on the table. For an environmental state to exist it would mean, overthrowing capitalism. But that would come at a cost. The pause in industrial and economic growth will only push us backward. Therefore, a conjugation of the ideas of industrialization and sustainable livelihood needs to happen for the betterment and achieving geo-sustainability.

Developing Green Urban Transport Policies (GUTP) has been undertaken by several local governments to curb $CO_2$ emissions. We have arrived at the precipice where we must develop green tech or cleantech transport. Powerful transnational corporations like automakers, their suppliers, and energy companies are central cleantech innovators in the transport and mobility domains; their research and development operations are spatially distributed, linked through organizational networks, knowledge exchange practices, temporary meetings, projects and platforms (e.g. van Winden et al. 2010). Sharing of knowledge and data on developing and utilizing resources to their fullest potential without depleting for future generations can be achieved through the enhanced coordination and cooperation between organizations and institutions that are already invested in cleantech research. Knowledge about new scientific developments and experiments is shared (and sometimes co-created) in global conferences and meetings before codification in patents or articles (e.g. World Biofuels conference) and new global business and development opportunities emerge under these settings. Simultaneously, many transnational research programs fund technology experimentation in different places (e.g. hydrogen-powered buses trials in European cities), as well as the set-up of international teams working on the development of concrete clean transport solutions (e.g. long duration car batteries, anti-fuel-corrosion materials, fuel catalyzers) (Carvalho et al. 2012).

## 27.5 Digital and Technological Transformation

In the realm of digital transformation, the term refers to the utilization of novel digital technologies to facilitate significant enhancements in business operations and markets, such as improving customer experience, streamlining operations, or establishing new business models (Paavola et al. 2017). In the contemporary digital era, businesses have increasingly faced scrutiny for unethical practices that have contributed to environmental harm. Consequently, they have been compelled to reassess their business models to align with current environmental safety standards. Li et al. (2018) examined digital transformation in the context of the influence of information technology (IT) on organizational structure and its alignment with business within Small and Medium-Sized Enterprises (SMEs). Digital transformation represents a process catalyzed by digital technologies, leading to organizational disruptions and significantly impacting organizational value creation, strategy, and structural mechanisms (Vial 2019). The advent of digital transformation has engendered the emergence of new business models and has induced disruptions in global markets and industries. The reverberations of digital transformation have upheaved traditional businesses, precipitating the entry of digitally proficient firms (Feroz et al. 2021).

Digital transformation in the wake of our changing times and changing needs is essential for further progress and development. An advanced digital transformation like artificial intelligence will help maintain our sustainable goals. The faster we develop our technologies the sooner we will have our answers for fighting our generation's vital problems. Starting with resource depletion and climate change, we must take digital help in recognizing and organizing our resources or thereof the lack of it. Digital transformation is referred to as "the use of new digital technologies to enable major business improvements in operations and markets such as enhancing customer experience, streamlining operations, or creating new business models" (Paavola et al. 2017). The advancement in the process of digital transformation is going to bring forth immense change in the market and impact consumer behavior. The business models that will be developed based on the demands of the customer will have a direct impact on the environment.

To achieve digital transformation, digital agility is needed to recombine digital assets with other organizational resources to change the way of doing business. By continuously sensing and seizing market opportunities, digital agility fosters the recombination and development of new products, services, and business models that enhance the value created for the customer (Karimi and Walter 2015; Sambamurthy et al. 2003; Teece 2010). Businesses have to keep in mind future needs and changing demands of the consumer and accordingly build their business models. It is necessary now more than ever that all businesses big or small start developing sustainable procedures for meeting their needs. The waste created by the retail markets needs immediate focus. Huge retail companies like Alibaba, and Amazon leave behind a trail of waste which ultimately harms the planet. The reuse, recycle hoax created by the plastic-producing businesses has yielded no measurable results. Therefore, it is time we combine data and analytical techniques to quantify the needs and demands of the consumer, to create less waste.

## 27.6 Conclusions

The discussions surrounding sustainable transitions have broadened over the past decade, marking a significant paradigm shift toward a holistic approach to address manifold challenges ranging across social, economic, political, environmental, and technological domains. The theme of sustainable transitions remains relatively underexplored, but recent research efforts have begun to throw light on its importance.

The study of sustainable transitions, particularly in the domains of aforesaid themes, has given impetus to various academic disciplines. Scholars have highlighted the importance of integrating geographical perspectives into the understanding and implementation of sustainable goals. Conferences and academic gatherings held across the world have provided platforms for dialogue and exchange of ideas, driving innovative theoretical developments at the intersection of different fields.

Social sustainability transformation underscores the peremptory of fostering inclusive communities, promoting social equity, and addressing vitally important issues such as growing inequalities and discrimination. Economic sustainability transformation entails balancing economic growth with social equity and environmental operation through practices such as circular economy models and investments in renewable energy. Environmental sustainability transformation is crucial for mitigating climate change impacts and ensuring the health and adaptability of ecosystems. Digital dexterity and advanced technologies like artificial intelligence big data, and blockchain technology offer opportunities to enhance resource management, consumer behaviour analysis, and business model orientation with environmental objectives.

As we traverse the complexities of sustainable transitions, it is evident that a collaborative effort is needed to address systemic challenges and foster meaningful change. Collaboration and coordination between academia, industry, policymakers, and civil society is essential to drive innovation, execute effective policies, and promote sustainable practices. By embracing a geospatial outlook and integrating insights from diverse fields, we can pave the way toward a more sustainable and resilient future for generations to come.

## References

Adloff F, Neckel S (2019) Futures of sustainability as modernization, transformation, and control: a conceptual framework. Sustain Sci 14:1015–1025

Asara V, Otero I, Demaria F et al (2015) Socially sustainable degrowth as a social–ecological transformation: repoliticizing sustainability. Sustain Sci 10(3):375–384. https://doi.org/10.1007/s11625-015-0321-9

Carvalho L, Mingardo G, Van Haaren J (2012) Green urban transport policies and cleantech innovations: evidence from Curitiba, Göteborg and Hamburg. Eur Plan Stud 20:375–396

Fastenrath S, Braun B (2018) Sustainability transition pathways in the building sector: energy-efficient building in Freiburg (Germany). Appl Geogr 90:339–349

Feroz AK, Zo H, Chiravuri A (2021) Digital transformation and environmental sustainability: a review and research agenda. Sustain 13(3):1530. https://doi.org/10.3390/su13031530

Hansen T, Coenen L (2015) The geography of sustainability transitions: review, synthesis, and reflections on an emergent research field. Environ Innov Soc Transit 17:92–109

Jonas M (2017) Transition or transformation?: a plea for the praxeological approach of radical socio-ecological change. In: Jonas M, Littig B (eds) Praxeological political analysis. Routledge, London, pp 116–133

Karimi J, Walter Z (2015) The role of dynamic capabilities in responding to digital disruption: a factor-based study of the newspaper industry. J Manag Inf Syst 32(1):39–81

Li L, Su F, Zhang W, Mao JY (2018) Digital transformation by SME entrepreneurs: a capability perspective. Inform Syst J 28:1129–1157

Paavola R, Hallikainen P, Elbanna A (2017) Role of middle managers in modular digital transformation: the case of Servu. In: Proceedings of the 25th European conference on information systems (ECIS 2017). Guimaraes, Portugal, pp 887–903

Sambamurthy V, Bharadwaj A, Grover V (2003) Shaping agility through digital options: reconceptualizing the role of information technology in contemporary firms. MIS Q 27:237–263

Teece DJ (2010) Business models, business strategy and innovation. Long Range Plan 43:172–194. https://doi.org/10.1016/j.lrp.2009.07.003

Van Winden W, Van den Berg L, Carvalho L, Van Tuijl E (2010) Manufacturing in the new urban economy. Routledge, Abingdon
Vial G (2019) Understanding digital transformation: a review and a research agenda. J Strateg Inf Syst 28:118–144

# Index




S. Anand et al. (eds.), *Emerging Geosustainability Transformations in India*, Advances in Geographical and Environmental Sciences,
https://doi.org/10.1007/978-981-97-7993-2

The manufacturer's authorised representative in the EU is Springer Nature Customer Service Centre GmbH, Europaplatz 3, 69115 Heidelberg, Germany. If you have any concerns regarding our products, please contact ProductSafety@springernature.com

Printed and bound by CPI Group (UK) Ltd, Croydon, CR0 4YY
01/10/2026
02225934-0017